AF545189

MICROELECTROMECHANICAL SYSTEMS AND DEVICES

Edited by **Nazmul Islam**

INTECHOPEN.COM

Microelectromechanical Systems and Devices
Edited by Nazmul Islam

Published by InTech

Janeza Trdine 9, 51000 Rijeka, Croatia

Edition 2014

Copyright © 2012 InTech

All chapters are distributed under the Creative Commons Attribution 3.0 license, which allows users to download, copy and build upon published articles even for commercial purposes, as long as the author and publisher are properly credited, which ensures maximum dissemination and a wider impact of our publications. After this work has been published by InTech, authors have the right to republish it, in whole or part, in any publication of which they are the author, and to make other personal use of the work. Any republication, referencing or personal use of the work must explicitly identify the original source.

As for readers, this license allows users to download, copy and build upon published chapters even for commercial purposes, as long as the author and publisher are properly credited, which ensures maximum dissemination and a wider impact of our publications.

Notice

Statements and opinions expressed in the chapters are these of the individual contributors and not necessarily those of the editors or publisher. No responsibility is accepted for the accuracy of information contained in the published chapters. The publisher assumes no responsibility for any damage or injury to persons or property arising out of the use of any materials, instructions, methods or ideas contained in the book.

Publishing Process Manager Vedran Greblo

Technical Editor Teodora Smiljanic

Cover Designer InTech Design Team

Additional hard copies can be obtained from orders@intechopen.com

Microelectromechanical Systems and Devices, Edited by Nazmul Islam
p. cm.
ISBN 978-953-51-0306-6

INTECH

open science | open minds

Contents

Preface

There have been numerous advances and inventions directly related to microelectromechanical systems (MEMS) and devices. The researchers have used the modified IC fabrication techniques at the beginning of MEMS era. Since then, MEMS researchers have continually advanced and augmented the capabilities of wafer-based fabrication technologies. These advances have been instrumental in the demonstration of new devices and applications, and even in the creation of new fields of research and development: bioMEMS, micro total analysis and system (μTAS), microfluidic devices, optical MEMS and RF MEMS. Experience indicates a need for a MEMS book covering biomedical materials as well as the most important process steps in bulk micro-machining and modeling. A special emphasis is put on silicon, the most important substrate in MEMS devices, and its material properties and measurement techniques as well as analytical methods used in the silicon material characterization. The primary aim of this book is to give a broader overview tailored for the needs of the MEMS industry rather than go too details in the scientific details. The other aims of this book are to disseminate practical knowledge about selective MEMS technology and its applications, to create a clear consciousness about the effectiveness of MEMS technologies, to stimulate the theoretical and applied research in these very important areas, and promote the practical using of these techniques in the industry.

As a particularly diverse and multidisciplinary area of research, the field of MEMS offers a vastly different set of challenges relative to typical IC fabrication and design. A wealth of knowledge exists in the MEMS community, but much of this expertise is most readily accessed by informal, nonmethodological means such as discussions with colleagues at conferences. We intend this book to provide the reader with the MEMS materials and processes, but beyond this goal, we intend for it to give practical insight into the workings and standard procedures carried out in research labs and production facilities on a daily basis. The chapters are meant to be a springboard of sorts, providing basic information about each topic, with a large number of classic and contemporary literature references to provide in-depth knowledge. We hope this book consolidates important information for readers and thereby spurs the creation of many new devices and processes. The authors of this book view the devices and processes as the fundamental building blocks for making complex systems. Keeping this in mind, the book is divided into four main sections: Chapters 1, 2, 3, and 4 covering materials related to bioMEMS devices; Chapters 5, 6, 7 and 8 covering materials on MEMS characterization and micromachining; Chapters 9, 10, 11, 12, and 13 covering RF and optical MEMS; Chapters 14, 15, 16, 17 and 18 covering MEMS based actuators.

Section I starts with the emerging field of bioMEMS, including MEMS coil for retinal prostheses, DNA extraction by micro/bio-fluidics devices, microcantilever integration with ACEO for lab-on-a-chip devices, and acoustic wave biosensors. Chapter 1 in the first section starts with the development of MEMS coils for retinal prostheses and those coils are fully micromachined in a way compatible with multielectrode arrays and the Parylene-based embedded chip packages. Chapter 2 presents overview of an integrated micro/bio-fluidic device capable of performing online cell lysis and DNA extraction. This device is a powerful tool for biological sample pre-treatment. More of the MEMS microfluidics devices are covered in Chapter 3. This chapter focuses on the MEMS microcantilever integration with ACEO and integrated micropump characterization with lab-on-a-chip applications. The entirety of Chapter 4 is devoted to the MEMS based acoustic wave biosensors characterization that has applications in emerging field of bioMEMS.

MEMS characterization, measurement, micromachining, macromodels are discussed in Section II. Chapter 5 of this section uses computer micro-vision and microscopic interferometry to carry out MEMS measurements, including dimensional (static) and moving (dynamic) properties analysis. Chapter 6 focuses on the description of issues and techniques in the interfacial adhesion for the MEMS devices. In chapter 7 the etch rate anisotropy in surfactant-modified etch solution is investigated. The surfactant in etching enables manufacturing of advanced and exciting structures for MEMS. Chapter 8 discusses the MEMS elements of different embedding systems and their macromodels for system-level simulation. Having System-level models of all MEMS components allow a fast and sufficiently exact simulation of entire MEMS. Analytical and numerical modeling is also covered in this chapter.

Section III focuses on the RF and Optical MEMS switches, microfabrication techniques and the new developments. RF MEMS switch has been demonstrated in Chapter 9 of this section. In addition, the experimental results and surface forces, charging contributions is also presented. RF MEMS switches are extended to Chapter 10. Two configurations of RF MEMS switches using electrostatic actuation, and several MIMs devices simulating the RF MEMS actuation pads, with various dielectric materials and electrodes, have been measured in this chapter. Chapter 11 provides a brief study of capillary forces based on a RF MEMS series switch. A six mask all metal fabrication process and fabrication of different novel switches is also presented. Chapter 12 presents the theory, basic physical design, and fabrication of various optical MEMS devices for display, and medical imaging. The optical MEMS technology promises to revolutionize broad categories by bringing together silicon-based microelectronics with micromachining technology, and optical components. That is the reason, understanding the thermal and optical response of laser-irradiated microsystems requires careful consideration of not only the individual thermal, optical, and mechanical parameters, but also the coupling that exists between them. All these are covered in Chapter 13.

Section IV emphasizes on MEMS based actuators. In Chapter 14 an exhaustive review on the preparation of PZT thick films have been carried out, taking specific focus in the effect of the infiltration in the preparation of high-quality films. Flexible MEMS Color Display on Polymer Materials is covered in Chapter 15. In order to design a flexible MEMS, special substrates and multilayer with high flexibility such as polymeric materials, ultra thin glass, thin metal foil, or even fabric should be considered and the authors have used the Printing Techniques to fabricated flexible MEMS. Chapter 16 discusses the design, fabrication, characterization, modeling, and reliability of thermal microactuators. Thermal, electrical, and mechanical measurements for bent-beam polycrystalline silicon thermal microactuators are also reported in this chapter. In Chapter 17, the major research activities that use a tensile testing method to study the mechanical behavior of thin film materials are reviewed. The final chapter of the book discusses mainly the characterization technique of diamond, Diamond-like carbon (DLC) and Diamond-like Nanocomposite (DLN) thin films and their application in MEMS devices.

Finally the editor would like to express his great appreciation to all the contributors for this unique window of opportunity to work with them, resulting in a comprehensive MEMS book, which illustrates a global cutting edge knowledge and expertise within the most vigorously growing industry today. The valuable contributions of the renowned researchers from different parts of the globe working on various aspects of MEMS devices deserve great appreciations. The entire credit goes to the InTech publishing group members for their tireless efforts to work on this project to publish the book in time. The editorial assistance of the process manager, Mr. Vedran Greblo needs special mention for the success of this book project.

Dr. Nazmul Islam
Director, NEMS/MEMS Lab
The University of Texas at Brownsville
USA

Part 1

BioMEMS Devices

1

Implantable Parylene MEMS RF Coil for Epiretinal Prostheses

Wen Li[1], Damien C. Rodger[2], James D. Weiland[2], Mark S. Humayun[2], Wentai Liu[4] and Yu-Chong Tai[3]

[1]Michigan State University,
[2]University of Southern California,
[3]California Institute of Technology
[4]University of California, Santa Cruz
USA

1. Introduction

Age related macular degeneration (AMD) and retinitis pigmentosa (RP) are two of the most common outer retinal degenerative diseases that have resulted in vision impairment and blindness of millions of people. Specifically, AMD has become the third leading cause of blindness on global scale, and is the primary cause of visual deficiency in industrialized countries (World Health Organization [WHO], 2011). In the United States, more than 500,000 people are suffering from RP and around 20,000 of them are legally blind (Artificial Retina Project, 2007). Whereas many treatment methods, including gene replacement therapy (Bennett et al., 1996), pharmaceutical therapy, nutritional therapy (Norton et al., 1993), photoreceptor and stem cell transplantations (MacLaren et al., 2006 & Tropepe et al., 2000), and dietary, have been explored to slow down the development of AMD and RP diseases in their early stages, vision impairment and blindness due to outer retinal degeneration currently remain incurable.

In the 1990's, researchers discovered that although the retinal photoreceptors are defective in RP patients, their optic nerves, bipolar and ganglion cells to which the photoreceptors synapse still function at a large rate (Humayun et al., 1999). Further studies showed similar results in AMD patients (Kim et al., 2002). These findings have created profound impact on the ophthalmology field, by providing the possibility of using artificial retinal prostheses to partially restore the lost vision function in AMD and RP patients. Two main retinal implant approaches are currently in development according to the layer of retina receiving the implanted device: subretinal (Chow et al., 2006; Rizzo, 2011; Zrenner et al., 1999) and epiretinal prostheses (Humayun et al., 1994; Weiland & Humayun 2008; Wong et al., 2009). Particularly, epiretinal implantation has received widespread attention over the last few years, for not only successful clinical trials demonstrating its efficacy in patients, but also its many advantages compared to others (Horch K.W. & Dhillon, G.S., 2004). First, the device implantation and follow-up examination only require standard ophthalmologic technologies, which can effectively reduce the risk of trauma during the surgery and also allow for the implant to be replaced easily. In addition, most of the implanted electronics are kept in the vitreous cavity so that potential thermal damage to the surrounding retinal tissues can be mitigated by taking the advantage of the heat dissipating properties of the vitreous. Finally, developments of modern microelectromechanical systems (MEMS) technologies enable complete intraocular retinal implants and high-density array stimulation.

Various epiretinal prosthetic technique options have been studied by a number of groups worldwide, by replacing the defective photoreceptors with a multielectrode array implanted on the surface of the inner retina between the vitreous and internal limiting membrane (Javaheri et al. 2006; Rizzo et al., 2004; Stieglitz et al. 2004). Although they have different system configurations, these systems generally utilize a pair of coils to transfer data and power wirelessly between an extraocular data acquisition system and intraocular electronics. In such implementations, the intraocular receiver coil, which resides inside the eye, has many constrains compared to the extraocular transmitter coil. First, the coil has to be mechanically durable to withstand the surgical procedure. It also needs to be flexible and small enough to facilitate the device implantation through small incisions. Finally, it should be chemically inert and biocompatible to prevent harmful interaction with surrounding tissue/cells. Currently, most systems are still using thick and stiff hand-wound coils as the receiver coils, which can cause notable degradation in the implant region. In addition, interconnections between the hand-wound coils and other components are usually formed by soldering, which will require additional hermetic package to ensure device biocompability. Planar microcoils that contain electroplated gold wires on flexible polyimide substrates have also been developed (Mokwa et al. 2008). While device functionality has been confirmed in animal models, several critical issues such as long-term reliability and biocompatibility need to be further addressed for human implantation.

To overcome these challenges, we have proposed a polymer-based MEMS technique for making microcoils. Devices typically consist of multi-layer conductive wires encapsulated by polymeric materials (Li et al., 2005; Li et al., 2006; Chen et al., 2008). In this design, Parylene C is selected as a main structural and packaging material because of its many unique properties, such as flexibility (Young's modulus ~ 4 GPa), chemical inertness, United States Pharmacopoeia (USP) Class VI biocompatibility (Rodger et al., 2006), and lower water permeability compared with other commonly used polymers (e.g. PDMS and polyimide) (Licari & Hughes, 1990). Microfabrication technology has several advantages over the hand-winding including miniaturized structures, precise dimensional control, and feasibility for system integration. The proposed coils will finally be integrated with other system components, such as circuitry, discrete electrical components, a flexible cable, and a high-density multielectrode array, to achieve a complete all-intraocular epiretinal prosthetic system (Rodger et al., 2008), as described in Fig. 1.

This chapter will concentrate on the design, microfabrication, and testing results of two types of Parylene-based intraocular MEMS coils for applications in epiretinal implantation. Specifically, Section 2 will introduce theoretical approaches for coil modeling and design. Coil's electrical properties, including self-inductance, effective series resistance (ESR), and parasitic capacitance, will be discussed theoretically with respect to their geometries. Section 3 describes a Parylene-metal-Parylene thin film technology for making the proposed microcoils. Two various types of microfabricated coil prototypes, including regular dual-layered planar coils and novel fold-and-bond coils, are implemented and characterized in Section 4. Successful data transmissions through such devices have been demonstrated using inductive coupling tests. Experiment also confirms that the quality factor (Q factor) of microcoils can increase proportionally with the increase of the number of metal layers.

Preliminary results suggest that the fold-and-bond technology is a very promising approach for making high Q MEMS coils in a simple, low-cost, and microfabrication compatible manner. Future direction for coil optimization will aim to increase the number of metal layers in order to enhance the Q factor and the power transfer efficiency.

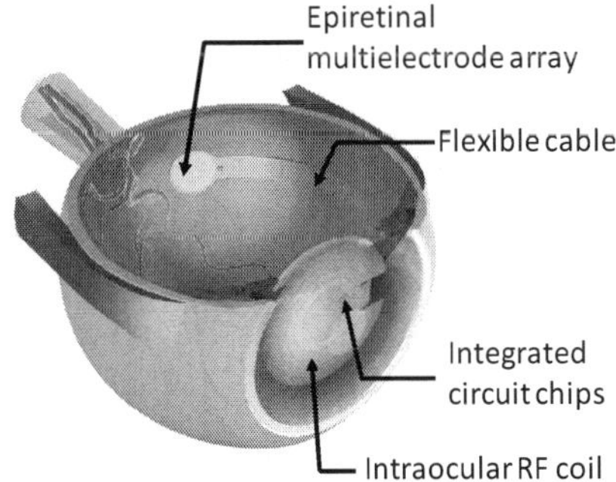

Fig. 1. System schematic of an all-intraocular retinal prosthetic system, which contains two MEMS radio-frequency (RF) coils for power and data transmission, circuitry integrated on a flexible Parylene cable for converting the signal to simulation pulses, and a high-density MEMS electrode array for simulating the neural cells.

2. Modeling of microcoils

Microfabricated coils usually suffer from low self-inductances and inevitable parasitic effects, namely parasitic resistances and capacitances, due to their small physical dimensions and technical constrains of surface micromachining. Particularly, for intraocular retinal implants in human subjects, planar coils with a maximal outer diameter of ~ 10 mm and a minimal inner diameter of ~ 3 mm are desired, which is limited by the space availability in the anterior chamber of human eyes. To better understand the electrical properties and parasitic effects in such small devices, we studied several analytical models and have now applied them to a simplified circular-shape planar coil as illustrated in Fig. 2. In this section, the coil self-inductance, ERS, and parasitic capacitance will be discussed separately with respect to the geometric parameters of devices.

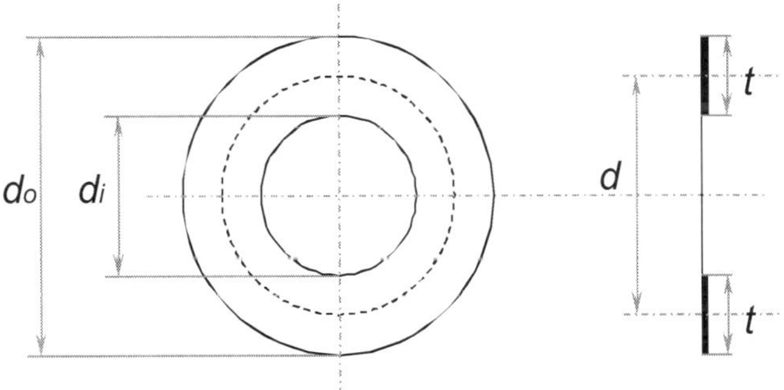

Fig. 2. Simplified model of a circular-shape planar MEMS coil.

2.1 Self-inductance

The self-inductance (L_s) of a multi-layer circular coil with rectangular cross-section can be calculated using the following equation (Dwight , 1945):

$$L_s = 2\pi d(nN)^2 \times 10^{-9}\left[\left(\ln\frac{4d}{t}\right)\left(1+\frac{t^2}{24d^2}\cdots\right)-\frac{1}{2}+\frac{43t^2}{288d^2}\cdots\right] \quad \text{(in Henry)}, \tag{1}$$

where d (in cm) is the mean diameter of the coil, t (in cm) is the coil width, n is the number of turns on each layer, and N is the total number of layers. This expression is valid only when the coil is operated at a low-frequency, i.e., no skin effect. The skin effect can be evaluated using a frequency-dependent factor, which is known as skin depth δ and can be calculated as

$$\delta = \sqrt{\frac{2\rho}{\omega\mu}} \quad \text{(in meter)}, \tag{2}$$

where ρ is the electrical resistivity of metal (in Ω·m), ω is the angular frequency (in rad/s), and μ is the permeability of metal (in H/m). In our proposed system, the data signal is modulated on a ~ 22 MHz carrier, and the power transfer is taken place at a frequency within 1-2 MHz. Therefore, the estimated skin depths at these low frequencies are much bigger than the thickness of metal thin films produced from physical vapor deposition (PVD). In this case, the skin effect can be negligible with an assumption of uniform current distribution in conductive wires.

2.2 Effective series resistance

ERS (R_s) is commonly used to estimate coil losses, which plays an important role in designing a power efficient inductive link. The ESR can be divided into two parts: DC resistance and frequency dependent resistance. Assuming the width of metal traces is much larger than the separation distance between adjacent turns that can be ignored, the DC resistance of the proposed coil can be calculated with the Ohm's law as given in (3),

$$R_s = \rho\frac{n^2 N\pi d}{t \times h} \quad (\Omega), \tag{3}$$

where ρ is the metal resistivity (in Ω·m), and h is the metal thickness (in m). As a result of the skin-effect, the frequency dependent part can be neglected at the low operating frequencies as mentioned earlier. Therefore, the equivalent ESR can be simply written as a DC resistance.

2.3 Parasitic capacitance

Parasitic capacitance (C_s) places a limit to the self-resonant frequency of the coil, above which the coil will not behave as an inductor any more. In a first-order approximation, the parasitic capacitance of a planar MEMS coil usually consists of two main components: the capacitance between turns and the capacitance between layers. A distributed model has been developed to estimate the equivalent parasitic capacitance, as discussed elsewhere (Wu et al. 2003 & Zolfaghari et al. 2001). In this method, a planar coil can be decomposed into equal sections by assuming consistent thickness and width of metal traces everywhere. The voltage profile can then be obtained by averaging the beginning and ending potential across the coil structure. With known voltage variations between the correlated sections of adjacent turns and layers, the total capacitive energy stored in the coil structure can be calculated using the distributed capacitance of each segment. The equivalent capacitance can then be approximated from the distributed capacitances, using the ideal double plate capacitor formula.

$$C_{eq-turn} = \sum_{k=1}^{n-1} \frac{1}{4} C_{ii} l_k [d(k+1) - d(k-1)]^2, \tag{4}$$

$$\begin{aligned} C_{eq-layer} &= \frac{1}{4} \sum_{k=1}^{N} (C_{m,m-1} + C_{m-2,m-3} + \ldots) \frac{A_k}{m^2} [4 - 2d(k-1) - 2d(k)]^2 \\ &+ \frac{1}{4} \sum_{k=1}^{N} (C_{m-1,m-2} + C_{m-3,m-4} + \ldots) \frac{A_k}{m^2} [2d(k-1) + 2d(k)]^2, \end{aligned} \tag{5}$$

$$C_{eq-total} = C_{eq-turn} + C_{eq-layer}, \tag{6}$$

Equations (4), (5), and (6) show the analytical formulas for calculating parasitic capacitance, where C_{ii} (in F) denotes the capacitance per unit length between adjacent metal turns, $C_{m,\,m-1}$ (in F) is the capacitance per unit area between the m-th and (m-1)-th metal layer, A_k (in m^2) is the trace occupied area of the k-th turn on each layer, and $d(k)=h_1+ h_2+\ldots+ h_k$, in which h_k is defined as the ratio of the wire length of the k-th turn (l_k) to the total wire length (l_{tot}). This simplified model neglects the second order parasitic capacitances between non-adjacent turns and layers, which are much less than the first order capacitances.

2.4 Quality factor

Q factor is an important metric for evaluating the efficiency of a coil, which is theoretically defined as the ratio of total stored energy to dissipated energy per cycle in a resonating system. With known L_s, R_s, and C_s, the Q factor of the coil can be derived from a 3-element circuit model (Fig. 3) (Wu, 2003). In order to obtain the coil Q factor mathematically, the total equivalent impedance (Z_s) is first studied, which can be written as the sum of a real resistance and an imaginary reactance (equation [7]).

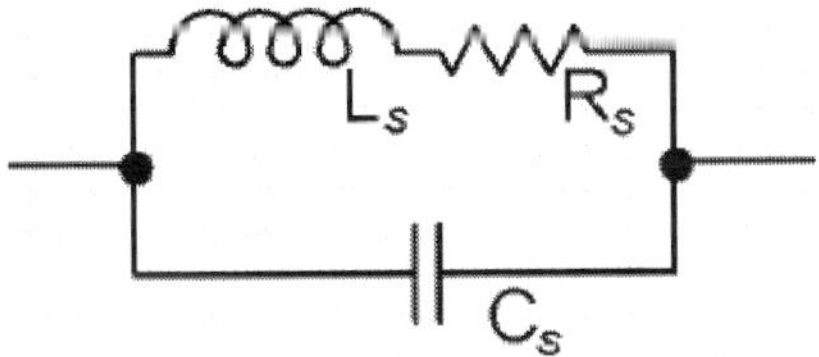

Fig. 3. Equivalent RLC circuit of a planar MEMS coil.

$$Z_s = \frac{R_s}{(1-\omega^2 L_s C_s)^2 + (\omega C_s R_s)^2} + j\frac{\omega(L_s - R_s^2 C_s - \omega^2 L_s^2 C_s)}{(1-\omega^2 L_s C_s)^2 + (\omega C_s R_s)^2}. \tag{7}$$

Then the self-resonant frequency ω_s can be expressed as:

$$\omega_s = \sqrt{\frac{1}{L_s C_s} - \frac{R_s^2}{L_s^2}} \approx \sqrt{\frac{1}{L_s C_s}}, \text{ when } R_s \ll \sqrt{\frac{L_s}{C_s}}. \tag{8}$$

For a retinal implant system, when both the external and internal units of the inductive link are tuned to a same resonant frequency ω_r, the maximal coupling energy can be delivered to the implanted system. In this case, the coil Q factor can be expressed with the following equation:

$$Q_r = \frac{\mathrm{Im}(Z_s)}{\mathrm{Re}(Z_s)} \approx \frac{\omega_r L_s}{R_s} \tag{9}$$

Combining with equations (1) and (3), equation (9) can be rewritten as

$$Q_r = \frac{2\omega_r Nth}{\rho}\left[\left(\ln\frac{4d}{t}\right)\left(1+\frac{t^2}{24d^2}\cdots\right)-\frac{1}{2}+\frac{43t^2}{288d^2}\cdots\right]\times 10^{-9} \tag{10}$$

Ideally, the Q factor of a coil should be as high as possible in order to minimize the power loss in the device as well as to maximize power transfer efficiency of the system. It can be seen from equation (10) that Q_r can be enhanced by increasing the number of coil layers (N), the coil width (t), and/or the thickness of the conductive layer (h). For an intraocular retinal implant, however, there is not much zoom to improve the coil width due to the small coil dimensions (inner diameter, outer diameter, etc.) confined by the eyeball size. Therefore, a more applicable way to increase a coil's Q factor is to increase the number of stacking layers as well as the thickness of conductive wires.

2.5 Finite element simulation

To validate the effectiveness of the theoretical models, finite element simulations (FES) are performed using a built-in package in CoventorWare (Coventor Inc., Cary, NC). As a demonstration, a coil with two layers of metal is designed, and its electrical characteristics are evaluated using both analytical models and FES, as summarized in Fig. 4 and Table 1.
During the simulation, an octagonal coil is used to approximate a circular shape due to memory constraint in CoventorWare. The coil self-inductance and the ESR are simulated over a frequency range from 10 kHz to 1 GHz. It can be seen that the self-inductance at 1 MHz shows only 2.2% deviation, and the ESR deviates by less than 6%, suggesting good agreement with the analytical models. The slight deviations might be introduced by the approximation of coil shape. Note that L_s and R_s both remain stable at frequencies below 10 MHz, indicating that skin effect or proximity effect is negligible at target frequencies of 1 or 2 MHz.

	OD (mm)	ID (mm)	Trace cross section (μm × μm)	Number of turns	L_s (μH)	R_s (Ω)	C_s (nF)	Q at 1MHz
Calculations	10	3	220 × 2	28 / Layer	5.0	28.9	65.5	1.1
FES	10	3	220 × 2	28 / Layer	4.9	27.4	--	1.12

Table 1. Coil characteristics estimated using both analytical models and FES.

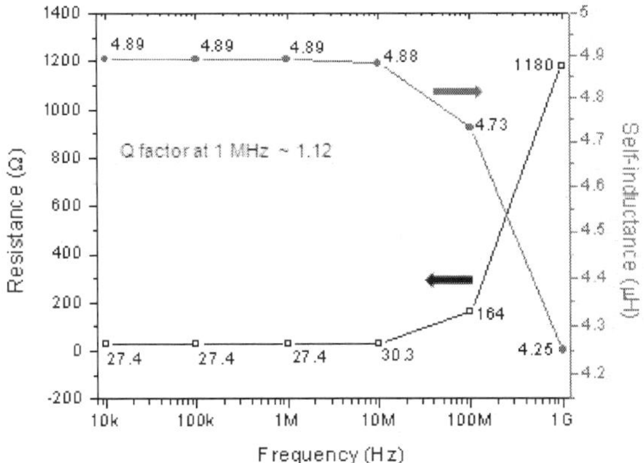

Fig. 4. Simulated self-inductance and ESR of the sample coil.

3. Parylene-metal-Parylene thin film technology

A multi-layer Parylene-metal thin film technology for making the proposed microcoils has been developed (Li et al., 2005). In this approach, thin-film metal conductive wires are sandwiched between multiple layers of Parylene C and interconnections between two adjacent layers are implemented using through holes in the Parylene insulation layer. Fig. 5 depicts a typical process flow for making a dual-metal-layer structure. Briefly, a layer of sacrificial photoresist is optionally spun on a standard silicon wafer, followed by Parylene C deposition (PDS 2120 system, Special Coating Systems, Indianapolis, IN, USA) (*Step 1*). A layer of metal is then deposited on top of the Parylene using an electron beam (e-beam) evaporator (SE600 RAP, CHA Industries, Fremont, CA, USA), and patterned using a wet etching process (*Step 2*). After that, a thin layer of Parylene C is deposited as an insulation layer between two metal layers, and the interconnection vias are selectively opened with oxygen plasma in a reactive ion etch system (RIE) (Semi Group Inc. T1000 TP/CC) using a photoresist mask(*Step 3*). After removing the photoresist mask, the second metal layer is evaporated and patterned, followed by another Parylene C deposition to conformally cover the exposed metal wire (*Step 4*). A photoresist mask is then patterned to expose the contact pads, as well as to define the coil shape (*Step 5*). Finally, oxygen plasma etch is performed to remove unwanted Parylene C, and the entire flexible device is released from the silicon substrate by dissolving the sacrificial photoresist (*Step 6*).

The reliability of the interconnections between nearby metal layers highly depends on the step coverage of the Parylene sidewall during metal evaporation, which can be improved by a slightly isotropic O_2 plasma etch (Meng et al., 2008). A special design of rotating wafer holder inside the e-beam evaporator also helps adjust the angle of attack of metal evaporant for best coverage. Microcoils comprising more than two layers of metal can be fabricated with similar procedure by alternating the Parylene C deposition, interconnection via fabrication, and metal evaporation process steps. Although it is specifically developed for microcoil fabrication, this technology can also be applied to the fabrication of other flexible, implantable devices with multi-layer Parylene-metal structures, such as dual-metal-layer electrode arrays (Rodger et al., 2008).

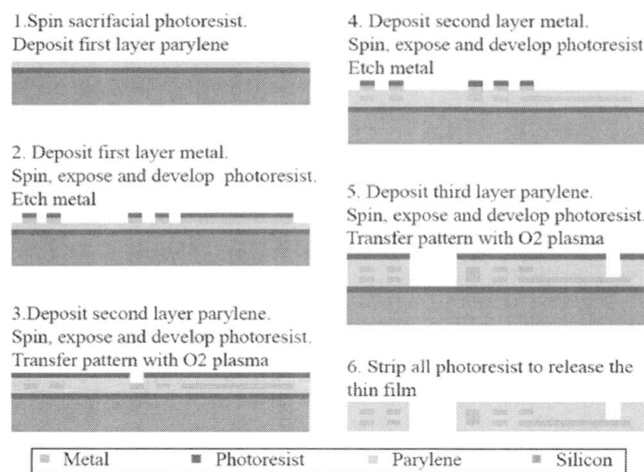

Fig. 5. Fabrication process flow of a dual-metal-layer Parylene-based MEMS structure.

For implantable devices, our Parylene-metal thin film technology has several unique advantages compared with conventional semiconductor-based microfabrication technologies. Using biocompatible Parylene directly as the actual substrate greatly simplifies the device integration and packaging procedures. Devices fabricated in this way are very flexible and foldable so that they can be implanted through small surgical incisions, allowing wounds to heal quickly. Moreover, the metal lines are completely padded by the Parylene material, and can therefore withstand repeated bending during surgical handling. Finally, a post-fabrication heat-molding process has been developed to modify the skins into various shapes that match the curvatures of the target implant areas (Tai et al., 2006).

4. Coil designs and fabrication results

4.1 Dual-layered MEMS coil

In this section, a planar coil is designed, which features: 1) dual-layer thin-film metal conductive wires sandwiched between multiple layers of Parylene C, and 2) interconnections between two adjacent layers that are formed by filling the Parylene through holes with PVD metal. Fig. 6 shows the microscope images of a fabricated coil and its interconnection via. This coil has totally 10 turns of wires made of approximately 2000 Å multiple layers of Ti/Au/Ti metallization. Titanium serves as an adhesion promoter to improve the bonding strength between gold and Parylene C. The device has overall dimensions of ~ 9.5 mm in outer diameter, ~ 5 mm in inner diameter, and ~ 11 μm in thickness, limited by the lens capsule size of the human eyes. The interconnection via occupies an area of ~ 0.06 mm^2 with a contact resistance of less than 1 Ω, which can be negligible compared with the total coil ESR. The device is proven to be very flexible and foldable (Fig. 7), easing the procedure of surgical insertion and lessening physical damages in the region of implantation.

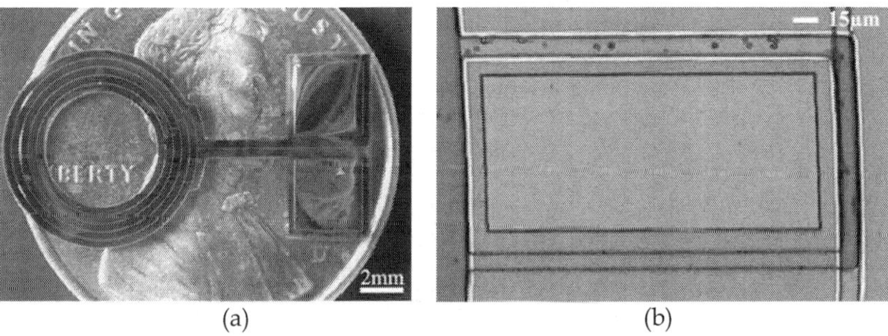

(a) (b)

Fig. 6. (a) A fabricated dual-metal-layer coil sitting on a penny. (b) The microscope image shows the interconnection via between two metal layers. (Li et al., @ 2005 IEEE)

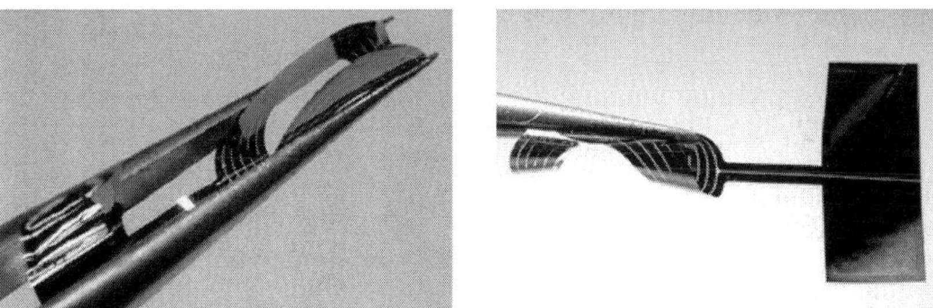

Fig. 7. Demonstration of the coil's flexibility and foldability. (Li et al., @ 2005 IEEE)

The electrical properties of the fabricated coil are characterized experimentally. Recall equation (7) in Section 2, by setting the derivation of the real part to zero and equating the imaginary part to zero, the self-inductance (L_s) and the parasitic capacitance (C_s) can be extracted using equations (11) and (12), where ω_0 is defined as the frequency at which the real part of the impedance is maximum, and ω_z is the zero-reactance frequency at which the imaginary part of the impedance is zero (Wu, 2003).

$$L_S = \frac{R_s}{\sqrt{2(\omega_0^2 - \omega_z^2)}}, \tag{11}$$

$$C_S = \frac{\sqrt{2(\omega_0^2 - \omega_z^2)}}{R_s(2\omega_0^2 - \omega_z^2)}. \tag{12}$$

For the coil in Fig. 6, the ESR (R_s) is measured to be around 72 Ω and the resistivity of e-beam deposited gold is calculated to be around 2.25×10^{-6} Ω·cm. This number agrees with the resistivity of bulk gold (2.2×10^{-6} Ω·cm), implying that the E-beam evaporated metal is void-free. The coil impedance is swept with an HP 4192A LF impedance analyzer over a frequency range from 5 Hz to 13 MHz. From the impedance versus frequency curves (Fig. 8), f_0 and f_z can be read with values of 7.5 MHz and 3.3 MHz respectively. Knowing R_s, ω_0, and ω_z, the coil self-inductance and capacitance are therefore calculated as L_s = 1.19 μH and

C_s = 201 pF. The theoretical numbers are also calculated using the abovementioned equations, and the fitting curves are plotted in Fig. 8, in comparison with the measured curves. The experimental data matches the theoretical calculations closely, with deviations of less than 19%. These errors may be attributed to the simplification of the 3-element model as well as interferences from the measurement instruments. The Q factor of the coil is obtained to be approximately 0.1 at the target frequency of 1 MHz, as expected from the design.

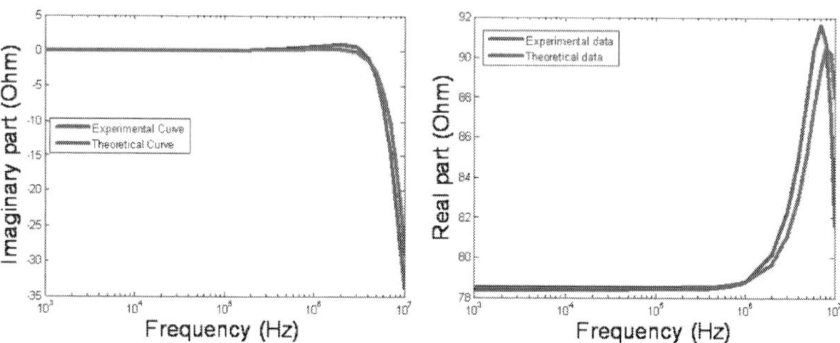

Fig. 8. Impedance measurement and curve fitting using the 3-element model: (a) Imaginary part; (b) Real part. (Red curves correspond to theoretical parameters of the fabricated coil: L_s = 1.0 µH, R_s = 67 Ω and C_s = 183 pF.)

The data and power transfer performances have also been verified using a custom designed data link at the University of California, Santa Cruz (UCSC). The testing waveforms are shown in Fig. 9, where the blue curve represents the data driving signal on the primary stage, the green curve represents the voltage across the primary coil, and the purple curve represents the receiving voltage across the secondary coil.

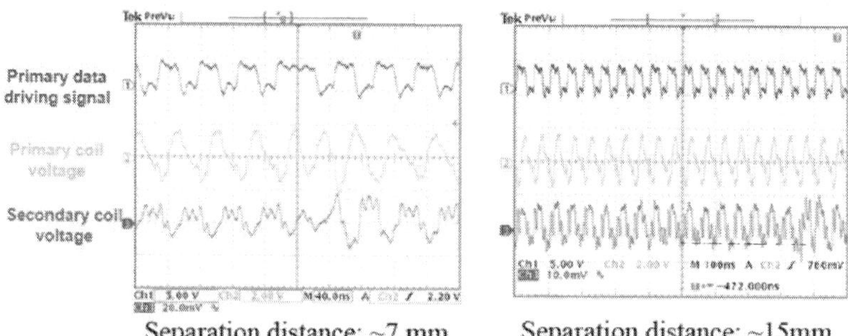

Fig. 9. Inductive coupling test waveforms: (a) received signal is 25 mV peak to peak; (b) received signal is 15 mV peak to peak.

While successful data transmission through our coils has been demonstrated, it is noted that this device has no driving capability due to its small Q factor (~ 0.1), meaning that the power cannot be delivered to the load. Therefore, enhancing coil's Q factor to achieve a higher power transfer efficiency is crucial for designing the next generation of coils.

Experiments have been done at UCSC to study the feasibility of using MEMS coils as the receiver coil for the current inductive link design. A rough estimate is that, in the worst case, a minimal Q factor of 10 will be needed in order to deliver ~100 mW for chip operation and stimulation. From the theoretical analysis herein, it is known that the Q factor can be enhanced by increasing the metal thickness and/or the number of metal layers. However, e-beam evaporated metals are usually limited in film thickness due to high process cost. Electroplated and sputtered metals can be thicker alternatives but their qualities, such as density and conductivity, are typically not as good as evaporated metals. This problem becomes more serious especially when devices are implanted inside harsh biological environments. From the device design aspect, increasing metal layers is more practical for the Q factor enhancement of MEMS coils, thus a fold-and-bond technology emerges as a good candidate.

4.2 Fold-and-bond coil

In the concept of fold-and-bond technology (Fig. 10), two or more thin-film planar spiral coil segments are fabricated from the same batch so that each segment has identical self-inductance and ESR, denoted by L_s and R_s, respectively. A new coil can then be formed by stacking *n* segments together in either parallel or series connections. For parallel stacking, particularly, the new coil will have identical inductance but *n*-times larger in equivalent metal thickness, resulting in an *n*-times lower ESR. As for series stacking, the resistance remains the same while the inductance increases by *n*-times because of the mutual inductance between adjacent layers. According to the definition of the Q factor, both approaches can achieve an *n*-times Q factor enhancement. In this section, the series stacking configuration is used to demonstrate the technology concept.

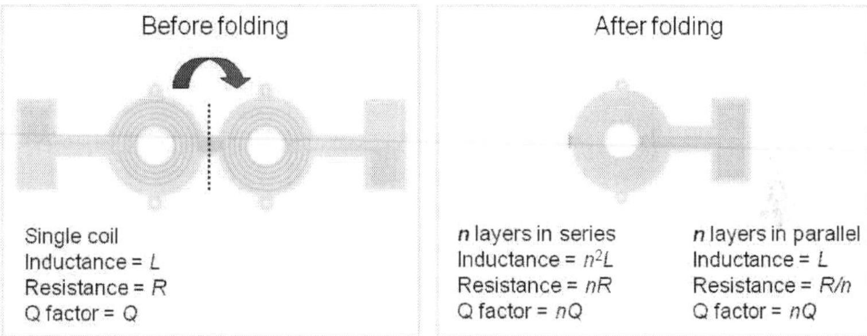

Fig. 10. Concept of the fold-and-bond technology for Q factor enhancement
A coil with one fold is depicted for representation.

Fold-and-bond coil's fabrication involves the dual-metal-layer Parylene/metal skin realization and a post-fabrication thermal bonding process (Li et al., 2008). The Parylene-metal skin with two buried layers of metal is first fabricated in the same manner as described in Fig. 5, in which one layer of metal is used to form the conductive wires of the coil, while the other layers is used to make the interconnections between the layers. This thin film skin can be folded and stacked into multiple layers because of the flexibility of Parylene C. While hand alignment under an optical microscope is used at the current stage, special alignment jigs can be custom designed in the future to achieve precise alignment of different layers. During the thermal bonding procedure, the folded device is sandwiched between two glass slides covered with aluminum sheets, which can avoid Parylene sticking on the glass. The whole unit is placed in a vacuum oven and bonded at 230 °C for two days. External pressure can be applied as needed to enable Parylene-Parylene bonding at moderate temperatures. The vacuum pressure is controlled at ~ 10 Torr to prevent Parylene C from unwanted oxidation in air at an elevated temperature.

Two Parylene-based skins with dual-layer embedded metal have been fabricated, as shown in Fig. 11. These prototypes are specifically designed for intraocular retinal prosthesis with the design parameters described in Table 2. The thickness of metal wires is increased to ~ 2 μm in order to further reduce the coil's DC resistance. The metal is covered with ~ 3.4 μm Parylene C on each side with the lead contact vias open. Fig. 12 shows the final devices after folding and thermal bonding. Misalignments of 10 μm to 30 μm have been observed, which is due to the lack of control with the hand alignment.

Type	Total number of turns	R_s (Ω)	L_s (μH)	C_s (pF)	Q at 1MHz
I	28	29	2.9	65.5	1.1
II	48	90	14.8	64.5	1

Table 2. Design specifications of two fold-and-bond coils.

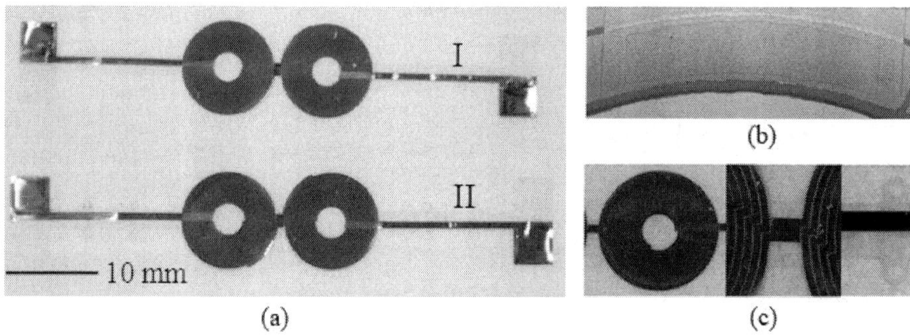

Fig. 11. Devices before thermal bonding: (a) Fabricated dual-metal-layer Parylene-based skins; (b) Microscope image of an interconnection via between two metal layers; (c) Photos of device details (from left to right): conductive wires of the coil, folding junction and suturing holes.

The devices still remain flexible after bonding (Fig. 13 (a)), indicating that annealing at a temperature below the melting point of Parylene C (290 °C) (Harder et al. 2002) will not alter the mechanical flexibility of the material. The DC resistances of the samples are measured before and after bonding with no significant change observed (Table 3), confirming the ductility and durability of metal traces. Stretching marks and Parylene cracks are found along the folding line after thermal treatment, which is caused by stress concentration during folding. Additional Parylene coating can be performed after thermal bonding to conformally cover these cracks in order to ensure a good sealing for final devices.

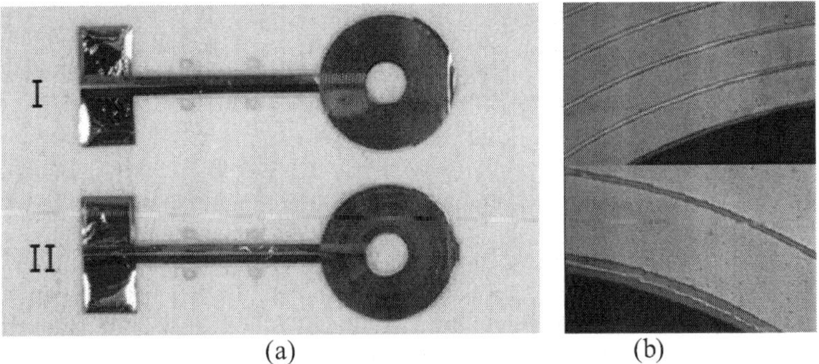

(a) (b)

Fig. 12. Devices after thermal bonding: (a) Fold-and-bond coils after thermal bonding. (b) Overlapping metal wires with misalignments of 10 μm to 30 μm.

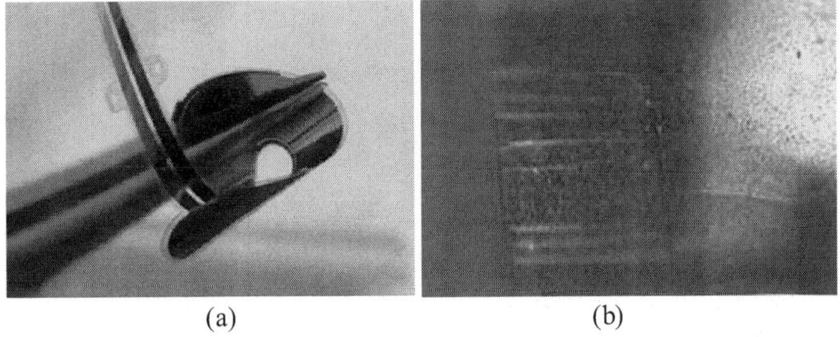

(a) (b)

Fig. 13. (a) Demonstration of device's flexibility; (b) Stretching marks along the folding line.

The electrical characteristics of the fabricated fold-and-bond coils are studied before and after thermal bonding. The lump parameters are extracted from the same 3-element model, as summarized in Table 3. As expected, the self-inductance and Q-factor of the coil are increased by more than 90% for both coil designs after folding. Changing the wire width has no significant impact on the parasitic capacitance, indicating the capacitance between adjacent layers is dominant over the capacitance between turns. Coil prototype I shows a much higher ESR compared to its theoretical value (29 Ω), which may be attributed to the non-uniformity of metal thickness. Overall, the measured values show good agreement with the theory predictions, demonstrating that the theoretical model can predict the coil properties effectively.

Type		R_s (Ω)	L_s (μH)	C_s (pF)	Calculated Q at 1MHz	Q Increase
I	Before	41.2	2.9	--	0.44	--
	After	41.8	5.7	62.1	0.85	95%
II	Before	91.3	8.6	--	0.58	--
	After	92.2	16.3	70.8	1.11	91%

Table 3. Measured electrical parameters of fold-and-bond coils using the 3-element model.

To study the power transfer efficacy, the fabricated devices are tested using a simplified inductive link, as shown in Fig. 14. In this setup, the transmitter coil is hand-wound with a self-inductance of ~ 23 μH and a series resistance of ~ 1.5 Ω. The inner diameter of the transmitter coil is optimized to be ~ 30 mm (Ko et al., 1977). The receiver coil is the fold-and-bond coil presented in Fig. 12. During the measurements, the primary stage is driven by an HP E3630A function generator with a sinusoidal input signal of 20 V peak to peak. Both the primary and secondary circuits are subject to parallel resonance with the same resonant frequencies of ~ 1 MHz. To minimize environmental interferences, both the transmitter and receiver coils are covered with aluminum foils as electromagnetic shielding.

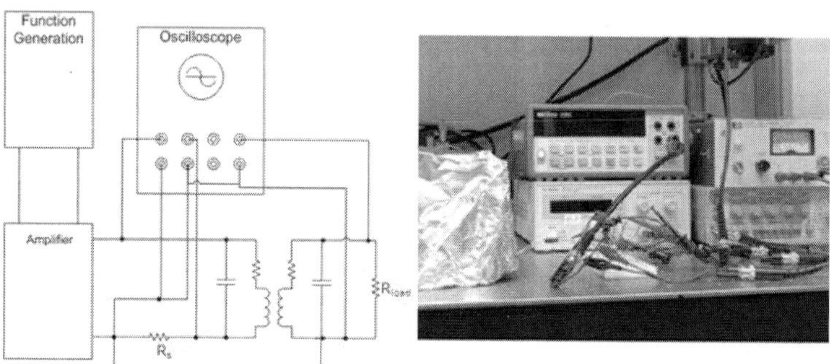

Fig. 14. Experimental setup of power transmission measurement and its circuit diagram.

Preliminary experiments have been performed. A 10 Ω series resistor (R_s) is incorporated in the primary stage to monitor the output current from the amplifier. The voltage across the transmitter coil is then calculated by subtracting the resistor voltage from the output voltage of the amplifier. The transferred power, which is defined as the power delivered to a 1 kΩ load resistor (R_{load}), can be obtained by directly measuring the voltage across the load resistor. The power transfer efficiency of two coil prototypes, which is calculated as the ratio of the transferred power to the total output power from the amplifier, is plotted in Fig. 15, as

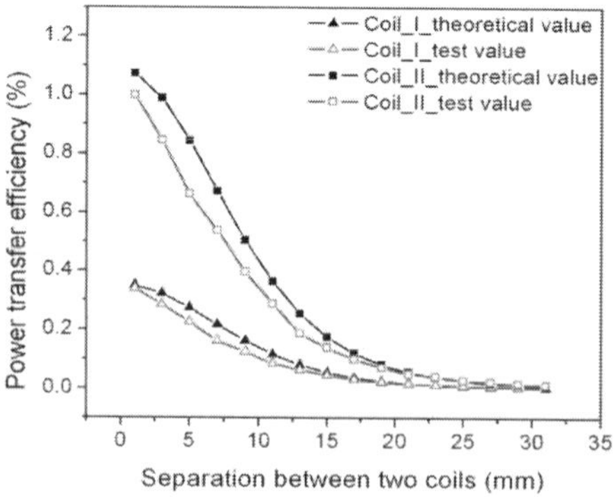

Fig. 15. Power transfer efficiencies of the fold-and-bond coils at 1 MHz, as functions of separation distances of the coil pair.

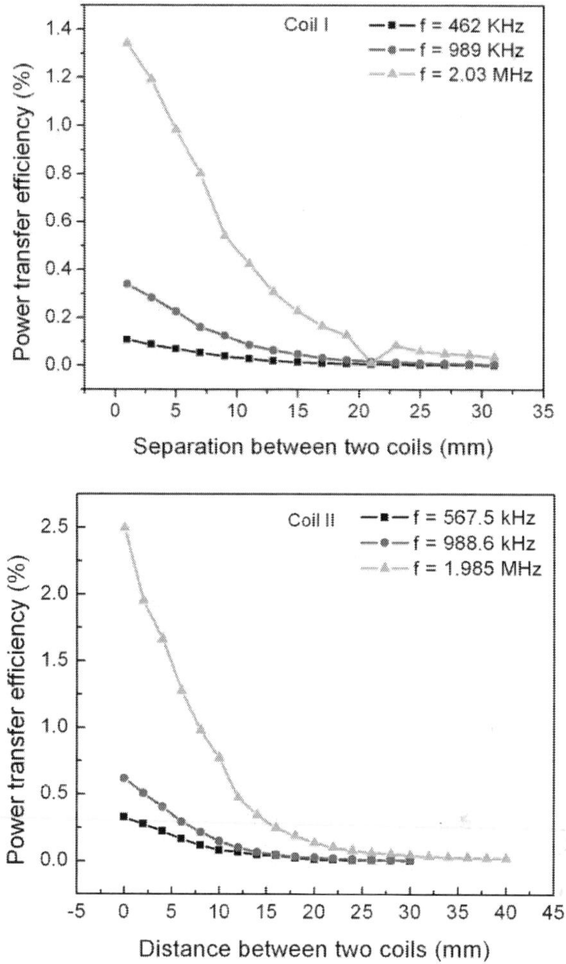

Fig. 16. The power transfer efficiency of difference devices vs. the separation distance of the coil pair.

functions of the separation distance between the coil pairs. For comparisons, theoretical data are also studied based on the models discussed elsewhere (Ko et al., 1977), which match the testing results within a reasonable range. As expected, the coil prototype II with higher self-inductance and Q factor exhibits higher power transfer efficiency at the same separation distances. The power transfer efficiencies at different operation frequencies have also been investigated, as plotted in Fig.16. The results show that the power transfer efficiency can be enhanced by more than 3 times when the operation frequency is changed from ~1 MHz to ~2 MHz. This can be attributed to the effective Q factor increases by almost two times as the frequency goes up.

Nevertheless, the power transfer efficiency of the current coil prototypes is still below 0.4% at the target implantation depth of ~ 15 mm, which is insufficient to drive electronics for high-density retinal stimulation. Whereas more power can be generated from the primary stage to compensate the low efficiency, the overheating issue of the coils becomes prominent. Therefore, future direction for coil optimization will mainly focus on increasing the number of metal layers to improve the Q factor. Theoretical evaluation predicts that at least 20 layers of metal will be required in order to achieve a reasonable Q factor of around 10, with a given gold thickness of ~ 2 μm.

4.3 Challenges in packaging

Device packaging for retinal prostheses presents several challenges. Electrical and fluidic isolation must be maintained to avoid device failures upon exposure to corrosive eye fluid. To ensure long-term use of implants, the interface between device and biological tissues should be stable and biocompatible in order to minimize inflammation and immune response. Packaging complication is also one of the significant challenges for high-density neural simulation/recording. Careful selection and evaluation of packaging materials and tools are necessary to address such challenges. We have studied the packaging performance of Parylene C using accelerated lifetime soak testing in heated saline. Preliminary results estimate that the lifetime of Parylene-coated metal at body temperature (37 °C) is more than 60 years, suggesting good packaging performance of Parylene C. Experimental details on Parylene packaging evaluation are discussed elsewhere (Li et al., 2010).

A chip-level integrated interconnect (CL-I^2) packaging method has also been explored for integration of microcoils with CMOS integrated circuit (IC) chips and high-density prosthetic electrodes (Rodger et al., 2005 & Li et al. 2010). In this method, all the IC components necessary for a retinal implant can be embedded in silicon cavities and functional Parylene-based MEMS devices (e.g. microcoil and high-density electrode array) can then be fabricated on the same platform using the abovementioned Parylene-metal thin film technology. Chip-to-microdevice interconnections can be constructed using standard microfabrication techniques such as photolithography and metal etching; therefore, eliminating wire-bonding, bump-bonding, or soldering steps. Whereas initial experiments show promising results, continuous investigations will be necessary to collect more data in order to optimize integration process and to further refine our knowledge of Parylene packaging behavior.

5. Conclusion

In this work, various types of MEMS coils have been designed and fabricated using the Parylene-metal-Parylene skin technology. Experiments have been performed to measure the electrical properties of the coils and the results show good agreement with the theoretical values. The data transfer effect has been successfully demonstrated with the telemetry link setup. However, the power transfer efficiency at the separation distance of 15 mm is below 0.4%, which is relatively low for high-density retinal stimulation. Given the constraints of device geometries, it is believed that increasing the number of metal layers will be the most effective and applicable way to enhance the Q factor of microcoils. According to the analytical models, at least 20 layers of metal will be needed in order to achieve a coil Q factor of approximately 10. While direct fabrication could be too complicated to be carried out, the fold-and-bond technology has proven itself as a very promising fabrication technology. Although specifically tailored to the needs of retinal prostheses, because our coils are fully micromachined in a way compatible with multielectrode arrays and the Parylene-based embedded chip packages, these devices can be easily integrated with various system components to achieve a new range of true system solutions for both biomedical and non-biomedical applications

6. Acknowledgment

This work is supported in part by the Engineering Research Center Program of the National Science Foundation under Award Number EEC-0310723 and by a fellowship from the Whitaker Foundation (D.R.). The authors would like to acknowledge Dr. Yang Zhi for his supports on the coil testing. We also want to thank Mr. Trevor Roper, Dr. Wen-Cheng Kuo, and other members at the Caltech Micromachining Laboratory for assistance with device simulation and fabrication.

7. References

Artificial Retina Project. (2007). Retinal Diseases: Age-Related Macular Degeneration and Retinitis Pigmentosa. Available from http://artificialretina.energy.gov/diseases.shtml.

Bennett, J.; Tanabe, T.; D. Zeng, Sun, Y.; Kjeldbye, H.; Gouras, P. & Maguire, A.M. (1996). Photoreceptor Cell Rescue in Retinal Degeneration (RD) Mice by In Vivo Gene Therapy. *Nature Medicine*, Vol. 2, (June 1996), pp. 649-654, ISSN 1078-8956.

Chen, P.J.; Kuo, W.C.; Li, W. & Tai, Y.C. (2006) Q-enhanced Fold-and-bond MEMS Inductors. *Proc. IEEE Int. Conf. on Nano/Micro Engineered and Molecular Systems*, ISBN 978-1-4244-1907-4, Sanya, Hainan Island, China, January 2008.

Chow, A.Y.; Chow, V.Y.; Packo, K.H.; Pollack, J.S.; Peyman, G.A. & Schuchard, R. (2006). The Artificial Silicon Retinamicrochip for the Treatment of Vision Loss from Retinitis Pigmentosa. *Arch Ophthalmol*, Vol. 122, (April 2004), pp. 460-469, ISSN 0003-9950.

Dwight, H.B. (1945). *Electrical Coils and Conductors*, McGraw-Hill, ASIN B0007IT6Z0, New York, United States.

Harder, T.; Yao, T.J.; He, Q.; Shih, C. Y. & Tai, Y.C. (2002). Residual Stress in Thin-film Parylene C. *Proc. IEEE Int. Conf. on Micro Electro Mechanical Systems*, ISBN 0-7803-7185-2, Las Vegas, United States, January 2002.

Horch K.W. & Dhillon, G.S. (2004). *Neuroprosthetics Theory and Practice (Series in Bioengineering & Biomedical Engineering-Vol.2)*, World Scientific Publishing Company, ISBN 9812380221, Singapore.

Humayun, M.S.; Propst, R.; Eugene de Juan Jr.; McCormick, K. & Hickingbotham, D. (1994). Bipolar Surface Electrical Stimulation of the Vertebrate Retina. *Arch Ophthalmol*, Vol. 112, (January 1994), pp. 110-116, ISSN 0003-9950.

Humayun, M.S.; Eugene de Juan Jr.; Weiland, J.D.; Dagnelie, G.; Katona, S.; Greenberg, R. & Suzuki, S. (1999). Pattern Electrical Stimulation of the Human Retina. *Vision Research*, Vol. 39, (July 1999), pp. 2569-2576, ISSN 0042-6989.

Javaheri, M.; Hahn, D.S.; Lakhanpal, R.R.; Weiland, J.D. & Humayun, M.S. (2006). Retinal Prostheses for the Blind. *Ann Acad Med,* Vol. 35, (March 2006), pp. 137-144, 2006, ISSN 0304-4602.

Kim, S.Y.; Sadda, S.; Pearlman, J.; Humayun, M.S.; Eugene de Juan Jr. & Green, W.R. (2002). Morphometric Analysis of the Macula in Eyes with Disciform Age-realted Macular Degeneration. *Retina,* Vol. 22, No. 4, (August 2002), pp. 471-477, ISSN 0275-004X.

Ko, W.H.; Liang, S.P. & Fung, C.D.F. (1977). Design of Radio-frequency Powered Coils for Implant Instruments. *Med. Bio. Eng. Comput.*, Vol. 15, (November 1977), pp. 634-640, ISSN 0140-0118.

Li, W.; Rodger, D.C.; Weiland, J.D; Humayun, M.S. & Tai, Y.C. (2005). Integrated Flexible Ocular Coil for Power and Data Transfer in Retinal Prostheses. *Proc.27th Ann. Int. IEEE-EMBS Conf.*, ISBN 0-7803-8741-4, Shanghai, China, January 2005.

Li, W.; Rodger, D.C.; Meng, E.; Weiland, J.D.; Humayun, M.S. & Tai, Y.C. (2006). Flexible Parylene Packaged Intraocular Coil for Retinal Prostheses. *Proc. 4th Int. IEEE-EMBS Special Topic Conf. on Microtechnologies in Medicine and Biology,* ISBN 1-4244-0338-3, Okinawa, Japan, May 2006.

Li, W.; Rodger, D.C. & Tai, Y.C. (2008). Implantable RF-coiled Chip Packaging. *Proc. IEEE Int. Conf. on Micro Electro Mechanical System*, ISSN 1084-6999, Tucson, United States, January 2008.

Li, W.; Rodger, D. C.; Meng, E.; Weiland, J. D.; Humayun, M. S. & Tai, Y.-C. (2010) Wafer-Level Parylene Packaging with Integrated RF Electronics for Wireless Retinal Prostheses. *Microelectromechanical Systems, Journal of,* Vol. 19, (June 2010) pp. 735-742, ISSN 1057-7157.

Licari J. J. & Hughes, L. A. (1990). *Handbook of Polymer Coating for Electronics: Chemistry, Technology, and Applications,* William Andrew Publishing/Noyes, ISBN 081551235X, Park Ridge, New Jersey, United States.

MacLaren, R.E.; Pearson, R.A.; MacNeil, A.; Douglas, R.H.; Salt, T.E.; Akimoto, M.; Swaroop, A.; Sowden, J.C. & Ali, R.R. (2006). Retinal Repair by Transplantation of Photoreceptor Precursors. *Nature,* Vol. 444, (November 2006), pp. 203-207, ISSN 0028-0836.

Meng E., Li P.Y. & Tai, Y.C. (2008) Plasma Removal of Parylene C, *Journal of Micromechanics and Microengineering,* Vol 18, (February 2008), pp. 045004, ISSN 1361-6439.

Mokwa, W.; Goertz, M. C.; Krisch, K. I.; Trieu, H.-K. & Walter, P. (2008) Intraocular Epiretinal Prosthesis to Restore Vision in Blind Humans. *Proc. 30th Ann. Int. IEEE EMBS Conf.,* ISBN 978-1-4244-1814-5, Vancouver, British Columbia, Canada, August 2008.

Norton, E.W.D.; Marmor, M.F.; Clowes, D.D.; Gamel, J.W.; Barr, C.C.; Fielder, A.R.; Marshall, J.; Berson, E.L.; Rosner, B.; Sandberg, M.A.; Hayes, K.C.; Nicholson, B.W.; Weigel-DiFranco, C.; Willett, W.; Felix, J.S. & Laties, A.M. (1993). A Randomized

Trial of Vitamin A and Vitamin E Supplementation for Retinitis Pigmentosa. *Arch Ophthalmol*, Vol. 11, (June 1993), pp. 1460-1466, ISSN 0003-9950.

Rizzo, J.F.; Wyatt, J.L.; Loewenstein, J.; Montezuma, S.; Shire, D.B.; Theogarajan, L. & Kelly, S.K. (2004). Development of a Wireless, Ab Externo Retinal Prosthesis. *Invest Ophthalmol Vis. Sci.*, Vol. 45, pp. 3399.

Rizzo, J.F. (2011). Update on Retinal Prosthetic Research: The Boston Retinal Implant Project. *Journal of Neuro-Ophthalmology,* Vol.32, (June 2011), pp. 160-168, ISSN 1070-8022.

Rodger, D. C.; Weiland, J. D.; Humayun, M. S. & Tai, Y.-C. (2005). Scalable Flexible Chip-level Parylene Package for High Lead Count Retinal Prostheses. *Proceedings of the 13th International Conference on Solid-State Sensors, Actuators and Microsystems*, ISBN 0-7803-8994-8, Seoul, South Korea, June 2005.

Rodger, D.C.; Weiland, J.D.; Humayun, M.S. & Tai, Y.C. (2006), Scalable High Lead-count Parylene Package for Retinal Prostheses. *Sensors and Actuators B: Chemical*, Vol. 117, (September 2006), pp. 107-114, ISSN 0925- 4005.

Rodger, D.C.; Fong, A.J.; Li, W.; Ameri, H.; Ahuja, A.K.; Gutierrez, C.; Lavrov, I.; Zhong, H.; Menon, P.R.; Meng, E.; Burdick, J.W.; Roy, R.R.; Edgerton, V.R.; Weiland, J.D.; Humayun, M.S. & Tai, Y.C. (2008). Flexible Parylene-based Multielectrode Array Technology for High-density Neural Stimulation and Recording. *Sensors and Actuators B: Chemical*, Vol. 132, (June 2008), pp. 449-460, ISSN 0925- 4005.

Stieglitz, T.; Haberer, W.; Lau, C. & Goertz, M. (2004). Development of an Inductively Coupled Epiretinal Vision Prosthesis. *Proc. Int. IEEE Eng. in Med. and Biol. Soc. Meet.*, ISBN 0-7803-8439-3, San Francisco, CA, USA, September 2004.

Tai, Y.C.; Rodger, D.C.; Li, W. & Tooker, A. (2006). *Method for Decreasing Chemical Diffusion in Parylene and Trapping at Parylene-to-parylene Interfaces,* US Patent Applicatio 11/408809. Available from
http://www.freepatentsonline.com/y2006/0255293.html

Tropepe, V.; Coles, B.L.K.; Chiasson, B.J.; Horsford, D.J.; Elia, A.J.; Mclnnes, R.R. & Kooy, D.V.D. (2000). Retinal Stem Cells in the Adult Mammalian Eye. *Science*, Vol. 287, (March 2000), pp. 2032-2036, ISSN 1934-7391.

Wong, Y.T.; Chen, S.C.; Seo, J.M.; Morley, J.W.; Lovell, N.H. & Suaning, G.J. (2009). Focal Activation of the Feline Retina via a Suprachoroidal Electrode Array. *Vision Research,* Vol 49, (May 2009), pp. 825-833, ISSN 0042-6989.

Weiland J.D. & Humayun, M.S. (2008). Visual Prosthesis. *Proceedings of the IEEE,* Vol. 96, (July 2008), pp. 1076-1004, ISSN 0018 9219.

World Health Organization. (2011). Prevention of Blindness and Visual Impairment. Available from http://www.who.int/blindness/causes/priority/en/index.html.

Wu, C.; Tang, C. & Liu, S. (2003). Analysis of On-chip Spiral Inductors Using the Distributed Capacitance Model. *IEEE J. of Solid-State Circuits,* Vol. 38, (June 2003), pp. 1040-1044, ISSN 0018-9200.

Wu J. (2003). *Inductive Links with Integrated Receiving Coils for MEMS and Implantable Applications* PhD thesis, University of Notre Dame. Available from
http://etd.nd.edu/ETD-db/theses/available/etd-09302003-162720/.

Zrenner, E.; Stett, A.; Weiss, S.; Aramant, R.B.; Guenther, E.; Kohler, K.; Miliczek, K.-D.; Seiler, M.J. & Haemmerle, H. (1999). Can Subretinal Microphotodiodes Successfully Replace Degenerated Photoreceptors? *Vision Research*, Vol. 39, (July 1999), pp. 2555-2567, ISSN 0042-6989.

Zolfaghari, A. Chan, A. & Razavi, B., Stacked Inductors and Transformers in CMOS Technology. *IEEE J. of Solid-State Circuits*, Vol. 36, (April 2001), pp. 620-628, ISSN 0018-9200.

2

MEMS-Based Microdevice for Cell Lysis and DNA Extraction

Xing Chen, Dafu Cui, Haoyuan Cai,
Hui Li, Jianhai Sun and Lulu Zhang
State Key Lab. of Transducer Tech., Inst. of Electronics, Chinese Academy of Sciences
China

1. Introduction

With the development of microelectromechanical system (MEMS) technology, micrototal analytical systems (μTAS) which has the potential for integrating sample pretreatment, target amplification, and detection, has been in progress. Micromachined analytical systems have several advantages over their large-scale counterparts, including low cost, disposability, low reagent and sample consumption, portability, and lower consumption. Many such devices have been demonstrated in the literature, including PCR microchips (Northrup et al., 1993; Copp et al., 1998; Panaro et al., 2005), DNA microchips (Fan et al., 1999), DNA biosensors (Kwakye et al., 2006), capillary electrophoresis (CE) microchips (Harrison et al., 1993; Backhouse et al., 2003; Liu et al., 2006), protein microchips (Yang et al., 2001; Wilson & Nie, 2006), etc. Most of these analytical processes need an effective yet simple method of obtaining high-quality DNA. Hence miniature devices for rapid sample pretreatment of DNA, including cell lysis and genomic DNA purification, are crucial for genetic application.

Traditional phenol extraction is a complex and time-consuming method for extracting DNA, and even some commercial purification kits require several centrifugal operations. The implementation of DNA purification on a microdevice is initially demonstrated based on the principle of solid phase extraction (SPE). The SPE on-microdevice can minimize sample loss and contamination problems as well as reduce analysis time, and besides, this SPE method can avoid problems of physical and biochemical degradation of DNA. For example, Tian et al. (Tian et al., 2000) established an SPE DNA purification microdevice in a capillary packing with silica resin matrix which could extract enough DNA for PCR reaction. Wolfe et al. (Wolfe et al., 2002) and Breadmore et al. (Breadmore et al., 2003) immobilized bare-silica beads matrix in microchannels by sol–gel technology for DNA purification. But a high packing density for larger surface area in the microfluidic device results in problems of backpressure and clogging of crude samples, and what is more, it is difficult to control the small particles in microdevices. A micropillar array fabricated by MEMS technology in a microchamber or channel increases the surface area available for DNA adsorption (Christel et al., 1999; Cady et al., 2003). However, the increasing surface area is limited and the problems of clogging could not be completely solved. Hence, a novel solid-phase matrix which should be easily integrated in microdevices is under demand.

It is well known that porous silicon with a relatively large specific surface area-to-volume (hundreds of square meters per cubic centimeter) can significantly increase the available interface area. Porous silicon can be easily obtained by electrochemical etching technology, and the pore geometry, the surface morphology, and the porosity of porous silicon can be very precisely controlled by the electrochemical etching conditions (Bisi et al., 2000; Schmuki et al., 2003). Porous silicon with several useful characteristics has already found applications in diverse fields such as solar cells (Bilyalov et al., 2003), RF (Park et al., 2001), (bio)chemical sensors (Massera et al., 2004; Björkqvist et al., 2004), etc.. Furthermore porous silicon with huge surface area as a kind of biocompatible solid support has been used to absorb enzyme, protein and other biologic molecules in fields of the enzyme micro reactors (Melander et al., 2005; Bengtsson et al., 2002), chromatography (Clicq et al., 2004), and antibody micro arrays (Steinhauer et al., 2005). Due to the fact that the technology for porous silicon fabrication is compatible with standard microelectronic and MEMS techniques, we have made use of porous silicon as the solid-phase matrix to extract DNA (Chen et al., 2006, 2007).

DNA purification from initial samples requires disrupting cells to liberate the nucleic acids before SPE process. Typical laboratory protocols for lysis steps include the use of enzymes (lysozyme), chemical lytic agents (detergents), and mechanical forces (sonication, bead milling). However, many such lysis techniques are not amenable to be implemented in a microfluidic platform. Miniaturization methods for lyzing cells are required so that cell lysis can be integrated into the μTAS. Therefore, a desirable lysis method should rapidly destroy cells while at the same time, it could not destroy nucleic acids and inhibit the following PCR reaction either, which would be amenable to be integrated with SPE on microdevices. Carlo et al. (Carlo et al., 2003) reported a mechanical cell lysis microdevice with nanostructural barbs which was used to disrupt sheep blood cells. However, the fabrication process of nanostructures is complex. Other miniaturization cell lysis methods include thermal (Lee & Tai, 1999), electrical (Gao et al., 2004; Taylor et al., 2001), ultrasonication (Belgrader et al., 2000; LaMontagne1 et al., 2002), and chemical treatments (Li & Jed Harrison, 1997). These approaches have been shown to be moderately successful. However, they all depend on the use of an external power supply and the devices may be quite complicated and costly to fabricate. Hence, chemical disruption methods are chosen due to the fact that they can be compatible with SPE on microdevices and do not require complex process of fabrication either. For example, Schilling et al. (Schilling et al., 2002) reported a relatively simple T-type microfluidic device that allowed the continuous lysis of bacterial cells using β-galactosidase. Sethu et al. (Sethu et al., 2004) established a continuous-flow microfluidic device for rapid erythrocyte lysis.

In this chapter, we have described a novel integrated MEMS-based microdevice capable of performing online cell lysis and genomic DNA purification during continuous flowing process. The method of chemical disruption was chosen for cell lysis, while the method of solid phase extraction was chosen for DNA purification. For cell lysis microdevice based on chemical disruption method, mixing is the key step due to the fact that each cell should be fully exposed to lysis buffer before the total lysis of cells, which is limited by the mixing speed. The mixing procedure was numerically simulated. Based on the results of the simulation, the mixing model for cell lysis was optimized to construct the microfluidic devices fabricated by MEMS technology. For DNA purification microdevice based on the SPE principle, solid-phase matrix is the key factor. The porous silicon was used as the solid-phase matrix to extract DNA. The porous silicon dioxide matrix was fabricated by electrochemical etching technology, and then were characterized by using SEM (scanning electron microscopy) and BET (Brunauer, Emmet, and Teller) nitrogen adsorption technique. The porous layer directly generated on the internal walls of the channels can greatly enhance the "active" surface area compared to the non-anodized one, and also avoid the problems of packing solid phase matrix.

In this integrated microdevice, cells were rapidly lyzed at first, and then genomic DNA was released and absorbed on a solid-phase matrix. Secondly, washing buffer was pumped through this microdevice for removing the proteins and other impurities which were also absorbed during the previous process. Finally, elution buffer was used to elute DNA and the desorbed DNA was collected at the outlet port. During continuous flow process, on-line rapid cell lysis and PCR amplifiable genomic DNA purification on the single MEMS-based microdevice have been implemented, which confirmed that the proposed microfluidic device was capable of directly providing genic analyte for the following molecular biology research or medical assay.

2. Theory

In microdevice channels, the flow is laminar in nature. Reynolds number, which is the ratio of the inertial forces to the viscous forces, can be used to characterize the flow. It is given by

$$Re = \rho \, v \, D_h / \mu \quad (1)$$

Where ρ is the fluid density, v is the flow velocity, and μ is the fluid viscosity. D_h represents the hydraylic diameter and, for rectangular channels, is given by

$$D_h = 2 \, a \, b/(a+b) \quad (2)$$

Where a and b are the cross sectional dimensions of the channel. The Reynolds numbers involved in the field of microfluidic device are usually below 100 (Gravesen et al., 1993), which prevent turbulent flow.

According the formula (1), the fluid density (ρ) and the fluid viscosity (μ) are constants. Re is depended on the flow velocity (v) and the hydraylic diameter (D_h). For our microdevice with a 200μm wide and 100μm deep rectangle channel, the hydraylic diameter (D_h) calculated by the formula (2) is 133μm. Note that the blood density (ρ) is 1.05g/cm^3, and the viscosity of blood (μ) is 3.115mPa.s. According to the formula (1), the Re in our experimental was 0.035~8.4, when the flow velocity (v) is about 0.0008~0.024m/s. Therefore a typical laminar flow is found in our experimental. Lack of turbulent flow limits the effective mixing of the fluids under investigation. On the other hand, for the cell lysis on the microdevice based on the chemical method, it requires that each cell should be fully exposed to lysis reagents in microchannels, which is determined by mixing performance of the microdevice.

In microdevice channels where flow is laminar in nature, the mixing is dictated by diffusion. Average time for a small portion of a fluid to diffuse a distance L can be estimated by

$$T_D = L^2 / D \quad (3)$$

Where D is the diffusion coefficient of the liquid. The above equation can be used to predict the order of time scale of mass diffusion. As it suggests, one can dramatically reduce the mixing time by reducing the diffusion length required for mixing or increasing the contact area between two different liquids while keeping the volume constant.

3. Experimental

3.1 Reagents

Triton-X 100, HNO_3, KOH, ethanol, Tris, EDTA, and HCl were purchased from Beijing Chemical Reagents Company (Beijing China). SYBR Green I dye was purchased from Molecular Probes (Leiden, The Netherlands). Guanidine thiocyanate, λ-DNA, nucleic acid extraction kit, and PCR kit were purchased from Tianwei (Beijing China). Primers for the 203-bp, -gapd gene and primers for the 250-bp, -action gene were purchased from Sangon (Shanghai China).

Diluted buffer (0.9% NaCl in PBS) was used to dilute whole blood. TE buffer (10mM Tris, 1 mM EDTA, titrated to pH 8), load buffer (4M GuSCN in TE buffer with 1% Triton-X 100, titrated to pH 6.7), wash buffer (70% ethanol with 10mM NaCl) and eluted buffer (namely TE buffer) were used for the DNA purification procedure. All solutions were prepared in distilled water.

3.2 MEMS-based microdevice design and fabrication

T-type mixing model and sandwich type mixing model were proposed. In the sandwich type model, line-type model and coil-type model of the mixing microchannels were designed. All the models (see Fig.2) were numerically simulated. The best model with the best mixing performance was used to construct the silicon substrate. The mixing condition of the best model was optimized by numerical simulation, and then it was verified by cell lysis experiments.

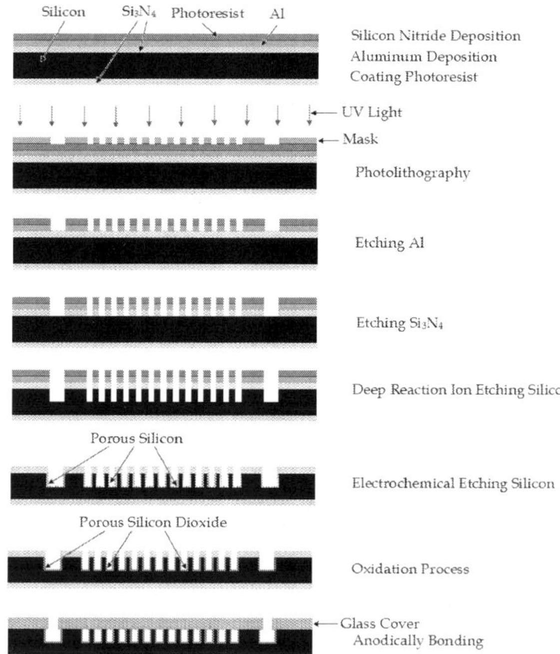

Fig. 1. Sequence for fabrication of the microdevice.

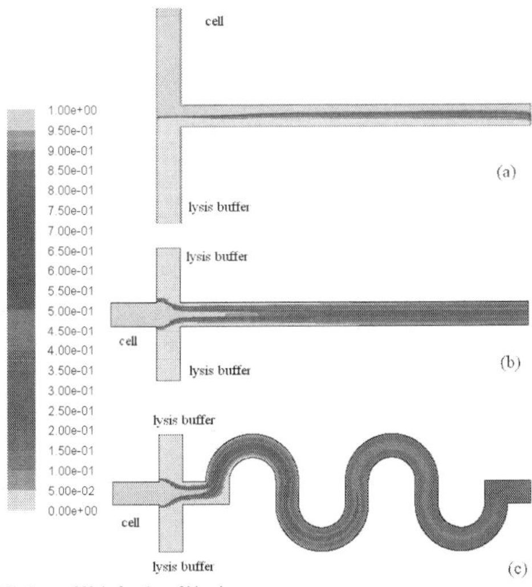

Fig. 2. Numerical simulation results for the species concentration distributions at T-type mixing model (a), sandwich type mixing model with lined channel (b) and sandwich type mixing model with coiled channel (c) when v_{cell} = 0.005m/s, v_{buffer} = 0.0025m/s. The color for "1.00e+00" and "0.00 e+00" in this figure refers to the concentration intensity of the pure cell sample and the pure lysis buffer, respectively. The software of FLUENT 6.2 was used to simulate.

The microdevice consisted of a silicon substrate and a glass cover. The fabrication process was shown in Fig. 1. A double-side polished n-type 0.01-0.1Ω·cm silicon wafer of (100) crystal orientation was first deposited with 0.3μm thick of silicon nitride (Si_3N_4) as the mask for fabricating porous silicon by a chemical vapour deposition (CVD) method, and then with 0.3μm aluminum as the mask for deep reaction ion etching (DRIE) by a electron beam vapour deposition technology. The wafer was spin coated with a positive photoresist (AZ1500) and patterned. After the exposed photoresist was developed, the exposed aluminum was removed by phosphoric acid and then the exposed silicon nitride was removed by plasma etching. The wafer was etched in a deep reaction ion etcher (Adixen, AMS100) to produce rectangle channels of 100μm in depth. After the process of removing the remained aluminum, a porous silicon layer on the internal walls of channels was anodized in 10% HF electrolyte at 20mA/cm^2 for 15min. After that, the silicon wafers were thoroughly rinsed in distilled water and oxidized at 1050°C for 1h. After oxidation process, the silicon nitride formerly patterned as the mask was removed by plasma etching for bonding. Finally glass covers (Corning Pyrex#7740), in which three holes had been drilled corresponding to the silicon substrate, were anodically bonded to the silicon wafers to form the closed channel by a bonder (Suss, SB6), thus microfluidic devices are fabricated.

3.3 Online cell lysis and DNA extraction procedure

The microdevice was mounted onto the stage of a microscope with a CCD camera and a video monitoring system. Blood was introduced through the cell inlet by a peristaltic pump, while lysis buffer (namely, load buffer) was pumped through the buffer inlet by another peristaltic pump. The value of the flow velocity was changed from 0.1μL/min to 25μL/min. When the lysis buffer and blood were blended in the channel during the continuous flow, cells were gradually lyzed online, and then genomic DNA was released and absorbed on the porous silicon matrix. DNA extraction procedure itself consisted of load, wash, and elution steps. In the load step, the released DNA was absorbed onto the porous silicon matrix in the presence of low pH (pH 6.7) and high concentration binding salt (4M GuSCN). And then proteins and possible PCR inhibitors were removed by passing 70% ethanol twice through the microdevice. Finally, DNA was eluted in TE buffer. 5 μL TE buffer was introduced in the microdevice for 10 min at 55°C and then 20 μL TE buffer was continuously passed through the microdevice.

3.4 Fluorescence detection and PCR amplification

DNA collected in the elution step was quantified by using SYBR Green I dye in a fluorometer using calibration curve of DNA which were generated using lambda DNA. Then DNA purified from whole blood was amplified by polymerase chain reaction (PCR). The 203-bp -gapd gene of rat's blood was amplified using the following primers: Forward Primer: 5'-AGAAGTACCTGCAACAGG- 3', Reverse Primer: 5'-GACGGACACATTGGGGGT- 3'. PCR reactions consisted of 2.5 μL standard 10×PCR buffer, 100 μmol dATP, dGTP, dCTP, and dTTP, 2.5 units of Taq polymerase, 50nmol of each primer, 5μL of the initial collected fraction, in a total volume of 25μL. These reactions were cycled in under the following conditions: 95°C denaturation for 5min, 35 cycles of 94°C for 1min, 68°C for 1min, 72°C for 1min, followed by a 10min extension at 72°C. DNA amplification was confirmed by gel electrophoresis.

4. Results and discussion

4.1 Design and simulation of mixing process

In general, comprehensive computational fluid dynamics (CFD) simulations are conducted before physical models are built and tested since these simulations enable the system parameters to be varied over a wide range of values and permit the simultaneous and instantaneous data collection of various aspects of models. Numerical simulation plays a key role in optimizing designs of microdevices and enables a reliable interpretation of the experimental results. T-type mixing model, sandwich-type mixing model with lined channel and coiled channel were designed and numerical simulated using FLUENT 6.2 software and all the simulations were two-dimensional.

The numerical simulation results are shown in Fig. 2. The mixing performance in sandwich-type mixing model is better than that in T-type one because of the increase of the contact area, and the mixing performance in coiled channel is much better than that in lined channel. Those numerical simulation results mentioned above provide a very clear understanding of the physical phenomena taking place in the two-dimensional microfluidic channels. In order to get a more accurate evaluation of the degree of mixing in the microchannel, a mixing index (σ) is used as follows (Erickson & Li, 2002):

$$\sigma = \left(1 - \frac{\int_0^h |c - c_\infty| dy}{\int_0^h |c_0 - c_\infty| dy}\right) \times 100\% \tag{4}$$

where C is the species concentration profile across the width of the microchannel (h), c_∞ is the completely mixed state (= 0.5) and C_0 is the completely unmixed state (= 0 or 1). Note that the confluent streams are completely mixed if σ = 100%. In contrast, they are completely unmixed if σ = 0%.

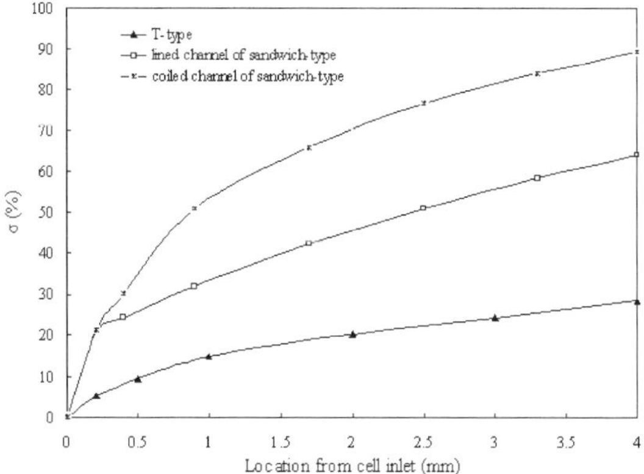

Fig. 3. Numerical evaluations of mixing efficiency index (σ) at different mixing models when v_{cell} = 0.005m/s, v_{buffer} = 0.0025m/s

Fig. 3 demonstrates the mixing efficiency of the proposed mixing models: T-type, sandwich-type with lined channel and sandwich-type with coiled channel. The use of coiled channel is effective in enhancing chaotic mixing, in which the mixing efficiency of 89.3% is obtained at the cross section located 4mm from inlet, by folding, stretching and reorienting fluid. Based on discussion above, the coiled channel of sandwich-type mixing model is chosen to construct the physical microdevice.

4.2 Optimization of cell lysis process

For cell lysis based on chemical disruption method, there are two steps which are mixing and chemical reaction respectively. In the microdevice, mixing is slow compared with the chemical reaction rate due to the fact that mixing in the microdevice is exclusively caused by diffusion because of laminar flow. The total lysis of cells requires that each cell should be fully exposed to lysis buffer, which is limited by the mixing speed. Thus the cell lysis efficiency is strongly depended on the mixing performance between the cell solution and the lysis buffer. From above results of simulation, a microdevice utilizing sandwich-type mixing model with coiled channel was designed and fabricated. As is shown in Fig. 4, the microdevice consists of a glass cover with two inlets and one outlet and a silicon substrate with an etched coiled channel of 200μm wide, 100μm deep and 20.16cm long.

The cell lysis performance is not only influenced by the design of the microchannels but also impacted by the velocity of the two solutions. With the increase of velocity, the time for cell lysis is reduced while the cell lysis efficiency decreases. In order to reduce the time for cell lysis and obtain high cell lysis efficiency, the velocity should be carefully examined. At first, the relationship between the mixing efficiency index and the inlet velocity was simulated using the coiled channel of sandwich-type mixing model.

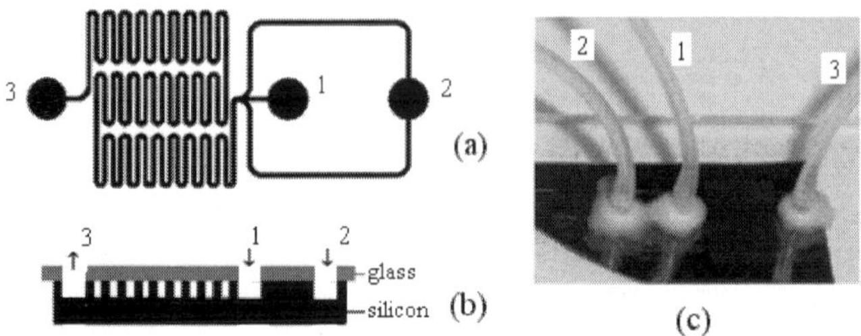

Fig. 4. Schematic top view (a), cross view (b) and photograph (c) of the microdevice. (1) Cell inlet; (2) Buffer inlet; (3) Outlet.

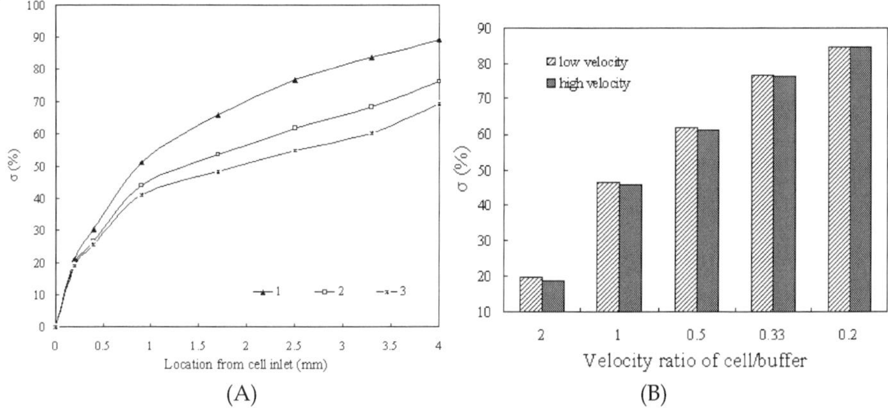

Fig. 5. Numerical evaluations of mixing efficiency index (σ) of coiled channel of sandwich-type mixing model. (Left) With different inlet velocities: "1" v_{cell} = 0.005m/s, v_{buffer} = 0.0025m/s; "2" v_{cell} = 0.02m/s, v_{buffer} = 0.01m/s; "3" v_{cell} = 0.2m/s, v_{buffer} = 0.1m/s. (Right) At the location near the inlet with different velocity ratio: "low velocity" v_{buffer} = 0.01m/s; "high velocity" v_{buffer} = 0.1m/s.

As is shown in Fig. 5A, the mixing efficiency is sharply decreased with the increase of the inlet velocity. This decrease of mixing efficiency can likely be attributed to the insufficient diffusion for the limited diffusion time. For the laminar flowing, the mixing time is proportional to the square of diffusion length so that decreasing diffusion length would reduce the mixing time sharply without deterioration of the mixing efficiency. The diffusion length can be changed by varying the cell and buffer velocity ratio. The lower cell/buffer velocity ratio leads to shorter diffusion length which can enhance mixing efficiency. Moreover at the location near the junction of three streams, the distribution of the laminar flow of the horizontal stream caused by the vertical streams enhances fluid mixing. And this enhancement is much stronger at the lower cell/ buffer velocity ratio. When the cell/buffer velocity ratio is 0.2, the mixing efficiency index is found to be more than 80% near the junction, which is insensitive to the value change of the flow velocity, shown in Fig. 5B.

In order to test the above results of simulation, the experiments of cell lysis were conducted by using the microfluidic device (Fig.4c). The cell sample was whole blood, while the lysis buffer was load buffer, 4M GuSCN in TE buffer with 1% Triton-X 100, titrated to pH 6.7. Triton X-100 is generic surfactants, with amphoteric chemical properties, which can react with water and can also react with lipid. High concentration guanidine salt is typical bonding salt for solid phase extraction nucleic acid and is also a generic strong denaturant. Moreover guanidine salt can quickly dissolve protein and can destroy the structure of cells. So the mixture of the traditional surfactant and the typical bonding salt was used as the lysis buffer for online cell lysis and DNA extraction.

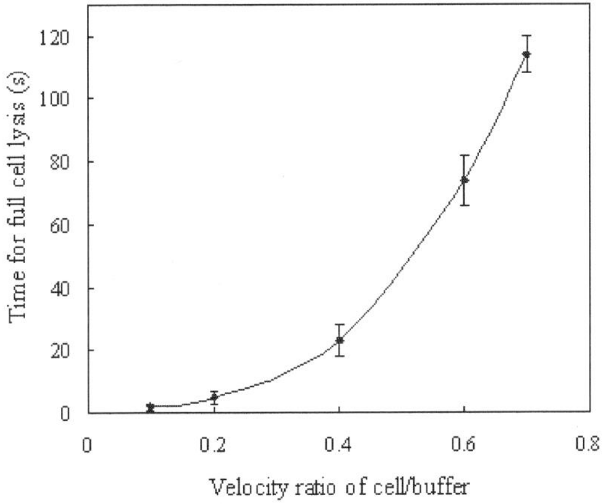

Fig. 6. Experimental evaluations of time for full cell lysis at variation of flow velocity ratio of cell/buffer when vbuffer = 1μL/min.

At first, the effect of flow velocity on the time taken for cell lysis was analyzed when the velocity ratio of cell/buffer was 1:1. For the flow velocity of 0.1, 0.2, 0.4 and 0.5μL/min, the time taken for complete lysis was tested to be 188s, 188s, 175s and 194s, at the location which was 3.92cm, 7.84cm, 14.56cm and 20.16cm from the inlet along the channel respectively. Note that the flow velocity of 1μL/min is equivalent to 0.833mm/s in the microchannel of 200μm wide and 100μm deep. The time taken for completely lysis at different flow rate with the same cell/buffer ratio was approximately the same; however cells could not be fully lyzed in the microdevice at more than 0.5μL/min. Then, the velocity ratio of cell/buffer was varied. As is shown in Fig. 6, the time required for full lysis is sharply reduced from approximate two minutes to several seconds with the decrease of the cell/buffer velocity ratio from 0.7 to 0.1. The photograph pictures of Fig. 7 show that the width of the cell stream is reduced with the decrease of the cell/buffer velocity ratio and is not affected by specific rate value. These experimental results are found to be consistent with the numerical results. Thus rapid cell lysis can be implemented under the condition of lower cell/buffer velocity ratio and higher specific rate value. In the end the cell velocity of 5μL/min and the lysis buffer velocity of 15~25μL/min were used for rapidly cell lysis.

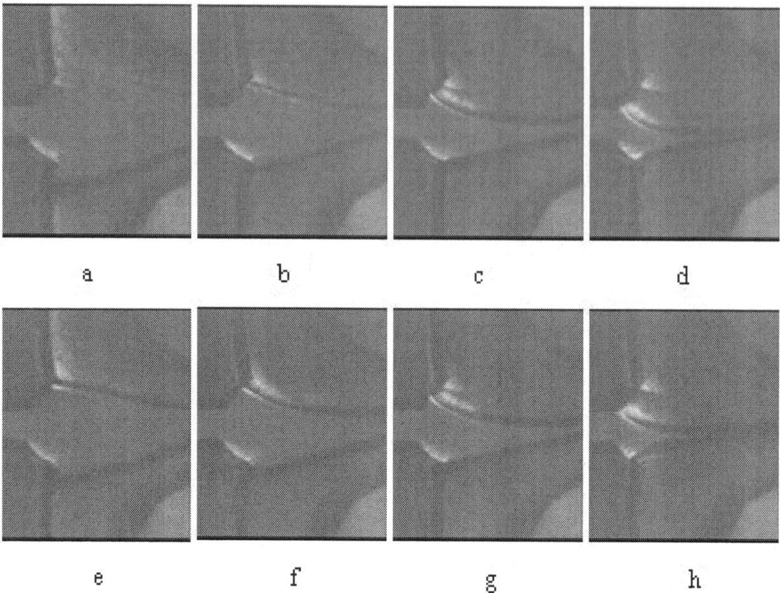

Fig. 7. Experimental results for the cell and buffer concentration distributions at location near the inlet with different velocity: "a" v_{cell} : v_{buffer} = 2:1, v_{buffer} = 1μL/min; "b" v_{cell} : v_{buffer} = 1:1, v_{buffer} = 1μL/min; "c" v_{cell} : v_{buffer} = 1:3, v_{buffer} = 1μL/min; "d" v_{cell} : v_{buffer} = 1:5, v_{buffer} = 1μL/min; "e" v_{cell} : v_{buffer} = 2:1, v_{buffer} = 10μL/min; "f" v_{cell} : v_{buffer} = 1:1, v_{buffer} = 10μL/min; "g" v_{cell} : v_{buffer} = 1:3, v_{buffer} = 10μL/min; "h" v_{cell} : v_{buffer} = 1:5, v_{buffer} = 10μL/min.

4.3 DNA purification using porous silicon matrix

DNA purification on-microdevice is based on SPE method: firstly adsorption of DNA onto a solid-phase matrix surface and then washing the impurities such as protein and finally elution of the purified DNA. The mechanism of DNA adsorption on a solid-phase matrix surface elucidated by Melzak et al. (Melzak et al., 1996) was that the adsorption of highly charged duplex DNA to negatively charged silica was controlled by three competing effects: weak electrostatic repulsion forces, dehydration and hydrogen bond formation. The surface area of the matrix significantly affects the adsorption of DNA onto the solid-phase matrix surface.

Porous silicon gotten on the surface of the tortuous rectangle channel in the microdevice by electrochemical etching technology can enhance the available surface area (see Fig. 8). The details of the preparation and optimization process for porous silicon were reported in our previous research (Chen et al., 2006; Chen et al., 2007). By using BET nitrogen adsorption experiment, pore size of the porous silicon is in the range of 20nm to 30nm. And the surface area of porous silicon is approximate 400m^2/g. Note that the volume of microfluidic device with a tortuous channel of 25cm long, 100μm deep and 200μm wide is 0.005cm^3. Thus the surface area to volume of microdevice with porous silicon is approximate 300m^2/cm^3 which is thousands of times higher than that of one without porous silicon since it is known that the surface area to volume of the microdevice without porous silicon is approximate 200cm^2/cm^3.

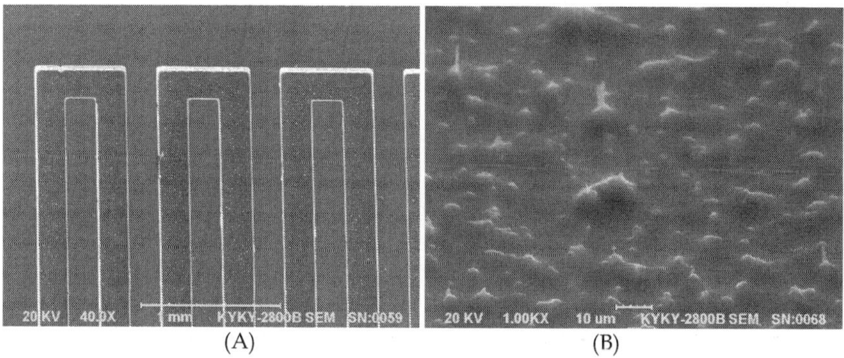

Fig. 8. SEM of porous silicon on the surface of the channel for DNA purification.
(A) ×40; (B) ×1000.

The microdevices with and without porous silicon were used to recover DNA from 200ng prepurified genomic DNA under the same experimental conditions, respectively. The performance of the microdevice with porous silicon is quite well, with an average of 83% (11.6%RSD) evaluated from five extractions, which is much higher than that of one without porous silicon with an average of 39.2% (8.7%RSD). It is important to note that 200ng DNA overloads the capacity of the microdevice without porous silicon due to the limitation of surface area. Therefore the binding capacity of the microdevice without porous silicon is approximately 75ng/cm^2 which agrees with the results of Cady et al. (Cady et al., 2003) who found that the binding capacity of nonporous micropillars was approximate 82ng/cm^2. The previous researches proved that the performance of DNA extraction microdevice was determined by the surface area of the matrix and the extracted DNA was found to increase linearly with the surface area (Fan et al., 1999; Kwakye et al., 2006). The binding capacity of porous silicon matrix would increase hundreds or thousands of times, while the extraction efficiency didn't improve so much. The reason could probably be that most of the internal pores and smaller pores might not be used to adsorb DNA, and the DNA adsorbed in these pores might not be easily eluted.

Comparison with the microdevice without the porous silicon layer, the microdevice with the porous layer achieved higher extracting efficiency of DNA without problems of clogging and high backpressure. Moreover this SPE microdevice with porous silicon can be easily integrated with other microdevices, since the porous silicon can be directly generated on the internal walls of channel in the microdevice.

4.4 Cell lysis and purification of DNA on microdevice

A microfluidic device capable of real time destroying cells and extracting DNA from the cell lysates during continuous flow was designed and fabricated. As is shown in Fig. 4, the microfluidic device with the size of 2cm×1.2cm consists of a silicon substrate within an etched coiled channel, and a glass cover with two inlets and one outlet according to the silicon substrate. The two inlets: cell inlet and buffer inlet are designed to introduce blood sample and lysis buffer simultaneously, leading to rapid lysis of blood cells near the location of cell inlet using the sandwich-type mixing model. Porous silicon layer fabricated on the surface of the internal wall of the coiled channel is designed to absorb DNA, which can strongly improve the extraction efficiency with the huge surface area. Thus this microfluidic device makes it possible to lyze cells and to purify DNA orderly.

Microfluidic devices with porous silicon matrix were used for lyzing blood cells and purifying genomic DNA, including three steps: load step, washing step, and elution step. In load step, blood cells were damaged, and the genomic DNA was released and adsorbed on the porous silicon matrix. In washing step, impurity such as protein was removed while the genomic DNA was still held on the porous matrix. In elution step, the purified DNA was desorbed from the porous matrix into the elution buffer. The process of online cell lysis and DNA extraction was in details as follows. Firstly, 10μL whole blood was introduced into the channel from the cell inlet, while 50μL lysis buffer was introduced from the buffer inlet. The lysis buffer was load buffer, namely 4M GuSCN in TE buffer with 1% Triton-X 100, titrated to pH 6.7, which included lysis reagent and bonding salt for DNA adsorption. GuSCN and Triton-X 100 were both the chemical reagents for cell lysis, while GuSCN is the bonding salt for DNA extraction. The flow velocity of whole blood sample was 5μL/min, whereas the flow velocity of lysis buffer was 25μL/min. The whole blood sample and the lysis buffer were mixed effectively at the location near the inlet, and blood cells were rapidly lyzed. After cell lysis, genomic DNA in blood cells was released and adsorbed on the porous matrix at the presence of low pH (pH 6.7) and high concentration binding salt (4M GuSCN). Secondly, 50μL washing buffer was pumped in the microchannel at 25μL/min from the buffer inlet, while the cell inlet was closed during washing step. Finally, 10μL of TE buffer was introduced in the microdevices and incubated in the microdevices for 5min at 55°C and then another 50μL of TE buffer was passed through continuously. As a whole, online cell lysis and genomic DNA purification could be implemented in less than 20min by using this integrated microdevice, while its large-scale counterparts commonly require more than several hours to finish the job.

In order to value the performance of the integrated microdevice, the elution buffer flowing out from the outlet port, in which the purified DNA was desorbed and eluted, was collected and subjected to fluorescence detection using SYBR Green I. From fluorescence detection, average 39.7ng genomic DNA was extracted from 1μL whole blood with 10.5%RSD by using three microdevices, respectively, at the same conditions. The extracted efficiency of DNA by the integrated microdevices is higher than commercial kit with silica resin which can only extract about 20~30ng DNA per microlitre blood. Moreover, it is only 10μl blood sample needed for the integrated microdevice to extract genomic DNA, while traditional phenol extraction or commercial kit requires several milliliters blood because of centrifugal operation.

The extraction of genomic DNA from a crude biological sample must be PCR-amplifiable. The lyzed cells are a complex mixture of proteins, peptides, lipids, carbohydrates, and other low molecular weight compounds that are known to inhibit DNA amplification by PCR. The genomic DNA extracted from whole blood in two separate experiments, respectively, was submitted for PCR to ensure that no inhibitory compounds were present. Shown in Fig. 9, a 203-bp fragment of -gapd gene is successfully amplified, which is identified by gel electrophoresis separation. This illustrates that the eluted DNA using the microdevice has been purified effectively for subsequent enzymatic reactions. Also shown from Fig. 9, the repeatability of this system is good enough for the downstream analytical steps.

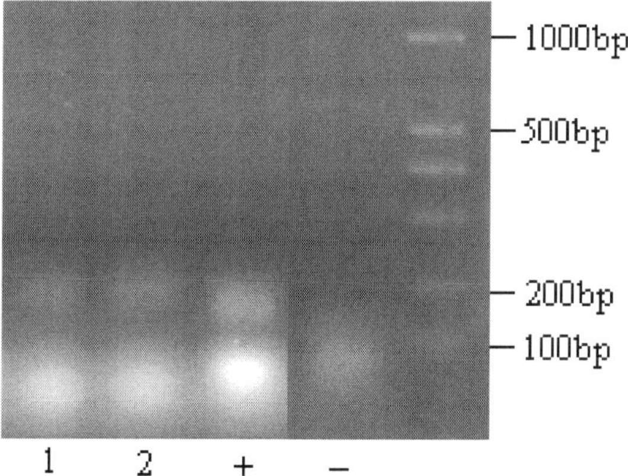

Fig. 9. Gel electrophoresis analysis of PCR products.

5. Conclusion

This chapter has demonstrated an integrated microdevice capable of performing online cell lysis and DNA extraction. Firstly micro total analytical systems (μTAS) base on MEMS technology was briefly introduced in Section 1. Then microdevices both for DNA extraction and cells lyzing were reviewed. The theory of flowing in microdevice was introduced in Section 2. In microdevice channels, the flow is laminar in nature. Then a novel MEMS-based microdevice capable of performing online cell lysis and DNA extraction was reported in Section 3 and 4. The fabrication procedure of the MEMS-based microdevice was presented in detail, while the experimental of cell lysis and DNA extraction were described respectively. Then the properties of this microdevice were studied by numerical simulation and experimental. According to the results of the simulation, the mixing performance in sandwich-type coiled channels was better than that in the T-type ones. According to the results of the experimental, a high performance of cell lysis was obtained in the sandwich-type microdevice at the optimized conditions of the cell/buffer velocity ratio <1/5. 83% DNA was recovered by the porous matrix, strongly contrary to the 39.2% DNA extracted by the non-porous one. And 39.7ng PCR-amplifiable genomic DNA was obtained from whole blood samples by using an integrated microdevice with sandwich-type coiled channels and porous silicon matrix. In general, the developed integrated microdevice providing a powerful tool for biological sample pre-treatment is shown to exhibit numerous advantages over its large-scale counterparts, including rapidness, much lower blood and reagent consumption, disposability, and portability and so on, which has the potential to integrate into μTAS for point-of -care medical diagnose.

6. Acknowledgement

The authors greatly acknowledge the financial support from the National Science Foundation of China under Grant number 60701019, 60427001 and 60501020. The authors are grateful to Mr. Feng Shen of Institute of Mechanics, Chinese Academy of Sciences for his assistance.

7. References

Backhouse, C.J.; Gajdal, A.; Pilarski, L.M. & Crabtree, H.J. (2003). Improved Resolution with Microchip-based Enhanced Field Inversion Electrophoresis, *Electrophoresis*, Vol.24, pp.1777-1786

Belgrader, P.; Okuzumi, M.; Pourahmadi, F.; Borkholder, D.A. & Northrup, M.A. (2000). A Microfluidic Cartridge to Prepare Spores for PCR Analysis, *Biosens. Bioelectron.*, Vol.14, pp.849–852.

Bengtsson, M.; Ekström, S.; Marko-Varga, G. & Laurell, T.; (2002). Improved Performance in Silicon Enzyme Microreactors Obtained by Homogeneous Porous Silicon Carrier Matrix, *Talanta*, Vol.56, pp.341-53

Bilyalov, R.; Stalmans, L.; Beaucarne, G.; Loo, R.; Caymax, M.; Poortmans, J.; & Nijs, J. (2001). Porous Silicon as an Intermediate Layer for Thin-Film Solar Cell, *Solar Energy Materials and Solar Cells*, Vol.65, No.1-4, pp.477-85

Bisi, O.; Ossicini, S. & Pavesi, L. (2000). Porous Silicon: a Quantum Sponge Structure for Silicon based Optoelectronics, *Surface Science Reports*, Vol.38, pp.1-126

Björkqvist, M.; Salonen, J.; Paski, J.; & Laine, E. (2004). Characterization of Thermally Carbonized Porous Silicon Humidity Sensor, *Sensors and Actuators A*, Vol.112, pp.244-7

Breadmore, M.C.; Wolfe, K.A.; Arcibal, I.G.; Leung, W.K.; Dickson, D.; Giordano, B.C.; Power, M.E.; Ferrance, J.P.; Feldman, S.H.; Norris, P.M. & Landers, J.P. (2003). Microchip-based Purification of DNA from Bilological Samples, *Anal. Chem.*, Vol.75, pp.1880-1886.

Cady, N.C.; Stelick, S. & Batt, C.A. (2003). Nucleic Acid Purification Using Microfabricated Silicon Structures, *Biosens. Bioelectron.*, Vol.19, pp.59-66.

Carlo, D.D.; Jeong, K.H. & Lee, L.P. (2003). Reagentless Mechanical Cell Lysis by Nanoscale Barbs in Microchannels for Sample Preparation, *Lab chip*, Vol.3, pp.287-291.

Chen, X.; Cui, D. F. Liu, C. C. & Li, H. (2007). Microfabrication and Characterization of Porous Channels for DNA Purification, *J. Micromech. Microeng.*, Vol.17, pp.68-75

Chen, X.; Cui, D. F.; Liu, C. C. & Cai, H. Y. (2006). Fabrication of Solid Phase Extraction DNA Chips based on Bio-Micro-electron-Mechanical System Technology, *Chin. J. Anal. Chem.*, Vol.34, pp.433−436

Christel, L.A.; Petersen, K.; McMillan, W. & Northrup, M.A. (1999). Rapid Automated Nucleic Acid Probe Assays Using Silicon Microstructures for Nucleic Acid Concentration, *Transactions of the ASME* , Vol.121, pp.22-27.

Clicq, D.; Tjerkstra, R. W.; Gardeniers, J. G. E.; van den Berg, A.; Baron, G. V. & Desmet, G.; (2004). Porous Silicon as a Stationary Phase for Shear-Driven Chromatography, *Journal of Chromatography A*, Vol.1032, pp.185-91

Copp, M.U.; Luechinger, M.B. & Manz, A. (1998). Chemical Amplification: Continuous-Flow PCR on a Chip, *Science*, Vol.280, pp.1046-1048

Erickson, D. & Li, D. (2002). Influence of Surface Heterogeneity on Electrokenetically Driven Microfluidic Mixing, *Langmuir*, Vol.18, pp.1883-1892.

Fan, Z.H.; Mangru, S;. Granzow, R.; Heaney, P.; Ho, W.; Dong, Q. & Kumar, R. (1999). Dynamic DNA Hybridization on a Chip Using Paramagnetic Beads, *Anal. Chem.*, Vol.71, pp.4851-4859

Gao, J.; Yin, X.F. & Fang, Z.L. (2004). Integration of Single Cell Injection, Cell Lysis, Separation and Detection of Intracellular Constituents on a Microfluidic Chip, *Lab chip*, Vol. 4, pp.47-52.

Gravesen, P.; Branebjerg, J. & Jensen, O.S. Microfluidics-a Review, (1993). *J. Micromech. Microeng.*, Vol.3, pp.168-182.

Harrison, D.J.; Fluri, K.; Seiler, K.; Fan, Z.; Effenhauser, C.S. & Manz, A. (1993). Micromaching a Miniaturized Capillary Electrophoresis-based Chemical Analysis System on a Chip, *Science*, Vol.261, pp.895-897

Kwakye, S.; Goral, V.N. & Baeumner, A.J. (2006). Electrochemical Microfluidic Biosensor for Nucleic Acid Detection with Integrated Minipotentiostat, *Biosens. Bioelectron.*, Vol.21, pp.2217–2223

LaMontagne1, M.G.; Jr., F.C.M.; Holden, P.A. & Reddy, C.A. (2002). Evaluation of Extraction and Purification Methods for Obtaining PCR-Amplifiable DNA from Compost for Microbial Community Analysis, *J. Microbiol. Meth.*, Vol.49, pp.255–264.

Lee, S.W. & Tai, Y.C. (1999). A Micro Cell Lysis Device, *Sens. Actuat. A*, Vol.73, pp.74–79.

Li, P.C.H.& Jed Harrison, D. (1997). Transport, Manipulation, and Reaction of Biological Cells On-Chip Using Electrokinetic Effects, *Anal. Chem.*, Vol.69, pp.1564-1568.

Liu, D.Y.; Shi, M.; Huang, H.Q.; Long, Z.C.; Zhou, X.M; Qin, J.H. & Lin, B.C. (2006). Isotachophoresis Preconcentration Integrated Microfluidic Chip for Highly Sensitive Genotyping of the Hepatitis B Virus, *J. Chromatogr. B*, Vol.844, pp.32-38

Massera, E.; Nasti, I.; Quercia, L.; Rea, I.; & Di Francia, G. (2004). Improvement of Stability and Recovery Time in Porous-Silicon-based NO_2, *sensor Sensors and Actuators B*, Vol.102, pp.195-7

Melander, C.; Bengtsson, M.; Schagerlöf, H.; Tjerneld, F.; Laurell, T. & Gorton, L. (2005). Investigation of Micro-Immobilised Enzyme Reactors Containing Endoglucanases for Efficient Hydrolysis of Cellodextrins and Cellulose Derivatives, *Analytica Chimica Acta*, Vol. 550, pp.182-90

Melzak, K.A.; Sherwood, C.S.; Turner, R.B. & Haynes, C.A. (1996). Driving Forces for DNA Adsorption to Silica in Perchlorate Solutions, *J. Colloid Interf. Sci.*, Vol.181, pp.635–644.

Northrup, M.A.; Ching, M.T.; White, R.M. & Watson, R.T. (1993). DNA Amplification with a Microfabricated Reaction Chamber,*Transducer*, pp.924-926

Panaro, N.J.; Lou, X.J.; Fortina, P.; Kricka, L.J. & Wilding, P. (2005). Micropillar Array Chip for Integrated White Blood Cell Isolation and PCR, *Biomolecular Engineering*, Vol.21, pp.157–162

Park, J. Y. & Lee, J. H. (2003). Characterization of 10 µm Thick Porous Silicon Dioxide Obtained by Complex Oxidation Process for RF Application, *Materials Chemistry and Physics*, Vol.82, pp.134-9

Schilling, E.A.; Kamholz, A.E. & Yager, P. (2002). Cell Lysis and Protein Extraction in a Microfluidic Device with Detection by a Fluorogenic Enzyme Assay, *Anal. Chem.*, Vol.74, pp.1798-1804.

Schmuki, P.; Schlierf, U.; Herrmann, T. & Champion, G. (2003) Pore Initiation and Growth on n-InP(100), *Electrochimica Acta*, Vol.48, pp.1301-8

Sethu, P.; Anahtar, M.; Moldawer, L. L.; Tompkins, R. G. & Toner, M. (2004). Continuous Flow Microfluidic Device for Rapid Erythrecyte Lysis, *Anal. Chem.*, Vol.76, pp.6247–6253.

Steinhauer, C.; Ressine, A.; Marko-Varga, G.; Laurell, T.; Borrebaeck, C. A. K. & Wingren, C. (2005). Biocompatibility of Surfaces for Antibody Microarrays: Design of Macroporous Silicon Substrates, *Analytical Biochemistry*, Vol.341, pp.204-13

Taylor, M.T.; Belgrader, P.; Furman, B.J.; Pourahmadi, F.; Kovacs, G.T.A. & Northrup, M.A. (2001). Lysing Bacterial Spores by Sonication through a Flexible Interface in a Microfluidic System, *Anal. Chem.*, Vol.73, pp.492-496.

Tian, H;. Hühmer, A.F.R. & Landers, J.P. (2000). Evaluation of Silicon Resins for Direct and Efficient Extraction of DNA from Complex Biological Matrices in a Miniaturized Format, *Anal. Biochem.*, Vol.283, pp.175-191.

Waters, L.C.; Jacobson, S.C.; Kroutchinina, N.; Khandurina, J.; Foote, R. S. & Ramsey, J.M. (1998). Microchip Device for Cell Lysis, Multiplex PCR Amplification, and Electrophoretic Sizing, *Anal. Chem.*, Vol.70, pp.158-162.

Wilson, M.S. & Nie, W. (2006). Multiplex Measurement of Seven Tumor Markers Using an Electrochemical Protein Chip, *Anal. Chem.*, Vol.78, pp.6476-6483

Wolfe, K.A.; Breadmore, M.C.; Ferrance, J.P.; Power, M.E.; Conroy, J.F.; Norris, P.M. & Landers, J.P. (2002). Toward a Microchip-based Solid-Phase Extraction Method for Isolation of Nucleic Acids, *Electrophoresis*, Vol.23, pp.727–733.

Yang, T.; Jung, S.; Mao, H. & Cremer, P.S. (2001). Fabrication of Phospholipid Bilayer-Coated Microchannels for on-Chip Immunoassays, *Anal. Chem.*, Vol.73, pp.165-169

MEMS Microfluidics for Lab-on-a-Chip Applications

Nazmul Islam and Saief Sayed
MEMS/NEMS Lab, The University of Texas at Brownsville
USA

1. Introduction

Micro-/Nano- fluid devices are becoming more prevalent, both in commercial applications and in scientific inquiry. Microfluidics, a branch of MEMS (Micro-Electro-Mechanical Systems) is key enabling factor in the miniaturization and integration of multiple functionalities for chemical analysis and synthesis in handheld microdevices, which require efficient methods for manipulating ultra small volumes of liquid as well as the contents in the fluid within the fluid networks. For biomedical applications, microfluidic chip arrays are being used to identify multiple bioparticles [1]. Recent developments in micro-fabrication technologies enabled different types of microfluidic functions such as micro-pumps [2, 3], micro-mixers [4], particle concentrator [5, 6], and various s types of injection systems (nano-needles). At the very beginning of microfluidics, people thought that microfluidic devices could just be a miniaturized version of macro- fluidic devices. The technological advancement on microfluidic systems has proven that the problem is far more complicated than scaling down a device geometrically. Therefore, a better understanding of the micro/nano scale properties is in order.

A dominant difference of microfluidic devices from their macro-scale counterparts is the increased surface/volume ratio, hence dominant surface force effects/friction. Micro channel needs high pressure for pressure driven flow to produce sufficient flow rate. The formula below relates the applied pressure with the conduit radius for a constant flow rate.

$$\Delta P = \frac{8\mu L Q}{\pi a^4} \qquad \mu : \text{viscosity}; \quad a : \text{conduit radius}$$

Every time we try to reduce the conduit radius into half, we need to have sixteen times of larger pressure to sustain the same flow rate. So at microscale, surface forces start to dominate due to the large surface/volume ratio. Therefore electroosmosis (as a type of surface forces) becomes the prime candidates for fluidic manipulation at micro scale. Direct Current Electroosmosis (DCEO) has a long history of being applied in miniaturized biochemical devices. However, DCEO has many undesirable effects, such as high voltage operation, electrolysis and resulting bubble generation, and pH gradient. In this chapter, we examined a new type of EO phenomena, ACEO (Alternating Current Electroosmosis), and how it can be employed to integrate with the microcantilever particle trapping.

2. MEMS microfluidics: Past, present and future

Microfluidics is an interdisciplinary area that focuses on the miniaturization of fluid-handling systems. The concept of complete lab-on-a-chip devices or micro-total analysis systems (μ-TAS), has recently generated great interest in a variety of industries, where transport and processes (including mixing, reaction, separation, and manipulation of chemicals and particles) are being applied on much smaller scales than traditional engineering technologies [7, 8]. This interest has led to tremendous growth in microfluidic technologies over the past decade.

A functional microfluidic chip should be able to realize certain functions, such as transporting, mixing (with reactants), sample treatment (concentrating, sorting). Figure 1 gives an example of a generic microfluidic chip. Unlike the microelectronics industry, where the current emphasis is on reducing the size of transistors, the field of microfluidics is focusing on investigating new fluid phenomenon at micro/nano-scale with more sophisticated fluid-handling capabilities. One of the promising types of micropumps is driven by electroosmosis (EO). EO pumps are purely driven by electric fields and have no moving parts. The central concept is utilizing the surface force. As the surface to volume ratio of the microchannel is high, the dominant surface force is a good choice for pumping the liquid in microchannel.

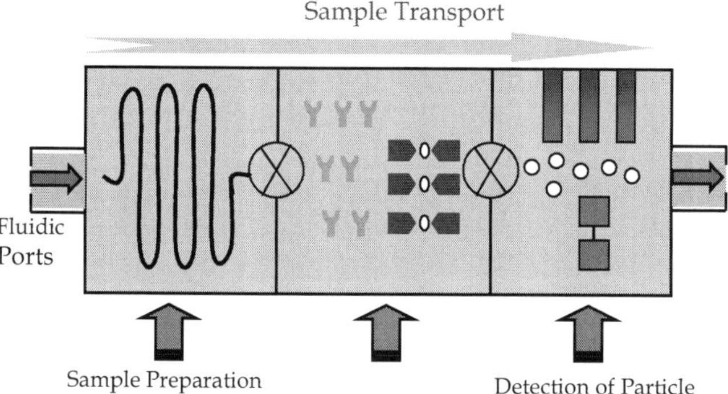

Fig. 1. A Generic Microfluidic Chip with the sample preparation, selection and detection of bio/nano/micro-particles.

As mentioned before, the field of microfluidics has far more complexity than people first expected. The dominant/efficient mechanism to manipulate fluid or biochemical samples will change with sample conductivity, pH value, and sample sizes. Also, we need to be concerned with side effects from those actuation mechanisms. So microfluidic industry did not develop in a similar way as the microelectronics industry. Lab-on-a-chip (LOC) devices have shown commercial success in biological applications such as electrophoretic separations and DNA sequencing, where DC electrokinetics or DC electric field is used to manipulate fluid/particles (here surface force is used versus pressure driven). However, because of its operation with high voltage, there are obstacles to extend DC electrokinetics to more fluidic functions. On the other hand, AC electrokinetics has important potentials in the field of life science. With the capability to manipulate particle and fluid motion at the microscale with low voltage, it meshes well with the requirements of lab-on-a-chip systems.

In this chapter a combination of sample transportation and particle detection in a microfluidic chip. As shown in figure 1, we need to transport the sample, for which I have experimentally validate the biased ACEO micro-pumping. The developed micropump is operated with smaller AC voltage, which is compatible with the lab-on-a-chip. For particle detection as shown in figure 1, we need high effective techniques to detect micro/nano-scale particulate. We also have envisioned to develop the validation technique for particle trapping. That is the reason we have interface the microcantilever with our microfluidic device. The applications of this integration can greatly benefit the advancement of AC electrokinetics.

3. Electrokinetics

Electrokinetics is the combination of "electric" plus "kinetics". Generally speaking, electrokinetics is the motion of liquid or particle under the influence of electric field. According to Probstein [7], the electrokinetic effects have been first observed by F. F. Reuss in 1809 via experiments on porous clay diaphragms. He has shown that in capillary, fluid moves from anode to cathode in the presence of an external electric field. In the mid 19th century, Wiedemann repeated this experiment and described the fundamentals of electrokinetics. This was followed later by the seminal work of Helmholtz in 1879 on the electric double layer theory, which related the electrical and flow parameters for electrokinetic transport. DC electrokinetics, including electroosmosis and electrophoresis, has almost two hundreds years' history and has been rather thoroughly investigated. Electroosmosis is the fluid motion caused by the electrical force acting on the double layers next to a charge surface. The ions in a double layer are moved by a tangential electric field, giving rise to movement of the whole double layer along the surface, which in turn puts the bulk fluid into motion through the viscous interaction.

On the other hand, AC electrokinetics has been studied for just a few years. AC electrokinetics can be classified into three categories, dielectrophoresis (DEP), the electrothermal effect (ET) and ac electroosmosis (ACEO). DEP is the force acting on the particles due to the difference in polarizability between the particles and the fluid. The electrothermal effect refers to the fluid motion caused by the interaction of electric fields and gradients of conductivity and permittivity of the fluid through Joule heating. AC electroosmosis is the fluid motion induced by moving double layers. We will focus on the AC electroosmosis, while we will discuss the effects from other electrokinetic phenomena. Basically, AC electroosmosis (ACEO) works by the same principle as DC EO. However, in ACEO, the surface charges are induced by externally applied voltages, rather than naturally occurring charges in DCEO, and consequently hundreds of time stronger. As a result, ACEO can generate flow velocity of several hundreds microns per second with a couple of volts.

AC techniques are more favorable over DC ones for following reasons: (1) low operating voltage makes it superior in terms of device portability; (2) avoids electrolysis and the resulting bubble generation; (3) minimizes pH gradient; (4) miniaturization and integration with other devices on lab-chips. We have extended the scope of ACEO by including electrochemical reactions (i.e. Faradaic charging), and then developed a new ACEO technique—asymmetric polarization ACEO. Asymmetric polarization of electrodes is achieved by combining the DC bias into AC signals over electrode pairs. Biased ACEO breaks the reflection symmetry to produce net flow in a symmetric pair of electrode. This technique adds more flexibility to the faster manipulation of bio-particulate.

3.1 AC Electrokinetics

AC electrokinetics provides a means to effectively control and manipulate particles and fluids at micro-scale. Switching electric field of AC elctrokinetics can suppress the electrolysis and hence the change of pH value at electrodes, which is inevitable in dc electrokinetics. Different from DC EO which relies on naturally induced charges, ACEO induces surface charges by applying voltage, which can be hundreds of times higher lower voltage is required to generate sufficient flow velocity. Additionally small spacing of electrodes makes it possible to reach high electric field (E) with low applied voltage (V). AC electrokinetics can operate low voltages, which is suitable for lab-on-a-chip operation.
AC electrokinetics comprises of dielectrophoresis, the thermal effect and ac electroosmosis. Dielectrophoresis (DEP) is the response of particles to the applied electric fields, and the electrothermal effect and ac electroosmosis are fluid motion caused by AC fields. Ramos et al have provided a rather comprehensive review [9] on various forces acting on micro-size particles on microelectrode arrays when electrodes are energized with ac voltages over a wide range of frequency. In subsequent work, Ramos et al [10] presented a RC model describing the frequency dependence of ACEO flow velocity by capacitive charging of the electrodes. Recently, Bazant and Squires [25, 28] also presented that AC electrokinetic phenomena can also occur for conducting particles, which they termed as Induced charged Electroosmosis (ICEO). They also predict that the velocity will scale as E^2, where E is the applied electric field. Vortices will also occur around a spherical metal because of their geometry. Bhatt [5] also reported that electrohydrodynamic effect arising from the application of alternating electric fields to patterned electrode surfaces.

3.2 AC Electroosmosis

ACEO was first investigated using a pair of planar electrodes as shown in figure 2. In figure 2, a pair of electrodes is placed parallel in an electrolyte. The first half cycle of the applied signal is shown in figure 2(a). The double layer is produced on the electrode surface like dc electroosmosis. Therefore there is a nonzero tangential component of electric field acting on the double layer. The interaction of this tangential electric field with the surface charge creates the force according to the Coulomb's law. This force is along the electrode, which in turn puts the fluid in motion. In the half cycle of the applied AC signal, the sign of the surface potential becomes reversed, as shown in figure 2(b). The ions move accordingly, keeping the sign of the double layer opposite to the potential. At the same time, the electric field also reversed its direction. So in this case both surface charge polarity and the electric field direction changed. Subsequently, the force acting on the double layer is still the same direction as in the first half cycle, keeping the fluid motion unchanged.
AC electroosmosis is much more complicated than DC electroosmosis due to two reasons: 1) The applied signal is oscillating; 2) The surface charge and the tangential field are coupled. An excellent experiment and theoretical study of AC electroosmosis can be found in [9, 10, 38]. The governing equations for the fluid motion inside the double layer is represented by following equation coupled with the continuity equation.

$$\rho_m \frac{D\vec{u}}{Dt} = -\nabla p + \mu \nabla^2 \vec{u} + \rho \vec{E} \tag{1}$$

Neglecting the term on the left hand side of equation (1) leads to,

$$-\nabla p+\mu\nabla^{2}\vec{u}+\rho\vec{E}=0 \qquad (2)$$

The fluid velocity can be obtained after substituting the charge density (ρ) and the electric field distribution into equation (2). The fluid velocity at the outer boundary of the double layer is given by,

$$\vec{u}_f=-\frac{\varepsilon}{4\mu}\Lambda\nabla_s\left(|\Delta V|^2\right) \qquad (3)$$

where Λ is a factor of the double layer structure, Δ_s represents a gradient across the surface, and ΔV is the potential drop across the double layer. According to the equation (3), fluid tends to move from higher electric field region to the lower electric field region. The fluid velocity at the electrode surface is,

$$u_s=\frac{1}{8}\frac{\varepsilon V_0 \Omega^{?}}{\mu s\left(1+\Omega^2\right)^2} \qquad (4)$$

where V_0 is the voltage amplitude and s is the distance from the electrode center. Ω is the nondimensional frequency,

$$\Omega=\omega\frac{\varepsilon}{\sigma}\frac{\pi}{2}\frac{s}{\lambda} \qquad (5)$$

where ω is the angular frequency and λ is the debye length.
Equation (4) gives a bell-shaped velocity profile with respect to the frequency. At high and low frequencies the velocity approaches to zero. We experimentally verified this bell-shaped velocity profile in our research. Our research emphasizes on the charging processes of electrodes. By studying surface EO flows with respect to AC potential, we identified ACEO induced by electrochemical reactions (i.e. Faradaic charging), and we have developed a new ACEO technique – asymmetric polarization ACEO. Here we also explain the competition between the Faradaic and Capacitive charging process.

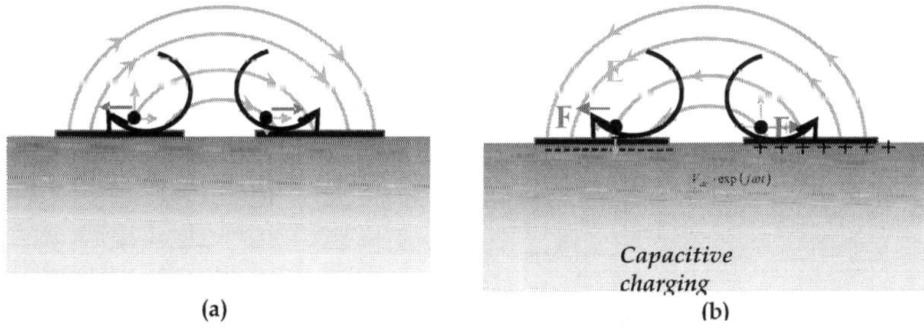

Fig. 2. ACEO fluid motion and induced charge at electrode surface. (a) during the half cycle when the left electrode has positive polarity; (b) during the next half cycle with opposite electrical polarity.

Faradaic charging generates co-ions from electrochemical reactions. When the electrode is positively charged, it goes through the reaction according to the Faradaic's law. On the other hand, capacitive charging attracts counter-ions from the electrolyte to screen the electrode potential. Our contribution in this field is that we have developed an "asymmetric polarization (A-P) ACEO" technique by adding a DC offset to the AC signal, and we used this A-P ACEO for trapping particles on the electrodes [11]. Asymmetric polarization of electrodes in a pair is achieved by combining the DC bias into AC signals over electrode pairs. For adding DC bias the reflection symmetry of electrode charging is broken, leading to asymmetric surface flow and net-flow. By adjusting the amplitude and frequency of AC signals, a variety of directed surface flows are produced on electrodes to manipulate and transport particles.

Capacitive charging and Faradaic charging coexist and compete for dominance when the electrodes are energized under biased conditions. The biased ACEO scheme is built on the fact that the two electrodes in a pair undergo the two distinct polarization processes. Biased AC EO is implemented by energizing electrode pairs with biased AC signals so that electrodes in a pair undergo polarizations different from each other. The advantage of such a scheme is twofold. First, it breaks the mirror symmetry of electric fields and, consequently, that of surface flows. If Faradaic charging occurs on one electrode while capacitive charging takes place on the other, then ions of the same sign populate the electrode surface and they migrate in the same direction under the influence of the electric fields. As a result, a unidirectional flow is produced at the electrode surface. So, non-uniform electric field has a good application in pumping action.

3.3 Fabrication of microfluidic devices

The processes developed for microelectronics, such as standard photolithographic methods, can be applied to silicon and glass substrates producing channel networks in two dimensions for sample transport, mixing, separation, and detection systems on a monolithic chip (Fig 1). A mask is made that has transparent and opaque regions that are patterned as a negative image of the desired channel layout. A UV-light source transfers the layout from the mask to the photoresist, which has been previously deposited on the substrate by spin-coating. The photoresist is then developed in a solvent that selectively removes either the exposed or the unexposed regions.

After developing the photoresist, it has a small amount of "hydrocarbon" material still on the wafer. If this is not removed it can affect the geomarty. Removing this very thin layer is something we call "descum". To descum we used an oxygen plasma generated in a parallel-plate etcher (Reactive Ion Etching - RIE). The oxygen plasma interacts with our undesirable hydrocarbon layer and burns it away. Then we deposit gold using the E-beam evaporator. As an adhesive layer we have used Cromium (Cr) between the gold and the wafer. After that photoresist is removed by lift-off process, and interdigitated electrode pattern is created for microfluidic experiment. Typical fabrication sequence is shown in the figure 3. Here we have shown the negative photolithography, where the image reversal is used.

Polymer based microfluidic devices are also getting much attention recent days. Polymer is resistant to chemicals and some are biocompatible for implantation by FDA [37]. PDMS has shown a number of advantages over other polymer materials (eg. SU-8, PMMA): PDMS microchips can be easily replicated and produced by rapid prototype approaches with low cost [31]. The excellent optical transparency of PDMS has been exploited to integrate different elements in the optical detection.

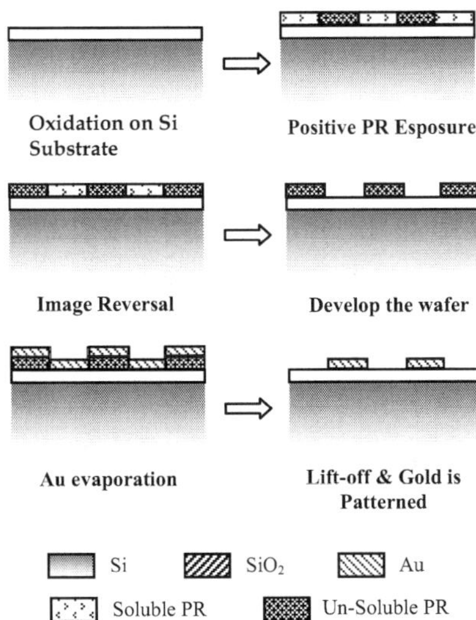

Fig. 3. Fabrication sequence for silicon, and glass microfluidic devices

For our experiments we have used PDMS to fabricate our microchannel for microfluidics particle trapping and micropumping. The first step of fabricating microchannel is to have the mold. We have fabricated the channel dimensions as low as 100 μm X 100 μm [cross-section]. Following is the steps of fabricating microchannel for our experiments.

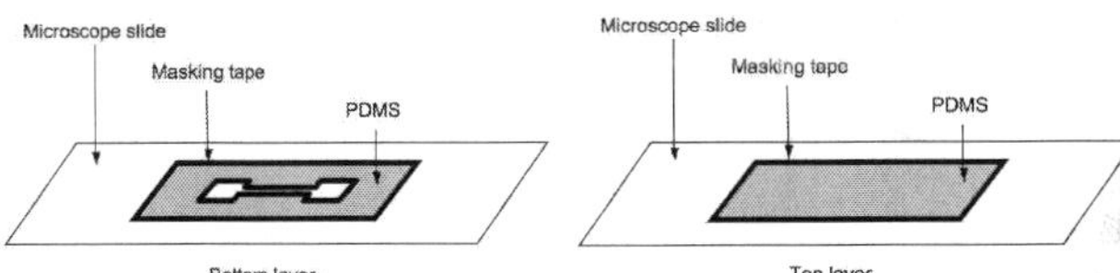

a. PDMS on two substrates to create a bottom layer and a top layer of the microchannel pump

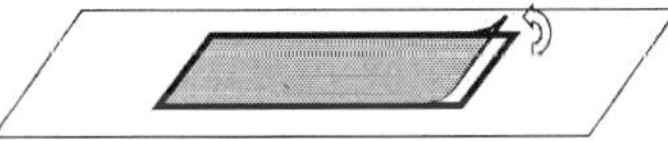

b. Top layer PDMS was peeled from the substrate after thermally cured

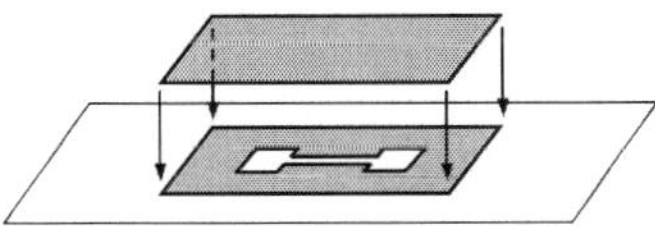

c. Top layer PDMS was placed on top of the patterned bottom layer

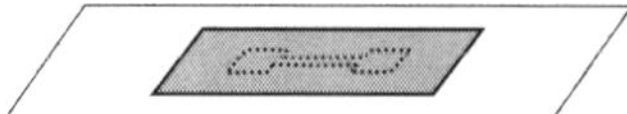

d. Top view of the sealed microchannel pump

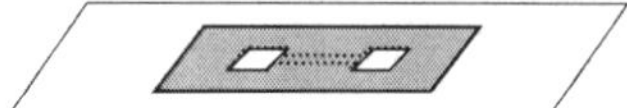

e. Two rectangular cuts were made for reservoirs

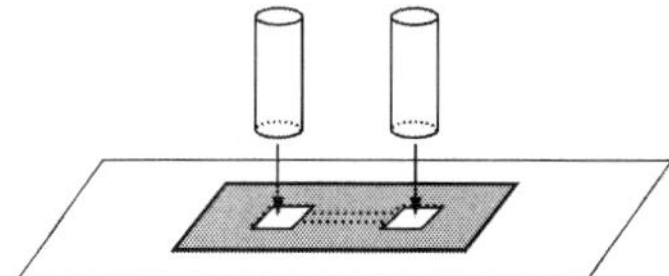

f. Integrating the reservoirs onto the pump

Fig. 4. Steps of fabricating microfluidic chamber

1. PDMS was prepared by thoroughly mixing the silicone Base and Curing Agent (SYLGARD 184 Silicone Elastomer Kit, Dow Corning, Midland, MI) at ratio 10:1 base:curing agent using a portable mixer. The mixture was then set in a small plastic container to reduce the air bubbles from the mixing process. This was done in approximately 30 minutes or until there were no visible bubbles. The PDMS was then poured onto two different glass substrates (microscope glass slides).
2. The first substrate functioned as the bottom layer, which was patterned using a piece of masking tape. The pattern defined the profile of the microfluidic pump. It contained a microchannel that connected two reservoirs at the end of the channel. The second substrate created the top layer, which would be used as the channel sealer. (See Figure 4a). Both substrates were thermally cured on a hot plate for 10 minutes at 150°C.
3. The PDMS for the top layer was then peeled from the glass substrate and was placed on top of the PDMS for the bottom layer. A careful placement was done to avoid any air gap between the two layers. Hence, to prevent a leak of the working liquid. (Figure 4-b, c, and d).
4. Two rectangular cuts were made on the top layer outlining the reservoir contours of the bottom layer. (Figure 4e).
5. Two reservoirs were made by cutting a glass pipette into two two-centimeter tubes. They were integrated on the pump using a 5-minute epoxy glue. (Figure 4f).

One of the most important advantages of PDMS is that it is easy to fabricate and allows a simple sealing with the planar substrates. However, PDMS has three significant problems in practical use.

- Dissolution or swelling by organic solvents.
- Absorption of chemical materials.
- Adhesion between PDMS and metal layer is not intact/perfect

To overcome these problems, we have coated PDMS with perfluoro amorphous polymer on the PDMS micro structures, which minimizes the problems of PDMS [36]. Perfluoro amorphous polymer (CYTOP, CTX-809A, Asahi Glass Company, Japan) is coated by spin coating on the PDMS structure. Since natural PDMS surface repels CYTOP, so we used O2 plasma pre-treatment on the surface before CYTOP coating. Our preliminary result shows that CYTOP coating was achieved without any deformation of the micro structure. This polymer based substrate with CYTOP can be used in different biological, chemical and lab-on-a-chip (LOC) applications.

The need for innovative fabrication methods to integrate higher levels of functionality into microfluidic and lab-on-a-chip devices is growing almost as rapidly as the number of potential applications for these miniature devices. The ability to make fully-integrated, multi-level fluidic systems with functional valves, pumping systems, electrical and electronic components, and other microeletromechanical system (MEMS) components is essential in order for this relatively new field to reach its full potential.

3.4 Electric field analysis

We have used Comsol Multiphysics (formerly FEMLab) to simulate the electric field distribution above a pair of planar electrodes (160micron width and 40 micron separation between the electrodes, with infinitesimal thickness). As shown in Figure 5a, tangential electric fields change directions over one electrode, which indicates that two counter-rotating vortices exist on one electrode, as schematically drawn in Figure 5b, countering to one vortex reported in the litereature.

The fluid velocity on the electrode surface is given [5] as

$$u = -(\varepsilon / \eta)(\xi - \varphi_b)E_t \tag{6}$$

where ε is the permittivity, η is the viscosity of bulk solution, $(\xi - \varphi_b)$ is the difference of potential between the double layer and the bulk solution, Et is the tangential component of electrical field. According to Equation 6, the velocity on the surface of double layer is proportional to the tangential field and potential difference between the double layer and fluid, which corresponds to the normal component of electrical field. Therefore, normalized boundary conditions on both electrodes are given as, $u = -E_x E_y$.

To model the fluid motions, the 2D Incompressible Navier-Stokes module is used in FEMLAB. In this case, the hydrodynamic property in the chamber is given as density = 1000 kg.m^{-1} and viscosity = 10^{-3} kg.m^{-1}.sec. The fluid velocity distribution is then obtained by solving Navier-Stokes equation with calculated field profile.

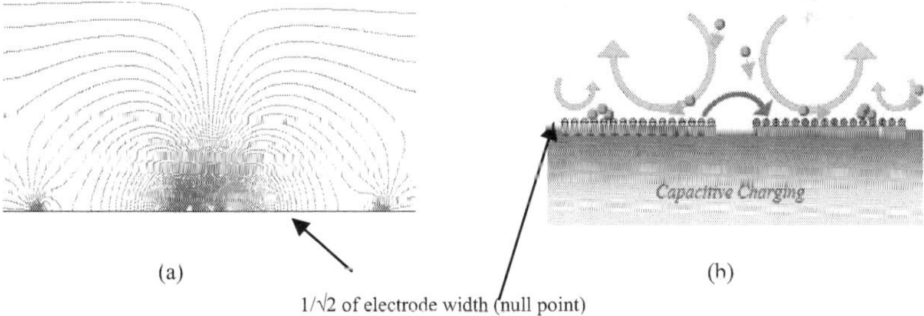

Fig. 5. (a) Comsol simulation for the Electric field distribution above a pair of planar electrodes with voltage of +1V & -1V in two electrodes (160/40micron). (b) Four counter-rotating vortices are formed above the electrodes due to changes in tangential electric fields, which facilitate particles aggregation on electrodes.

3.5 Impedance analysis for optimization

The charging and ion migration process at electrodes can be represented by an equivalent circuit element. To analyze the effect of ACEO, it is necessary to develop an RC equivalent circuit for the pair of electrodes. Figure 6(a) is the equivalent circuit for the planar electrode in an aqueous environment. There are two paths that the planar interdigitated electrodes are connected as shown in the figure. One path goes through Ccell, which stands for the direct dielectric coupling between electrodes and it jumps the dielectric coupling through the fluid, and the environment. The other path goes through the fluid, which can be treated as resistance since it obeys Ohm's law. It is in series with capacitance for double layer charging at the interface of electrolyte and electrode on both ends. Figure 6(b) is the simplified RC equivalent circuit model. At low frequency, the reactance from double layer capacitance is high. Thereby a large portion of voltage drop happens within the double layer, suitable for the ACEO phenomenon to take place. At high frequency, time is limited for double layers to form, consequently inducing a great amount of surface charge in the double layer.
So, for the high frequency case the reactance gets smaller and more voltage drop across the resistive bulk fluid. As there is little voltage drop, hence little surface charge on the electrode, so ACEO becomes negligible at high frequency. By using the equivalent circuit model we can theoretically analyze the frequency range for the ACEO mechanism.

$$Z_{\text{mod}\,eled} = \left[j\frac{1}{2*pi*f*C_{cell}} \;\; || \;\; \left(R_{sol} + j\frac{1}{2*pi*f*C_{tot_dl}} \right) \right] + 2R_{lead}$$

The values of the above mentioned components in the equivalent circuit can be extracted by impedance measurements. 5 mV_{rms} excitation level was used for impedance measurement. For 5 mV excitation level in can be very well assumed. In this condition faradaic charging will not take place, so we neglected the faradaic charging (Fig. 6b) for this analysis. The simplified equivalent circuit is shown in Fig 6(b). Following is the modeled impedance value,

$$Z_{\text{mod}\,eled} = \left[j\frac{1}{2*pi*f*C_{cell}} \;\; || \;\; \left(R_{sol} + j\frac{1}{2*pi*f*C_{tot_dl}} \right) \right] + 2R_{lead}$$

(a) (b)

Fig. 6. Equivalent impedance for interdigitated pattern; (a) RC equivalent circuit for planar electrode configuration; (b) Simplified RC equivalent circuit for the modeling.

C_{cell} and C_{dbl} can be extracted from the experimental plot. At smaller frequency the C_{dbl} is dominated and at higher frequency $C_{cell/dielectric}$ dominants. Our extracted parameters are,

C_{cell} = 305 pF; C_{dbl} = 28.85 nF; R_{lead} = 18 Ω; R_{sol} = 966.234 Ω. We put all these values to our $Z_{modeled}$ and compare the plot with the experimental impedance plot from the impedance analyzer. The two plots matched fairly, so we can conclude that our extracted modeled parameter is correct.

The impedance measurements have been done for two different situations, 1) the electrode pair in DI water, and 2) the electrode pair in the DI water seeded with particles. We use information from impedance measurement to optimize the operating condition (signal frequency, magnitude etc.) of ACEO. Impedance plot for the two different excitation levels are also compared in figure 7. As seen from the figure, for high excitation voltage the impedance goes down. This is because the charge, Q is constant in the double layer, and the double layer capacitance is inversely proportional to the applied oscillator voltage (Q=CV). That is the reason the impedance plot for the two different excitation levels are different.

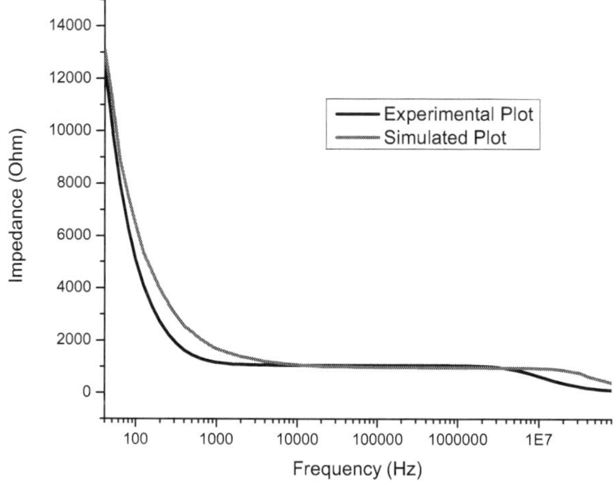

Fig. 7. Comparison of impedance plot between the experimental and the modeled data

Nyquist plot shows the frequency response for the linear system. Figure 8 shows the Nyquist plot for the equivalent circuit. The low frequency area of Nyquist plot denotes the double layer capacitive effect. Point "a" denotes the solution of resistance, which we extracted from figure 7 is 966.234 Ω. The graphical representation (figure 8) using a Nyquist plot represents the two parallel path of the microfluidic system, so that we can analyze the pre-dominant of the impedance in smaller and higher frequency range. The impedance analysis is very important to distinguish the effect of different electrokinetic forces.

From the figure 7 we also can see that the difference of the impedance with and without the particle is more in frequency below 1 KHz. So we adopted signal frequency range between 100 Hz to 1 kHz for particle trapping. The experiment was done at 500 mV_{rms} oscillation level. Same characteristics were obtained at 1 V_{rms}. Our goal is to determine a higher velocity electrode pattern with the polystyrene particle, so both experiments and calculation were performed to determine the optimum signal magnitude for these four sizes of electrode pair (160/40, 160/20, 80/40, 80/20).

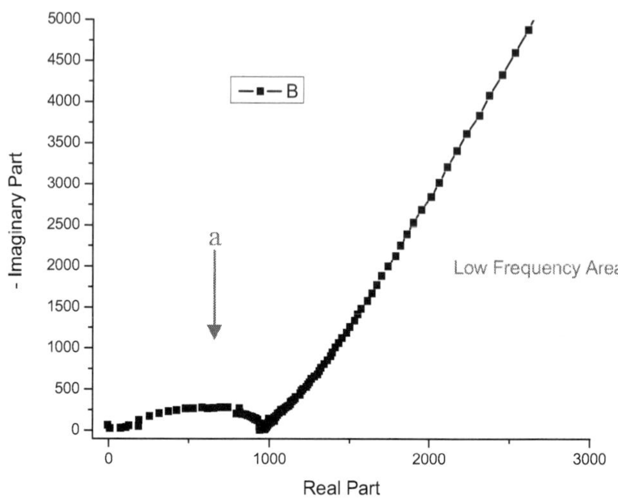

Fig. 8. Nyquist Plot for real vs Imaginary plot at 5mV Osc Level

4. Biased AC electroosmosis micropump

Micropump is critical to transport small amounts of fluid for many microfluidic applications, ranging from drug delivery, bio-fluid analysis, to microelectronics cooling. With the development of MEMS technology, micropumps have been designed and fabricated to integrate with lab-on-a-chip (LOC). Due to the large surface to volume ratio in microchannels, surface tension and viscous forces play an important role in the flow characteristics. So for micropumping action we need to choose the electroosmosis technique which benefits from the higher surface to volume ratio. For this reason electroosmotic action is suitable for miropumping action.

Electroosmotic (EO) pumping is the motion of bulk liquid caused by the application of an electric field to a channel. Electroosmotic (EO) pumps (*a.k.a.* electrokinetic pumps) have no moving parts and are capable of generating high flow rate per device volume. People use electroosmotic pumping to achieve both significant flow rates and pressures, and a fairly wide range of working electrolytes may be used (including deionized water, buffered aqueous electrolytes). These devices have significant pressure capacity in a compact structure. We have achieved flow rates in excess of 400 micron/second. EO pumps offer some advantages over other miniature pumps for microchannel cooling applications and integrated bio-analytical systems.

High-pressure capacity, millimeter-scale, porous-media based EO pumps have been demonstrated [32, 33], and most of the micropumps which are presented in the literature are DCEO micropumps. However, DCEO micropump suffers from high voltage operation (several kVs) and consequently excessive electrochemical reactions and electrolysis at the electrodes. This high voltage operation also creates the pH gradients and bubble which is not favorable for micropumping [28-32]. Again, many of the current fabrication techniques of porous-media EO pumps are not compatible with standard microfabrication processes and this poses a significant obstacle to the chip-level integration of EO pumps into microsystems. Our developed micropump is operated with smaller AC voltage and microfabrication compatible.

Reliability, compatibility, and cost are also criterions for selecting or designing micropumps. Micropumps must perform proposed functions without being damaged (re-usable), and at the same time must not bring changes in the medium. Compatibility with microsystems requires precisely pumping the desired range of fluid volumes and proper overall size of the micropump. Micropump can be used to manipulate the fluid volumes ranging from a few picoleters to hundreds of microliters for different biomedical applications, such as single molecule detection, species separation, antigen-antibody binding. The pump size is important for integrating the compact microsystem. The simplicity in design and fabrication of micropumps is also desirable. By using our fabricated micro-electrode array the problems of the electrokinetic micropumps can be solved.

4.1 The features of AC electroosmosis micropumps

Novel pumps based on ac electroosmosis are also investigated in this research. Here the AC electroosmosis is the main mechanism to drive fluids, which are quantitatively investigated by both experiments and Comsol simulations. Compared to other micropumps without moving parts, this AC electroosmotic micropump has following unique features.

1. Low operating voltage makes it superior in terms of device portability. In our design the AC electric fields are applied to pump fluids as small as 500mV bias voltage.
2. Avoids electrolysis and the resulting bubble generation. Currently EHD and MHD micropumps use DC fields to produce fluid motion, generating bubbles at the electrodes. In this research the applied AC fields has a frequency range of 100Hz to 5 KHz, allowing no time for bubble generation.
3. The dimension of our micropump is 5 mm long, 500 μm wide and 100 μm height. These pumps can be scaled down linearly to submicron/nanoscale or up by expanding laterally.
4. Minimizes pH gradient for using the smaller voltage.
5. The designed micropump is simple in structure. The channel and electrodes array are combined to achieve electroosmotic pumping. The designed ACEO micropump can also be integrated with lab-on-a-chip for miniaturization.

These features of the AC electroosmotic pump provide higher reliability, higher compatibility and lower cost. It can be possible for mass production of the electrode array on the wafer. The channel can also be fabricated using PDMS by using Si mold.

The designed AC electroosmotic pumps do not require open channels and hence are very useful for a sealed lab-on-a-chip system, where the fluid is circulating in closed channels. The use of this type of AC electroosmotic pump does not need open reservoir, avoiding undesired pressure due to different fluid heights in the reservoir.

4.2 Design of AC electroosmotic micropumps

Electroosmotic micropumps have been used widely in broad applications [33, 34]. Novel micropumps based on ACEO have been designed to drive fluids. AC electroosmotic micropumps can operate at lower voltage to avoid undesirable electrolysis and pH gradients. The main concept of getting the pumping action is to get the uni-directional flow by breaking the reflection-symmetry in the geometry or applied signal. Figure 9 shows the pumping action breaking the symmetry by changing the geometry [35]. By using the asymmetric pattern of electrode bigger vortices are generated on the larger electrode, which eventually dominants and produce uni-directional flow. The smaller electrode still produces the counter-rotating vortices that reduce the net flow.

Our design of AC electroosmotic micropump is based on the biased AC electroosmosis technique for symmetrical electrode array. The separation of the in-pair symmetrical electrodes (80 micron size) are 40 μm. We repeated the pair of electrodes to make the array of electrodes. The separation of the elctrode pair is 100 μm. For this particular configuration we name it 80/40/100 configuration. The corresponding numerical simulations using the finite element software FEMLAB are also conducted to verify the concept. The good agreement between the simulations and the experimental data regarding the uni-directional flow is also demonstrated.

4.2.1 Capacitive and faradaic charging effect

Biased ACEO is realized by applying biased AC signals over electrode pairs, leaving the electrolyte floating; therefore, two electrodes have different electrical potentials relative to the electrolyte. With a biased AC signal, $V_{applied}=V_0(1+\cos\omega t)$ over the electrodes, the left electrode is always positive and more prone to Faradaic charging, while the other is always negative and subject to capacitive charging. For the biased signal one electrode has a positive offset with the potential always greater than zero, while the other one lower than zero (Fig. 9). When the voltage exceeds the threshold for reaction, asymmetric vortices are formed above two electrodes as faradaic reactions take place at the positively biased electrodes. Faradaic reaction generates co-ions following Faraday's law. For the other electrode with a negative offset, counter ions are attracted to the electrode. Therefore, for two electrodes, same polarity of ions is induced. A unidirectional fluid loop is consequently formed by tangential fields, as shown in Figure 9. For the negative particles in an aqueous environment, they adhere to the stagnation line on the positively biased electrode. Henceforward, biased ACEO exhibits directional particle assembly.

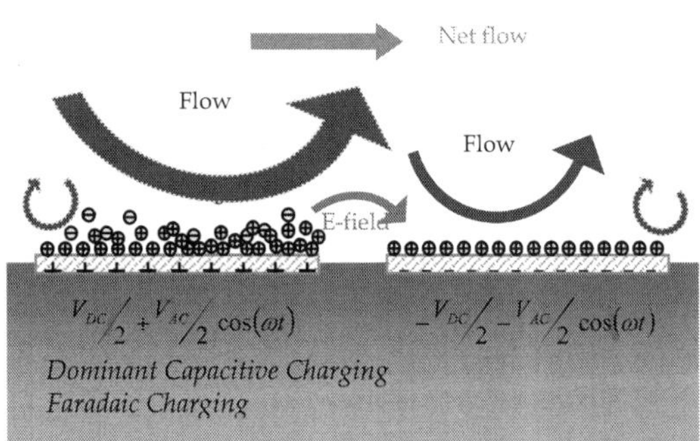

Fig. 9. Asymmetric polarization with appropriate magnitude can produce uni-directional micropumping.

Because most bioparticles are negatively charged, the DC bias can provide a synergy of AC and DC electrokinetics for more efficient particle collection. Electrophoretic/electrostatic force is exerted simultaneously with ACEO to move bioparticles towards positively-biased electrodes, as shown in Figure 9. Figure 9 already explained the electric field direction and net fluid flow for the asymmetric biased electrode pattern. Also in next section we have experimentally proved unidirectional particle flow. Breaking the charging symmetry on electrodes are main concept of producing uni-directional flow. With the biased AC signal, one electrode is at a higher voltage and hence undergoes Faradaic charging with cations, and the other electrode is at a lower voltage and hence is polarized by capacitive charging with the cations as well. This combination of the two polarization on the two electrodes of an electrode pair produces a uni-directional flow on each of the electrode pairs on electrode array. Which eventually produce a continuation of fluid flow and show the pumping action.

4.2.2 Unidirectional ACEO micropump

When using the asymmetric electrode pattern the direction of the pumping can not be reversed. For several biomedical applications there is a need for bi-directional flow directions. A common medical treatment procedure makes the use of a bidirectional flow control of one or more fluids to and from a patient. For these applications we have developed the biased AC electroosmotic micropump which can operate in both directions. Here we have broken the symmetry by applying asymmetric voltage on the symmetric electrode pattern, which eventually breaks the symmetry of the pattern. Figure 10 shows the electrode array, where L_{pair} is the separation between the two electrode pairs and L_{array} is the separation within the electrode pair. The mechanism of the designed pump is explained in Fig 10. As shown in the picture, both positive biased electrode (faradic charging) and the negative biased electrode (capacitive charging) generate the same positive charges on the electrode. The electrode surface has an excess of positive charges that creates a uni-directional flow in electrode pair, which eventually produce net flow on the array of electrodes. However, the coupling between neighboring electrode pairs will produce counterflows to the net flow produced within the pair and reduce the pumping efficiency. To reduce the undesirable coupling between the two adjacent pairs, L_{array} is kept larger than L_{pair}, but not so large that the flow loses its momentum.

Figure 10 is the schematic of pumping action for symmetrical electrode pattern with the biased applied voltage. The spacing Lpair and Larray will be the controlling factor for the pumping velocity. The rule of thumb is $L_{array} > L_{pair}$, so that the consecutive electrodes of Lpair and Larray do not form the EO flow in the reverse direction. From our earlier experiment we already have got L_{pair} of 20μm for the 80/20 configuration of our electrode pattern, which produces the highest fluid velocity.

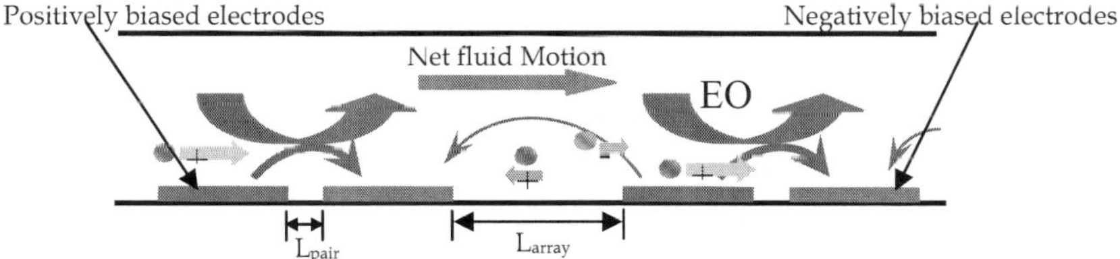

Fig. 10. Biased ACEO can produce uni-direction fuid motion, which also imparts differential velocities to particles with various charge/mass ratio.

Biased ACEO pump will be investigated by manipulating the faradic charging effect. We will also study the improvement in the pumping velocity. The pumping setup is shown in figure 11. The channel dimension is 350μm X 1.5mm X 5mm. The highest measured velocity for this pump is 150 μm/sec.

Fig. 11. Experimental setup of first version of the micropump.

4.3 Experimental results

ACEO flow is examined using microfabricated arrays of electrode pairs on silicon substrate. Au/Cr (90nm/10nm thickness) electrodes were fabricated by lift-off procedure in IC processing. Cr is the adhesion layer between the substrate and Au, and Au is in contact with electrolytes. The electrodes were 20 mm long, 0.1 μm think, 80 μm wide with a 20 μm separation (denoted as 80/20). Microfluidic chambers were formed by sealing silicone microchambers (PC8R-0.5, Grace Bio-Labs, Inc.) over the wafer, which have a height of 500 μm. Polystyrene spheres (1 μm diameter; Fluka Chemica) seeded in DI water was used to track fluid motion.

Figure 12 gives the comparative analysis of the microfluidic velocity generated using four types of patterned Au electrodes. The 80/20 configuration gives the highest velocity, which is suitable for the pumping application.

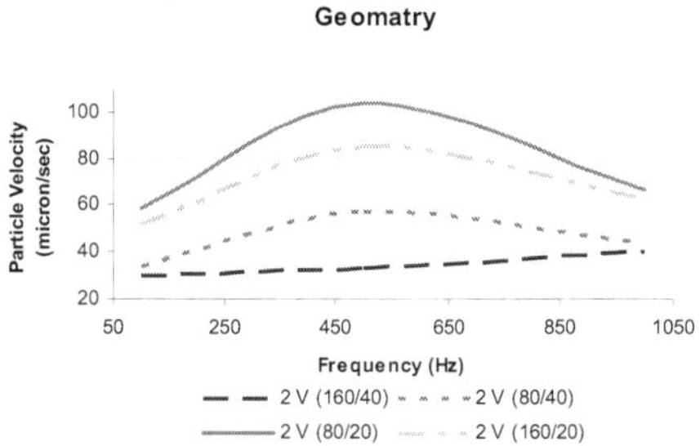

Fig. 12. Microfluidic Experiment result & comparison of four types of electrode geometry

4.4 Optimization of biased ACEO micropump

The goal of the research is to minimize the micropump reverse flow velocity. Obviously we increase the applied voltage it will increase the pumping velocity, and it will also increase the reverse pumping velocity (for the vortices). So, here we have focused on decreasing the reverse direction flow. Figure 13 explains the optimization concept by decreasing the channel height of the microfluidic chamber. Later on we also have presented the numerical simulation using Comsol Multiphysics for three different channel heights and verified the concept. The experimental result demonstrates that the thinner channel height ~ 100 μm increases the velocity of the micropump. In this case we also have demonstrated the bulk fluid flow, as here the surface to volume ratio is high.

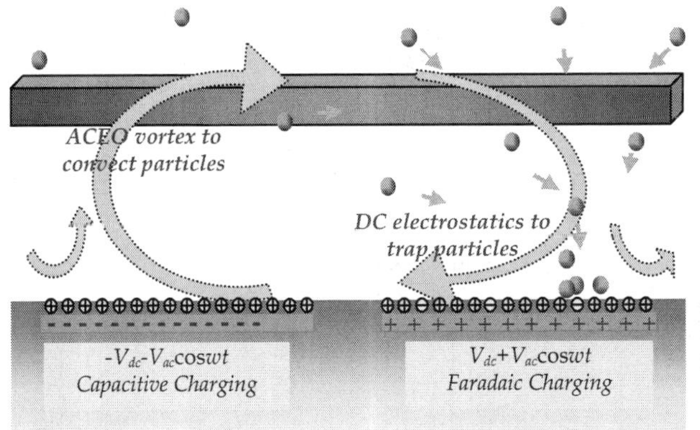

Fig. 13. Optimization Concept: to decrease the channel height to reduce the reverse flow

As seen from figure 13, the big vortex is suppressed / obstructed by the top wall of the channel. According to our analysis the vortex size is depended on the electrode geometry and the channel height. For the fixed electrode geometry the size of the vortex is only depended on the channel height, which is explained by Comsol Multiphysics simulation. The boundary condition for inflow and outflow was the key factor to run the simuation. Our findings show that smaller channel height will increase the surface flow, which will be described more in the next section.

Simulation of ACEO micropump has been performed prior to experiments. The simulation model, consists of two symmetrical electrodes of 80 μm width and the separation between the electrodes are 40 μm. The height of the chamber is 200 μm. From simulation we found that at the edge of the electrode the electric field is maximum. After simulating the electric field analysis the Navier-Stokes simulation is done to calculate the velocity field in the chamber. Initial fluid velocity is set to zero for the Navier-Stokes (NS) simulation. The fluid motions are generated by applied electric signals and a large vortex is formed for biased applied signal.

5. ACEO sensor Integrated with microcantilever

We have investigated another microstructural design for ACEO devices, which is similar to a pair of parallel plates (Fig. 15). In this configuration, electrodes are facing each other, similar to a pair of parallel plates. The two face-to-face electrodes are asymmetric in design, so they produce non-uniform electric field. For mechanism identical to that of planar electrodes, surface EO flows are generated from the electrode edges inwards, and slow down to stagnation at the center, where particles are expected to trap.

5.1 ACEO particle trap

ACEO can transport the particles from a large region in the bulk fluid to the electrode surface. The flow velocity is important for optimizing the micropump and particle transportation. In contrast to electrophoretic and dielectrophoretic (DEP) velocity, which are typically limited to less than 20 microns per second [9], the ACEO velocity exceed 100 micron/sec. Figure 14(a) shows the initial distribution of particles when no signals are applied over the electrodes. Figure 14(b) shows that particles accumulated from both sides into lines at approximately $1/\sqrt{2}$ of electrode width [12]. This corroborates the theoretic prediction, since fluid velocity reduces at the null points of electric fields, and particles become trapped to the electrodes due to surface forces of between particles and electrodes. For the biased ACEO experiments, applied voltage exceeds the threshold for reactions at V0=1.5V (i.e. high level & low level of biased voltage is 3V & 0V respectively). At the same voltage, the maximum flow velocity shifts to higher frequency compared with symmetric AC signals. This is because Faradaic polarization becomes suppressed at high frequency. Beyond 500Hz, microflows from capacitive charging are much stronger than those from Faradaic charging, so that the stagnation point on the left electrode disappears. At 100Hz, streamlines from capacitive charging and Faradaic charging become connected, forming a large vortex over the electrode pair and the particles aligned on the right electrode. Figure 14b also validates the null point formation by Comsol Multiphysics simulation in Figure5.

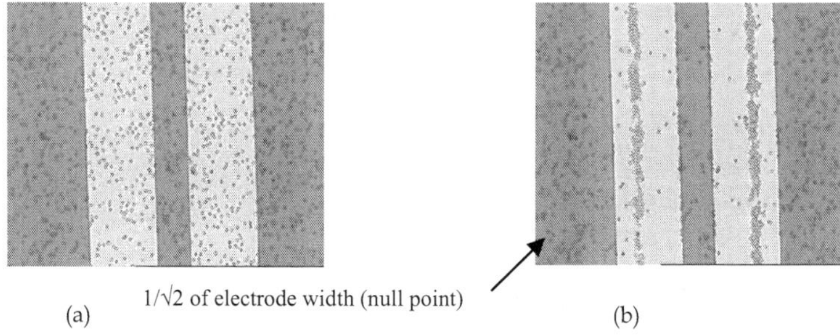

Fig. 14. (a) Particle without the supply voltage (b) Experimental picture of the particles accumulating at the $1/\sqrt{2}$ electrode width;

5.2 Electric field analysis of parallel plate particle trap

Most ACEO devices reported so far adopt a side-by-side (interdigitated) configuration. To integrate such design on microcantilevers would call for sophisticated microfabrication. We use a face-to-face configuration, very much like a pair of parallel plates, with one plate having smaller electrode area than the other, as shown in Figure 15. As the top and bottom electrodes are asymmetric, the tangential electric fields are generated which induces electro-osmotic fluid motion. For the first half cycle as in Figure 15a, the bottom electrode is at positive potential, and negative ions are induced at the interface of the electrode and fluid. These negative ions interact with the electric field and produce two counter-rotating vortices from electrode edges inward on the electrode surface, creating null point at the center of the electrode.

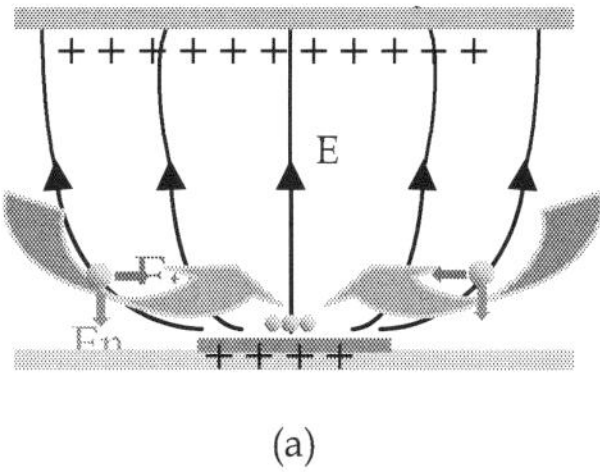

(a)

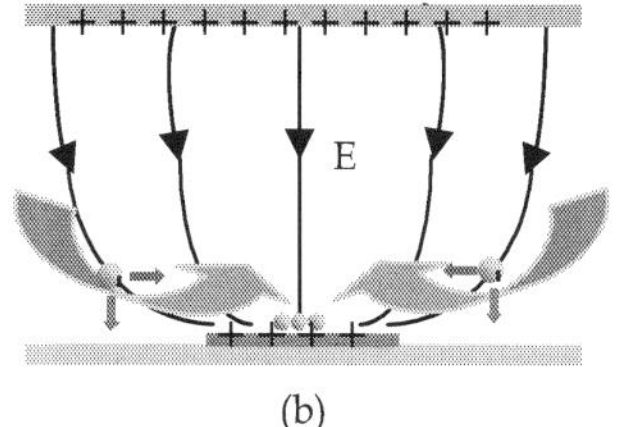

(b)

Fig. 15. Concept of Parallel plate particle trapping for an AC cycle; (a) during the half cycle when the bottom electrode has positive polarity and ITO coated top electrode is negative; (b) next half cycle with opposite polarity. Flow motion and induced charges is also shown

For the next half cycle as in Figure 15b, applied potentials switches polarity and the bottom electrode is at negative potential. Here the induced positive ions will interact with the electric field and again produces two counter-rotating vortices from electrode edges inward, and fluid motions are sustained thus the particles are trapped at the centre of the electrode. As the bottom plate is smaller than the top plate, the electric field is almost always normal to the top electrode, hence tangential force can be neglected.

The tangential electric field for the asymmetric electrode pattern induces electro-osmotic fluid motion in the bottom plate. It is the microfluidic flow that conveys particles from the bulk of the fluid onto the fluid surface. The stagnation point created at the centre of the bottom plate. The particles are trapped at the stagnation point of the fluid.

5.3 Microcantilever particle trapping using ACEO

This section explains the novel particle trapping method using microcantilever. Here we have presented the first integration of the microcantilever with the ACEO particle trapping mechanism. Recently microcantilever sensor technology has boomed and become a promising sensor technology. Microcantilever sensors have several advantages over many other sensor technologies, including faster response time, lower cost of fabrication, the ability to explore microenvironments, and improved portability. Cantilever resonance responses, such as frequency, deflection, Q-factor, and amplitude, undergo changes due to adsorption or changes in environment. Resonance frequency of a microcantilever can be used to detect particles. When the target is loaded on the microcantilever, the resonance frequency of microcantilever is going to change. That means for the mass loading on the cantilever the resonance frequency is supposed to go down.

The parallel plate design has been used to attract particles onto cantilevers for high sensitivity detection [14]. Because particle trapping/concentrating effect is more obvious with the smaller electrode, in our design the metal-coated cantilevers are facing one large electrode (covering a whole fluid chamber), so that particles will aggregate on the cantilevers. The tangential electric field of parallel plate configuration is generated for the asymmetric electrode pattern, which induces electro-osmosis fluid motion. In our design of microcantilever trap, the metal-coated cantilevers substitute the patterned bottom electrode, so that particles will aggregate on the cantilevers [15]. Figure 16 shows the experimental setup of cantilever particle trap.

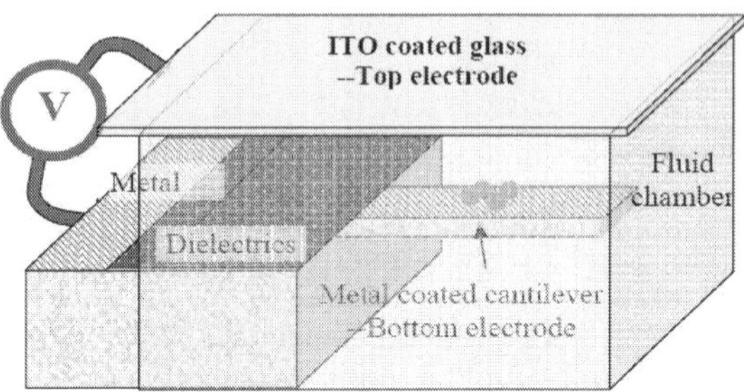

Fig. 16. Experimental Setup of Cantilever particle trap

As shown in Figure 16, photoresists (dielectrics) are coated on the conductive areas other than the cantilevers to suppress unwanted local EO flows. The ITO glass slide works as the top electrode, which is covering the whole fluid chamber. We have used Au-coated AFM probes as the MC, which has the dimension of 125μm X 30μm X 4μm. Tipped MC side was not used to avoid sharp electric fields.

Figure 17 shows the experimental results of trapping 200nm fluorescent particles on MC. After applying the AC signal (100Hz, 400mVp-p), suspended particles accumulate at the center of the cantilever from all directions. As time passes, more fluorescent particles from the surrounding area accumulated and formed bright object pattern. After the particle trapping on the surface, the MC was dried with AC signals applied, so that particles will not get dispersed by diffusion, surface tension, etc. Then the particle trapping effect was verified with AFM resonance measurement.

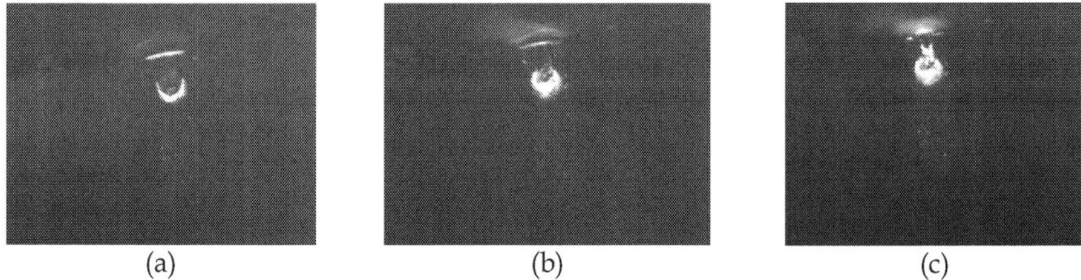

(a) (b) (c)

Fig. 17. Image sequence of 200 nm fluorescent particles trapped on the micro-cantilever;

5.3.1 Microcantilever particle trapping validation

To verify the concentrated particle trapping on MC, we also measure the resonance frequency of MC before and after trapping experiments. MC resonance frequency is inversely proportional to the differential mass of cantilever [21]. The sensitivity of a cantilever to mass loading is mainly determined by the excited cantilever resonance frequency.

$$\Delta m = \frac{K}{4\pi^2}\left(\frac{1}{f'^2} - \frac{1}{f^2}\right)$$

Where, Δm is the mass change;
K=sensor spring constant;
f = resonance frequency before mass adsorption;
And f' is the resonance frequency during mass adsorption.

Changes in the mass and surface properties of the microcantilever through binding or hybridization of analytes to receptor molecules will directly influence its surface stress. This causes the microcantilever to deflect and the deflection is proportional to the analyte concentration and inversely proportional to mass loading. The more the particle concentration on ACEO-cantilever, the more is the bending. So the more mass on the cantilever means the lower resonance frequency. From our experimental result (Fig. 18) we have got the resonance frequency of the MC goes down to 276.07 KHz after particle trapped on the MC for ACEO, which translates to a change of mass. For a change in mass we get,

$$\frac{f_1}{f_2} = \frac{\sqrt{m_{eff} + \Delta m}}{\sqrt{m_{eff}}}$$

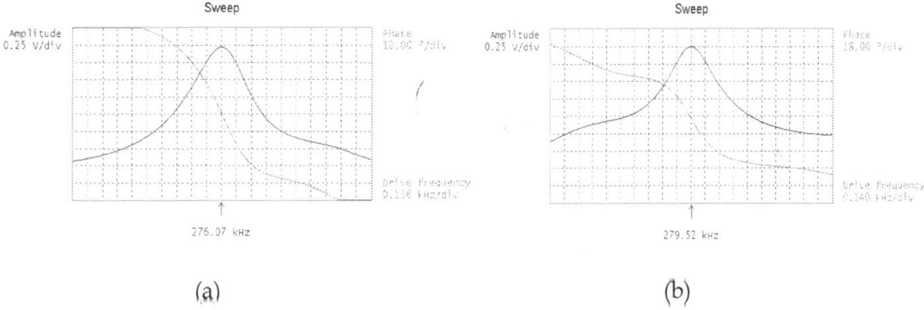

Fig. 18. Resonance frequencies of the MCs (measured with multimode AFM.)
(a) After the particle trapping by ACEO, 276.07 kHz, (b) Control experiment with no electric signal applied, 279.52 kHz.

Microcantilever dimension is 125μm x 30μm x 4μm. The volume of the MC is 1.5e-14 m^3. Microcantilever is Si based, and the density of Si is 2330 kg/m^3. So the mass of microcantilever is 3.495e-8 gm. By putting the frequency and mass values in equation (4.2), we have got an increase of 2.52% increase of mass for the frequency change from 279.52 KHz to 276.07 KHz. The change in resonant frequency as a function of the particle mass binding on the cantilever beam surface forms the basis of the particle detection scheme.

5.3.2 Analysis of particle trapping using SEM

Our research also shows that applying the electric field creates a certain crystal shape of the concentrated particles. Without applying the electric field, the particles are accumulating in layers, and formed no crystal shape (Fig. 19a). But when the electric field was applied for ACEO, the particles formed the close-packed layer of colloidal particles, and take the shape of crystalline structure, as shown in Figure 19(b). Compare with the Figure 19(a), the crystalline structure (Fig. 19b) increases the number of particles in the bottom layer [22].

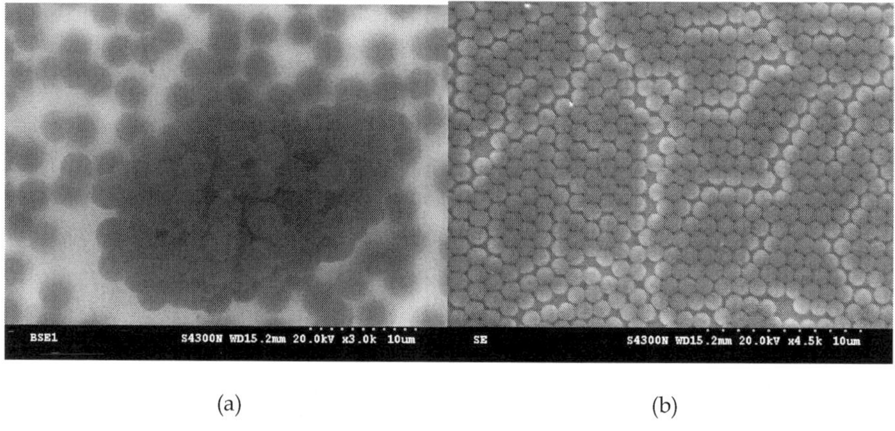

(a) (b)

Fig. 19. SEM image of the particles; (a) Particles dried on the surface, no electric field is applied (b) For ACEO particle trapping, particles take the Crystal Shape.

6. Conclusion

AC electroosmosis, with its advantages of low power, low voltage and higher velocity, has been used to develop pumps. Such AC electrokinetic micropumps are presented in literature which have fast pumping (velocities » mm/s) velocity with low driving voltage of a few volts. But the literature review used the asymmetric pattern of electrodes. This chapter presents the novel micropump which utilizes the symmetric electrode pattern with the bias AC voltage. The chapter also focused on the MEMS microcantilever integration with ACEO.

The novel DC-biased, AC electroosmotic micropump operates in low voltage to avoid undesirable electrolysis and pH gradients. The developed uni-directional micropump prototype breaks the symmetry of the electrode array by applying the DC-biased AC signal, which has proven to be advantageous when compared to those pumps that use only asymmetric electrode patterning. In this work, a novel technique also has been developed to trap the particles and pump the fluid. Interfacing the micro-cantilever with ACEO mechanism has expanded its capability for biological, physical and chemical detection and makes the whole system ultra sensitive. The research work substantially enriches the portfolio of transducers, lab-on-a-chip (LOC) and MEMS that can be used in high performance miniaturized analytical systems.

7. Acknowledgment

This work is supported in part by the Nanotechnology for Undergraduate Education Program of the National Science Foundation (NSF-NUE) under Award Number EEC-1138205. The work is also supported from the internal fellowship grant from the University of Texas at Brownsville. We also want to thank Arizona State University (ASU) Nanofab Laboratory for assistance with microfabrication.

8. References

[1] Helene Andersson, Albert ven den Berg, " Microfluidic devices for cellomics : A review", Sensors and Actuators, B 92, (2003), 315-325.
[2] Vincent Studer, Anne Pépin, Yong Chen, and Armand Ajdari, "An integrated ac electrokinetic pump in a microfluidic loop for fast and tunable flow control", Analyst, vol. 129, pp. 944-949, 2004.
[3] C.-H. Chen and J.G. Santiago, "A planar electroosmotic micropump," J. Microelectromech. Syst., vol. 11, pp. 672-683, 2002.
[4] Lastochkin, D., Zhou, R., Wang, P., Ben, Y. and Chang, H.-C., "Electrokinetic Micropump and Micromixer Design Based on AC Faradaic Polarization", J. of Applied Physics , 96 , 1730 (2004).
[5] K. H. Bhatt, S. Grego and O. D. Velev, "An AC Electrokinetic Technique for Collection and Concentration of Particles and Cells on Patterned Electrodes," *Langmuir* 21, pp. 6603-6612, 2005.
[6] M. R. Brown, and C. D. Meinhart, "AC electroosmotic flow in a DNA concentrator," Microfluidics & Nanofluidics, ISSN:1613-4982 (Print) 1613-4990 (Online), 2006.
[7] Ronald F. Probstein, Physicochemical Hydrodynamics : An Introduction, 2nd Editon, New York : John Wiley and Sons, c 1994
[8] Stone, H.A. and S. Kim, "Microfluidics: Basic issues, applications, and challenges," AIChE J., 47(6), 1250-1253 (2001).
[9] A. Ramos, H. Morgan, N. G. Green and A. Castellanos, "AC electrokinetics: a review of forces in microelectrode structures", J. Phys. D: Appl. Phys. 31(1998) 2338–53
[10] A. Ramos, H. Morgan, N. G. Green and A. Castellanos, "AC electric-field-induced fluid flow in microelectrodes", J. Colloid Interface Sci. 217(1999) 420–2
[11] J. Wu, "Electrokinetic Microfluidics for On-Chip Bioparticle Processing," IEEE Trans. Nanotech., Mar. 2006.
[12] J. Wu, Y. Ben and H.-C. Chang, "Particle Detection by Micro- Electrical Impedance Spectroscopy with Asymmetric-Polarization AC Electroosmotic Trapping," Microfluidics & Nanofluidics, 1(2), pp. 161-167, 2005

[13] Julio M. Ottino and Stephen Wiggins, " Introduction: mixing in microfluidics", Phil. Trans. Royal Society Lond. A (2004) 362, 923-935.

[14] N. Islam, M. Lian, S. Swaminathan and J. Wu, "Micro/Nano- Particulate Fluid Manipulation in AC Electro-Kinetic Lab-on-a-Chip," 2nd ASM - IEEE EMBS Conf. Bio, Micro & Nanosyst., Jan. 15-18, 2006, San Francisco, CA, USA.

[15] J. Wu, N. Islam and M. Lian, "High Sensitivity Particle Detection By Biased AC Electro-Osmotic Trapping on Cantilever," 19th IEEE Int'l Conf. Micro Electro Mechanical Systems (MEMS 2006), Jan. 22-26, 2006, pp. 566 - 569, Istanbul, Turkey.

[16] N. Islam, M Lian, and J. Wu, "Enhancing Microcantilever Sensitivity with Integrated AC Electroosmotic Trapping," submitted to Bio-microfluidics Journal.

[17] J. Wu and H.-C. Chang, "Asymmetrically Biased AC Electrochemical Micropump," AIChE annual meeting 2004, Nov. 7 – 12, Austin, TX.

[18] Rosenthal A, Voldman J, "Dielectrophoretic traps for single-particle patterning," Biophysical Journal, 88(3), pp. 2193-2205, 2005.

[19] Meinhart C D, Wang D and Turner K, "Measurement of AC Electrokinetic Flows," J. Biomedical Microdevices, 5(2), 139-145, 2003

[20] Morgan H, Green N G, "Electrokinetics: Colloids and Nanoparticles," Research Studies Press Ltd: UK 2002

[21] Sepaniak M, Datskos P, Lavrik N, Tipple C, "Microcantilever Transducers: A New Approach in Sensor Technology," Analytical Chemistry, November 1, pp. 568A-575A, 2002.

[22] Green N G, Ramos A, Gonzalez A, Morgan H, and Castellanos A, "Fluid flow induced by nonuniform ac electric fields in electrolytes on microelectrodes. III. Observation of streamlines and numerical simulation," Physical Review E 66, 026305, 2002.

[23] Taylor, M. T., " Simulation of microfluidic pumping in a genomic DNA blood-processing cassette," Journal of Micromechanics and Microengineering, 2003. 13(2): pp 201-208.

[24] Laser, D. and J. Santiago, "A Review of micropumps," J. Micromechanics and Microengineering, 2004. 14:pp R34-R64.

[25] T.M. Squires and M.Z. Bazant, "Induced-charge Electro-osmosis," J. Fluid Mech., 509, pp. 217-252, 2004.

[26] Vincent Studer, Anne Pépin, Yong Chen, and Armand Ajdari, "An integrated ac electrokinetic pump in a microfluidic loop for fast and tunable flow control", The Analyst Journal, 129, 944-949, 2004.

[27] J. Wu, N. Islam and M. Lian, "High Sensitivity Particle Detection By Biased AC Electro-Osmotic Trapping on Cantilever," 19th IEEE Int'l Conf. Micro Electro Mechanical Systems (MEMS 2006), Jan. 22-26, 2006, pp. 566 - 569, Istanbul, Turkey.

[28] M. Z. Bazant, T. M. Squires, "Induced-charge Electrokinetic Phenomena: Theory and Microfluidic Applications", Phys. Rev. Letter, 92(6), 066101 (2004)

[29] A. Castellanos, A. Ramos, A. Gonzalez, N. G. Green, and H. Morgan, "Electrohydrodynamics and dielectrophoresis in microsystems: Scaling laws," J. Phys. D, Appl. Phys., vol. 36, pp. 2584–2597, 2003.

[30] Josh H. M. Lam, Raymond H. W. Lam, Kin Fong Lei, Winnie W. Y. Chow and Wen J. Li, "A Polymer-based Micro fluidic Mixing System Driven by Vortex Micropumps", Proceedings of the 5th World Congress on Intelligent Control and Automation, June 15-19, 2004.

[31] C. Liu, D. Cui, H. Cai, X. Chen, Z. Geng, "A rigid poly(dimethylsiloxane) sandwich electrophoresis microchip based on thin-casting method", Electrophoresis 2006, 27, 2917–2923

[32] N.G. Green, A. Ramos, A. Gonzalez, A. Castellanos, and H. Morgan, "Electrothermally induced fluid flow on microelectrodes," Journal of Electrostatics, 53 (2001) 71-87.

[33] Lastochkin, D., Zhou, R., Wang, P., Ben, Y. and Chang, H.-C., "Electrokinetic Micropump and Micromixer Design Based on AC Faradaic Polarization", J. of Applied Physics , 96 , 1730 (2004).

[34] Iki, N., H. Hoshino, and T. Yotsuyanagi, " A capillary electrophoretic reactor with an electroosmosis control method for measurement of dissociation kinetics of metal complexes," Analytical Chemistry, 2000. 72(20): pp. 4812-4820.

[35] A. Ramos, A. Gonzalez, A. Castellanos, N.G. Green, and H. Morgan, "Pumping of liquids with ac voltages applied to asymmetric pairs of microelectrodes," Phys. Rev. E, 67, 056302 (2003).

[36] Shaorong Liu, Qiaosheng Pu, and Joann J. Lu, "Electric field-decoupled electroosmotic pump for microfluidic devices", Journal of Chromatography A, 1013 (2003), 57-64.

[37] Ping Wang, Zilin Chen, and Hsueh-Chia Chang, "A new electro-osmotic pump based on silica monoliths", ELSEVIER-Sensors and Actuators B113 (2006), 500-509.

[38] A. Ramos, H. Morgan, N. G. Green, A. Castellanos. "AC electrokinetics: a review of forces in microelectrode structures". J. Phys. D: Appl. Phys. Vol 31. pp 2338-2353.

[39] N.G. Green, A. Ramos, A. Gonzalez, H. Morgan, and A. Castellanos, "Fluid flow induced by nonuniform ac electric fields in electrolytes on microelectrodes. III. Observation of streamlines and numerical simulation," Physical Review E, 2002, Vol.66, 026305

[40] Shuhuai Yao and Juan G. Santiago, "Porous glass electroosmotic pumps: theory", ELSEVIER- Journal of Colloid and Interface Science, 268 (2003), 133-142.

[41] Chuan-Hua Chen and Juan G. Santiago, "A Planar Electroosmotic Micropump", Journal of Microelectromechanical Systems, Vol.11, No.6, December 2002.

[42] P. H. Paul, D.W. Arnold, and D. J. Rakestraw, "Electrokinetic generation of high pressures using porous microstructures," in _-TAS 98, Banff, Canada, 1998.

[43] S. Zeng, C. H. Chen, J. C. Mikkelsen Jr., and J. G. Santiago, "Fabrication and characterization of electroosmotic micropumps," Sensors Actuat. B, vol. 79, pp. 107–114, 2001

[44] A. Ramos, A. Gonzalez, A. Castellanos, N.G. Green, and H. Morgan, "Pumping of liquids with ac voltages applied to asymmetric pairs of microelectrodes," Phys. Rev. E, 67, 056302 (2003).

[45] N. Islam and J. Wu, "Microfluidic Transport by AC Electroosmosis", Journal of Physics: Conference Series, 34, pp. 356 - 361, 2006s

4

Acoustic Wave Based MEMS Devices, Development and Applications

Ioana Voiculescu and Anis N. Nordin
[1]City College of New York
[2]International Islamic University
[1]USA
[2]Malaysia

1. Introduction

Acoustic waves based MEMS devices offer a promising technology platform for a wide range of applications due to their high sensitivity and the capability to operate wirelessly. These devices utilize an acoustic wave propagating through or on the surface of a piezoelectric material, as its sensing mechanism. Any variations to the characteristics of the propagation path affect the velocity or amplitude of the wave.

Important application for acoustic wave devices as sensors include torque and tire pressure sensors (Cullen et al., 1980; Cullen et al., 1975; Pohl et al., 1997), gas sensors (Levit et al., 2002; Nakamoto et al., 1996; Staples, 1999; Wohltjen et al., 1979), biosensors for medical applications (Andle et al., 1995; Ballantine et al., 1996; Cavic et al., 1999; Janshoff et al., 2000), and industrial and commercial applications (vapor, humidity, temperature, and mass sensors) (Bowers et al., 1991; Cheeke et al., 1996; Smith, 2001; N. J. Vellekoop et al., 1999; Vetelino et al., 1996; Weld et al., 1999).

This chapter is focused on two important applications of the acoustic-wave based MEMS devices; (1) biosensors and (2) telecommunications. The technological advancement of the micro-electromechanical systems (MEMS) facilitated the development of biosensors and various devices for telecommunications.

There has been increasing interest to develop miniature, portable and low-cost biosensors fabricated using MEMS technologies. For biological applications the acoustic wave device is integrated in a microfluidic system and the sensing area is coated with a biospecific layer. When a bioanalyte interacts with this sensing layer, physical, chemical, and/or biochemical changes are produced. Typically, mass and viscosity changes of the biospecific layer can be detected by analyzing changes in the acoustic wave properties such as velocity, attenuation and resonant frequency of the sensor. An important advantage of the acoustic wave biosensors is simple electronic readout that characterizes these sensors. The measurement of the resonant frequency or time delay can be performed with high degree of precision using conventional electronics.

Currently, a limitation of acoustic wave devices for biological applications is that they reuire expensive electronic detection systems, such as network analyzers. A final product aimed at the end user market must be small, portable and packaged into a highly integrated cost effective system. For acoustic wave biosensors integrated in a lab-on-chip device, sample pre-treatment, purification and concentration, as well as a good interface between the user and the integrated sensing system also need to be developed in the future

Historically, acoustic wave devices are widely used in telecommunications industry, primarily in mobile cell phones and base stations. Surface Acoustic Wave (SAW) devices are capable of performing powerful signal processing and have been successfully functioning as filters, resonators and duplexers for the past 60 years. Although SAW devices are technological mature and have served the telecommunication industry for several decades, these devices are typically fabricated on piezoelectric substrates and are packaged as discrete components. The wide flexibility and capabilities of the SAW device to form filters, resonators there has been the motivation to integrate such devices on silicon substrates (Nordin et al., 2007; M. J. Vellekoop et al., 1987; Visser et al., 1989). Standard Complementary Metal Oxide Semiconductor (CMOS) technology with additional MEMS post-processing was used for the fabrication of a CMOS SAW resonator in 0.6 μm AMIs CMOS technology (Nordin et al., 2007). The advantage of using standard CMOS technology for the fabrication of a SAW resonator is that active circuitry can be fabricated adjacent to the CMOS resonator on the same electronic chip.

Telecommunication devices based on acoustic waves have different requirements compared to biosensors. The biosensors operates at frequencies in the range of MHz where acoustic wave devices operating as a filter or resonator are expected to operate at high frequencies (GHz) and have high quality factors and low insertion losses. With the advancement in lithographic techniques, the acoustic wave based devices have the advantage of meeting the stringent requirement of telecommunication industry of having Qs in the 10,000 range and silicon compatibility.

A simple, robust, cheap packaging method is also critical for the commercialization of the acoustic wave devices. The integration of the acoustic wave based MEMS biosensor in the microfluidic system is a complex matter. The integration technique is influenced by the sensor fabrication process and the type of the biological applications. In some applications the sensor could be embedded in a microfluidic reservoir. In the case when the biological application requires different biological solutions to be introduced on the sensor sensitive area the biosensor could be embedded in a microfluidic channel. The packaging of the acoustic wave devices used for telecommunication is less complicated since these devices are embedded in the package and do not need to be in contact with liquid.

2. Acoustic wave MEMS devices as biosensors

There has been increasing interest to develop miniature, portable and low-cost biosensors fabricated using MEMS technologies. MEMS technology has been adopted from the integrated circuit (IC) industry and applied to the miniaturization of a large range of systems including acoustic wave based devices. Recent technological advancement of MEMS processes allows the fabrication of thin piezoelectric films and the integration of acoustic wave based devices, and electronics on a common silicon substrate. The acoustic wave MEMS biosensors presented in this chapter could be categorized in two main groups; resonators and delay lines.

For biological applications the Acoustic Wave Based MEMS devices are integrated in a microfluidic system and the sensing area is coated with a biospecific layer. When a bioanalyte interacts with this sensing layer, physical, chemical, and/or biochemical changes are produced. Typically, mass and viscosity changes of the biospecific layer can be detected by analyzing changes in the acoustic wave properties such as velocity, attenuation and resonant frequency of the sensor. An important advantage of the acoustic wave biosensors is simple electronic readout that characterizes these sensors. The measurement of the resonant frequency or time delay can be performed with high degree of precision using conventional electronics.

The Sauerbrey equation correlates the changes of the resonant frequency of an acoustic wave resonator with the mass deposited on it. The acoustic wave propagating on a piezoelectric substrate is generated and received using IDTs. In the case of a biosensor resonator, the cell to be analyzed or the antibody layer for protein marker detection are added on the IDTs. This will cause a shift of the resonant frequency due to the increasing of mass, where f_i and f_o are the resonant frequencies before and after loading the sensor.

The Sauerbrey equation is defined as;

$$\Delta f = -\frac{2 f_0^2 \Delta m}{A\sqrt{\rho_q \mu_q}} = -2.26 \cdot 10^6 f_0^2 \frac{\Delta m}{A} \tag{1}$$

where

$$f_o - f_i = \Delta f \tag{2}$$

From (1) the change Δf of the resonant frequency of the piezoelectric crystal is directly proportional to the mass loaded on the acoustic wave resonator, where Δm is expressed in g and Δf and f_0 in Hz (Skládal, 2003).

Generally, the acoustic wave MEMS resonators employed for biosensing applications are FBAR and acoustic wave based delay lines. For FBAR type biosensor the excitation electrodes are fabricated at both sides of the piezoelectric substrate and the acoustic waves propagate through the volume of the substrate. The detection mechanisms occur at the opposite surfaces of the piezoelectric substrate.

The acoustic wave based delay lines reported in the literature as MEMS biosensors are surface acoustic wave (SAW) delay lines that consists of two sets of interdigitated transducers (IDT)s fabricated on the same side of a thin layer of piezoelectric material. The acoustic wave is produced by one set of IDTs and the second set of IDTs is used to detect the acoustic wave. In the case of a biosensor, the surface between these two sets of IDTs is covered with a biological layer sensitive to the analyte to be detected, as illustrated in Fig. 1.

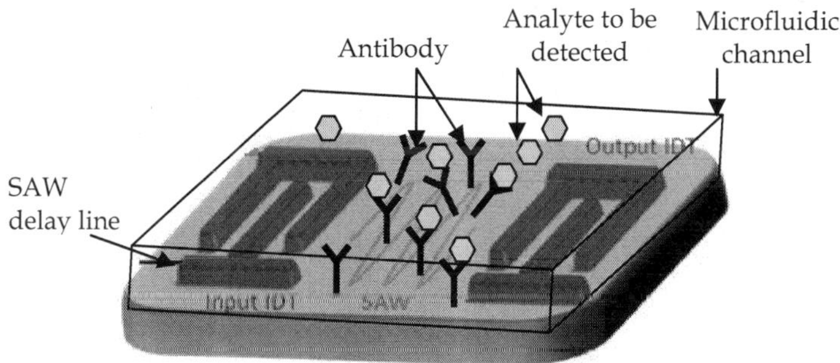

Fig. 1. SAW delay line biosensor integrated in a microfluidic channel. The surface between the IDTs is coated with antibodies sensitive to the analyte to be detected. The analyte

molecules binding to the immobilized antibodies on the sensor surface influence the velocity of the SAW and hence the output signal generated by the driving electronics.

The absorption of the analyte on the sensitive layer will produce a time delay in the acoustic wave propagation. The main disadvantage of the acoustic wave based devices when used as biosensors is the degradation of performance due to liquid damping. In liquid the quality factor Q drops (usually more than 90% reduction) and negatively affects the device sensitivity. Since most of the biological applications are performed in liquid only few types of acoustic wave devices could be integrated in microfluidic channels, without significant degradation of the sensor performance.

2.1 Film Bulk Acoustic wave Resonators (FBAR)

In recent years the thin film fabrication technology has made substantial progress particularly in view of high frequency resonators. In the case of MEMS-based FBAR resonators the expensive single crystalline substrates used for quartz crystal microbalance (QCM) resonators, could be replaced with a large range of thin piezoelectric films. FBAR resonators that are fabricated from a thin piezoelectric substrate and the excitation electrodes are fabricated at both sides of the piezoelectric substrate. The FBAR resonators could be integrated in a microfluidic system and successfully used for biosensing applications because of low damping of the acoustic wave in the liquid (Gabl et al., 2003; Weber et al., 2006; Wingqvist et al., 2005).

A MEMS FBAR biosensor with aluminum nitride (AlN) as piezoelectric film is illustrated in Fig. 2 (Wingqvist et al., 2005). The thickness of the AlN film is 2μm. Bottom and top Al electrodes were patterned with standard lithography and etching processes. The overlap of the top and bottom electrode defines the active area where the acoustic wave is generated. In order to fabricate a shear acoustic wave FBAR, the AlN thin films were grown with the crystallographic axis inclined with an angle of 30° relative to the surface normal. The shear mode is preferred for liquid application instead of the longitudinal mode because of low-loss operation in liquid and small reduction of the quality factor Q. The silicon wafer was etched from the back side to fabricate a free standing membrane used to isolate the resonator acoustically from the substrate and define a cavity. This cavity was further connected to a microfluidic transport system for analyte delivery to the bottom electrode of the resonator. For this type of FBAR, the bottom electrode is the sensing electrode. An Au layer was thermally evaporated onto the bottom Al electrode to create a biochemically suitable surface for subsequent tests. The sensor was tested with different concentration of albumin in solution and the detection limit was 0.3 ng/cm^2.

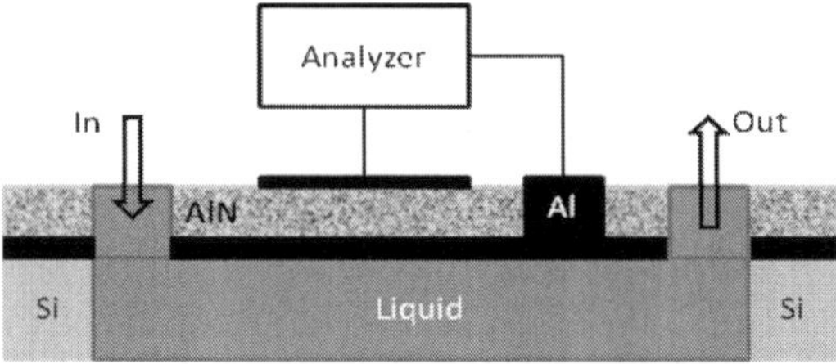

Fig. 2. Schematic of a shear wave FBAR with a microfluidic transport system.

Another FBAR resonator used for biological applications, illustrated in Fig. 3, is fabricated from a thin film of ZnO (Zhang et al., 2009). This FBAR resonator was fabricated from a <100> silicon wafer. A thin silicon dioxide layer was thermally grown on the wafer, followed by the low pressure chemical vapor deposition (LPCVD) of a thin SiN layer. An Al film was evaporated and patterned on the SiN layer as the bottom electrode and determines the effective area of the FBAR. The ZnO piezoelectric layer with the thicknesses from 0.55 µm to 4.2 µm was sputtered on the Al bottom electrode. Layers of Cr/Au representing the top electrode were sputtered and patterned by lift-off over the ZnO layer. Au was chosen as the top electrode due to its excellent conductivity and good affinity for biomolecular binding. The microfluidic channel was fabricated from a 3 µm thick parylene film deposited on the top electrode. The silicon wafer was backside etched using deep reactive ion etch (DRIE) to release the SiN membrane. The SiO_2 layer underneath the SiN layer was removed by wet etching, see Fig. 3a. The top Au electrode of the FBAR represents the sensing electrode and in the case of biological applications will be coated with a thin layer of biological material. This sensor is demonstrated a quality factor Q in liquid of 120. Q is improved by integrating a microfluidic channel on FBAR, which confines the liquid to a height comparable to the acoustic wavelength. However, this FBAR biosensor is sensitive to temperature variations. Increasing the temperature degrades Q, resulting in degradation of resolution.

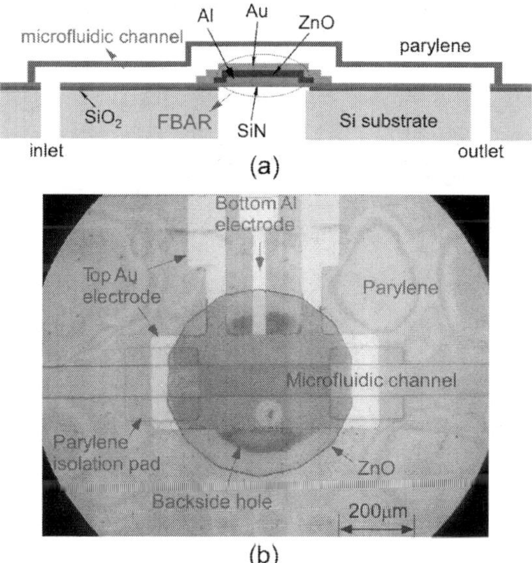

Fig. 3. (a) Schematic of the FBAR sensor integrated with a microfluidic channel and (b) Top view of the fabricated FBAR sensor; a microfluidic channel run across the FBAR sensor (Zhang et al., 2009). (© [2009] IEEE.) Used with permission.

Typical MEMS FBAR's have high quality factor Q in the air and vacuum which range from a few hundreds to a few thousands. High Qs are possible since the acoustic wave generated in the FBAR is well confined by the very large acoustic impedance mismatch between the solid materials and air. However, for biological applications the FBAR resonators are immersed in liquid and the solid-liquid interface becomes leaky for acoustic waves since the impedance mismatch is small, resulting in the degradation of *Q* of FBAR. To improve *Q* of FBAR in liquid, it is recommended that the microfluidic channels on top of the FBAR have the heights comparable to the acoustic wavelength in FBAR. This minimizes the dissipation in the liquid and hence improves the resonator's *Q*.

An interesting design of an FBAR without significant reduction of *Q* factor in liquid environments is illustrated in Fig. 4 (Pottigari et al., 2009). This device employs a thin vacuum gap between the FBAR and the sensing interface to prevent acoustic energy loss in liquid. This vacuum gap acts as an acoustic energy loss isolation layer of the FBAR in the aqueous viscoelastic media and reduces the direct contact area at the interface between FBAR and liquid. This vacuum separation is achieved by using micro-posts between the top electrode and the sensing diaphragm, which is in contact with the liquid. When the liquid is loaded on the sensing diaphragm, the mass is directly transferred onto the FBAR through the microposts. The vacuum gap protects the FBAR surface from the liquid, prevents the acoustic energy loss from the liquid, and contributes to maintenance of a high *Q* factor in the liquid without losing mass sensitivity.

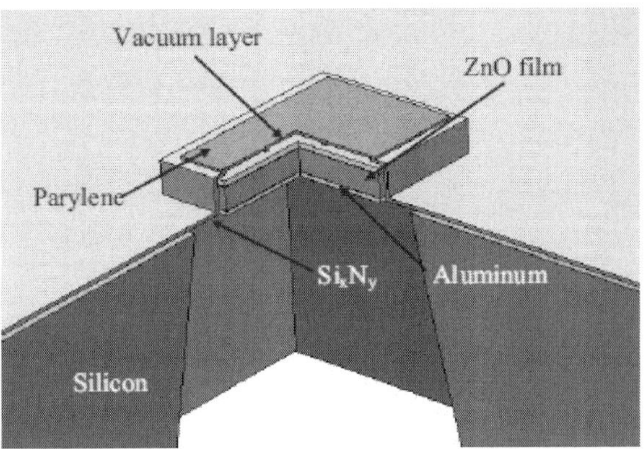

Fig. 4. Cross-section view of the V-FBAR device (Pottigari et al., 2009).
(© [2009] IEEE). Used with permission.

This vacuum (V)-FBAR uses a 0.7 µm thick ZnO film as piezoelectric material. The ZnO piezoelectric film is deposited by a radio-frequency (RF) sputtering system. A 1.6 µm thick parylene layer forms the top sensing diaphragm. Fig. 5 shows the V-FBAR integrated with a parylene sensing diaphragm that is supported by several parylene microposts. The height of the vacuum space is 2 µm and the active sensing area is 200 µm x 200 µm (Pottigari et al., 2009). The sensing diaphragm fabricated over a vacuum gap and the microposts are enhancing the sensitivity of the FBAR by separating the liquid damping effect from the operating frequency of the device. This device was conceived for biological applications.

Another recently developed MEMS based FBAR-type biological sensor is illustrated in Fig. 6 (Gabl et al., 2003). This FBAR is formed by two electrodes, fabricated at both sides of a thin a piezoelectric layer and integrated above a silicon substrate. The active vibrating region of the resonator is coated with a receptor layer which is sensitive to the biological analyte to be detected. The attachment of the analyte molecules leads to an increase of the resonator mass load and a decrease of the resonant frequency which can be electrically determined.

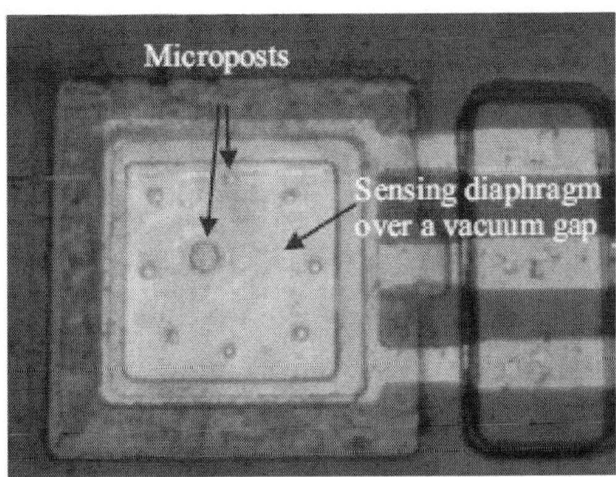

Fig. 5. Photo of fabricated V-FBAR (Pottigari et al., 2009).
(© [2009] IEEE). Used with permission.

The FBAR sensors have been fabricated on silicon substrates employing reactive magnetron sputtering of ZnO (Fig. 7). The 100 nm Au top-electrode provides a low electrical series resistance as well as a common chemical base for the binding of bioreceptor molecules. The quarter wavelength thick bottom electrode acts acoustically as an efficient reflection layer and ensures a high mass sensitivity as well as a low ohmic series resistance. A 3-fold ZnO/Pt mirror is fabricated below the FBAR sensor in order to isolate the sensor from the silicon substrate, see Figs. 6 and 7.

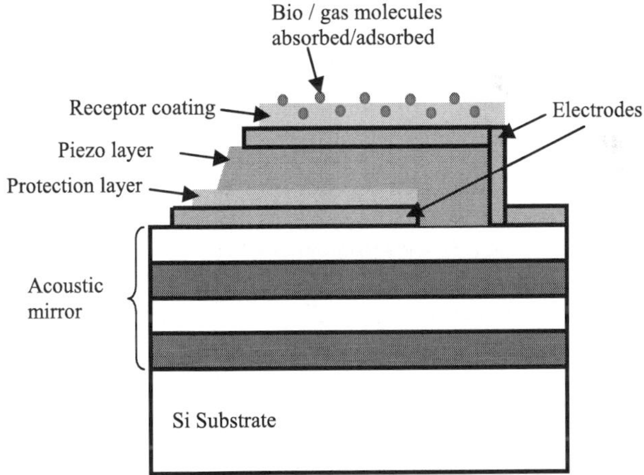

Fig. 6. Illustration of FBAR type biological sensor.

To verify the use of this type of FBAR in biosensing, a receptor assay of biotin-labeled DNA oligos coupled to the gold surface and streptavidin as the target molecule, were used. This FBAR biosensor is characterized by a sensitivity three orders of magnitude larger than for typical QCM.

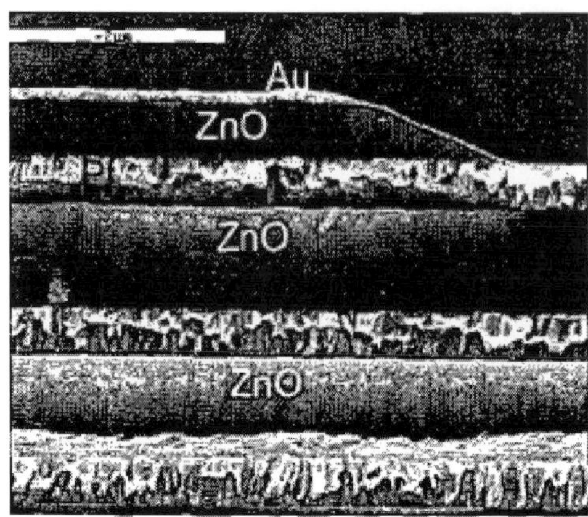

Fig. 7. SEM image of the acoustic mirror and ZnO thin films on a Si substrate. (Gabl et al., 2003). (© [2003] IEEE). Used with permission.

A nano-FBAR biosensor that uses nanomaterials to increase its sensitive area and it is based on $Mg_xZn_{1-x}O$ as piezoelectric material was reported in the literature (Chen et al., 2009). This device is built on Si substrates with an acoustic mirror consisting of alternating quarter-wavelength silicon dioxide (SiO_2) and tungsten (W) layers to isolate the FBAR from the Si substrate, see Figs. 8 and 9. High-quality ZnO and $Mg_xZn_{1-x}O$ thin films are achieved using RF sputtering technique. Tuning of the device's operating frequency was realized by varying the Mg composition in the piezoelectric $Mg_xZn_{1-x}O$ layer. The ZnO nanostructures were grown on Au electrode situated on the TFBAR's top surface using metalorganic

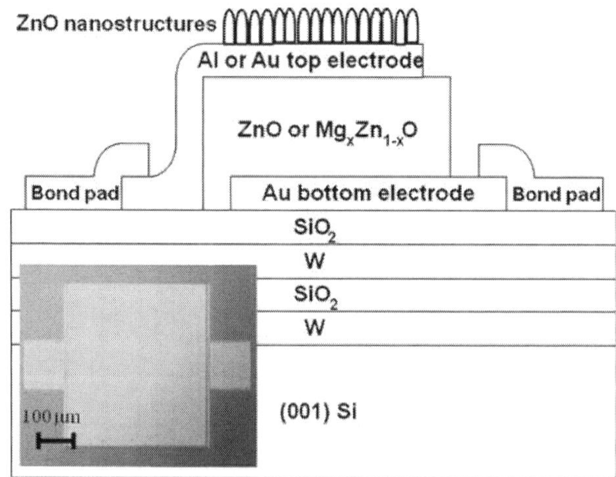

Fig. 8. Schematic diagram of the MgxZn1-xO FBAR structure; the inset shows a micrograph of a FBAR device. For the FBAR nanosensor, Au top electrode is used to facilitate MOCVD growth of ZnO nanostructures (Chen et al., 2009). (© [2009] IEEE). Used with permission.

chemical vapor deposition (MOCVD). These nanostructures offer a very large sensing area, faster response, and higher sensitivities over the planar sensor configuration. Because of the large sensitive area, the mass sensitivity of this biosensor is higher than 103 Hz cm²/ng. In order to employ this nano-FBAR for biosensing, the nanostructured ZnO surface was functionalized to selectively immobilize DNA. The device sensitivity was 16.25 ng of hybridized DNA and linker molecules combined.

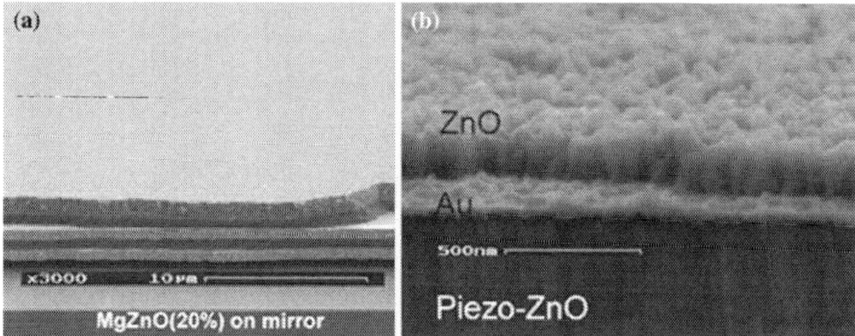

Fig. 9. Cross-sectional SEM images of (a) Mg0.2Zn0.8O film deposited on the mirror/Si structure and (b) ZnO nanostructures deposited on Au electrodes (Chen et al., 2009). (©[2009] IEEE). Used with permission.

The FBAR can be modeled into an equivalent circuit using a constant "clamped" capacitance C_0 connected in parallel with an acoustic (motional) arm that consists of motional capacitance C_m, motional inductance L_m, and motional resistance R_m, as illustrated in Fig. 10. The formulas for each component are given below:

$$C_0 = \varepsilon_r \varepsilon_0 \frac{A}{d}, \tag{6}$$

where A is the area of overlap of the two electrodes, ε_r is the relative static permittivity of the material between the electrodes, ε_0 is the electric constant and d is the separation between the electrodes.

$$C_m = \left[\left(\frac{f_p}{f_s}\right)^2 - 1\right] \tag{7}$$

where f_p is the parallel resonant frequency and f_s is the series resonant frequency.

$$L_m = \frac{1}{(2\pi f_s)^2 C_m} \tag{8}$$

$$R_m = \frac{1}{2\pi f_s C_m Q} \tag{9}$$

where Q is the quality factor of the resonator.

The FBAR biosensor is operated based on the dependency of the resonator frequency on its mass. The biomolecules attached on the sensing electrode increase the sensor mass. The sensitivity S (frequency shift per mass attachment) is described as:

$$S = \frac{f_0}{M} \quad (10)$$

where f_0 is the operating frequency and M is the resonator mass [11].

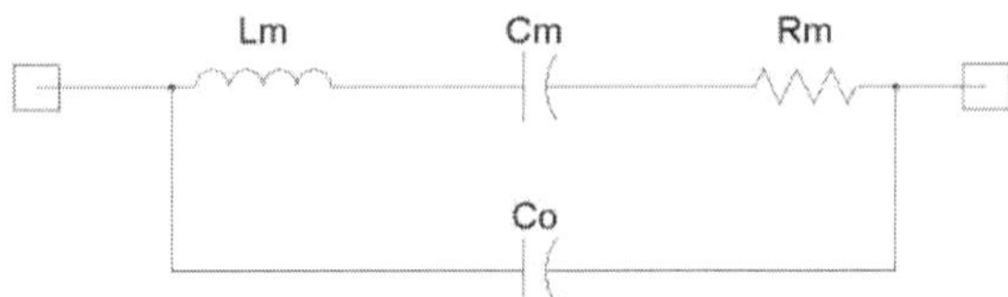

Fig. 10. FBAR equivalent circuit, where L_m is the motional mass, C_m is the motional capacitance, R_m is the motional resistance and C_0 is the static capacitance.

2.2 SAW delay lines as biosensors

Surface acoustic wave (SAW) delay lines were also studied for biosensing and could be integrated in microfluidic systems. A SAW delay line consists of two IDTs that are electrode pairs fabricated on the same side of a thin piezoelectric layer via photolithography. A sinusoidal voltage applied to the input IDTs translates into oscillating mechanical strain that forms a SAW that propagates along the surface of the piezoelectric thin film. The SAW is then

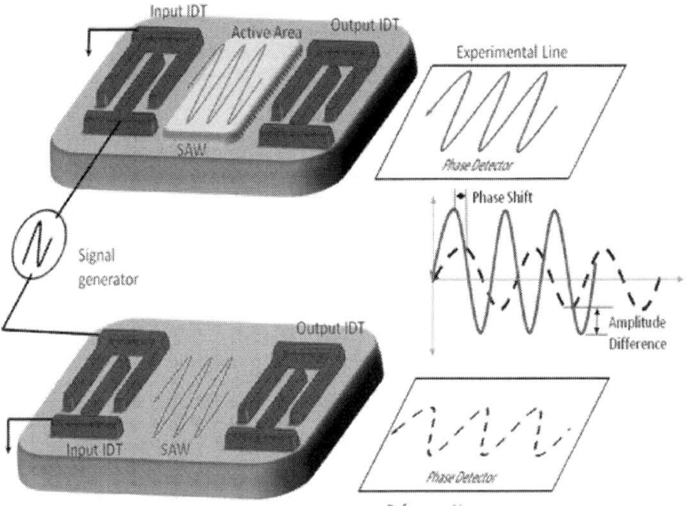

Fig. 11. A SAW delay line consists in input IDT and output IDT fabricated on a piezoelectric substrate. Two delay lines operate in parallel, with one line acting as a reference line and the other acting as an experimental line. A sinusoidal voltage is applied to the input IDT, which develops an alternating electric field that is translated into a mechanical SAW by the piezoelectric effect. The velocity of the SAW is affected by the mass loading, the liquid viscosity and the temperature of the substrate surface. Any difference in velocity between the two delay lines will be reflected as a phase shift and amplitude difference.

converted back into a sinusoidal voltage of different frequency (phase) and amplitude at the output IDTS. These differences are related to changes in the velocity of the SAW and can be correlated to changes in the mass loading, viscosity and temperature of the substrate. The effect of the temperature on the substrate could be compensated by using a dual delay line configuration. In this case only one of the SAW devices will be functionalized with biological molecules. The reference SAW delay line is not functionalized with biological molecules and it is used only for temperature compensation. Both SAW delay lines will operate at the same temperature. Figure 11 illustrates a SAW biosensor in a dual delay line configuration (Arruda et al., 2009). Measurements comparing the experimental delay line with the reference delay line are used to compensate the effect of temperature on the biosensor.
A layer of bio-molecules consisting of protein cross-linkers and antibodies is coated on the sensitive surface of the sensing device in the path of the traveling waves, between the two sets of IDTs, as illustrated in Fig. 11. If specific target proteins (antigens) are present, they bind to the antibodies, creating the mass loading on the surface of the substrate. As a result, a time delays in the propagation of the SAWs will occur.

2.3 Guided Surface Acoustic Wave resonators

One subcategory of SAW, the Love propagation mode, is especially promising for chemical and biosensing applications due to its high sensitivity to mass loading and the ability to make measurements in liquid environments with minimal propagation losses. Shear-horizontal waves can be guided by placing a thin guiding layer on a SH-SAW sensor. The bare SH-SAW resonator has a lower sensitivity because the acoustic wave goes deeper into the substrate. The sensitivity to surface perturbations could be increased when a waveguiding layer fabricated on top of the IDTs is used. The waveguiding layer also provides protection from chemicals in the liquid. Waves that propagate through the guiding layer are known as Love waves. Dielectrics such as silicon dioxide, silicon nitride and most polymers can be used as waveguide materials. Polymers have a lower shear wave velocity and therefore they are recommended for Love mode SAW sensors. Acoustic efficiency is improved when the waveguide layer is thin and does not load or attenuate the traveling acoustic wave. A typical Love mode SAW sensor is illustrated in Fig. 12.

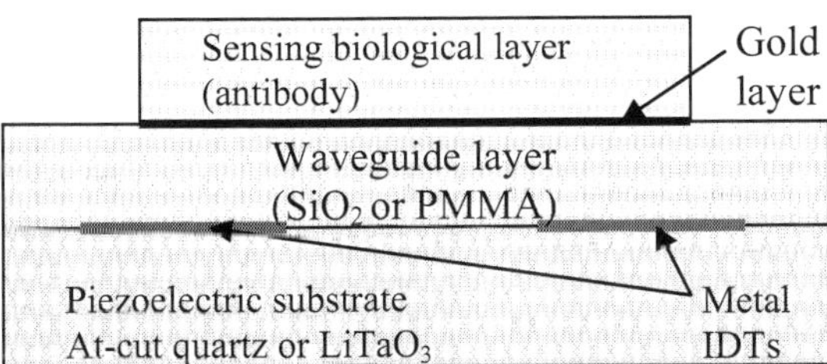

Fig. 12. Illustration of a Love mode surface acoustic wave sensor formed by a piezoelectric substrate, the waveguide layer, IDTs, and the sensing layer. The waveguide layer could be a SiO_2 film or a polymer layer and its function is to minimize propagation losses.

An example of a MEMS Love wave SAW device was implemented on $LiTaO_3$ substrates with Cr/Au IDTs and poly(methyl methacrylate) (PMMA) waveguide (Bender et al., 2000;

Branch et al., 2004). The waveguide is used to trap the SAW along the surface of the piezoelectric substrate to minimize energy losses and to protect the IDTs from corroding in a liquid-sensing environment. The surface of the waveguide is in contact with the sensing layer and it is often covered with a 50 nm thick gold layer, as illustrated in Fig.12. This gold layer provides better adherence of the sensing layer (antibodies) and prevents the nonspecific binding of proteins to the active area. Fabrication of the Cr/Au IDTs used standard lithographic techniques and the PMMA layer was spin-coated on the device. The thickness of the PMMA layer was optimized to avoid acoustic attenuation. These devices were tested with goat immunoglobulin G (IgG) and indicated minimum mass sensitivity of 17 pg/mm^2.

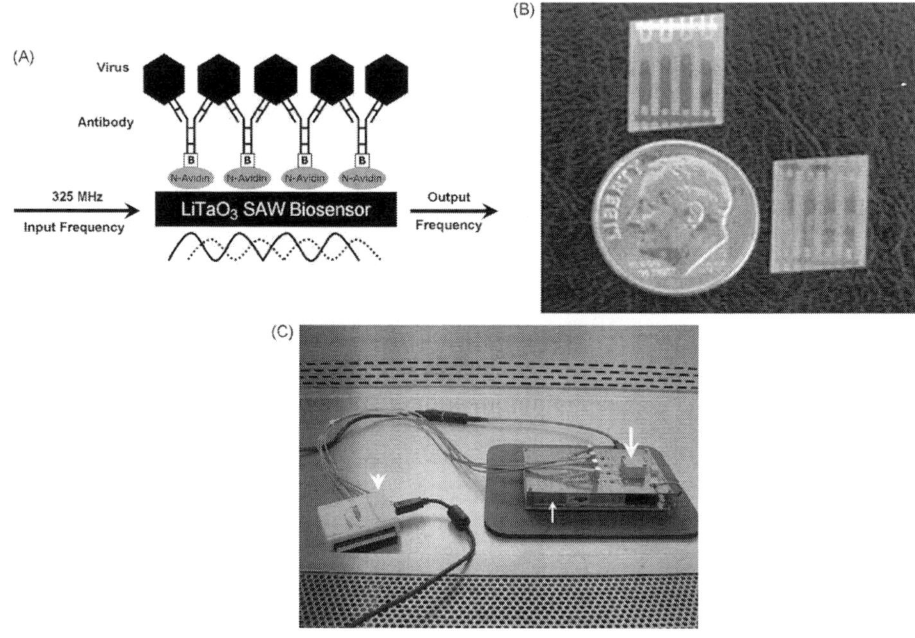

Fig. 13. (A) Concept of the antibody-based virus surface acoustic wave (SAW) biosensor. The lithium tantalate (LiTaO3) sensor surface was coated with NeutrAvidin Biotin Binding Protein (N-Avidin) and coupled to biotinylated (B) monoclonal anti-JVB antibody for Coxsackie virus. Anti-SNV-G1 glycoprotein scFv antibody for SNV detection was coupled directly. Molecular interaction between virus and antibody elicits an acoustic wave leading to a change in the input frequency of 325 MHz. (B) The sensor wafer is shown in scale compared to a dime coin; four aluminum delay lines are visible; one serves as the reference and three as the test delay lines. (C) The SAW detection board (thin arrow) with the fluidic housing (thick arrow) and the output interface device (arrowhead) to a laptop computer is shown. Reprinted with permission from Ref. (Bisoffi et al., 2008).
(© [2008] Biosensors and Bioelectronics). Used with permission.

Love wave SAW biosensors have also been used to test specific binding of different concentrations of Immunoglobulin G in the range of 0.7- 667 nM using a sensing surface

popular in the communications industry. To improve silicon compatibility, efforts have been made to implement the SAW resonator using standard CMOS process with minimal post-processing. Results indicate that while this device shows promise, significant improvement is required before the CMOS SAW resonator can meet the stringent communication requirements. In this aspect, FBARs have shown better performance in terms of quality factor (6000) and low insertion losses. However, complete CMOS-compatibility has not yet been achieved and the device still requires bond wires for connections to the circuitry.

5. References

Andle, J. C., & Vetelino, J. F. (1995).*Acoustic wave biosensors.* Proceedings of IEEE Ultrasonics Symposium, ISBN 1051-0117, 7-10 Nov 1995

Arruda, D. L., Wilson, W. C., Nguyen, C., Yao, Q. W., Caiazzo, R. J., Talpasanu, I., et al. (2009). Microelectrical sensors as emerging platforms for protein biomarker detection in point-of-care diagnostics. *Expert Review of Molecular Diagnostics,* Vol. *9,* No. 7, (2009), pp. 749-755.

Ballantine, D. S., White, R. M., Frye, G. C., Martin, S. J., Ricco, A. J., Zellers, E. T., et al. (1996). *Acoustic wave sensors: theory, design, and physico-chemical applications*: Academic Press San Diego.

Bender, F., Cernosek, R. W., & Josse, F. (2000). Love-wave biosensors using cross-linked polymer waveguides on LiTaO 3 substrates. *Electronics Letters,* Vol. *36,* No. 19, (2000), pp. 1672-1673.

Bisoffi, M., Hjelle, B., Brown, D. C., Branch, D. W., Edwards, T. L., Brozik, S. M., et al. (2008). Detection of viral bioagents using a shear horizontal surface acoustic wave biosensor. *Biosensors and Bioelectronics,* Vol. *23,* No. 9, (2008), pp. 1397-1403,.

Bowers, W. D., Chuan, R. L., & Duong, T. M. (1991). A 200 MHz surface acoustic wave resonator mass microbalance. *Review of scientific instruments,* Vol. *62,* No. (1991), pp. 1624.

Branch, D. W., & Brozik, S. M. (2004). Low-level detection of a Bacillus anthracis simulant using Love-wave biosensors on 36°YX LiTaO3. *Biosensors and Bioelectronics,* Vol. *19,* No. 8, (2004), pp. 849-859.

Buff, W., Rusko, M., Goroll, E., Ehrenpfordt, J., & Vandahl, T. (1997).*Universal pressure and temperature SAW sensor for wireless applications.* Proceedings of IEEE Ultrasonics Symposium.

Campanella, H., Cabruja, E., Montserrat, J., Uranga, A., Barniol, N., & Esteve, J. (2008). Thin-Film Bulk Acoustic Wave Resonator Floating Above CMOS Substrate. *Electron Device Letters, IEEE,* Vol. *29,* No. 1, (2008), pp. 28-30.

Campbell, C. (1998). *Surface acoustic wave devices for mobile and wireless communications.* San Diego. Academic Press.

Cavic, B. A., Hayward, G. L., & Thompson, M. (1999). Acoustic waves and the study of biochemical macromolecules and cells at the sensor-liquid interface. *Analyst,* Vol. *124,* No. 10, (1999), pp. 1405-1420, ISSN 0003-2654.

Cheeke, J. D. N., Tashtoush, N., & Eddy, N. (1996).*Surface acoustic wave humidity sensor based on the changes in the viscoelastic properties of a polymer film.* Proceedings of IEEE Ultrasonics Symposium, ISBN 1051-0117, 3-6 Nov 1996.

Chen, Y., Reyes, P. I., Duan, Z., Saraf, G., Wittstruck, R., Lu, Y., et al. (2009). Multifunctional ZnO-Based Thin-Film Bulk Acoustic Resonator for Biosensors. *Journal of Electronic Materials,* Vol. *38,* No. 8, (2009), pp. 1605-1611.

Cullen, D. E., & Montress, G. K. (1980).*Progress in the Development of SAW Resonator Pressure Transducers.* Proceedings of IEEE Ultrasonics Symposium, 1980.

Cullen, D. E., & Reeder, T. M. (1975).*Measurement of SAW Velocity Versus Strain for YX and ST Quartz.* Proceedings of IEEE Ultrasonics Symposium, 1975.

Gabl, R., Green, E., Schreiter, M., Feucht, H. D., Zeininger, H., Primig, R., et al. (2003).*Novel integrated FBAR sensors: a universal technology platform for bio- and gas-detection.* Proceedings of IEEE Sensors, 22-24 Oct. 2003.

Gizeli, E., Bender, F., Rasmusson, A., Saha, K., Josse, F., & Cernosek, R. (2003). Sensitivity of the acoustic waveguide biosensor to protein binding as a function of the waveguide properties. *Biosensors and Bioelectronics,* Vol. *18,* No. 11, (2003), pp. 1399-1406.

Gorla, C. R., Emanetoglu, N. W., Liang, S.,. Mayo, W. E., Lua Y., Wraback, M., & Shen, H. (1999) Structural, optical, and surface acoustic wave properties of epitaxial ZnO films grown on [011-2] Sapphire by metalorganic chemical vapor deposition", *J. Appl. Phys.* 85(51), pp. 2595, 1999.

Hara, M., Kuypers, J., Abe, T., & Esashi, M. (2003).*MEMS based thin film 2 GHz resonator for CMOS integration.* Proceedings of Microwave Symposium Digest, 2003 IEEE MTT-S International, ISBN 0149-645X.

Harrington, B. P., Shahmohammadi, M., & Abdolvand, R.*Toward ultimate performance in GHZ MEMS resonators: Low impedance and high Q.* Proceedings of IEEE 23rd International Conference on Micro Electro Mechanical Systems (MEMS), ISBN 1084-6999, 24-28 Jan. 2010

Hikita, M., Takubo, C., & Asai, K. (2000). New high performance SAW convolvers used in high bit rate and wideband spread spectrum CDMA communications system. *IEEE Transactions on Ultrasonics, Ferroelectrics and Frequency Control,* Vol. *47,* No. 1, (2000), pp. 233-241.

Hunter, I. C., Billonet, L., Jarry, B., & Guillon, P. (2002). Microwave filters-applications and technology. *IEEE Transactions on Microwave Theory and Techniques,* Vol. *50,* No. 3, (2002), pp. 794-805.

Janshoff, A., Galla, H. J., & Steinem, C. (2000). Piezoelectric Mass-Sensing Devices as Biosensors-An Alternative to Optical Biosensors? *Angew Chem Int Ed Engl,* Vol. *39,* No. 22, (2000), pp. 4004-4032, ISSN 1521-3773.

Länge, K., Rapp, B. E., & Rapp, M. (2008). Surface acoustic wave biosensors: a review. *Analytical and Bioanalytical Chemistry,* Vol. *391,* No. 5, (2008), pp. 1509-1519.

Lavasani, H. M., Wanling, P., Harrington, B., Abdolvand, R., & Ayazi, F. A 76 dB Ohm 1.7 GHz 0.18 um CMOS Tunable TIA Using Broadband Current Pre-Amplifier for High Frequency Lateral MEMS Oscillators. *Solid-State Circuits, IEEE Journal of,* Vol. *46,* No. 1, pp. 224-235.

Levit, N., Pestov, D., & Tepper, G. (2002). High surface area polymer coatings for SAW-based chemical sensor applications. *Sensors and Actuators B: Chemical,* Vol. *82,* No. 2-3, (2002), pp. 241-249.

Macchiarella, G., & Stracca, G. B. (1982).*SAW Devices for Telecommunications: Examples and Applications.* Proceedings of IEEE Ultrasonics Symposium, 1982

Mason, W. P., Thurston, R. N., (1972) *Physical Acoustics*, Academic Press, Inc., New York, 1972.

Morgan, D. P. (1985). *Surface-wave devices for signal processing*. Amsterdam ; New York: Elsevier.

Nakamoto, T., Nakamura, K., & Moriizumi, T. (1996).*Study of oscillator-circuit behavior for QCM gas sensor*. Proceedings of IEEE Ultrasonics Symposium, ISSN 1051-0117, 3-6 Nov 1996

Nomura, T., Takebayashi, R., & Saitoh, A. (1998). Chemical sensor based on surface acoustic wave resonator using Langmuir-Blodgett film. *IEEE Transactions on Ultrasonics, Ferroelectrics and Frequency Control*, Vol. *45*, No. 5, (1998), pp. 1261-1265.

Nordin, A. N., & Zaghloul, M. E. (2007). Modeling and Fabrication of CMOS Surface Acoustic Wave Resonators. *IEEE Transactions on Microwave Theory and Techniques*, Vol. *55*, No. 5, (2007), pp. 992-1001.

Otis, B. P., & Rabaey, J. M. (2003). A 300-/spl mu/W 1.9-GHz CMOS oscillator utilizing micromachined resonators. *IEEE Journal of Solid-State Circuits*, Vol. *38*, No. 7, (2003), pp. 1271-1274.

Pohl, A., Ostermayer, G., Reindl, L., & Seifert, F. (1997).*Monitoring the tire pressure at cars using passive SAW sensors*. Proceedings of IEEE Ultrasonics Symposium, 1051-0117, 5-8 Oct 1997.

Pottigari, S. S., & Jae Wan, K. (2009).*Vacuum-gapped film bulk acoustic resonator for low-loss mass sensing in liquid*. Proceedings of International Solid-State Sensors, Actuators and Microsystems Conference, 21-25 June 2009

Rebeiz, G. M. (2003). *RF MEMS: theory, design, and technology*: John Wiley and Sons.

Reindl, L., Scholl, G., Ostertag, T., Ruppel, C. C. W., Bulst, W. E., & Seifert, F. (1996).*SAW devices as wireless passive sensors*. Proceedings of IEEE Ultrasonics Symposium, 1051-0117, 3-6 Nov 1996

Ruby, R. C., Bradley, P., Oshmyansky, Y., Chien, A., & Larson, J. D., III. (2001).*Thin film bulk wave acoustic resonators (FBAR) for wireless applications*. Proceedings of IEEE Ultrasonics Symposium,

Sadek, A. Z., Wlodarski, W., Shin, K., Kaner, R. B., & Kalantar-zadeh, K. (2006). A layered surface acoustic wave gas sensor based on a polyaniline/In2O3 nanofibre composite. *Nanotechnology*, Vol. *17*, No. 17, (2006), pp. 4488, ISSN 0957-4484

Skládal, P. (2003). Piezoelectric quartz crystal sensors applied for bioanalytical assays and characterization of affinity interactions. *Journal of the Brazilian Chemical Society*, Vol. *14*, No. (2003), pp. 491-502, ISSN 0103-5053

Smith, A. L. (2001). Mass and heat flow measurement sensor: Google Patents.

Springer, A., Huemer, M., Reindl, L., Ruppel, C. C. W., Pohl, A., Seifert, F., et al. (1998). A robust ultra-broad-band wireless communication system using SAW chirped delay lines. *IEEE Transactions on Microwave Theory and Techniques*, Vol. *46*, No. 12, (1998), pp. 2213-2219.

Staples, E. J. (1999).*Electronic nose simulation of olfactory response containing 500 orthogonal sensors in 10 seconds*. Proceedings of IEEE Ultrasonics Symposium, 1051-0117, 1999.

Vellekoop, M. J., Nieuwkoop, E., Haartsan, J. C., & Venema, A. (1987). *A Monolithic SAW Physical-Electronic System for Sensors*. Proceedings of IEEE Ultrasonics Symposium, 1987

Vellekoop, N. J., Jakoby, B., & Bastemeijer, J. (1999).*A Love-wave ice detector*. Proceedings of IEEE Ultrasonics Symposium, ISBN 1051-0117, 1999

Vetelino, K. A., Story, P. R., Mileham, R. D., & Galipeau, D. W. (1996). Improved dew point measurements based on a SAW sensor. *Sensors and Actuators B: Chemical,* Vol. *35,* No. 1-3, (1996), pp. 91-98,

Visser, J. H., Vellekoop, M. J., Venema, A., Drift, E. v. d., Rek, P. J. M., & Nederhof, A. J. (1989).*Surface Acoustic Wave filter in ZnO-SiO2-Si layered structures*. Proceedings of IEEE Ultrasonics Symposium.

Weber, J., Albers, W. M., Tuppurainen, J., Link, M., Gabl, R., Wersing, W., et al. (2006). Shear mode FBARs as highly sensitive liquid biosensors. *Sensors and Actuators A: Physical,* Vol. *128,* No. 1, (2006), pp. 84-88.

Weigel, R., Morgan, D. P., Owens, J. M., Ballato, A., Lakin, K. M., Hashimoto, K., et al. (2002). Microwave acoustic materials, devices, and applications. *IEEE Transactions on Microwave Theory and Techniques,* Vol. *50,* No. 3, (2002), pp. 738-749.

Weld, C. E., Sternhagen, J. D., Mileham, R. D., Mitzner, K. D., & Galipeau, D. W. (1999).*Temperature measurement using surface skimming bulk waves*. Proceedings of IEEE Ultrasonics Symposium, 1051-0117, 1999.

White, R. M., & Voltmer, F. W. (1965). Direct piezoelectric coupling to surface elastic waves. *Applied Physics Letters,* Vol. *7,* No. (1965), pp. 314-316,

Wingqvist, G., Bjurstrom, J., Liljeholm, L., Katardjiev, I., & Spetz, A. L. (2005).*Shear mode AlN thin film electroacoustic resonator for biosensor applications*. Proceedings of IEEE Sensors, Oct. 30 2005-Nov. 3 2005

Wohltjen, H., & Dessy, R. (1979). Surface acoustic wave probe for chemical analysis. I. Introduction and instrument description. *Analytical Chemistry,* Vol. *51,* No. 9, (1979), pp. 1458-1464.

Zhang, X., Xu, W., Abbaspour-Tamijani, A., & Chae, J. (2009).*Thermal Analysis and Characterization of a High Q Film Bulk Acoustic Resonator (FBAR) as Biosensors in Liquids*. Proceedings of IEEE 22nd International Conference on MEMS, pp. 939-942.

Zhou, W. (2009). *Integration of MEMS Resonators within CMOS Technology*. Cornell.

Zuo, C., Van der Spiegel, J., & Piazza, G. (2010). 1.05-GHz CMOS Oscillator Based on Lateral-Field-Excited Piezoelectric AlN Contour-Mode MEMS Resonators. *Chengjie Zuo,* Vol., No. (2010), pp. 15.

Part 2

MEMS Characterization and Micromachining

5

MEMS Characterization Based on Optical Measuring Methods

Tong Guo[1], Long Ma[2] and Yan Bian[3]
[1]Tianjin University
[2]Civil Aviation University of China
[3]Tianjin University of Technology and Education
P.R. China

1. Introduction

Micro Electro Mechanical Systems (MEMS) is developed based on the semi-conductor technology, however, relative material, design, fabrication, simulation, packaging and test are more complex than those in semi-conductor technology. In the primary stage, MEMS technology focused on the design and development, now on the commercialization and improving reliability and decreasing cost and price. So test is increasingly important to MEMS technology and testing cost is about 1/3 of the whole cost of MEMS. In order to improve the production and decrease the cost, producers and researchers pay more attention to MEMS test to solve all the testing problems from design to packaging process.

There are a number of methods to carry out these measurements, such as scanning electron microscopy (SEM), atomic force microscopy (AFM), stylus profiler, and optical profiler, etc. Every method has its advantages and disadvantages.

- SEM is one of the most common measurement tools. However, nearly all nonconductive specimens examined using SEM need to be coated with a thin film of conducting material. This may result in bending or distortion of the device, especially where free structures such as cantilever beams. SEM tests are also time consuming and not suitable for a production environment.
- AFM has been suggested as a MEMS measurement tool. As with SEM, analysis may be slow (about 20 min/device), and the limited measurement range of an AFM (100 μm×100 μm×5 μm, Veeco multimode AFM) means that it is unable to investigate large samples or out-of-plane devices such as the cantilevers. It is also difficult to examine packaged devices using an AFM.
- Mechanical stylus surface profilers are commonly used for dimensional measurements in MEMS. While these can measure step heights with a high accuracy, they are not suitable for the analysis of freestanding structures, where the stylus may break the device under test. Deep, high aspect ratio devices also pose problems, as the stylus probe may be too large to accurately reproduce the surface profile.

If MEMS devices need to fit the large-scale production, it is essential that these measurements must be cheaply and easily made at the wafer level, without the need for large space, expensive packaging or destructive test methods. Optical techniques can offer solutions to many of these problems. This chapter uses computer micro-vision and microscopic interferometry to carry out MEMS measurements, including dimensional (static) and moving (dynamic) properties analysis. The moving properties can be classified into in-plane (lateral) movements and out-of-plane (vertical) movements. The techniques involved are simple, fast, non-destructive, requiring no sample preparation and may be carried out at wafer level - all important requirements for high volume production.

2. System set-up

2.1 Microscopic interferometer

The interferometer is the device which can generate the interferogram patterns. The typical microscopic interferometers used in MEMS measurement include Michelson-type, Mirau-type and Linnik-type. The scheme of optical structures in these interferometers is shown in figure 1.

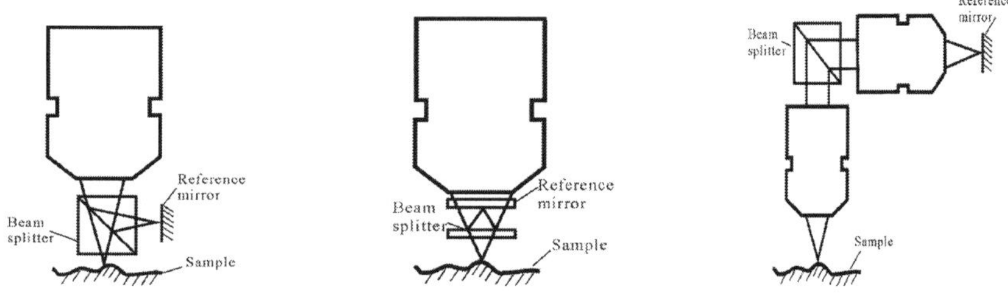

(a) Michelson interferometer (b) Mirau interferometer (c) Linnik interferometer

Fig. 1. Schematic layout of the three types of interferometers

Inside the interferometer the light is divided into two beams, one beam is guided on the sample surface as the test beam, and the other one goes to the reference mirror as the reference beam. When the OPD is within the coherence length, the reflected test beam and reference beam will interfere with each other and produce interference fringe patterns. Figure 1(a) presents the Michelson interferometer, where a beam splitter is placed in front of the objective. Hence, the working distance of the objective must be relatively large which makes the magnification of Michelson interferometer the lowest of the three types of interferometers shown in figure 1. Figure 1 (b) is the Mirau interferometer, which is widely equipped in lots of commercial optical profilers. Inside this interferometer, the light is split by a very thin and flat optical mirror. As result, the mechanical structure of this interferometer is more compact than that of Michelson type. Besides, the light path between the test beam and the reference beam is very similar, which can also minimize optical disturbance. Figure 1 (c) illustrates the layout of the Linnik interferometer, from where one can see the beam splitter is located behind the objective. Therefore, Linnik interferometer can use the objective with shorter working distance and higher magnification. However, the unconformity between the two objectives may cause measurement errors, and that is why in the application of Linnik interferometer the two objectives must be matched very well.

2.2 Measurement system

Two sets of measurement systems were developed in our laboratory. The first one is for static characterization, the other one is for dynamic characterization.

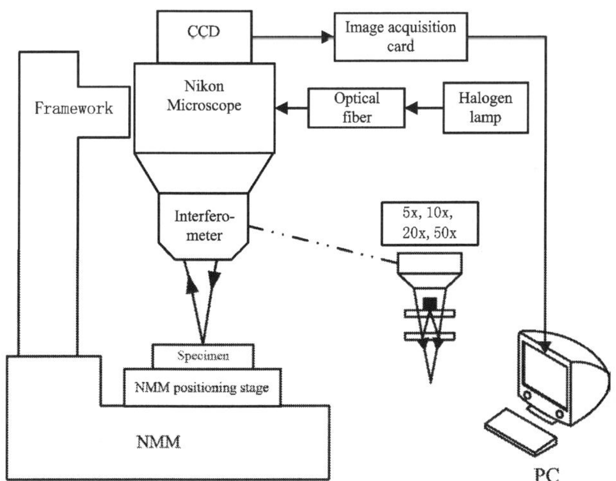

Fig. 2. Scheme of the static characterization system

As a foundation of our system, the nano-measuring machine (NMM) was used, which was developed in Ilmenau University of Technology of Germany for surface measurement within a measuring volume of 25mm×25mm×5mm. Three commercial interferometers inside the NMM read the stage position in real time so that the positioning control loop can assure a movement resolution of 0.1 nm. The sensor arrangement provides an Abbe error-free measurement on all three NMM axes. Based on the NMM, various measuring systems could be integrated.

The layout of the first experimental system used in this context is illustrated in figure 2. To increase the optical system robustness, we chose a commercial interferometer system with halogen lamp fibre illumination from NIKON. A CCD camera recorded the interferogram at equidistant positions and transfered them into a computer for data processing. In the laboratory, the system was assembled inside a cover where a vibration isolation table was provided. Additionally, in order to further decrease the influence of vibrations, we placed the entire equipment on a special independent foundation.

The second system is called micro motion analyzer (MMA). The structure of MMA is shown in figure 3. MMA was equipped with several long working distance optical objectives from Zeiss company, including 5×, 10×, 20× and 50×. There were two kinds of light sources: one was a high performance LED (Nichia NSPG 500S LED, central wavelength was 525nm, optical bandwidth was 40nm) to be used in the in-plane motion measurement; the other was a LD (Hitachi HL6501MG, the power was 50mW, the wavelength was 658nm) to be used in the out-of-plane motion measurement. A nano-positioner (PI P-721.CL, capacitor sensor feedback control in close loop) was used to shift the phase of the interferogram and adjust the optical path with sub-nanometer resolution. The bright field and interference field were switched by putting a stop plate in the reference optical path. Images were captured by a digital CCD camera (Microvision, 1280×1024 pixels, 9μm pixel distance, 100% filling factor, 10 bits grayscale resolution), then transferred to the server to be post-processed. The user can transfer and receive instructions and data through TCP/IP protocol on a client PC. The system was also equipped with a signal generating unit and a high voltage amplifying unit to output high voltage signal and stimulate the tested MEMS device. The system was placed on a vibration isolation table (TMC Lab Table) to decrease the effect of outside vibration. The system was also equipped with a probe station (Karl Suss, Germany) to test the unpackaged MEMS devices. Here digital phase lock loop (PLL) made the stimulating signal, illuminating signal and the image capturing synchronized. The frequency range of Arbitrary Waveform Generator (AWG) was from 1 Hz to 10 MHz, and the sampling frequency was 40 MHz.

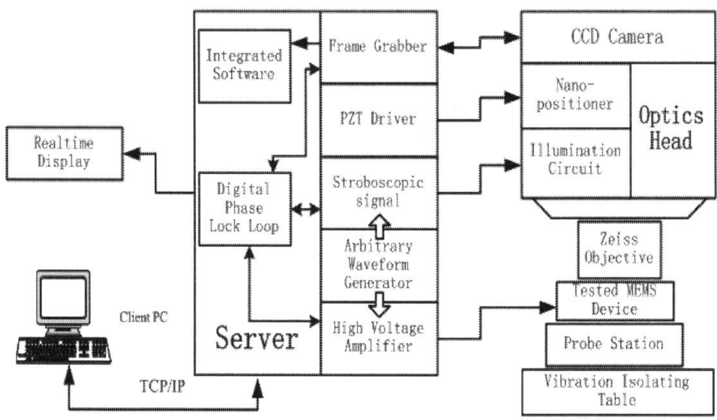

Fig. 3. Structure diagram of the MMA

3. Measurement principles

3.1 Static characterization

Static characterization is a very important aspect in MEMS technology, especially for the design and manufacture of micro- or nano-sensors and actuators. In general case, the goal of static characterization is to derive the geometrical parameters, such as surface topography, profile, waviness, roughness and depth-width ratio, etc.

3.1.1 Phase-shifting interferometry (PSI)

If we denote the coordinates in the plane of the CCD array as (x, y), we can write the intensity of the interference pattern I(x, y) as:

$$I(x,y) = A(x,y) + B(x,y)\cos[\phi(x,y) + \varphi] \quad (1)$$

Where $A(x, y)$ is the function representing the mean background intensity of the interferogram; $B(x, y)$ is the function determining the modulation of the interferogram; $\phi(x, y)$ is the function dependent on the unwrapped phase $\theta(x, y)$ which represents the height change on the sample; phase φ is a constant, which can be adjusted by changing the position in the axial direction between the objective and the sample.

Using a series of I measured at different phase φ value, Hariharan phase-shifting algorithm is used to extract height information on devices' surface in I. The intensities are I_1, I_2, I_3, I_4 and I_5 respectively. The formula of calculating wrapped phase ϕ is as follows.

$$\phi(x,y) = \arctan\left[\frac{2(I_2 - I_4)}{2I_3 - I_1 - I_5}\right] \tag{2}$$

This algorithm need change п/2 between two successive image capturing. This algorithm is repressive to the calibration error of the phase-shifter and the nonlinear errors of the detector, so it improves the measurement accuracy.

Measurements using single-wavelength interferometry have an inherent 2п phase ambiguity, which means phase maps must be unwrapped by removing artificial 2п jumps before they can be properly interpreted in terms of surface height. This process is called phase unwrapping. Now there are many kinds of phase unwrapping algorithms. In the experiments, considering about the testing speed and the real quality of the interferogram, quality-guided path following algorithm is used to retrieve the actual phase map. The surface height can be calculated by the following formula.

$$h = \frac{\theta}{4\pi}\lambda \tag{3}$$

Here, λ is the wavelength of the light source.

Figure 4 shows the wrapped phase map and the unwrapped map in the process of dealing with the captured images. The tested sample is a membrane.

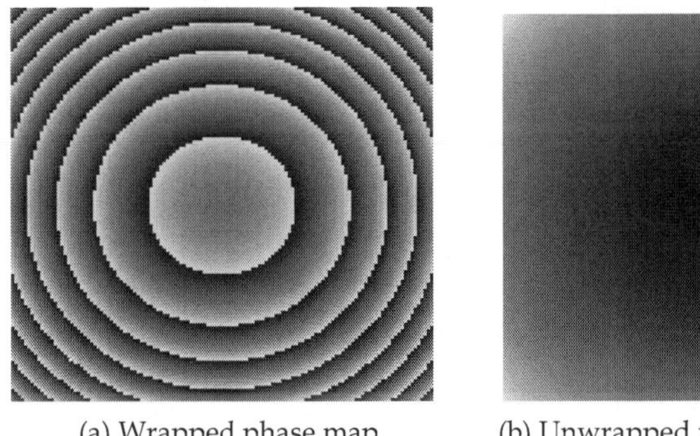

(a) Wrapped phase map (b) Unwrapped phase map

Fig. 4. Pictures of processed results

3.1.2 White light scanning interferometry (WLI)

PSI is widely used to test smooth surfaces and is very accurate, resulting in vertical measurements with sub-nanometer resolution. However, PSI cannot obtain a correct profile for objects that have large step changes because it becomes ineffective as height discontinuities of adjacent pixels exceed one quarter of the used wavelength (λ/4), which is also called "phase ambiguity".

White light scanning interferometry (WLI) provides a good solution to overcome "phase ambiguity". Here we track point P and Q in figure 5 to illustrate the scan process of WLI. When Q stays outside the coherence length of the light source, the intensity detected is nearly the same with the background irradiance, corresponding to t_1 to t_2. Afterwards, from time t_2 when Q starts to move into the coherence length zone, the signals extracted of point Q will be kept modulating by the interferogram until it travels out at t_5. During the scan, the maximum visibility will occur at t_4 where the testing beam matches the reference beam. The above process is roughly similarity for point P. Once retrieved the correlogram, the peak position of the intensities can then be used as an indicator of the surface relative height.

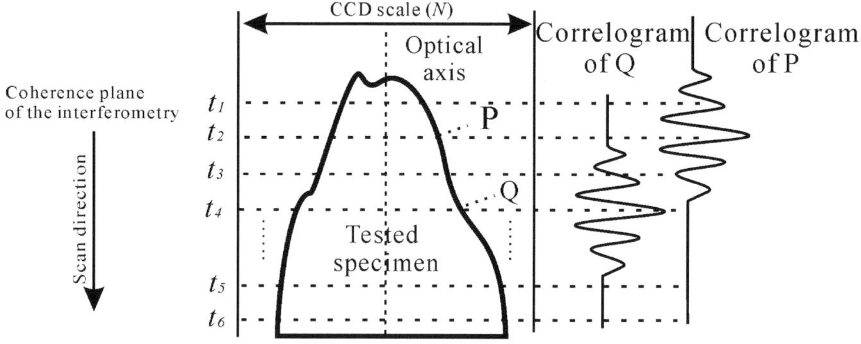

Fig. 5. Schematic diagram of white light vertical scanning interferometry

The white light correlogram recorded by the CCD camera in the white light interferometry can be expressed as:

$$I = I_0\left[1 + \gamma g(z)\cos(\phi + \alpha)\right] \tag{4}$$

Where I is the measured intensity, I_0 is the background irradiance value, γ is a constant, $g(z)$ is the fringe visibility in the form of Gaussian function, ϕ is the phase value depended on the optical path difference and the α is the additional phase term due to different reflections. Generally, the maximum location of $g(z)$ is usually extracted for the height evaluation. This process can be done either in spatial space or in frequency domain.

So far, the white light signal demodulation techniques can be basically categorized into two main groups: spatial domain algorithms and frequency domain algorithms. The first group consists of polynomial interpolation, centroid method and Hilbert transform, whereas the latter has Fourier transform, wavelet transform and spatial frequency domain analysis, etc. In general case, algorithms from the second group can perform the surface characterization with a higher resolution. Nevertheless, because of the transform procedures from the spatial domain to frequency domain (various types of convolutions), the second group algorithms are usually time consuming; on the contrary, the first group is efficient but with lower resolution. The other methods are generally based upon the above algorithms and will not be mentioned here.

3.1.3 White light phase-shifting interferometry (WLPSI)

3.1.3.1 Phase extraction

Carré proposed a new phase shifting interferometry in 1966. Unlike the conventional PSI, Carré's method does not require several fixed phase steps (for example, λ/8), but only equal ones, which makes it much easier for most PZTs to fulfill. Suppose the phase step is set to be 2δ, Carré's method can be expressed as:

$$I_1(x,y) = I_0\{1+\gamma\cos[\phi(x,y)-3\delta]\} \tag{5}$$

$$I_2(x,y) = I_0\{1+\gamma\cos[\phi(x,y)-\delta]\} \tag{6}$$

$$I_3(x,y) = I_0\{1+\gamma\cos[\phi(x,y)+\delta]\} \tag{7}$$

$$I_4(x,y) = I_0\{1+\gamma\cos[\phi(x,y)+3\delta]\} \tag{8}$$

Where I_1 to I_4 are the recorded intensities, I_0 is the background irradiance, γ is a constant indicated the visibility and ϕ is the phase term needed to be extracted.
Then the phase term ϕ can be computed as:

$$\phi = tg^{-1}\frac{\sqrt{|[(I_1-I_4)+(I_2-I_3)][3(I_2-I_3)-(I_1-I_4)]|}}{|[(I_2+I_3)-(I_1+I_4)]|} \tag{9}$$

Here ϕ falls within (-π, +π] and the surface information can be written as:

$$h = \frac{\bar{\phi}}{4\pi}\lambda \tag{10}$$

Where h is the relative height of the surface, $\bar{\phi}$ is the unwrapped phase term, λ is the wavelength used. The relative heights on every point together can then give a surface 3D map of the tested sample.
However, in the white light correlogram, the existence of $g(z)$ will make g be a function of the spatial position along the scanning direction, which means error occurs when we employ the Carré method to carry out the phase computation from a white light interferogram. In the next, we will use computer simulation to study how much this kind of error affects the measuring accuracy.
The generated discrete white light signal $I(n)$ is as follows:

$$I(n) = 200 + 200\exp\{-\frac{[(n-20)\times\Delta]^2}{\sigma^2}\}\cos[\frac{4\pi}{\lambda}(n-20)\times\Delta+\frac{\pi}{7}] \tag{11}$$

Where Δ is the scanning step 60 nm, σ is set to be 500 nm($\sigma = l_c/2\pi$, l_c is the coherence length of the light source), the additional phase term due to different reflections is set to be π/7. Figure 6 is the phase computation simulation, which indicates that: phase extraction error owing to the visibility variety is minimum at the position of the coherence peak; on the other hand, the phase error has a trend of becoming bigger while the phase position travels away from the centre. The minimum phase error is less than 0.02 rad and the equivalent measurement error is smaller than $\lambda/500$. Assumed the wavelength is 600 nm, this error could be about 1nm and is negligible in comparison with the environment disturbance. This can be explained as the weak modulation effect of the Gaussian function around the zero order fringe of the white light signal, where the value of the Gaussian function can be treated as 1(for normalized Gaussian function).

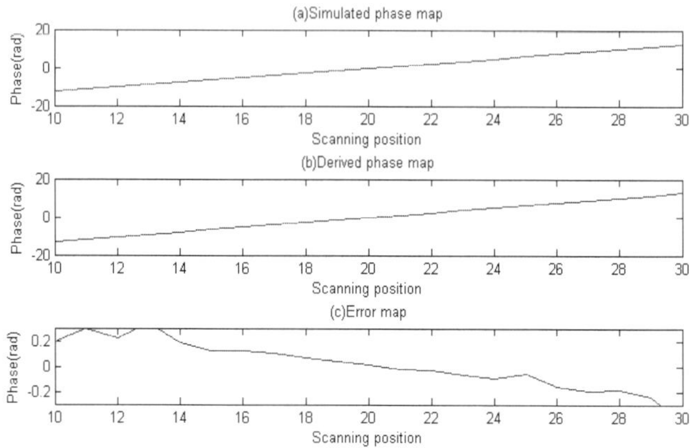

Fig. 6. Phase extraction simulation

3.1.3.2 Height evaluation

Based on the phase computation simulation, the relative height of the tested profile h can then be determined as:

$$h = (stepnumber - peakstep) \times \Delta - t \times [\frac{(\phi + 2k\pi)}{4\pi}\lambda] \tag{12}$$

Where *stepnumber* is the scanning numbers, *peakstep* corresponds to the phase term which comes out of Carré method, Δ is the scanning step and t is the NA parameter. Concerning the NA parameter, both Katherine Creath and C.J.R.Sheppard have given lots of detailed research on the relationship between the objective NA and the fringe width. Here we take this relationship into account and use the results from Ingelstam's equation to calculate this parameter.

$$t = 1 + \frac{(NA_{eff})^2}{4} \tag{13}$$

Where NA_{eff} is the effective numerical aperture.

3.2 Dynamic characterization

The MMA is a highly integrated video microscope, using stroboscopic techniques to capture images of small, fast moving targets. The MMA uses both bright field and interference field based illumination modes combined with sophisticated machine vision algorithms to quantify micro motions. The MMA server and optics head combine the video microscopy with interferometry. However, successful measurements demand several characteristics in the target. The MMA needs two things: a target that can be moved in a periodic manner, and a target whose video image has contrast or structure when illuminated.

3.2.1 In-plane motion measurement using video microscopy

The MMA calculates the motion of a selected region in a sequence of images using machine vision algorithms. The MMA algorithms are hybrids constructed from a broad class of algorithms (gradient based) that exploit changes in brightness (grayscale values) between images and are capable of measuring motions smaller than the individual pixel size.
Fast moving targets appear blurry. One way of looking at fast moving targets is to slow down their apparent motion using a strobed light source. The scheme shown in figure 7 outlines how carefully timed pulses of light can capture snapshots (samples) of an object motion. These snapshots or images can then be used to reconstruct the displacement trajectory of the target. This is the case for displacements in the focal plane of the microscope or for displacements along the optical axis. It turns out that sometimes different types of images are better for measuring different types of displacements. The MMA optics module provides illumination for two types of images, both of which are acquired in a similar manner.

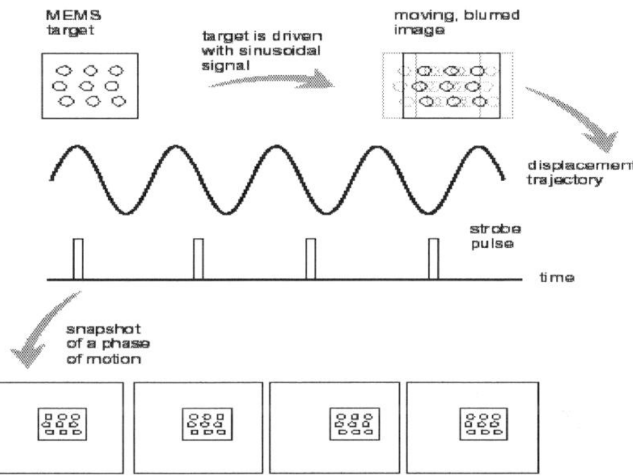

Fig. 7. Freezing rapid movement by stroboscopic illumination

3.2.1.1 Bright field image

By shining a monochromatic incoherent beam of light through the objective lens and uniformily illuminating the target, a bright field image is produced. This is the familiar view through the microscope eyepiece. A bright field image of a MEMS device is shown in figure 8(a).
Optical flow is used to describe the measured motion of brightness patterns between images. The algorithms used by the MMA are optical flow algorithms and are based on two important assumptions or constraints: 1) The brightness of a target region is constant over time, 2) The target region moves as a rigid body.
Motion of the target modulates brightness in the image. The image of a MEMS target moves across an array of pixels. The position of the image is shown at two consecutive timesteps and the registers at the bottom of each image represent the nominal brightness of the corresponding pixel in row 6 (in figure 9). Even sub-pixel motion has measureably changed the brightness in the pixels.

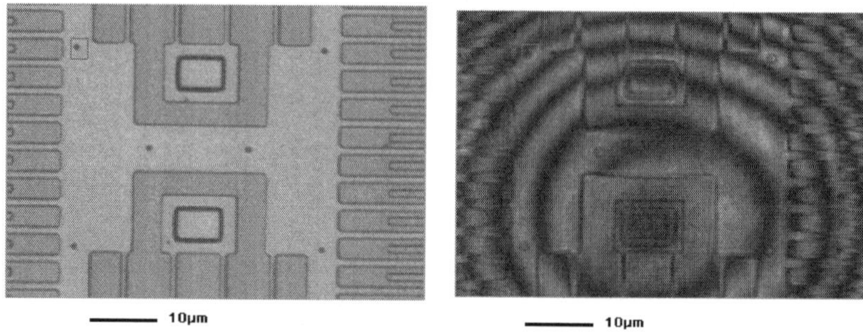

(a) Bright field image (b) Interferometric field image

Fig. 8. Images of the tested micro-resonator

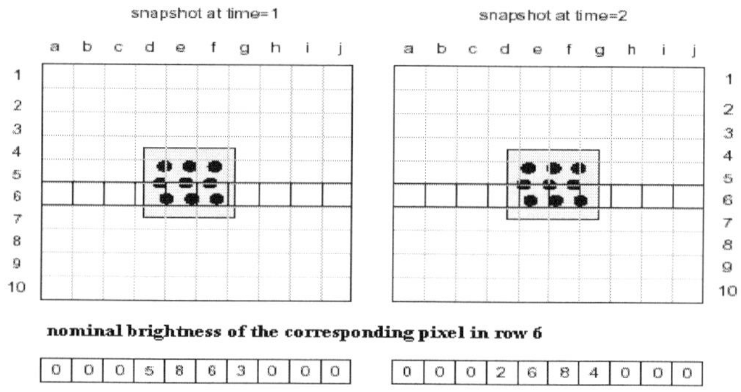

Fig. 9. In-plane motion measurement using optical flow algorithm

The optical flow algorithms used by the MMA work for in-plane or out-of-plane measurements. For out-of-plane measurements, image sequences can be collected at different planes of focus (optical sectioning). In this way, brightness gradients can be sampled along the optical axis and then used in a manner identical to in-plane sequences.

3.2.1.2 Interference field image

Figure 8(b) is the interference field image, which is formed by the reflected light from the tested sample and the reference mirror in the optics module. The interferogram in the picture is sensitive to the optical path change of the sample, which can be used to calculate the height of the sample and the out-of-plane motion between images.

3.2.2 Sub-pixel displacements measurement

The final result of the computation is an extremely powerful tool capable of measuring motions of magnified targets well below the resolution of human vision. Although, the resolution of a single static image is limited in a theoretical sense by the wavelength of the light used to generate the image, the resolution of a motion measurement is limited by the sensitivity of the CCD camera as shown in figure 9. Sub-pixel motion of an object or region (larger than a pixel) can be measured by using the modulation of the brightness of relevant pixels. It is also clear from figure 9 that changes in the measured brightness can be produced by factors other than motion of the target such as noise in the CCD camera's electronics or fluctuations in the illumination intensity.

3.2.3 Qualifying displacements

In order to quantify in-plane or out-of-plane displacements, the algorithms need a length scale. It means that different measurements need corresponding length scale.

For bright field images (in-plane measurements and optical sectioning), an in-plane length standard such as a grating (a target with a ruling of known spacing) is used to calibrate the length per pixel for a given magnification. The algorithms then convert the calculated displacements from pixels to micrometers.

Interference images have a more convenient length scale available. In this case, measurements are quantified using the known wavelength of the illumination source. This length standard is independent of magnification.

3.2.4 Noise floor analysis

After the data acquisition process is finished, the stimulating system is turned off and the measurement is done again. These two kinds of data are analyzed using the same processing method. Then repeating it five times, the noise floor is described by the RMS value.

4. Experimental results

4.1 Step structure measurement

4.1.1 10 μm standard step height measurement

A 10 μm standard step height fabricated by VLSI (9.976 μm+0.028 μm) was measured, as shown in figure 10, the mean height is 9.984 μm, while the standard deviation is 0.010 μm. The result comparison was presented in figure 11. We can clearly see the step height derived from WLPSI stays in the middle of all the results. Since the algorithms from the second group can achieve higher measurement resolution, the WLPSI can improve the measurement resolution compared with that of the algorithms from the first group.

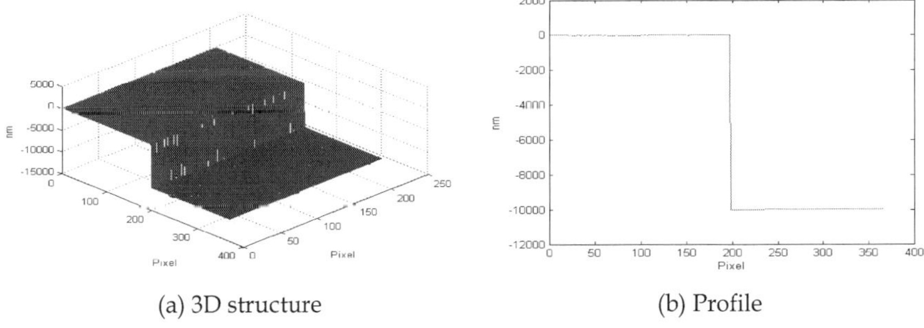

(a) 3D structure (b) Profile

Fig. 10. Measurement of the 10 μm standard step height

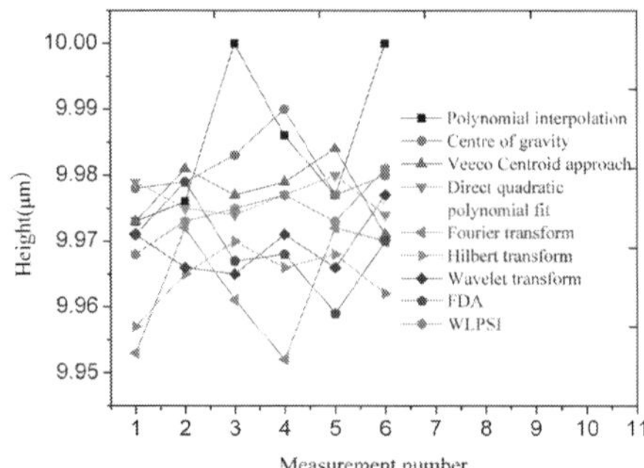

Fig. 11. 10 µm standard step height measurement comparison

4.1.2 44 nm standard step height measurement

Many coherence peak detecting algorithms in white light interferometry perform the height evaluations through either recovering the envelop function or locating the centroid of the correlogram. It works quite well for most surfaces, however, when it comes to some step height alike structures that lower than the coherence length of the illuminator, overshoot will be observed at these discontinuities. This overshoot is also known as batwings, which comes out in form of the high frequency information in the measurement results. Normally, the larger magnification objective always has the higher cutoff spatial frequency, so it will also bring stronger batwings. The batwings is shown in figure 12, where a 44 nm standard step height manufactured by VLSI (43.2 nm±0.6 nm) was measured by Fourier transform.

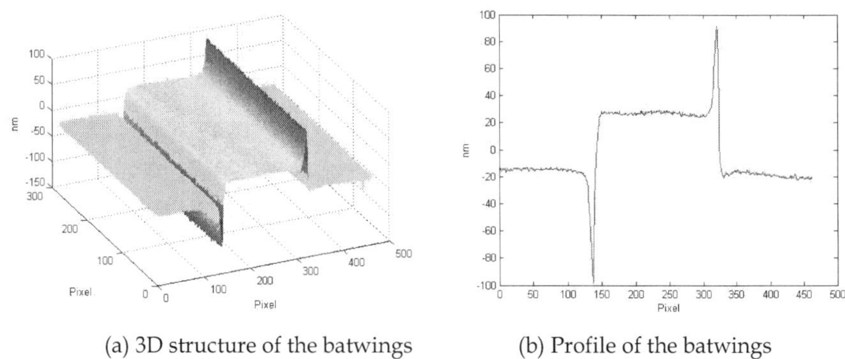

(a) 3D structure of the batwings (b) Profile of the batwings

Fig. 12. Measurement of the 44 nm standard step height with batwings

During the vertical scanning, on condition that the step discontinuities is lower than the coherence length, the light diffracted by the top edge and the light reflected from the bottom will interfere with each other and then travel back to the interferometer. It is because of this diffraction that deforms the correlogram and produces batwings. However, the batwings never occurred in PSI measurements, which gives us a clue to find out the solution for this problem: intensities are much easier to be affected than the phase information in the white light signals. In this work, we repeated the measurements with WLPSI on the above 44 nm standard step height in figure 12. As illustrated in figure 13, it gave out correct results without batwings.

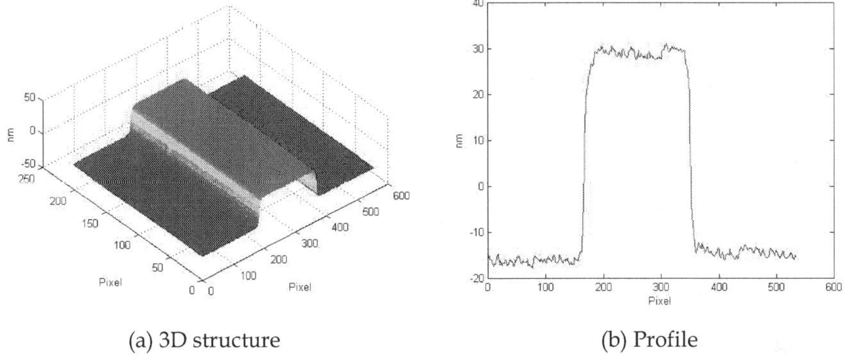

(a) 3D structure (b) Profile

Fig. 13. Measurement of the 44 nm standard step height with WLPSI

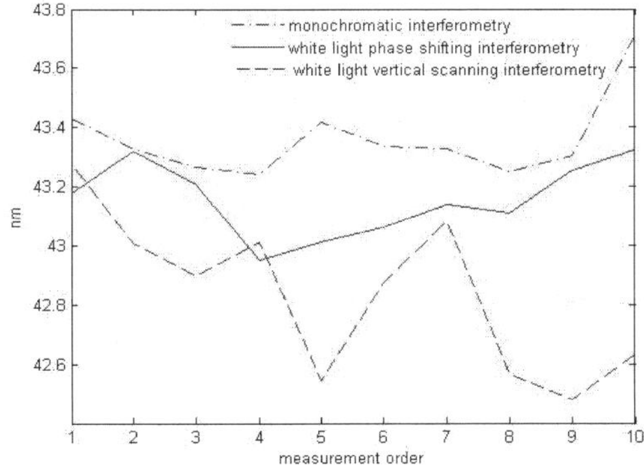

Fig. 14. Comparison of the results from the 44 nm standard step height calibrations

	PSI	WLPSI	WLI
Mean height(nm)	43.36	43.16	42.84
Standard deviation(nm)	0.14	0.13	0.27

Table 1. Measurement results of the 44 nm standard step height using different methods

Figure 14 and table 1 show the 44 nm standard step height evaluations comparison between PSI, WLPSI and WLI. We can clearly see that the results from WLPSI ended up basically in the middle of these three methods and it is close to the results from PSI, which also shows the ability of WLPSI in improving the measurement accuracy compared with the traditional WLI.

4.2 Micro pressure sensor measurement

Experiments were done on a micro pressure sensor using PSI under the air condition. From figure 15(a), a deformed membrane can be seen to show the air pressure. The deform value shows the change of the pressure. From figure 15(b), the height difference between the highest position and the lowest position is 1345.7 nm, which is matched with the designed value.

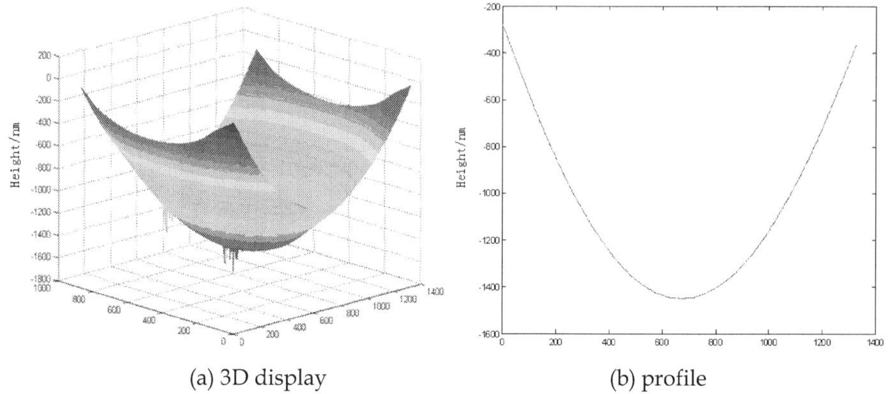

(a) 3D display (b) profile

Fig. 15. Results of micro pressure sensor measurement

4.3 Film structure measurement

Combining with Otsu method from image segmentation technique, we measured a film thickness standard in WLI measurement (centre average thickness: 1052.2 nm±0.9 nm, refraction index: 1.46, model number: FTS4-10100, VLSI) which was calibrated by an ellipsometer in 632.8 nm wavelength. The system equipped 10× Mirau objective to perform a set of repeated measurements". The thickness and the surface topography were successfully extracted, which were shown in table 2 and figure 16. The R_a values of the upper and lower surfaces were 7.30 nm and 7.32 nm, respectively.

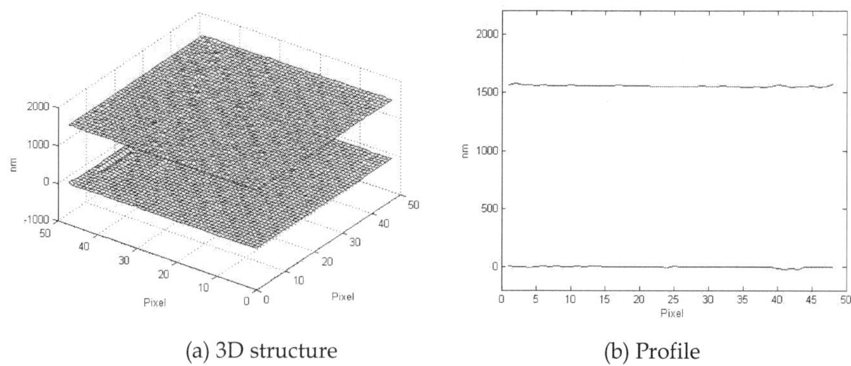

(a) 3D structure (b) Profile

Fig. 16. Film thickness standard measurement

No.	Thickness(nm)	No.	Thickness(nm)	Mean value(nm)	Standard deviation(nm)
1	1053.1	6	1050.9	1051.8	1.43
2	1053.1	7	1052.3		
3	1050.0	8	1052.1		
4	1048.8	9	1052.1		
5	1050.9	10	1053.1		

Table 2. Measurements on film thickness standard

4.4 Micro-resonator measurement

With WLPSI, a micro resonator manufactured by Microelectronics Center of North Carolina (MCNC) was measured. With a 20× Mirau objective, the micro resonator was illuminated by a white light illuminator (central wavelength is 600 nm), the scanning range was configured to be 9 μm and the scanning step was 45 nm. The result is shown in figure 17. The height of the comb-finger profile is approximately 3.8 μm.

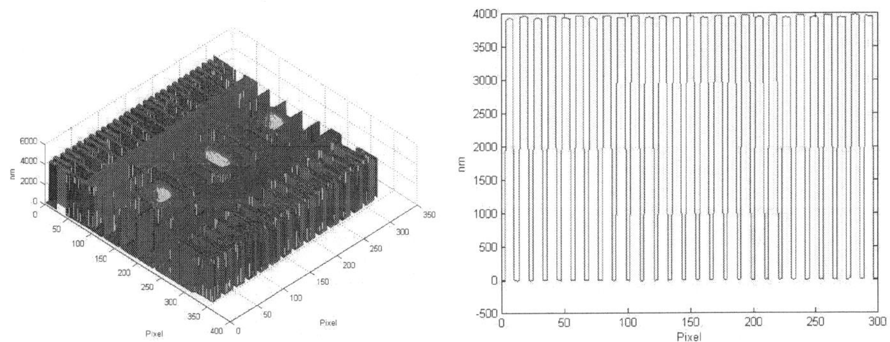

(a) 3D structure of the resonator (b) Comb-finger profile of the resonator

Fig. 17. Measurement of the resonator

4.5 Dynamic characterization

4.5.1 In-plane motion measurement

A 20× objective was used in the dynamic experiments. The stimulating signal was sine waveform with 10V amplitude, 20V offset voltage. The exposing time of CCD was 20ms for LED illumination and 100μs for LD illumination. The strobed pulse percent was 100%. The strobed phase number was 8 for in-plane motion measurement and 16 for out-of-plane motion measurement.

In order to get the resonant frequency of the tested device, the sweep-frequency measurement in a large range was firstly needed. The frequency increases in the logarithm mode, so the range of resonant frequency can be decided through one time frequency sweeping measurement. Then another frequency sweeping measurement can be done in a small range. In summary, smaller range can achieve more times measurement, and the gotten resonant frequency will be more accurate. Figure 18 shows the experimental results in the range from 100 Hz to 100 kHz, the sweeping number is 21. After that, the frequency sweeping range can be decreased to 22 kHz~25 kHz, the sweeping number was 100 (see in figure 19). From that, the resonant frequency of the microstructure can be obtained. From the amplitude-frequency and phase-frequency curves, the device can be fitted into a second order system, the resonant frequency of the device is about 23.41 kHz, the maximum moving amplitude is around 764.52 nm, so the quality factor can be calculated as Q=23.41/(24.04-22.87)=20. Because of the air damp during the moving process of the device, the quality factor is not high. All data will be fed back to the designer and the quality factor of the device can be increased through improving the structure of the micro-resonator. Then five times of experiments without stimulating signal were done. The noise floor was calculated as 0.56 nm.

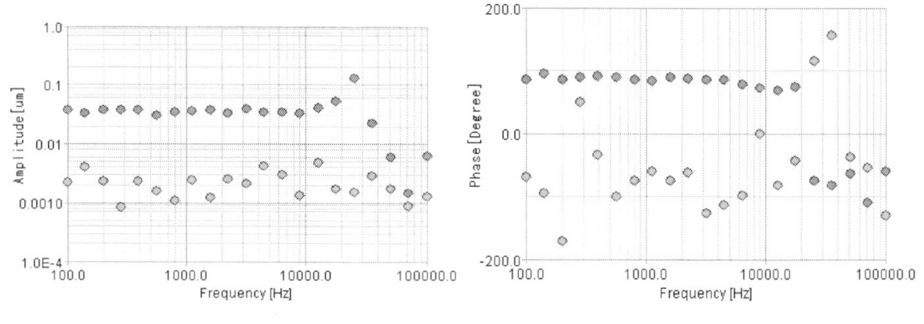

(a) Amplitude-frequency curve (b) Phase-frequency curve

Fig. 18. Large range sweep-frequency measurement of in-plane motion. The blue curve is the result in x direction. The yellow curve is the result in y direction.

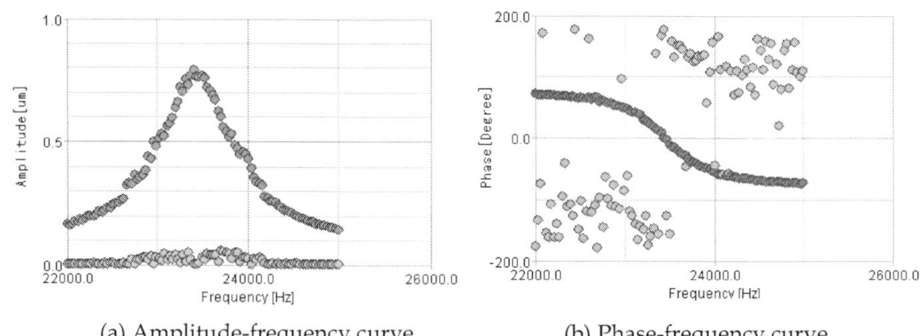

(a) Amplitude-frequency curve (b) Phase-frequency curve

Fig. 19. Small range sweep-frequency measurement of in-plane motion. The blue curve is the result in x direction. The yellow curve is the result in y direction.

In order to study the relationship between the stimulating voltage and the movement, amplitude sweeping experiments were performed. The stimulating frequency was 23 kHz with the starting amplitude 0% and the ending amplitude 100% (10 V). The sweeping number was 21 and the amplitude increased linearly. The experimental results were shown in figure 20. The curves show that the relationship between the moving amplitude and the stimulating voltage is linear which is consistent with the computer simulated results. It draws the conclusion that the design and fabrication process of the device are valid and the device has good dynamic behaviors.

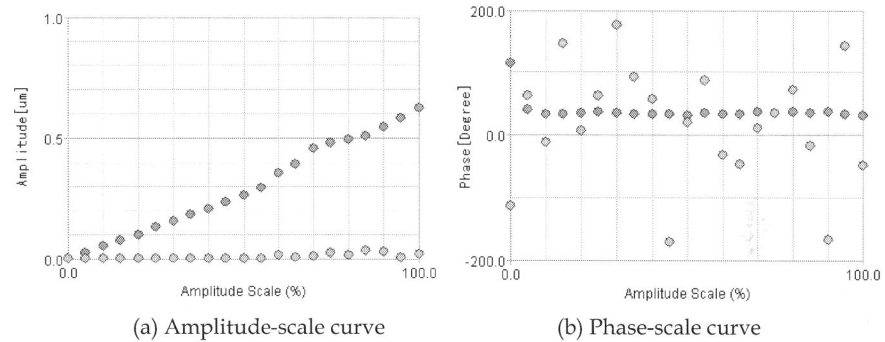

(a) Amplitude-scale curve (b) Phase-scale curve

Fig. 20. Sweep-amplitude measurement of in-plane motion. The blue curve is the result in x direction. The yellow curve is the result in y direction.

4.5.2 Dynamic profile measurement

A region of interest (ROI, see in figure 21) on the device was tested to study the dynamic profile change at a certain stimulating signal. The dynamic profiles were shown in figure 22.

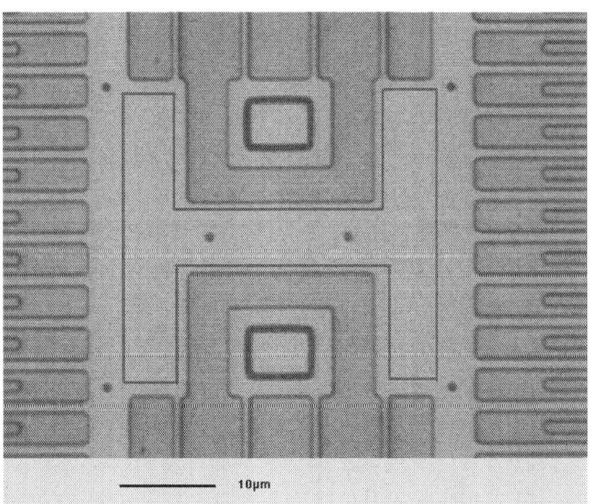

Fig. 21. Choice of ROI

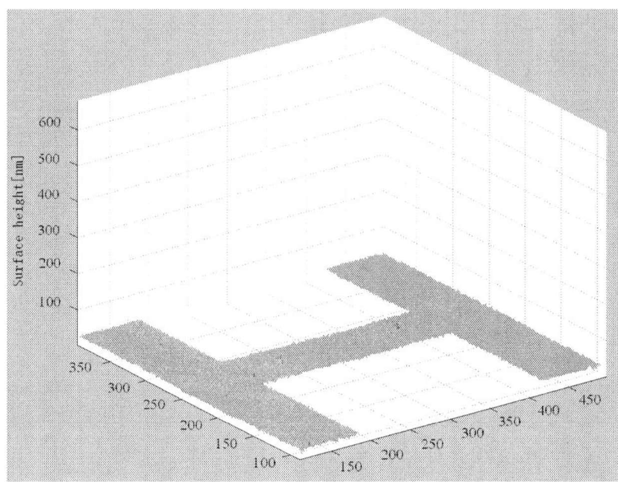

(a) Profile at the second phase

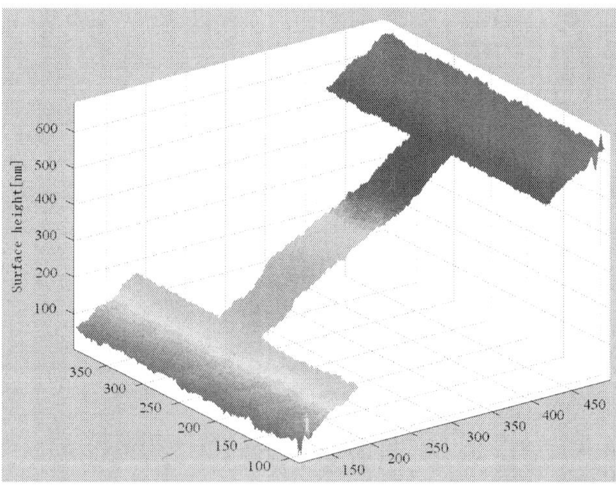

(b) Profile at the ninth phase

Fig. 22. Dynamic profile measurement

The stimulating voltage frequency was 23 kHz, the amplitude was 10 V, and the offset voltage was 20 V. The motion can be classified into three kinds of cases: 1) one reference plane; 2) the first phase of the device movement; 3) the reference mirror in the interferometer. Here the second case is chosen. From figure 22, it can be shown that the movement of the device is like sinusoidal waveform and there is a distort phenomena in the process of the movement.

5. Conclusion

This chapter has discussed several optical methods to realize MEMS characterization, including dimensional (static) and moving (dynamic) properties analysis. Two measuring systems are introduced. NMM based system combines nano-measuring machine with high positioning accuracy and microscopic interferometers. PSI, WLI, WLPSI are applied in the system to measure the dimensional parameters. MMA combines video microscopy, stroboscopic illumination and special algorithms. It employs two kinds of nondestructive methods - computer microvision for in-plane motion measurement and phase shifting interferometry for out-of-plane motion measurement. This system can test the three-dimensional motions and dynamic profiles with nanometer accuracy.

6. Acknowledgment

The authors gratefully acknowledge the support of National Natural Science Fund (91023022), International Cooperation Project of MOST (2008DFA71610) and Natural Science Fund of Tianjin (09JCYBJC05300).

7. References

Bosseboeuf, A., Gilles, J. P., Danaie, K. et al.(1999). A versatile microscopic profilometer-vibrometer for static and dynamic characterization of micromechanical devices. *Proc. SPIE*, Vol. 3825, pp. 123-133

Burdess, J. S., Harris, A. J., Wood, D. et al. (1997). A system for the dynamic characterization of microstructures. *J. Microelectromechan. Syst.*, Vol.6, No.4, pp. 322-328

Fu, X., Liu, Y. Q., Hu, X. D. et al. (2004). Micro Laser Doppler Vibrometer Technology for MEMS Dynamic Measurement. *J of Optoelectronics·Laser*, Vol.15, No.11, pp.1357-1360

Guo, T., Chang, H., Chen, J.P. et al. (2009). Micro-motion Analyzer used for Dynamic MEMS Characterization, *Optics and Lasers in Engineering*, Vol. 47, No.3-4, pp. 512-517

Guo, T., Wu, Z. C., Ma, L. et al.(2010). Dynamic MEMS characterization system using differential phase measurement method, *Proceedings of SPIE*, Vol. 7544, pp. 75444W

Guo, T., Ma, L., Zhao, J. et al. (2011). A nanomeasuring machine based white light tilt scanning interferometer for large scale optical array structure measurement, *Optics and Lasers in Engineering*, Vol. 49, No.9-10, pp.1124-1130

Guo, T., Ma, L., Chen, J.P. et al.(2011). MEMS surface characterization based on white light phase shifting interferometry, *Optical Engineering*, Vol. 50, No.5, pp. 053606

Hartzell, A. L., Woodilla, D. J. (2001). MEMS reliability, characterization, and test. *Proc. SPIE*, Vol. 4558, pp. 1-5

Ma, L., Guo, T., Yuan, F. et al.(2009). Thick film geometric parameters measurement by white light interferometry, *Proceedings of SPIE*, Vol. 7507, pp. 75070G

6

Surface Characterization and Interfacial Adhesion in MEMS Devices

Y. F. Peng and Y. B. Guo
Xiamen University
China

1. Introduction

The characteristic size of MEMS is ranging from atomic and molecular scales to micrometer and several millimeters scales. Components that reach micro-scale size have a high surface to volume ratio, which leaves them be highly subjected to micro-scale effect and susceptible to surface forces. Devices that utilize MEMS technology will often having mating surfaces. Adhesion force can arise from any number of phenomenon such as van der Waals, capillary, ionic and molecular forces. The components used in MEMS structures are very light (on the order of a few micrograms) and operate under very light loads (on the order of a few micrograms to a few milligrams). Surface forces between the adjacent surfaces are becoming dominant over the inertial force in MEMS devices. Because of the micro-sized component, the adhesion forces can pull the adjacent compliant structure into contact and result the interfacial adhesion, which may cause the device-malfunction to a great extent. The operation and performance of lightly loaded micro/nano components in MEMS are highly dependent on the adhesive interactions between mating surfaces. In a word, it is important that the mechanisms of interfacial adhesion should be explained, and separating techniques should be added to the design of MEMS scale components to ensure there is no unwanted contact. Furthermore, the interfacial adhesion between two adjacent mating surfaces is determined by the interaction of rough surfaces. The surface is all rough though in different range. The interactions among different asperities are complicated because the surface topography is consisting of so many asperities. It stands to reason that the proper surface characterization is necessary to elucidate the interfacial adhesion.

The interfacial adhesion is the science and technology of two interacting surfaces in relative motion and of related subjects and practices. It is also valuable in the fundamental understanding of interfacial phenomena to provide a bridge between science and engineering in MEMS. The differences between the conventional or macro-contact and micro/nano-adhesion are contrasted in Table 1. In macro-contact mechanics, tests are conducted on components with relatively large mass under heavily loaded conditions. In these tests, contacting between mating surfaces is inevitable and the bulk properties of mating components dominate the contacting performance. In micro/nano-adhesion, measurements are made on components, at least one of the mating components, with relatively small mass under lightly loaded conditions. The interaction is not limited only to the contacting condition. In this situation, though without contact, the attractive interaction between mating opposite surfaces at small approaching distance can't be neglected. Some of the smaller asperities on the micro-sized surface will be stretched, while some of the taller ones will be compressed through contact. The classical contact mechanics is no longer valid in analyzing the interaction of mating micro/nano-sized surface. It is necessary to explore methods to solve the interfacial adhesion problems in MEMS devices.

Macro-contact mechanics		Micro/Nano adhesion	
Condition	Contact (Inevitable)	Condition	Contact or non-contact
	Large mass		Small mass
	Heavy load		Light or zero load
Method	Hertz theory	Method	Need to consider the adhesion of surface forces
	Linear elastic mechanics		
Target	Bulk material	Target	Surface (Few atomic layers to several μm depths)

Table 1. Comparison between macro-contact and micro/nano-adhesion

In this chapter, we will take a close look at surface geometric structure, or surface topography, and surface forces to elucidate the adhesion problems between mating micro-sized MEMS surfaces. Firstly, the complexities of the surface microstructure devices are discussed. Secondly, several typical surface-measurement instruments are introduced. Thirdly, the techniques to characterize the complex micro-scale surfaces are presented. Finally, the surface forces are described in a summary form, and then the adhesion models are given to interpret the adhesive interaction of MEMS devices.

2. Characterization and modeling of microstructure surface

2.1 Complexities of surface microstructure

Whether a surface is rough or smooth, the answer is – it depends on a roughness sensors used (Bushan, 1999)! The problem of scale-dependent roughness is very intrinsic to solid surfaces. For most solid surfaces it is observed that under repeated magnification, more and more roughness keeps appearing until the atomic scales are reached where roughness occurs in the form of atomic steps (Williams & Bartlet, 1991) as shown graphically in Fig.1. The roughness often appears random and disordered, ranges from around 10^{-4}m (0.1 mm) to about 10^{-9}m (1 nm) and does not seem to follow any particular structural pattern (Thomas, 1982). The randomness and the multiple roughness scales both contribute to the complexity of the surface geometric structure.

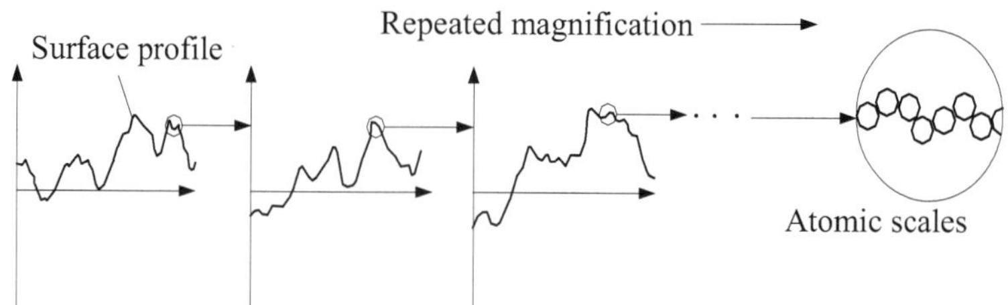

Fig. 1. Scale-dependent of surface roughness

2.2 Surface measurement techniques

Because of the complexities of surface microstructure, the measuring techniques and instruments are important to achieve the surface information to characterize and model the surface microstructure in MEMS devices. The accuracy of traditional contact (probing) as well as noncontact techniques has been perfected to a level allowing measurement of roughness in the nanometer range (Fig.2). The most accurate profilometer probes allow measurement of summit heights of several Angstroms (Bennet & Dancy, 1981, Bhushan et al., 1988). Yet, the comparatively poor lateral (horizontal) resolution significantly limits application of these techniques to the nanometer topographies when the distance between asperities is much less than the solution or 0.1-1μm. The development of techniques using probes smaller than the radius of the probing needle or the light wavelength makes it possible to extend the spectrum of surfaces studied (Myshkin et al., 2003). The scanning electron microscope (SEM) technique can be used to gauge topography with a comparable resolution both vertically and laterally by interpreting the emission intensity of the secondary electrons the topographic pattern (Myshkin et al., 1992). The scanning tunneling microscope (STM) has a still finer probe, which is the electron flux tunneled between the target surface and the needle tip. In this case the surface topography resolution is 0.01 and 0.1 nm in the vertical and lateral directions, respectively (Binnig & Rohrer, 1982). Hence, the STM technique and others resulting from its progress make it possible to use this approach for more accurate topographic investigations of solids on the nanoscale. Significant prospects are connected with the application of atomic force microscope (AFM) (Sarid, 1991) in which atomic-molecular surface effects are registered.

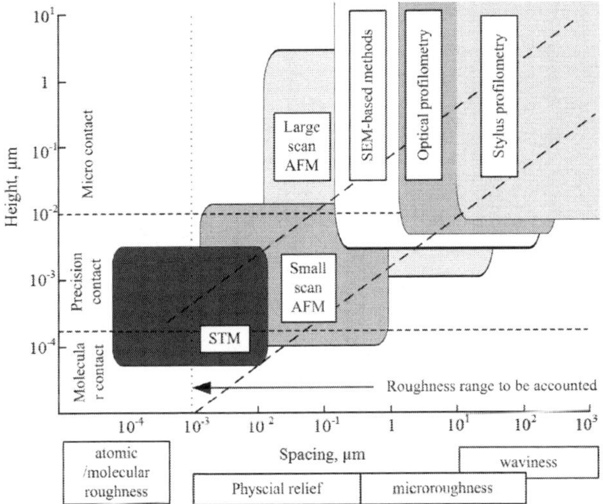

Fig. 2. Diagram of the height and spacing parameters and ranges of vertical-lateral resolution for different methods of roughness measurement (Myshkin et al., 2003).

The scanning tunneling microscope (STM) developed by Dr. Gerd Binnig and Heinrich Rohrer has revolutionized the study of surfaces and is rapidly becoming a required tool in almost every surface characterization laboratory. It is the first instrument capable of directly obtaining three-dimensional images of solid surfaces with atomic resolution (Binnig et al., 1982). Today's STMs can be used in the ambient environment for atomic-scale imaging of surfaces. Generally, samples to be imaged with STM must be conductive enough to allow a few nanoamperes of current to flow from the bias voltage source to the area to be scanned. In many cases, nonconductive samples can be coated with a thin layer of a conductive material to facilitate imaging. AFM can be used for measurement of all engineering surfaces which may be either electrically conducting or insulating. AFM has now become a main surface profiler for topographic measurements on micro- to nanoscale (Bhushan & Blackman, 1991; Oden et al., 1992; Bhushan et al, 1997). STMs, AFMs, and their modifications can be used at extreme magnifications ranging from 10^{-3} to $10^{-9}\times$ in x-, y-, and z-directions for imaging macro- to atomic dimensions with high-resolution information and for spectroscopy (Bushan, 1999). These instruments can be used in any sample environment such as ambient air (Binnig & Smith, 1986), various gases (Burnham et al., 1990), liquid (Marti et al., 1987; Binggeli et al., 1993), vacuum (Binnig et al., 1982), low temperatures (Hug et al., 1993), and high temperatures. To decrease the wear of brittle tip and extend its application in biological research, the carbon nanotube (CNT) has been used to probe the sample instead by adhered it on top of a tip (Fang et al., 2008) in AFM. The resolution ratio can reach about 3nm with functional single-walled CNT in scanning the grease double molecular membranes, while it is about 15nm for conventional Si and of Si3N4 (Yamachika et al., 2004). Fig.3 shows such a tip-CNT probe and the captured image.

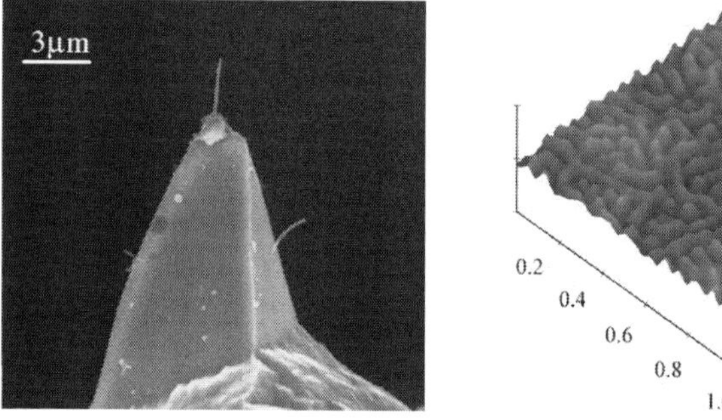

Fig. 3. AFM CNT probe and captured image (Fang et al., 2008); (a). SEM images of CNT probe (b). Images of a styrene-ethylene/butylene-styrene copolymer

2.3 Surface characterization techniques

The characterization of the surface roughness on the micro/nanoscale needs more thorough investigation. This is essential for solving interfacial adhesion phenomena. The randomness suggests that the statistical methods of roughness characterization should be adopted to determine the average dimensions of topographical elements forming the surfaces of solids. In addition, a rough surface involves so many length scales ranges from atomic/molecular level to nano or micro scale, then the characterization techniques must be independent of any length scale. In this section, both the statistical and fractal method to characterize surface roughness are presented in a way that is suitable to model adhesion of mating MEMS devices.

2.3.1 Probability height distribution

One of the characteristics of a rough surface is the probability distribution (Papoulis, 1965). It is often found that the normal or Gaussian distribution fits the experimentally obtained probability distribution quite well (Thomas, 1982; Bhushan, 1990). In addition, it is simple to use for mathematical calculation (Greenwood & Williamson, 1966; Chang et al., 1987). The bell-shaped normal distribution (Papoulis, 1965) which has a variance of unity is given as

$$g(\bar{z}) = \frac{1}{\sqrt{2\pi}} \exp\left[-\frac{(\bar{z} - \bar{z}_m)^2}{2} \right] \qquad -\infty < \bar{z} < \infty \tag{1}$$

where $\bar{z}_m = \bar{z}/\sigma$ is the nondimensional mean height, σ is the standard deviation. The mean height and the standard deviation can be found from a roughness measurement $z(x,y)$ as

$$z_m = \frac{1}{L_x L_y} \int_0^{L_x} \int_0^{L_y} z(x,y)dxdy = \frac{1}{N_x N_y} \sum_{i=1}^{N_x} \sum_{j=1}^{N_y} z(x_i, y_j) \tag{2}$$

$$\sigma = \sqrt{\frac{1}{L_x L_y} \int_0^{L_x} \int_0^{L_y} \left[z(x,y) - z_m \right]^2 dxdy} = \sqrt{\frac{1}{N_x N_y} \sum_{i=1}^{N_x} \sum_{j=1}^{N_y} \left[z(x_i, y_j) - z_m \right]^2} \tag{3}$$

Here, L_x and L_y are the lengths of surface sample, whereas N_x and N_y are the number of points in the x and y lateral directions, respectively. The integral formulation is for theoretical calculations, whereas the summation is used for calculating the values from finite experimental data.

Although used extensively, the normal distribution has limitations in its applicability. The normal distribution near the tail is not an accurate representation of real surfaces. This is an important point since it is usually the tail of the distribution that is significant for calculating the real area of contact (Bushan, 1999). The inverted chi-squared (ICS) distribution fit the experimental data much better near the tail of the distribution (Brown & Scholz, 1985). This is given for zero mean and in terms of nondimensional height, $\bar{z}$, as

$$g(\bar{z}) = \frac{(\nu/2)^{\nu/4}}{\Gamma(\nu/2)} (\bar{z}_{\max} - \bar{z})^{(\nu/2)-1} e(\bar{z} - \bar{z}_{\max})\sqrt{\nu/2} \qquad -\infty < \bar{z} < \bar{z}_{\max} \tag{4}$$

which has a variance of 2ν and a maximum height $\bar{z}_{\max} = \sqrt{\nu/2}$. The advantage of the ICS distribution is it has a finite maximum height, as does a real surface, and has a controlling parameter ν, which gives a better fit to the topography data. It is found that as ν increases, the ICS distribution tends toward the normal distribution (Bushan, 1999; Brown & Scholz, 1985).

Berry and Hannay (Berry & Hannay, 1978) suggested that the variance can be represented as follows:

$$\sigma^2 \approx L^n \tag{5}$$

where L is the length of the sample and n varies between 0 and 2.

If the exponent n in Equation (5) is equal to zero, then the rough surface is generally to be a statically stationary process. This means that the measured roughness sample is a true statistical representation of the entire rough surface. However, n is not equal to zero in the general cases. Then a rough surface is assumed to be a nonstationary random process and the standard deviation is scale dependent, which arises from the probability distribution of a small surface region may be different from that of the larger one. The gathered roughness measurements of a wide range surfaces by Sayles and Thomas (Sayles & Thomas, 1978) have shown that the variance of the height distribution is a function of the sample length and in fact suggested that the variance varied as $\sigma^2 \approx L$. This behavior implies that the surface is a nonstationary process and any length of the surface cannot fully represent the surface in a statistical sense.

Other statistical parameters, such as rms slope σ' and rms curvature σ'' proposed by Nayak (Nayak, 1971, 1973) are also used in surface roughness characterization (Greenwood & Williamson, 1966; Nayak, 1973) and to model the elastic-plastic contact of isotropic and anistropic solid bodies (McCool, 1986). The question is that the determination of σ, σ' and σ'' depends on the sample size, instrument resolution, and experimental filter used to acquire the topography data (Yan & Komvopoulos, 1998), that is whether the rms parameters vary with the statistical sample size or the instrument resolution. Given a rough surface, an instrument with resolution τ will measure the surface height of points that are separated by a distance τ. If τ is reduced, new locations on the surface are accessed. Due to the multiple scales of roughness present, a reduction in τ makes the measured profile look different for the same surface. It is thus necessary to obtain some scale-independent techniques for roughness characterization.

2.3.2 Fractal techniques

It is found that the power spectra of engineering surfaces produced by random processes, such as cleavage, solidification, vapour deposition, and directionally unbiased machining, have been obaserved to follow inverse power laws over a wide range of length scales (Majumdar & Tien, 1990). This is an inherent property of fractal geometry illustrating its potential to represent surface features from the microscale down to the nanoscale (Yan & Komvopoulos, 1998). Fractal geometry, pioneered by Manderbrot (Mandelbrot, 1967) when studying the problem of the length of Britain coastline, can be observed in various natural phenomena, such as precipitation, turbulence, and surface topography, and is characterized by continuity, nondifferentiablity, and self-affinity. Recent works (Kardar et al., 1986; Gagnepain, 1986; Majumdar & Bhushan, 1990) have shown that the fractal geometry can be utilized to develop a scale-independent characterization technique of the fractallike behavior for a rough surface. The mathematical properties of fractal geometry can be satified by the Weierstrass-Mandelbrot (W-M) function given by (Berry & Lewis, 1980)

$$w(x) = \sum \gamma^{(D-2)n}\left(1 - e^{i\gamma^n x}\right)e^{i\phi_n} \tag{6}$$

where w is a complex function of the real variable x. A fractal profile $z(x)$ can be obtained as the real part of $w(x)$

$$z(x) = \mathrm{Re}\left[z(x)\right] = \sum_{n=-\infty}^{\infty} \gamma^{(D-2)n}\left[\cos\phi_n - \cos(\gamma^n + \phi_n)\right] \tag{7}$$

where D ($1 < D < 2$) is the fractal dimension of the profile, is a frequency index, ϕ_n is a random phase, and γ ($\gamma > 1$) is a parameter that determines the density of frequencies in the profile, which is often chosen to be 1.5. The right hand side of Equation is a superposition of cosine function with geometrically increasing frequencies. The random phase ϕ_n is used to prevent the surface profile. The approximate continuous power spectrum, $P(\omega)$, of the profile $z(x)$ given be Equation

$$P(\omega) = \frac{1}{\omega^{(5-2D)}\ln\gamma} \tag{8}$$

is an inverse power function of the spatial frequency, ω, and has been observed to hold for many engineering surfaces.

The two-variable function developed by Ausloos and Berman (Ausloos & Berman, 1980) can be used to model fractal surfaces exhibiting corrugations in all directions. The height function of a fractal surface can be expressed

$$z(\rho,\theta) = \left(\frac{\ln\gamma}{M}\right)^{\frac{1}{2}} \sum_{m=1}^{M} A_m \sum_{n=-\infty}^{\infty} \left(\kappa\gamma^n\right)^{(D-3)} \cdot \left\{\cos\phi_{m,n} - \cos\left[\kappa\gamma^n \rho\cos(\theta - \alpha_m) + \phi_{m,n}\right]\right\} \tag{9}$$

where D ($2 < D < 3$) is the fractal dimension of the surface. The physical significance of D is the extent of space occupied by the rough surface, with larger D values corresponding to denser profiles. For isotropic surfaces, the value of D can be determined from the slope of the log-log plot of power spectrum (Wang & Komvopoulos, 1994; Gagnepain & Rogues-Carmes, 1986). The parameter M denotes the number of supposed ridges used to construct the surface. The anisotropy of the surface geometry is controlled by the magnitude of A_m. For isotropic surface, $A_m = A$ for all m values; for anisotropic surfaces, A_m varies with m. The arbitrary angle α_m is used to offset the ridges in the azimuthal direction. The parameter κ is a wave number related to the sample size, $\kappa = \pi m/M$. The frequency index n is a finite value. The lowest frequency of index $n_{\min}$ is equal to $1/L$ and it can also be set equal to zero. The upper limit of n is

$$n_{\max} = \mathrm{int}\left[\frac{\log(L/L_s)}{\log\gamma}\right] \tag{10}$$

where $\mathrm{int}[...]$ denotes the maximum integer value of the number in the brackets, L_s is the cut-off length of sample.

By introducing a new length parameter G such that G, the surface height function of 3D isotropic surfaces can be obtained

$$z(x,y) = L\left(\frac{G}{L}\right)^{D-2}\left(\frac{\ln\gamma}{M}\right)^{\frac{1}{2}}\sum_{m=1}^{M}\sum_{n=0}^{n_{max}}\gamma^{(D-3)n}\cdot\left\{\cos\phi_{m,n} - \cos\left[\frac{2\pi\gamma^n(x^2+y^2)^{\frac{1}{2}}}{L}\cdot\cos\left(\tan^{-1}\left(\frac{y}{x}\right)-\frac{\pi m}{M}\right)+\phi_{m,n}\right]\right\} \quad (11)$$

The Equation (13) can be used to represent a 3D isotropic fractal surface. This function of surface height provides a deterministic means of generating stochastic rough surfaces. The only unknown variables in Equation (13) are the scale dependent fractal parameters G and D, which can be determined experimentally. Therefore, this fractal approach has the inherent capability of representing surfaces at various length scales, different from those at which the measurements were made (Yan & Komvopoulos, 1998).

3. Surface forces and adhesion mechanics

Surface microstructures typically range from 0.1 to several µm in thickness with lateral dimensions of 10-500µm, and lateral and vertical gaps to other structures or to the substrate of around 1µm (Maboudian & Howe, 1997). The large surface area and small offset from adjacent surfaces makes these microstructures especially vulnerable to adhesion upon contact. The causes of strong adhesion can be traced to the interfacial forces existing at the dimensions of microstructures. These include capillary, electrostatic, van der Waals, and chemical forces.

3.1 Surface forces and adhesion work

There are a wide variety of surface forces (Israelachvili, 1992). Capillary, electrostatic and van der Waals forces can each contribute to adhesion under different circumstances in MEMS devices.

3.1.1 van der Waals forces

Van der Waals force is the force acting between atoms or small molecules, which includes dispersion force， Debye force and dipole-dipole force. The interaction potential between atoms or molecules of each force is a function of $1/r^6$, which r is the separation between atoms. For two flat parallel surfaces, and for separations less than a characteristic distance, $r_0 \approx 20$ nm (nonretarded regime), the attractive force per unit area is given by (Israelachvili, 1985)

$$F_{vdW}(r) = \frac{A}{6\pi r^3} \quad (12)$$

where A is the Hamaker constant to reflect the strength of the van der Waals interaction for two bodies in medium. However, for separations larger than r_0, the attraction is retarded. Taking retardation into account, it is proposed (Cheng & Cole, 1988) that the van der Waals force per unit area is represented more accurately as

$$F_{vdW}(r) = \frac{A}{6\pi r^3}\frac{r_0}{r^3(r+r_0)} \quad (13)$$

The work of adhesion between two surfaces interacting with each other via van der Waals interaction can be obtained by integration from contact, $r_{0,vdW}$, to infinity (Maboudian & Howe, 1997)

$$W_{vdW}(r) = \frac{A}{12\pi r_{0,vdW}^2} \tag{14}$$

3.1.2 Electrostatic forces

Electrostatic forces are the forces between charged bodies. Charges are known to accumulate from the ambient and migrate across insulating surfaces on silicon chips. Early in the development of integrated-circuit (IC) technology, charge migration was the source of device instabilities. Transport of both positive and negative ionic species has been observed in the presence of lateral electrical fields (Shockley, 1964). Electrostatic attraction may also arise due to a difference in the work function of the approaching surfaces. Neglecting the internal space charge regions, the force per unit area acting between surfaces with potential difference V separated by an air gap with permittivity ε_0 is given by

$$F_{el}(r) = \frac{\varepsilon_0 V^2}{2r^2} \tag{15}$$

and the associated energy is given by

$$W_{el}(r) = \frac{\varepsilon_0 V^2}{2r} \tag{16}$$

3.1.3 Capillary force

With the presence of a thin liquid film, such as a lubricant or adsorbed water layer at the contact interface, menisci will form around the contacting and noncontacting asperities due to surface energy effects (Israelachvili, 1985). Fig.4 (a) shows the condition that the amount of liquid film volume was large enough to immerse the rough surface. When the mating surfaces are pulled apart, meniscus will formed underneath the microstructures. Then the liquid meniscus will create a pressure because of the pressure difference across the curved liquid-air interface, which is called the capillary pressure (Laplace pressure), and is given by

$$P_l = \gamma_l \left(\frac{1}{r_1} + \frac{1}{r_2} \right) \tag{17}$$

The liquid surface tension is denoted by γ_l, and the two radii of curvature of the liquid surface are termed by r_1 parallel to the surface normal of the substrate and r_2 (in the plane of the substrate) (Adamson, 1990; Israelachvili,1985). Since in micromechanical structures lateral dimensions are often much larger than the vertical spacing, $r_2 \gg r_1$, and in this case, Equation (17) simplifies to

$$P_l = \frac{\gamma_l}{d}\left(\cos\theta_1 + \cos\theta_2\right) \tag{18}$$

where θ_1, θ_2 is the upper and lower contact angle of liquid bridge, and d is the separation distance between the two surfaces, equal to $r_1(\cos\theta_1 + \cos\theta_2)$. On a hydrophilic surface ($\theta < 90^\circ$), such as the native oxide of silicon, the meniscus shape will be concave underneath a structure shown in Fig.4 (a). This creates an attractive capillary force that may sufficiently strong to pull the compliant structures into contact.

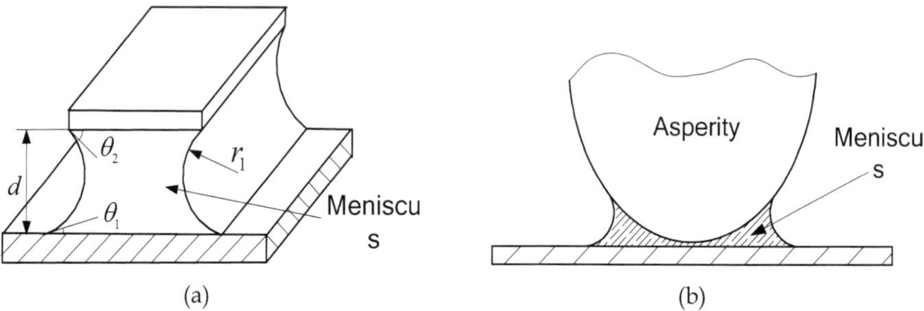

Fig. 4. Wetting and contact angle

The isolated micromenisci would occur at the contact interface, if the amount of meniscus volume were not large enough to immerse all asperities of the rough surface, as shown in Fig.4 (b), the meniscus radius at equilibrium is equal to the so-called Kelvin radius r_k. It is related to the Kelvin equation value and controlled by the relative vapor pressure relative humidity (Admoson, 1990). At equilibrium, the meniscus curvature is related to the relative vapor pressure (H_r) by the Kelvin equation:

$$\gamma_l \left(\frac{1}{r_1} + \frac{1}{r_2} \right)^{-1} \equiv r_k = \frac{\gamma_l v}{RT \log H_r} \tag{19}$$

where r_1 and r_2 are the two radii of curvature of the meniscus, r_k is the Kelvin radius, γ_l is the surface tension of the liquid, and v is its molar volume ($\gamma_l v / RT \approx 0.54$ nm for water at 20°C). As two hydrophilic surfaces approach each other in a humid environment, the liquid undergoes capillary condensation as soon as the separation equals

$$d_0 \approx r_k(\cos\theta_1 + \cos\theta_2) \tag{20}$$

If, after the condensation has occurred, the two surfaces are pulled apart, the volume of the condensate is essentially constant and is given by $V = S_w d_0$, where S_w is the wetted surface area (Maboudian & Howe,1997). The effect of a liquid condensate on the adhesion force per unit area between two parallel plates is then given by

$$F_{cap}(d) = \frac{\gamma_l d_0}{d^2}(\cos\theta_1 + \cos\theta_2) \tag{21}$$

If we assume that, as the two surfaces are pulled apart, the meniscus breaks at a separation much larger than d_0, then integrating Equation (21) from d_0 to infinity yields the work of adhesion due to capillary forces

$$W_{cap} = \gamma_l(\cos\theta_1 + \cos\theta_2) \quad (22)$$

3.2 Adhesion models of single asperity

To better understand the interfacial adhesion of MEMS devices, it is important to provide an adequate background of the prior work performed in the area of adhesive rough surface contact. For determining the interfacial adhesive behavior, several solutions have been developed and many of these theories idealize the asperity in contact with a half rigid flat as a spherical shape.

Hertz theory is the famous continuum contact mode to predict the contact area for various geometries. It relates the radius of the circle of contact a_{H} to the load P, the spherical indenter radius R, and the equivalent elastic modulus of the contacting materials K by:

$$P_{\mathrm{H}} = \frac{Ka_{\mathrm{H}}^3}{R} \quad (23)$$

and between the contact radius a_{H} and the indentation depth δ,

$$\delta = \frac{a_{\mathrm{H}}^2}{R} \quad (24)$$

In the presence of surface forces, Hertz theory can underestimate the contact area, especially when the load diminishes to zero. Considering the contact between a rigid sphere with half rigid flat, the adhesion force P_a, between then is given be Bradely theory (Bradley, 1932) as

$$P_a = 2\pi\omega R \quad (25)$$

DMT theory was then proposed by Derjaguin, Muller and Toporov to account for the long-ranged attraction around the periphery of the contact area. The DMT model gives the contact radius a_{DMT} related to the work of adhesion, ω, by

$$P_{\mathrm{DMT}} = \frac{Ka_{\mathrm{DMT}}^3}{R} - 2\pi\omega R \quad (26)$$

$$\delta = \frac{a_{\mathrm{DMT}}^2}{R} \quad (27)$$

It is apparent that DMT is Hertz with an offset due to surface forces. Therefore, DMT theory applies to rigid systems, low adhesion and small radii of curvature. JKR theory, described by Johnson, Kendall and Roberts, takes the short-ranged attractive forces among the contact area into account. It related the contact radius, a_{JKR}, to the work of adhesion, ω, as

$$P_{\mathrm{JKR}} = \frac{Ka_{\mathrm{JKR}}^3}{R} - \sqrt{6\pi\omega Ka_{\mathrm{JKR}}^3} \quad (28)$$

$$\delta = \frac{a_{\mathrm{JKR}}^2}{R} - \frac{2}{3}\sqrt{\frac{6\pi\omega a_{\mathrm{JKR}}^3}{K}} \tag{29}$$

JKR theory applies well to highly adhesive systems that have large radii of curvature and low stiffness. To bridge the DMT theory and JKR theory, by following the analysis of Tabor (Tabor, 1977; Muller et al., 1980) pointed out that the two theories represented the opposite extremes of a dimensionless parameter μ given as

$$\mu = \left(\frac{R\omega^2}{E'^2\varepsilon^2}\right) \tag{30}$$

where ε is the equilibrium spacing in the lennard-Jones potential. μ can be interpreted as the ratio of elastic deformation resulting from adhesion to the effective range of surface forces.

A more complex, yet more accurate, description of sphere–flat adhesion mechanic, which is referred as MD model, was formulated by Maugis (Maugis, 1992). By analogy with the plastic zone ahead of a crack, the adhesion is represented by a constant additive traction acting over an annular region around the contact area. The ratio of the width of the annular region to the radius of the contact area is denoted by m. The set of equations relating the dimensionless load, approach is

$$1 = \frac{\lambda A^2}{2}\left[\sqrt{m^2-1} + (m^2-2)tg^{-1}\sqrt{m^2-1}\right] + \frac{4\lambda^2 A}{3}\left[1 - m + \sqrt{m^2-1}\,tg^{-1}\sqrt{m^2-1}\right] \tag{31}$$

$$\overline{P} = A^3 - \lambda A^2\left(\sqrt{m^2-1} + m^2 tg^{-1}\sqrt{m^2-1}\right) \tag{32}$$

$$\Delta = A^2 - \frac{4\lambda A}{3}\sqrt{m^2-1} \tag{33}$$

where λ is another dimensionless number, called transition parameter λ, and is related to μ by $\lambda = 1.157\mu$. The dimensionless parameters that appear in the above equations are defined as follows: $\lambda = \frac{2\sigma_0}{\left(\pi\omega K^2/R\right)^{1/3}}$, $\overline{P} = \frac{P}{\pi\omega R}$, $A = \frac{a_{MD}}{\left(\pi\omega R^2/K\right)^{1/3}}$, $\Delta = \frac{\delta}{\left(\pi\omega^2 R/K^2\right)^{1/3}}$, where the adhesion work ω is defined as $\sigma_0 h_0$, σ_0 is the adhesive attraction equals Dugdale stress and h_0 is the effective range of Dugdale stress.

For each previous mentioned theories were presented, there may be cases when assumptions made for a given approach do not exactly describe the materials combinations or the geometry, which are depicted physically in Fig.5.

Following the analysis of Maguis, Kim et al (Kim et al., 1998) offers an extension of the MD solution by adding to the solution regime when the asperity might not be in physical contact but still in the range of adhesion. The KMJ extensions explained by Kim is

$$1 = \frac{\pi}{4}\lambda c^2 + \frac{2}{3}c(\pi - 2)\lambda^2 + \xi \tag{34}$$

$$\overline{P} = -\frac{\pi\lambda c^2}{2} \tag{35}$$

$$\Delta = -\frac{4C\lambda}{3} - \frac{2}{\pi}\frac{\xi}{\lambda} \tag{36}$$

where $C = \frac{c}{\left(\pi\omega R^2/K\right)^{1/3}}$, c represents the adhesive contact zone radius, ξ is ratio of h_g/h_0, and h_g, h_0 are the gap between the deformed asperity at $r=0$ and $r=c$ respectively shown in Fig.5(e.)

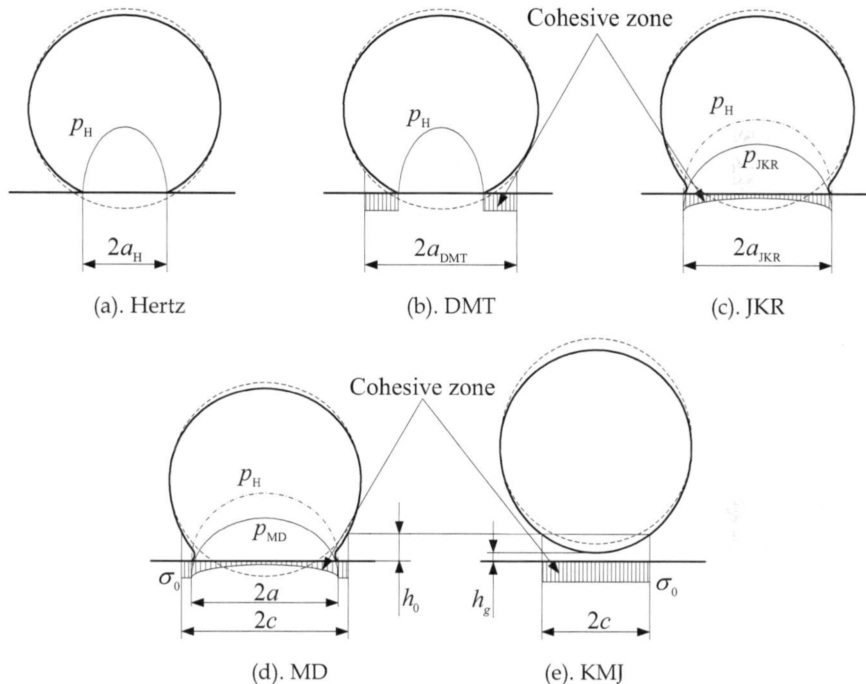

Fig. 5. Interactive forces for the (a). Hertz, (b). DMT, (c). JKR, (d). MD, and (e) KMJ.

3.3 Adhesion for micro-sized rough surface

In practice, the contact at the surface interface is governed by asperity interaction. Since the surfaces are not smooth, contact of two multiscale rough surfaces will occur only at discrete points which sustain the total compressive force. The typical contact interface which is formed of contact spots of different sizes that are spatially distributed randomly over the interface. The size of contact spots ranges from nanometers to micrometers, making adhesion a multiscale phenomenon.

3.3.1 Statistical adhesion theories for micro-sized rough surface

Most contact theories of the rough surface thus far are mostly based upon the conventional statistical parameters such as standard deviation of asperity heights, slope and radius of curvature (Greenwood & Williamson, 1966; Fuller & Tabor, 1975). The Greenwood-Williamson (GW) model assumes the surface to be composed of hemispherical asperities all having the same radius of curvature R. The summit heights or asperity peaks are distributed randomly about a mean summit plane and follow a Gaussian distribution with a standard deviation, σ. If there exists a probability density function $\varphi(z)$ of asperity heights, then it is possible to find the probability that an asperity will be greater than a certain height, d. The distance d represents the length from the mean plane of asperity heights to the smooth surface. The probability that an asperity height is greater than d is given by:

$$\int_d^{\infty} \varphi(z)\mathrm{d}z \tag{37}$$

Therefore, it follows that the number of asperities in contact is represented by

$$n = N\int_d^{\infty} \varphi(z)\mathrm{d}z \tag{38}$$

where N represents the total number of asperities.

Having the numerical expressions for the non-dimensional contact radius (A_i) and load ($\bar{P}_i$) for a single asperity as a function of Δ, the total contact area and load can be formed (Morrow et al., 2003)

$$A_{total} = N\left(\frac{\pi\omega R^2}{K}\right)^{2/3} \int_d^{\infty} \pi A_i(\Delta)^2 \varphi(z)\mathrm{d}z \tag{39}$$

$$P_{total} = N\pi\omega R\int_d^{\infty} \bar{P}_i(\Delta)^2 \varphi(z)\mathrm{d}z \tag{40}$$

The asperities that have a height than d greater than are deformed by a distance $\delta = z - d$. Assuming a Gaussian distribution and to have a relationship between Δ and z, the following equation can be obtained:

$$\frac{P_{total}}{N\pi\omega RP_c(\lambda)} = \int_d^{\infty} \bar{P}_i\left(\frac{\delta}{\bar{\delta}}\right)^2 \exp\left(-\frac{(\delta+d)^2}{2\sigma^2}\right)\mathrm{d}\delta \tag{41}$$

where $\bar{\delta}$ is $\left(\pi^2\omega^2 R/K^2\right)^{1/3}$. The above equation is only valid when the smooth surface progressively approaches the rough surface until a minimum d is reached [Fuller and Tabor]. Because of the existence of δ_c ($-\left(3\pi^2\omega^2 R/4K^2\right)^{1/3}$) to abrupt rupture, or pull-off, the asperities will no longer contribute to the adherence force when asperities were extended above δ_c. Therefore, Morrow et al made the adjustment of the lower integration limit by the amount $\delta_c(\lambda)$, the adhesion model given above then takes the form:

$$\frac{P_{total}}{N\pi\omega RP_c(\lambda)} = \frac{1}{P_c(\lambda)\sqrt{2\pi}} \int_{-2\delta_c^*(\lambda)}^{\infty} \overline{P}_i \left(\frac{\delta}{\overline{\delta}}\right)^2 \exp\left(-\frac{\left(\delta^* + d^*\right)^2}{2}\right) \mathrm{d}\delta^* \tag{42}$$

$$\frac{A_{total}}{N\left(\pi\omega R^2/K\right)^{2/3} A_c(\lambda)} = \frac{\pi}{A_c(\lambda)\sqrt{2\pi}} \int_{-2\delta_c^*(\lambda)}^{\infty} A\left(\frac{\delta^*}{\overline{\delta}^*}\right)^2 \exp\left(-\frac{\left(\delta^* + d^*\right)^2}{2}\right) \mathrm{d}\delta^* \tag{43}$$

Any term in above equation that has a superscript has been divided by σ. The improvement equation by Morrow is similar in form to the rough surface integral of Fuller and Tabor, but has some important difference. The mort important is that the lower integration limit δ_c and P_c are functions of λ, which gives the solution validity over the entire range of the transition parameter. P_c represents the force at which the system becomes stable under force control. The normalization factor $P_c(\lambda)$ is determined by finding the point at which the tangent for load deflection curve becomes zero. $\delta_c(\lambda)$ can also be determined in a similar manner. The critical step in obtaining an adhesive rough surface solution is to find the load at which the system becomes unstable, i. e. the minimum pull-off force ($P_{\min}$). The d^* can be solved by set the derivative of Eq. (42) equals to zero.

3.3.2 Fractal adhesion theories for micro-sized rough surface

The statistically based adhesive theory can be used with confidence as long as the length scale is known before hand (Morrow, 2003). It is well documented that surfaces exhibit roughness on many different length scales (Majumdar & Bhushan, 1990; Majumdar, 1989; Majumdar & Bhushan, 1991). The topography of any surface can be thought of as roughness surperimposed on top of roughness. Majumdar et al (Majumdar & Bhushan, 1990; Majumdar & Tien, 1990) have proven that the multi-scale nature for surface roughness can be represented by fractal geometry. Then it is reasonable to establish the adhesion model of rough interface based on the fractal parameters. Majumdar (Majumdar & Bhushan, 1991) has argued that the size distribution of contact spots can be given as:

$$n(s) = \frac{D}{2s}\left(\frac{s_l}{s}\right)^{D/2} \tag{44}$$

where s is the contact area and s_l represents the largest spot contact area.

3.3.2.1 Fractal model for adhesive contact of JKR type

By assuming that the plasticity plays a minor role in the asperity contact due to the light loading conditions, Morrow et al proposed a fractal model for adhesive contact of JKR type (Morrow & Lovell, 2003). The model follows the example set forth by Majumdar and Bhushan (Majumdar & Bhushan, 1991). It was assumed that the interference which a spherical has with a rigid plane is given by:

$$\delta = G^{(D-1)} l^{(2-D)} \tag{45}$$

where l is the length scale of the asperity as shown in Fig. 6.

A relationship between the truncated and real contact area is developed by equaling the interference distance for both the Hertizan truncated and JKR conditions:

$$s_{JKR} = 9.25\left(\frac{s'^{D}G^{2(1-D)}\omega}{K}\right)^{2/3}\left(1.68 - 0.61\sqrt{5.95 - \frac{\sqrt{(s'\pi)/2}}{\left(\left(s'^{D}G^{2(1-D)}\omega\right)/K\right)^{1/3}}}\right) \tag{46}$$

Following the work of Majumdar and Bhushan, the expression for the interference, δ, in terms of truncated area s':

$$\delta = G^{(D-1)}s'^{(2-D)/2} \tag{47}$$

The radius of curvature at the asperity tip in terms of fractal parameters (Majumdar & Bhusha, 1991)

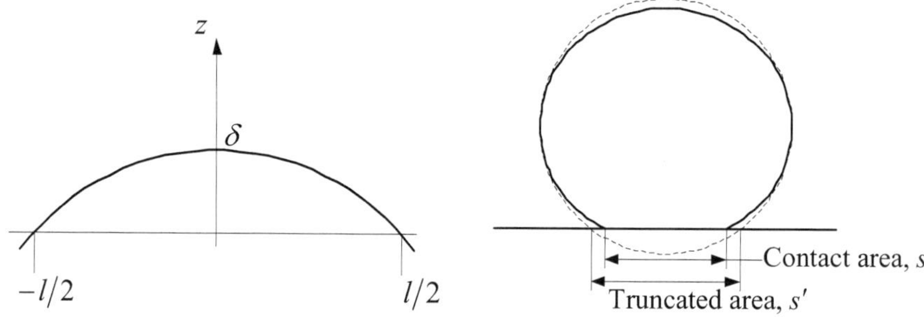

(a) Geometry of contact spot with a given interference (b) Truncated area and contact area

Fig. 6. Fractal approximation of asperity contact;

$$R = \frac{s'^{D/2}}{\pi^2 G^{D-1}} \tag{48}$$

Because of the surface attraction, the asperities are stretched when the contacting surfaces are pulled away. The critical interference δ_c is given

$$\delta_c = -\left(\frac{3\omega^2 s'^{D/2}}{4K^2 G^{D-1}}\right)^{1/3} \tag{49}$$

The critical contact area s'_c for adhesion is broken is:

$$s'_c = \left(\frac{3\omega^2}{4K^2}\right)^{1/(3-2D)} G^{4(1-D)/(3-2D)} \tag{50}$$

Supposing that the size distribution of the asperities, $n(s)$, is known, the real area of contact can be integrated

$$s'_{real} = \int_{s'_c}^{s'_l} S'n(s)ds' = \frac{D}{2-D}\left(s'_l - s'^{1-D/2}_c s'^{D/2}_l\right) \tag{51}$$

where s'_l represents the truncated area of the largest contact spot and is related with the total truncated area S' with

$$s'_l = \left(\frac{3-D}{D-1}\right)S' \tag{52}$$

Then the pull-off force is

$$P = \int_{s'_c}^{s'_l} P(s')n(s)ds' \tag{53}$$

The above Equation can be numerically integrated and solved (Morrow, 2003).

3.3.2.2 Fractal elastic-adhesive model

Based on the work of Yan and Komvopolous (Yan & Komvopoulos, 1998), Morrow proposed a 3D fractal elastic adhesive rough surface solution methodology (Morrow, 2003). In order to develop easy to use expressions for the asperity interference, Yan and Komvopoulos developed a two-dimensional form of Equation (11) to account for the M number of ridges by introducing a multiplicative factor which was eventually set equal to 1. This implied that a two-dimensional W-M function could be used to approximate a fractal function in three dimensions. Yan states in (Yan & Komvopoulos, 1998) that since the radius of curvature of each asperity is much greater than the height of the asperity then this relationship can be assumed to be

$$a' = 2R\delta \tag{54}$$

where a' is the truncated contact radius. Then the radius of curvature is:

$$R = \frac{s'^{(D-1)/2}}{2^{5-D}\pi^{(D-1)/2}G^{D-2}(\ln\gamma)^{1/2}} \tag{55}$$

where the truncated contact area s' can be expressed as

$$s' = \left(\frac{\delta}{2^{4-D}\pi^{(D-3)/2}G^{D-2}(\ln\gamma)^{1/2}}\right)^{2/(3-D)} \tag{56}$$

The fractal relationship for the radius of curvature, R can be used to modify the transition parameter λ which is related to fractal dimensions

$$\lambda = 1.16\left(\frac{\omega s'^{(D-1)/2}}{2^{5-D}K^2\pi^{(D-1)/2}G^{D-2}\sqrt{\ln\eta z_0^3}}\right) \tag{57}$$

where z_0 is the intermolecular distance. From the fractal adhesion model, it can be found that transition parameter λ is no more a constant, but a variant with the approaching distance δ of the asperity during the adhesive contact process. Then the truncated area for each ith asperity can be determined based on a given interference δ_i :

$$\frac{s_i'}{\pi} = 2\left(\frac{s'^{(D-1)/2}}{2^{5-D}\pi^{(D-1)/2}G^{D-2}(\ln\gamma)^{1/2}}\right)\delta_i \tag{58}$$

Equation 60 can be solved using a fixed point iteration scheme to determine the truncated area. It should be noted that the truncated area is used for the first iteration of the algorithm only. To solve for the total load and contact area, Morrow proposed a novel numerical algorithm. In this algorithm the Maugis-Dugdale solution is used to model the micro-sized contact of each asperity. The first step is to generate the surface topography from the fractal parameters using W-M function. Next the surface is offset to introduce an initial penetration into the rigid plane. The interference, δ_i, of each asperity is determined. Based on this interference, the truncated area is subsequently computed for only the asperities that are physically interfering with the rigid plane and is then used to compute the radius of curvature and transition parameter λ for each asperity. Once the truncated area is initially determined, the main iteration scheme is started to determine the real area of contact for each asperity. Using the values for λ and R, the contact radius, m and δ_{com} are computed. These values are computed based on the adhesive contact solution of Maugis given in Equations (31-33). Once the iterations have converged, the values of the load, P_i, and area, S_i are added to the totals (P_{total} and S_{total}). All interfering asperities are iterated on in this fashion and then the surface is moved to the next separation locatio1n and the procedure starts once again.

3.3.2.3 Fractal elastic-plastic adhesion model

According to Majumdar-Bhushan model (Majumdar & Bhushan, 1991), the truncated area $s' > s_c'$ are elastically deformed since they satisfy the condition of $\delta < \delta_c$, whereas asperities with $s' \le s_c'$ satisfy the plastic flow criterion and are thus considered to be in fully plastic deformation state. This result is in disagreement with that derived from the GW model. The reason for this disagreement is that the present analysis accounts for the dependence of the curvature radius on microcontact area, whereas in the GW model the curvature radius of asperity is considered to be invariant. The critical microcontact area for plastic flow of the entire asperity is

$$s_c' = \left(\frac{2^{11-2D}}{9\pi^{4-D}}G^{2D-4}\ln\gamma\left(\frac{E}{H}\right)^2\right)^{1/(D-2)} \tag{59}$$

The elastic adhesion of asperities can be analyzed by Morrow method (Morrow, 2003). Assuming the asperity is at a fully plastic microcontact, the contact pressure within the contact zone is the hardness, H . Then with the Maugis-Dugdale approximation to the adhesive interaction, the adhesive contact pressure for the microcontact can be approximated as (Peng & Guo, 2007)

$$p_p = \begin{cases} H - \dfrac{2\sigma_0}{\pi}\tan^{-1}\sqrt{\dfrac{c^2 - a^2}{a^2 - x^2}} & |x| \ll a \\ -\sigma_0 & a < |x| < c \end{cases} \tag{60}$$

The adherence force can be obtained by integration over the contact and cohesive zones

$$P_p = Hs - 2\sigma_0 a^2 \left[\sqrt{\frac{c^2}{a^2} - 1} + \frac{c^2}{a^2} \tan^{-1} \sqrt{\frac{c^2}{a^2} - 1} \right] \tag{61}$$

where $s = 2\pi a^2$ is the real contact area. When plastically deformed, the contact area s of the microcontact is just the truncated area, which means $a = a'$ and is

$$a = \frac{1}{2} \left(\frac{\delta}{2G^{D-2} (\ln \gamma)^{1/2}} \right)^{1/(3-D)} \tag{62}$$

The radius of cohesive zone c can also be determined by geometrical consideration

$$c = \frac{1}{2} \left(\frac{\delta + h_0}{2G^{D-2} (\ln \gamma)^{1/2}} \right)^{1/(3-D)} \tag{63}$$

If $s_l' > s_c'$, both elastic and fully plastic microcontacts exists. Thus the total adherence force for the fractal surface includes the elastic adhesion and plastic adhesion forces, which is

$$F_t = \sum_j P_{ej} + \sum_j P_{pj} \tag{64}$$

where P_{ej} is the adherence force of j th asperity in elastic contact which can be determined by MD theory utilizing the Morrow method, and P_{pj} is the adherence force of j th asperity in plastic contact which can be calculated by Equation (63).

3.3.2.4 Adhesion model for meniscus stiction

The issue of meniscus force is of critical importance for the microsized interfacial interaction mechanism, such as in the magnetic storage hard-disk drives (Bhushan, 1996). For a given interacting system and environmental parameters, the Laplace pressure and the meniscus height can be assumed to be a constant, whose character is similar to that of the Dugdale stress in linear elastic fracture mechanics. To solve the adhesive problem of the capillary force due to meniscus, the effective work of adhesion can be defined by the product of the meniscus height and Laplace pressure (Xue & Polycarpou,200; Peng et al., 2009)

$$\omega_l = h_c p_l \tag{65}$$

Then by substituting the σ_0 with p_l, and h_0 with h_c, the above methodology, mentioned in section 3.3.2.2 and 3.3.2.3, can be adopted to solve the sitiction problem in presence of meniscus.

4. Summary

In this chapter, we have attempted to present a description of issues and techniques in the interfacial adhesion for the MEMS devices. We firstly discus the complexity of the surface microstructure. Then we present the techniques to characterize the micro-scale surfaces. Finally, we introduce the adhesion models to interpret the adhesive interaction of MEMS devices.

It is hoped that the introductions in this chapter can gain the rational understanding leading to the design of better MEMS structures in the technologically field. The interpretation of interfacial adhesion is challenging for the application of MEMS technology. It is further complicated by the inability to observe the interfacial interactions directly, resulting in conclusions from inference. The gap between theoretical research of rough surface adhesion and the real world where thousands or millions of asperities are involved remains enormous. Then it is clear that there is still a great deal of research necessary to obtain a comprehensive understanding of adhesion at the microscale. The high-resolution instrument should be developed and well calibrated, with which one can measure both the microstructure topography and adhesion, especially the biological sample and hydrophilic surface. It is essential that the proper data processing method should be presented to reflect the intrinsic characters more accurately, and helps to understand the sources of error.

5. References

Adamson, A. W. (1990). *Physical Chemistry of Surfaces*, 5th ed., Wiley, New York.

Ausloos and Berman. (1980). A multivariate Weierstrass-Mandelbrot function, *Proceedings of the Royal Society A* 370, 459.

Bennet, J.M. & Dancy, J.H. (1981). Stylus profiling instrument for measuring statistical properties of smooth optical surfaces, *Applied Optics* Vol. 20: 1785–1802.

Berry, M. V. & Hannay, J. H. (1978). Topography of Random Surfaces, *Nature* Vol. 271: 573.

Berry, M. V. & Lewis, Z. V. (1980). On the Weierstrass-Mandelbrot Fractal Function, *Proceedings of the Royal Society A* Vol. 370: 459-484.

Binggeli, M., Christoph, R., Hintermann, H.E., Colchero, J., Marti, O. (1993). Friction Force Measurements on Potential Controlled Graphite in an Electrolytic Environment, *Nanotechnology* Vol. 4: 59–63.

Binnig, G., Quate, C. F. and Gerber, C. (1986). Atomic Force Microscope, *Physics Review Letter* Vol. 56: 930–933.

Binnig, G. & Rohrer, H. (1982). Scanning tunneling microscopy, *Helvetica Physica Acta* Vol. 55: 726-735.

Binnig, G. & Smith, D.P.E. (1986). Single-Tube Three-Dimensional Scanner for Scanning Tunneling Microscopy, *Review of Scientific Instruments* Vol. 57: 1688.

Bhushan, B. (1999). *Handbook of Micro-Nanotribology*, Second Edition. Bharat Bhushan. CRC Press.

Bhushan, B. (1996). *Tribology and Mechanics of Magnetic Storage Devices*, 2nd ed., Springer, New York.

Bhushan, B. & Blackman, G.S. (1991). Atomic Force Microscopy of Magnetic Rigid Disks and Sliders and Its Applications to Tribology, *Journal of Tribology* Vol. 113: 452–458.

Bhushan, B. & Dugger, M. T. (1990). Real Contact Area Measurements on Magnetic Rigid Disks, *Wear* Vol. 137: 41–50.

Bhushan, B., Sundararajan, S., Scott, W.W., and Chilamakuri, S. (1997). Stiction Analysis of Magnetic Tapes, *IEEE Transactions on Magnetics* Vol. 33: 3211–3213.

Bhushan, B., Wyant, J.C., Meiling, J. (1988). A new three-dimensional non-contact digital optical profiler, *Wear* Vol. 122: 301–312.

Bradley, R. S. (1932). The coercive force between solid surfaces and the surface energy of solids, *Philosophical Magazine* Vol. 13: 853-862.

Brown, S. R. and Scholz, C. H. (1985). Closure of Random Elastic Surfaces in Contact, *Journal of Geophysical Research* Vol. 90: 5531–5545.

Burnham, N.A., Domiguez, D.D., Mowery, R.L., Colton, R.J. (1990). Probing the Surface Forces of Monolayer Films with an Atomic Force Microscope, *Physics Review Letter* Vol. 64: 1931–1934.

Chang, W. R., Etsion, I., and Bogy, D. B. (1987). An Elastic-Plastic Model for the Contact of Rough Surfaces, *Journal of Tribology* Vol. 109: 257–263.

Cheng, E. & Cole, M. W. (1988). Retardation and Many Body Effects in Multilayer Film Adsorption, *Physics Review B* Vol. 38: 987-995.

Fang, F. Z., Xu, Z. W. and Dong, S. (2008). Study on phase images of a carbon nanotube probe in atomic force microscopy, *Measurement Science & Technology* Vol. 19(No. 5) doi:10.1088/0957-0233/19/5/055501.

Fuller, K. & Tabor, D. (1975). The effect of surface roughness on the adhesion of elastic solids, *Proceedings of the Royal Society A* Vol. 345: 327-342.

Gangepain ' J. & Roques-Carmes. (1986). Fractal Approach to Two-Dimensional and Three-Dimensional Surface Roughness, Wear Vol. 109: 119–126.

Greenwood, J. A. & Williamson, J. B. P. (1966). Contact of Nominally Flat Surfaces, *Proceedings of the Royal Society A* Vol. 295: 300–319.

Hug, H.J., Moser, A., Jung, Th., Fritz, O., Wadas, A., Parashikor, I., Guntherodt, H.J. (1993). Low Temperature Magnetic Force Microscopy, *Review of Scientific Instruments* Vol. 64: 2920–2925.

Israelachvili, J. N. (1985). *Intermolecular and Surface Forces*. Academic, London.

Kardar, M., Parisi, G., and Zhang, Y. C. (1986). Dynamic Scaling of Growing Interfaces, *Physics Review Letter* Vol. 56: 889–892.

Kim, K.S., McMeeking, R.M. and Johnson. K.L. (1998). Adhesion, slip, cohesive zones and energy fluxes for elastic spheres in contact, *Journal of the Mechanics and Physics of Solids* Vol. 46:243-266.

Maguis, D. (1992). Adhesion of spheres: The JKR-DMT transition using a Dugdale model, *Journal of Collid and Interface Science* Vol. 150(No. 1): 243-269.

Maboudian, R. & Howe. R. T. (1997). Critical Review: Adhesion in surface micromechanical structures, *Journal of Vacuum Science & Technology B* Vol. 15(No.1): 1-20.

Majumdar, A. (1989). *Fractal surfaces and their Applications to Surface Phenomena.* PhD thesis, University of California, Berkley.

Majumdar, A. & Bhushan, B. (1991). Fractal model of elastic-plastic contact between rough surfaces, *Journal of Tribology* Vol. 113:1–11.

Majumdar, A. and Bhushan, B. (1990). Role of Fractal Geometry in Roughness Characterization and Contact Mechanics of Surfaces, *Journal of Tribology* Vol. 112: 205–216.

Majumdar, A. & Tien, C. L. (1990). Fractal Characterization and Simulation of Rough Surfaces, *Wear* Vol. 136: 313–327.

Mandelbrot, B. B. (1967). How Long is the Coast of Britain? Statistical Self-Similarity and Fractional Dimension. *Science* Vol. 155: 636–638.

Marti, O., Drake, B., Hansma, P.K. (1987). Atomic Force Microscopy of Liquid-Covered Surfaces: Atomic Resolution Images, *Applied Physics Letter* Vol. 51: 484–486.

McCool, J. I. (1986). Comparison of Models for the Contact of Rough Surfaces, *Wear* Vol. 107: 37–60.

Morrow, A. (2003). *Adhesive Rough Surface Contact*, Ph.D. thesis, University of Pittsburgh, Pittsburgh.

Morrow, C., Lovell, M.R., Ning, X. (2003). A JKR–DMT transition solution for adhesive rough surface contact, *Journal of Physics D Applied Physics* Vol. 36: 534–540.

Myshkin, N.K., Ya. A., Grigoriev, S.A., Chizhik, Choi, K.Y., Petrokovets, M.I. (2003). Surface roughness and texture analysis in microscale, *Wear* Vol. 254: 1001–1009.

Myshkin, N.K., Ya., Grigoriev, A., Kholodilov, O.V. (1992). Quantitative analysis of surface topography using scanning electron microscopy, *Wear* Vol.153 (No.1): 119–133.

Muller, V. M., Yushchenko, V. S., B. V. Derjaguin. (1980). On the influence of molecular forces on the deformation of anelastic sphere and its sticking to a rigid plane, *Journal of Colloid Science on Sciencc* Vol. 77: 91-101.

Nayak, P. R. (1971). Random Process Model of Rough Surfaces, *Journal of Lubrication Technology* Vol. 93: 398–407.

Nayak, P. R. (1973). Random Process Model of Rough Surfaces in Plastic Contact, *Wear* Vol. 26: 305–333.

Oden, P.I., Majumdar, A., Bhushan, B., Padmanabhan, A., and Graham, J.J. (1992). AFM Imaging, Roughness Analysis and Contact Mechanics of Magnetic Tape and Head Surfaces, *Journal of Tribology* Vol. 114: 666–674.

Papoulis, A. (1965). *Probability, Random Variables and Stochastic Processes*, McGraw Hill, New York.

Peng, Y. F. & Guo, Y. B. (2007). An Adhesion Model for Elastic-Plastic Fractal Surfaces, *Journal of Applied Physics* Vol. 102(No.5): 3510-7.

Peng, Y. F., Guo, Y. B., Hong, Y. Q. (2009). An Adhesion Model for Elastic-contacting Fractal Surfaces in Presence of Meniscus, *ASME Journal of Tribology* Vol. 131: 024504-1-5.

Sarid, D. (1191). *Scanning Force Microscopy*, Oxford University Press, New York.

Sayles, R. S. and Thomas, T. R. (1978). Surface Topography as a Nonstationary Random Process, *Nature* Vol. 271: 431–434.

Shockley, W., Hooper, W.W., Queisser, H.J. and Schroen, W. (1964). Mobile electric charges on insulating oxides with application to oxide covered silicon p-n junctions, *Surface Science* Vol. 2: 277-287.

Tabor, D. (1977). Surface forces and surface interactions, *Journal of Colloid Science on Science* Vol. 58: 2-13.

Thomas, T. R. (1982). *Rough Surfaces*, Longman, New York.

Wang, S. & Komvopoulos, K. (1994). A fractal theory of the interfacial temperature distribution in the slow sliding regime: Part I – Elastic contact and heat transfer analysis, *Journal of Tribology* Vol. 116: 812 - 823.

Williams, E. D. & Bartlet, N. C. (1991). Thermodynamics of Surface Morphology, *Science* Vol. 251: 393–400.

Xue, X. & Polycarpou A. (2007). An improved meniscus surface model for contacting rough surfaces, *Journal of Colloid and Interface Science* Vol. 311: 203–211.

Yamachika, R. et al. (2004). Controlled Atomic Doping of a Single C60 Molecule, *Science* Vol. 304:281-284.

Yan, W. & Komvopoulos, K. (1998). Contact analysis of elastic-plastic fractal surfaces, *Journal of Applied Physics* Vol. 84: 3617.

7

Advanced Surfactant-Modified Wet Anisotropic Etching

Bin Tang and Kazuo Sato
Nagoya University
Japan

1. Introduction

In the area of Micro Electro Mechanical Systems (MEMS), bulk micromachining and surface micromachining are two main technologies. Bulk micromachining defines structures by selectively etching inside a substrate while surface micromachining uses a succession of thin film deposition and selective etching on top of a substrate. The two technologies are quite different, resulting in different dimensions and different mechanical properties. Although bulk micromachining is usually considered to be the older technology, the two developments run parallel. This is due to the fact that the two different approaches have trade-offs. Taking an example of MEMS capacitance accelerometers, surface micromachined structures use smaller chip area, thus leaving more space for the electronics. On the other hand, in bulk micromachining the larger mass gives greater sensitivity for accelerometers and the larger area leads to larger capacitances for easy read out, which are extremely useful in the inertial device fabrication.

1.1 Wet anisotropic etching in MEMS

Bulk micromachining technology relies on isotropic or anisotropic liquid phase (wet) etching as well as by plasma phase (dry) etching in single crystalline silicon for the formation of functional shapes and patterns in many major applications. Although dry etching has penetrated the traditional territory of wet etchants, the high cost in dry etching and the difficulty in the etch rate uniformity on the whole wafer in wet isotropic etching still make wet anisotropic etching the most affordable method for the reliable production, if wet anisotropic etching can be used to deliver a similar intermediate or final structure.

The formation of crystal facets due to etching is referred to as faceting. When the primary flat of Si {100} is along the [110] direction, rectangular structures with concave corners are easily made with four (111) sidewalls and a (100) plane as the bottom. If the slow etching (111) planes meet, etching will be self-limiting to result in inverted pyramids. Etched grooves, trenches, wells and other basic structures of diaphragms (membranes), beams, and cantilevers exemplify the features of crystal plane-dependent etching. Combined with the use of mask patterns, the etch rate anisotropy becomes a most valuable property as it provides a low-cost, precise means for the production of three-dimensional shapes delimited by smooth, shiny facets, leading to complex structures with multiple functionalities, as shown in Fig. 1.

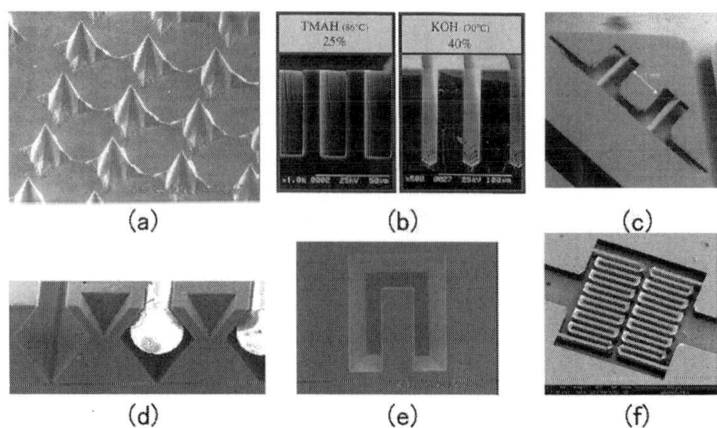

Fig. 1. Wet anisotropic etched structures for various kinds of applications: (a) microneedles for bio-medical applications, Shikida et al. 2006; (b) deep grooves, Sato et al. 1998; (c) mass-spring systems for accelerometers, Butefisch et al. 2000; (d) grooves for the optical fibre alignment, Hoffmann et al. 2002; (e) cantilever beam; (f) suspended filament beams as MEMS heaters, Lee et al. 2009.

A number of alkaline etchants have been tried for wet anisotropic etching of single crystal silicon. Some main features of wet etchants are compared in Table 1. EDP (also referred to as EPW for ethylene diamine, pyrocathecol and water) is not wildly used because of occupational safety and health hazards. Therefore, potassium hydroxide (KOH) and tetra methyl ammonium hydroxide (TMAH) are the most commonly used anisotropic etchants. Based on the ability to withstand the chemical attack by these etchants, silicon oxide (SiO_2), silicon nitride (Si_3N_4), and other metal layers (e.g. Cr, Au) have been used as masking materials. KOH is non-toxic, easy to use, provides excellent etching profiles and has a good selectivity between Si and Si_3N_4, although poor for SiO_2. In addition, KOH is incompatible with CMOS processing due to the presence of an alkali metal. Although TMAH has a lower etch rate, it has outstanding characteristics, such as a high selectivity between Si and SiO_2, and the absence of harmful ions. Hence, TMAH solutions are preferred in recent research and production.

Etchant	KOH (40 wt%)	TMAH (25 wt%)	EDP (80 wt%)
Rate (at 80 °C) µm/min	1	0.5	1 (at 115 °C)
Etching of SiO_2 mask	0	++	+
Compatibility for IC process	-	+	+
Handling (toxicity)	+	+	-
Cost	++	+	+

+:Good 0:Fair -:Poor

Table 1. Comparison of some main features of wet etchants.

1.2 Motivation of surfactant-modified etchants

With different etchants or in the conditions of different concentrations and temperatures, people could get different etching characteristics, which can be clarified as etch rate anisotropy, surface roughness and mask-corner undercut. Surface roughness improvement is important when considering the optical and tribological applications. On the other hand, the conventional design of MEMS structures fabricated by wet silicon bulk micromachining on Si {100} has sharp edge convex and concave corners. This design exhibits stress concentration at the concave corners when a load is applied, which may initiate micro cracks. By providing rounded concave corners instead of sharp ones, the stress can be reduced, thus improving the mechanical efficiency of the microstructures. In pure TMAH solutions, however, both surface roughness and mask-corner undercut are not good enough in regard to above applications. Fig. 2 illustrates anisotropic etched cantilever beam shaped patterns with mask-corners in 10 wt% TMAH at 60 °C. The etched surface is full of hillocks with an average roughness of 110nm. Therefore, many other factors have to be thought over.

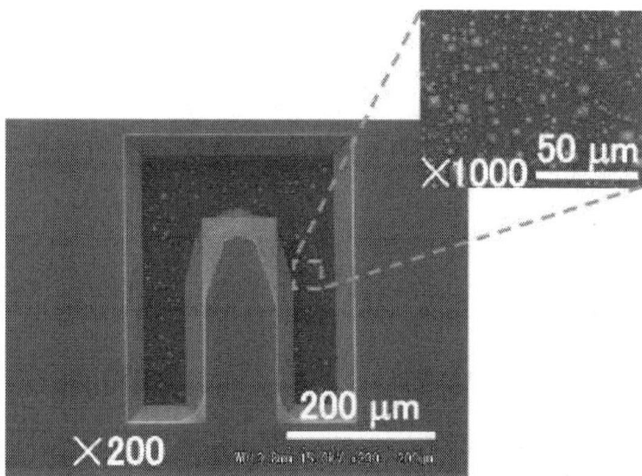

Fig. 2. Anisotropic etched cantilever beam shaped patterns with mask-corners in 10 wt% TMAH at 60 °C (etching depth = 34 μm). The upper right figure is 100 μm x 100 μm.

Among those methods for improving surface roughness and mask-corner undercut, for instance, metal impurities, alcohols, diffusion, light, pressure, microwave irradiation, corner compensation and ultrasonic irradiation et al., surfactant-modified wet anisotropic etching is more outstanding for the effects in both surface smoothness and undercut decrease, also for its stabilization of anisotropy change.

Various ionic (e.g. anionic SDSS, cationic ASPEG, etc.) and non-ionic (e.g. PEG, NC series, Triton X-100, etc.) surfactants have been investigated. Although anionic surfactants exhibit the highest etching rate, the cationic and non-ionic surfactants are suitable for TMAH solutions to improve the roughness of the etched surface owing to the excellent capacity to wet the silicon wafer. TMAH solutions with cationic and non-ionic surfactants are IC-compatible process. Furthermore, adding non-ionic surfactant to TMAH solutions can efficiently reduce undercutting at mask-corners. Such an addition is preferred when accurate profiles are required without very deep etching. Therefore, non-ionic surfactant-modified etching process attracts researchers' attention. With regard to easily handling and less toxicity, in this study, Triton X-100 is selected.

This chapter starts by providing the completely etch rate anisotropy in surfactant-modified wet etching in section 2; the mechanism behind of the change of etching characters when compared with pure etchants will be analyzed in section 3; several applications for the fabrication of new structures by using this advanced anisotropic wet etching will be presented in section 4.

2. Characterization of the etch rate anisotropy

Different etchants can give different etching properties. As one of the most important properties, the etch rate anisotropy clearly manifests the etching behavior in a concentrated solution. In this chapter, the characterization of the etch rate anisotropy is studied by using hemispherical silicon specimens, with and without surfactant in TMAH solutions. Especially, surfactant-modified etching process is analyzed in detail because of its benefit in MEMS applications.

2.1 Experimental details

The hemispherical specimen enables us to obtain the etch rate for all range of crystallographic orientations under the same etching conditions simultaneously, because all the orientations are placed on its surface. The P-type single-crystal hemispherical specimen with a diameter of 44 mm (resistivity: 6–12 Ω-cm) is used in the evaluation. The ingot of hemispheres is provided by Sumitomo Sitix Corporation. In order to produce hemispheres, it is mechanically ground, lapped and polished into mirrored surfaces with a sphericity of less than 10 μm, latitude from 0 to 90° and surface roughness of 0.005–0.007 μm in the arithmetical average by Okamoto Kogakukosakusho Corporation. The etch rate at each orientation is calculated by measuring the shape change before and after etching. The optimized etching depth should be in the range of 100–150 μm in order to avoid interference between neighboring orientations while maintaining the resolution of the geometry measurement. The shape is measured using a 3D profile machine UPMC550-CARAT (Carl Zeiss Co.) with an accuracy of less than 1.0 μm. Fig. 3 shows the locations of crystallographic orientations on the silicon hemispherical sample and a schematic view of the surface profile measurement. Area 'A' corresponds to the measurable area of the hemispherical silicon sample. The place outside the measurement zone and the bottom area are protected by a thermally grown oxide layer. The surface profile is probed every 2° of latitude ranging from 0° to 70°, and every 2° of longitude ranging from 0° to 360°. Supplements of de-ionized (DI) water into an etching bath every 2 h control the tolerance of etching temperature within 1 °C. The total numbers of probe points to be measured are 6480.

TMAH (Toyo Gosei Co. Ltd) and Triton X-100 (Amersham Biosciences) are used as the main etchant and surfactant, respectively. The Triton solution is used to prepare the surfactant-added TMAH solution. The fresh etchant is employed in every subsequent experiment.

2.2 Etch rate anisotropy

The photos of the hemispherical specimen before and after etching at 61 °C in TMAH + Triton are taken as one of the examples and presented in Fig. 4. The contour maps made by the 3D Anisotropic-Etching Simulator (FabMeister-ES) are shown in Fig. 5. Due to the crystallographic symmetry of the hemispherical specimen, only one quarter part with averaged etch rates among equivalent orientations is presented. The range of the planes affected by surfactant adding is clearly visible. The etch rates of exact and vicinal {100} planes are almost unaffected when the surfactant is added, while the etch rates of exact and vicinal {110} planes are reduced significantly.

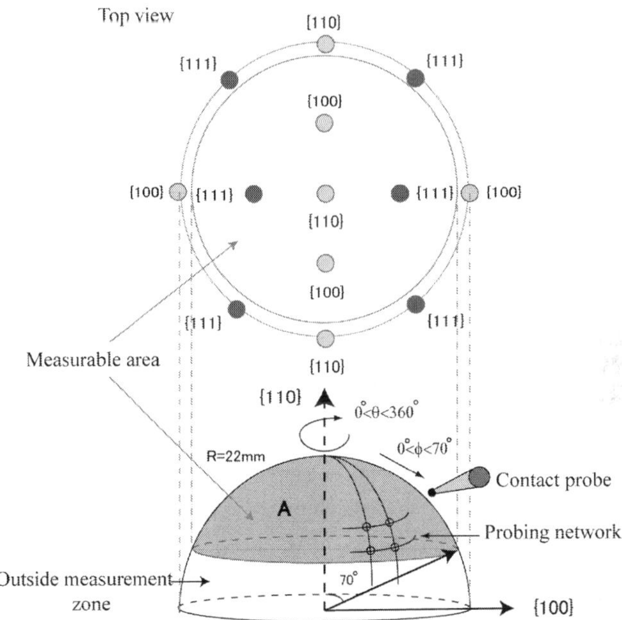

Fig. 3. Location of different crystallographic orientations on one silicon hemispherical sample and schematic view of surface profile measurement. (Reproduced with permission from IOP)

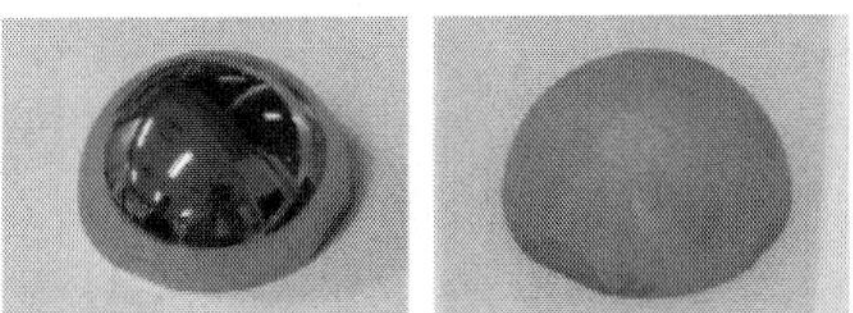

Fig. 4. Photos of the hemisphere specimen: (a) before etching and (b) after etching (etching temperature = 61 °C).

The effect of etching temperature on the etch rates is shown in Fig. 6. The etching anisotropy is influenced by temperature. The orientation of the highest etch rate is shifted toward {350} with the increase in temperature. Contrary to 71 and 81 °C, no single plane prominently appears as the highest etching rate plane at 61 °C. The etch rates of the orientations between {100} and {110} largely depend on the temperature; however, those in between {110} and {111} exhibit less dependence. It can obviously be concluded from these results that the etching anisotropy depends upon the etching temperature. This will result in different etched profiles due to the difference in etching temperatures even if the same mask is used. Examples of measured etch rates for nine different orientations are listed in Table 2. The etch rate ratios of {mnl}/{100}, where m, n, and l are integers, increase with temperature for the planes lying between {100} and {110} (excluding {100} and {110}). On the other hand, these ratios for the planes lying between {110} and {111} are almost the same for 61 °C; however, for other temperatures (71 and 81 °C), the values vary.

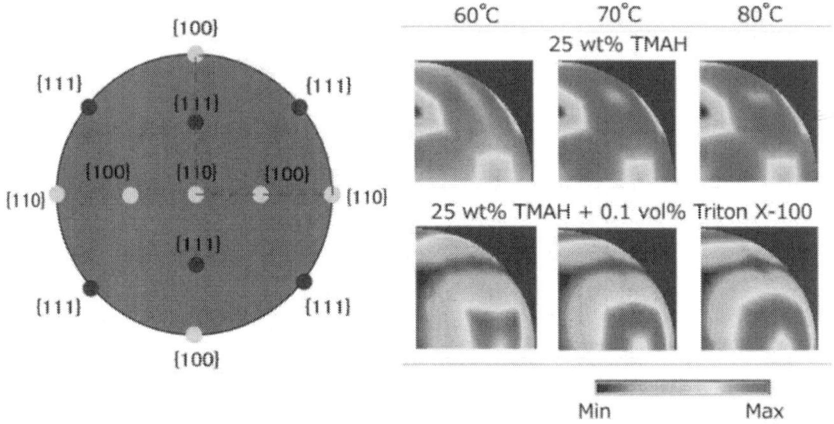

Fig. 5. Contour maps of etch rate in pure and surfactant-added TMAH solutions

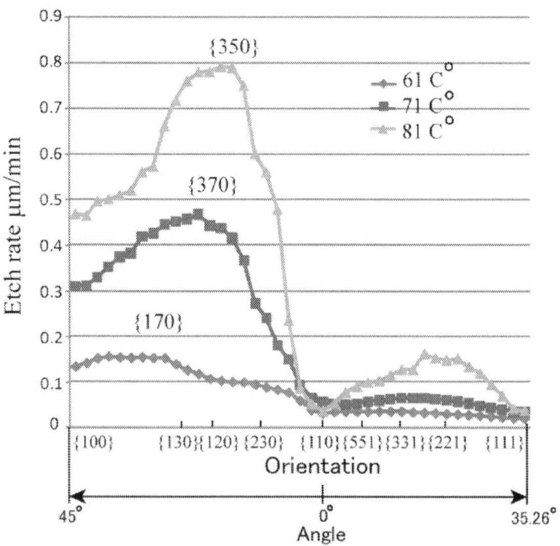

Fig. 6. Effect of temperature on the etch rate anisotropy in 25 wt % TMAH + 0.1 vol% Triton X-100.

Orientation	61°C		71°C		81°C	
	Etch rate (μm/min)	{mnl}/{100}	Etch rate (μm/min)	{mnl}/{100}	Etch rate (μm/min)	{mnl}/{100}
{100}	0.133	1.000	0.308	1.000	0.468	1.000
{130}	0.138	1.038	0.462	1.500	0.717	1.532
{120}	0.102	0.767	0.437	1.419	0.709	1.515
{230}	0.093	0.699	0.272	0.883	0.600	1.282
{110}	0.032	0.241	0.055	0.179	0.036	0.075
{551}	0.034	0.256	0.055	0.179	0.098	0.209
{331}	0.034	0.256	0.064	0.208	0.126	0.269
{221}	0.033	0.248	0.062	0.201	0.152	0.325
{111}	0.019	0.143	0.033	0.107	0.035	0.077

Table 2. Orientation- and temperature-dependent etch rates and etch rate ratios of {mnl}/{100} in TMAH + Triton solutions.

Temperature-dependent etching anisotropy is verified by fabricating 3D circular shape cavities on silicon {100} wafers. Fig. 7 shows the SEM images of circular cavities (etch depth = 100 μm) formed in Triton-added 25 wt% TMAH at 61 and 81 °C using circular shape mask opening, respectively. SEM images are taken after removal of the oxide masking layer. In the case of TMAH + Triton, the {110} and its vicinal planes emerge along <100> directions as these planes exhibit significantly low etch rates. The cavity etched at 81 °C provides better roundness than that of etched at 61 °C. This is because of improved anisotropy ({mnl}/{100}) at high temperature (Table 2). When the temperature is increased from 61 to 81 °C, the etch rate ratios of {110}/{100} and {111}/{100} are changed from 0.241 and 0.143 to 0.077 and 0.075, respectively.

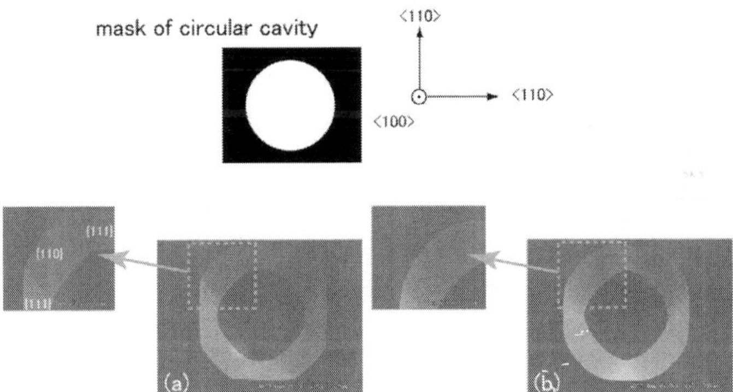

Fig. 7. Circle-like 3D microstructures in Si {100} wafers using one-step etching in 25 wt % TMAH + 0.1 vol% Triton X-100 at: (a) 61 °C and (b) 81 °C.

The Arrhenius-type dependence of etching rates for six planes {110}, {100}, {111}, {331}, {711}, and {221} is plotted in Fig. 8. The activation energy of TMAH + Triton (0.1-0.2 eV) is lower than that of pure TMAH (or KOH) solution (0.5-0.7 eV). The extreme resemblance of activation energy for {331} and {221}, which lie between {110} and {111}, may be associated with the almost same etch rates as described earlier. In particular, {110} plane has very low activation energy.

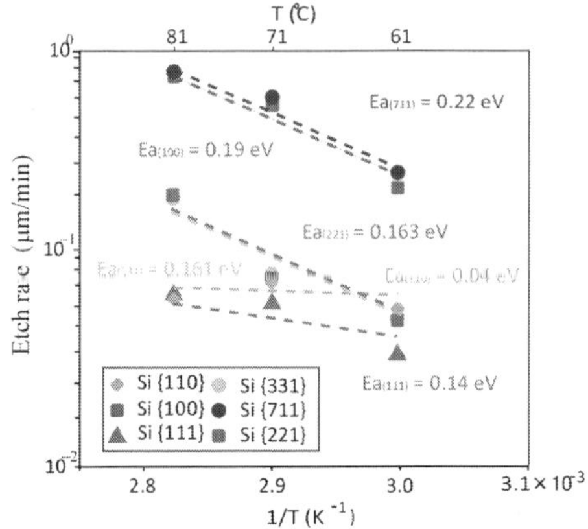

Fig. 8. The Arrhenius plot of etching rates for different planes.

3. Adsorption of surfactant molecules and its effect on etching

The change in the etching behavior of TMAH after the incorporation of the surfactant indicates that the reaction mechanism at the surface is affected by the surfactant molecules, which may block selectively some of the active surface sites, particularly those that appear on Si {110}. Triton X-100 has the following molecular structure:

C_8H_{17} — (benzene ring) — $O(CH_2CH_2O)_nH$

hydrophobic chain (head) hydrophilic chain (tail)

One end of the molecule is hydrophilic, and the other is hydrophobic.

3.1 Adsorption of surfactant molecules on silicon surfaces

Ellipsometry is utilized as a surface-sensitive technique in order to determine the thickness of thin films adsorbed on different surfaces at the subnanometer. Although the non-ionic surfactants do not take part in the etching reaction, their molecules are known to adsorb on the silicon surface by FT-IR observation. In this section, we report detailed ellipsometry measurements in order to determine and characterize the preferential adsorption of the non-ionic surfactant Triton X-100 on {110} and {100} surfaces before etching. The study focuses on the dependence of the thickness of the surfactant layer on various conditions such as the surfactant bath time (SBT), its temperature (T) and the use of ultrasonic agitation (UA), etc.

For experiments, we use three-inch diameter, p-type single crystalline {100} and {110} silicon wafers of 5-10 Ω-cm resistivity. These orientations are selected because of two reasons: (i) most widely used and commonly available, (ii) surfactant effects are quite different for Si {100} and Si {110}. Firstly, the wafers are diced into 13×13 mm^2 small samples. Thereafter, the samples are properly cleaned in chemical solutions followed by thorough rinse in de-ionized (DI) water. In this study all the containers are either glass or Teflon, depending on the type of solution. The samples are now stored in DI water. A surfactant bath consisting of 1 vol% Triton X-100 in DI water is prepared.

The density of adsorbed surfactant molecules in the layer increases as the surfactant concentration is increased, reaching a maximum value (adsorption saturation) at a concentration that is similar in magnitude to the critical micelle concentration (CMC), typically between 0.01 vol% and 0.1 vol% (i.e. 100–1000 ppm). Increasing the surfactant concentration further results in a larger number of micelles in solution but does not affect the adsorption density at the interface. The Triton concentration value of 1 vol% in DI water is a simple choice in order to study the effect of the pre-adsorbed layer as it ensures that enough surfactant is adsorbed on the surface according to the CMC argument.

The samples are dipped in the surfactant bath at room temperature and 60 °C for different times. Prior to immersion in the Triton bath, the samples are dipped in 5% hydrofluoric (HF) acid solution for 1 min and then thoroughly rinsed in DI water. This step is attempted to make the surface hydrophobic. After the surfactant bath, the samples are gently dipped in DI water for several times to remove the most weakly adsorbed surfactant molecules from the surface. These dippings are carried out in still water in a Teflon container. The number of dippings in DI water after the surfactant bath may affect the thickness of the adsorbed layer. Therefore, the effect of the number of gentle dippings is also studied. Moreover, the effect of ultrasonic agitation during the surfactant bath is also investigated. After several dippings in DI water, the samples are dried and used for surfactant layer thickness measurements by ellipsometry. Bare silicon samples are used as reference surfaces. For the ellipsometry measurements of the obtained Triton films we used a standard geometry where the sample is placed horizontally and visible light is reflected at a grazing angle and received by the detector. The spectral analysis is performed using a commercial spectroscopic ellipsometry analyzer (MARY-102).

Although the thickness of the Triton layer (h) increases with the Surfactant Bath Time (SBT) and the thickness can be reduced by performing one or several dippings (ND), we have observed that two Triton layers of equal thickness that have been prepared in different ways, namely, (i) by only the Triton bath and, (ii) by the Triton bath followed by several dippings, have different deviations for the measured thickness in the ellipsometry measurements. When the thickness is determined by focusing the light beam on different regions of a sample without rinsing, the measurements exhibit large fluctuations (±6 Å), indicating an uneven surfactant distribution (large salience). However, the scattering range of the thickness for the samples that have been dipped is less than ±2 Å, thus indicating a more homogeneous packing of the surfactant molecules. For improved repeatability and control, we conclude already at this stage that the Triton layer should be prepared by following the second procedure (Triton bath followed by several dippings).

In order to evaluate how the thickness of the Triton layer decreases with the number of dippings, we consider a Triton layer that is initially saturated, meaning that the silicon surface has been exposed to the Triton solution for a sufficiently long time (24 h), thus ensuring that the thickness has reached its maximum value. Fig. 9(a) shows the thickness of surfactant layer as a function of the number of dippings. Here, ND = 0 means that the sample is directly dried by air. The plot shows that, for both {100} and {110} oriented wafers, a layer of finite, non-zero thickness is measured, indicating that the surfactants are adsorbed on the silicon surface. Generally, {110} samples show a thicker Triton layer, indicating a larger ability to attract surfactant molecules. Below ND = 3, the difference between Si {110} and Si {100} at the same number of dippings is small, and the thickness is high for both orientations. For these particular measurements, the thickness changes significantly between ND = 2 and ND = 3, although it does not change much between ND = 1 and ND = 2. Above ND = 3, the thickness remains finite and constant for {110}, suggesting that the surfactant layers are strongly adsorbed -at least they cannot be easily removed by rinsing- while the thickness is vanishingly small for {100}, indicating a weaker adsorption as the surfactant layer can be easily removed. Thus, we conclude that ND = 3 is the optimal number of dippings and use it for all subsequent experiments. For ND = 3, the typical deviation in the layer thickness is roughly ±1 Å.

The surfactant layer thickness as a function of the Surfactant Bath Time (SBT) is presented in Fig. 9(b) (only ND = 3 is concerned) for two different surfactant bath temperatures (Room Temperature and 60 °C). The adsorption of the surfactant on both Si {100} and Si {110} are saturation processes, characterized by a saturation time (τ_{sat}) and a saturation thickness (h_{sat}). For non-ionic surfactants, the adsorption kinetics involve an initial fast depletion of the surfactant molecules immediately bordering the interface, followed by the diffusion of surfactant molecules from the bulk etchant to the interface and, finally, a rearrangement of the adsorbed surfactant molecules into a final packing structure. Although the actual shape of the monotonic increase of the thickness before saturation should contain valuable information about the manner how the surfactant is packed into a layer, such a study is out of the scope of the present work. From the figure, it is apparent that for the nearly same temperature {100} develops a thinner surfactant layer than {110}. The saturation thickness will become larger for smaller ND values. This observation is good in conformity with the less temperature-dependence of non-ionic surfactant adsorption.

Fig. 9(c) shows an Arrhenius plot of the saturation thickness against inverse temperature. The Arrhenius equation ($h \propto e^{-Ea/KT}$) gives the dependence of adsorption on absolute temperature and activation energy. We find that the apparent activation energy is Ea ~ 0.15 eV for {110} while Ea ~ 0.01 eV for {100}, showing that there is clearly bigger barrier to be overcome in the adsorption process of surfactant molecules on the {110} surface. The larger activation energy for {110} suggests that chemisorption may play a role on these surfaces, whereas the small activation energy for {100} indicates that only physisorption is involved for these surfaces.

From the Arrhenius plot, it is observed that adsorption is dominantly thermal for {110}, while nearly athermal for {100}. The increase in adsorption density with temperature is mostly due to the reduction in the size of the hydration shells surrounding the surfactant molecules, especially around the hydrophilic polyethylene oxide chains. The adsorption process is a trade-off between (i) the energy reduction obtained through adsorption by increasing the number of contacts between the hydrophobic alkyl chains and the surface, and (ii) the entropy increase obtained by remaining dissolved in a more disordered state. The hydration shells are smaller at higher temperatures, resulting in less water becoming ordered per adsorbed molecule, thus allowing a larger number of adsorbed surfactant molecules at higher temperature.

Agitation is a key method that can significantly affect the wet etching quality, including the etch rate and surface morphology. The etching properties of etched surfaces with ultrasonic agitation are satisfactory and superior to no agitation. Fig. 9(d) shows the surfactant layer thickness obtained using 110 W ultrasonic cleaner (VS-D100) during residence in the surfactant bath at room temperature. After adding ultrasonic agitation, the surfactant layer thickness for {110} increases with respect to no agitation (cf. figure 3.4), indicating that appropriate forced convection can improve the adsorption of surfactant molecules.

However, for {100} the thickness is only slightly larger than without agitation, indicating little change of the surface properties. The error bars are included in figure 3.6 to stress the larger variations as compared to no agitation.

Although our study shows that a larger adsorption of Triton molecules can be obtained by using ultrasonic agitation, especially on {110}, the oscillating force results in large fluctuations in the surfactant thickness. An inhomogeneously adsorbed layer is considered a disadvantage in terms of repeatability and surface roughness control.

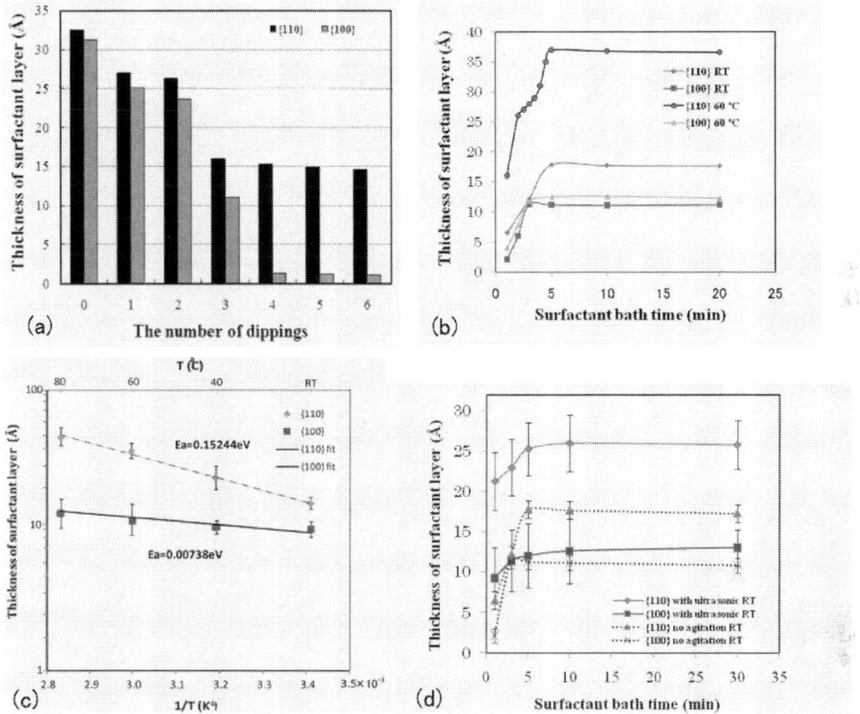

Fig. 9. Ellipsometry study: (a) thickness of the surfactant layer obtained in 1 vol% Triton as a function of the number of dipping in DI water at Room temperature; (b) surfactant layer thickness as a function of Surfactant Bath Time (SBT); (c) Arrhenius plot of the saturation thickness of surfactant; (d) comparison of the thickness of surfactant attached on the silicon surface with ultrasonic agitation at room temperature (Reproduced with permission from Elsevier).

3.2 Effect of the adsorbed surfactant layer on etching

In order to observe the effect of the adsorbed surfactant layer on the etched silicon surface morphology and etch rate, 10 wt% TMAH etchant is used at two different temperatures (room temperature and 60 °C, for both the Triton bath and the etching in TMAH) with ND = 3. The etch rates and etched surface morphology of {110} and {100} with pre-adsorbed Triton layer of different thicknesses are shown in Fig. 10 (etch depth ~ 30 ± 3 μm).

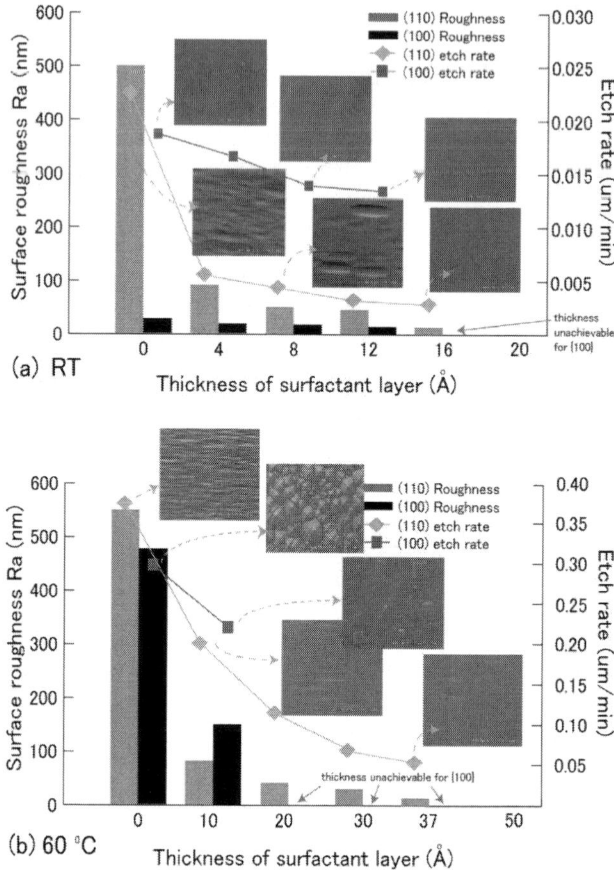

Fig. 10. Effect of pre-adsorbed surfactant layer (1 vol% Triton in DI water) on the surface roughness (Ra) and etch rate for Si {110} and Si {100} in 10 wt% TMAH: (a) room temperature; and (b) 60 °C.

For room temperature, the dramatic transformation in the surface morphology correlates directly with a strong reduction in the measured surface roughness. For {110}, it is found that typical zigzag structures emerge in pure TMAH while short pre-treatment in Triton (producing a layer thickness of 12 Å) drastically smoothens the surface, in spite of the reduced thickness of the surfactant layer. The saturated layer thickness for {110} (≈ 16 Å) produces a very smooth silicon surface (Ra ≈ 20 nm). For {100}, the surfactant pre-treatment provides some improvement in the morphology, even when the initial surface is already very smooth. Similar experiments carried out at 60 °C are shown in Fig. 10(b) (etch depth ~ 35 ± 3 μm). Although {110} shows similar surface morphologies at 60 °C and RT, the surface morphology of {100} is very different, characterized by the formation of pyramidal hillocks. Nevertheless, the roughness of both {110} and {100} is improved by the use of the surfactant layer.

In a similar manner, the thickness of the surfactant layer above the silicon surface has an effect on the etch rate in TMAH solutions after the surfactant pretreatment. For {110}, there is a significant reduction in the etch rate for thin surfactant layers. However, for {100} the reduction in the etch rate is only moderate. While {110} has a higher etch rate than {100} in pure TMAH in Fig. 10(b), the pre-treated samples show an opposite behavior, with {100} faster than {110}. This behavior is similar as for directly etching in solutions of Triton added TMAH, indicating that the dissolved surfactant is adsorbed on the surface during etching, as recently shown in FT-IR experiment. Compared with room temperature, there is a less sudden reduction in the etch rate of Si {110} wafers at 60 °C. In the case of {100}, a moderate etch rate reduction is observed at 60 °C (as for room temperature). The saturated layer thickness for {110} (≈ 37 Å) produces a very smooth silicon surface (Ra ≈ 20 nm or less).

4. Applications on MEMS

Significantly different etching behaviour of TMAH + Triton from that of traditionally used anisotropic etchants is very useful for MEMS applications in order to extend the range of 3D structures fabricated by wet etching because the surfactant is adsorbed at the silicon-etchant interface as a thin layer to act as a filter moderating the etching behaviour. In this chapter, we present three applications using surfactant-modified etchants and point out its great potential on advanced MEMS structures.

4.1 Conformal structures

The corner compensation method is the most widely used method for fabricating the sharp edge convex corners. The design and dimensions of the compensation structures require the knowledge of the undercutting ratio and its dependence on the etchant. If the design of MEMS structures does not include any rounded concave and/or sharp convex corners but a smooth etched surface is necessarily required, then the high concentration (20 - 25 wt% TMAH) should be selected for anisotropic etching. If the structures comprise rounded concave and sharp edge convex corners, the pure TMAH cannot be used due to severe undercutting.

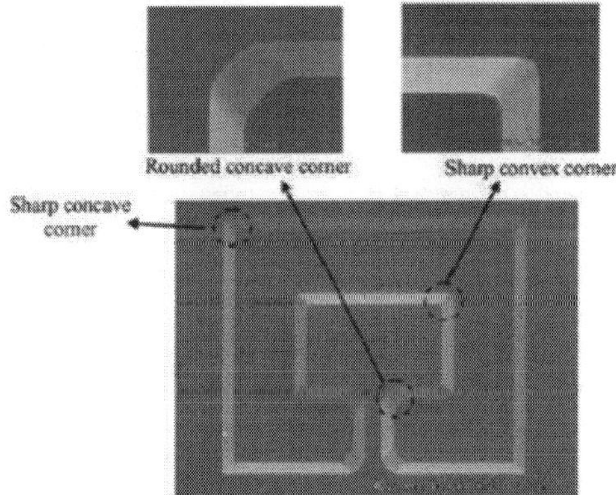

Fig. 11. Conformal mesa shape fabricated in surfactant-modified solutions.

The sharp convex corners and rounded concave corners etched in surfactant-modified TMAH solutions are shown in Fig. 11. Various kinds of etched patterns (etch depth about 35 μm) using 25 wt% TMAH without and with the surfactant are shown in Fig. 12. The undercutting at the convex corners is considerably reduced because of the changed etching anisotropy by the adsorption of surfactant molecules. This solution exhibits minimum undercutting and provides very smooth surfaces while keeping a reasonable etch rate.

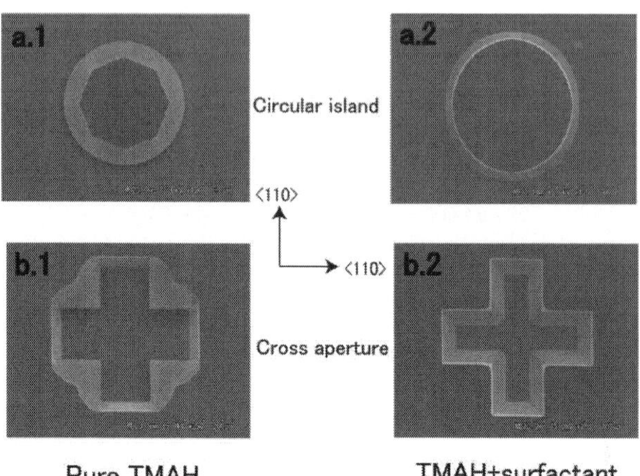

Fig. 12. Two kinds of etched patterns in pure and surfactant added 25 wt% TMAH at 60 °C (etch depth = 35 μm).

4.2 Scalloping removal for vertical micro-lens

Recently, the optical elements perpendicular to the substrate, for instance the cylindrical lens for micro-laser-scanning module applications, have been required as they can be integrated to a desired position on the same substrate using standard semiconductor processes. In order to fabricate some class of devices such as a micro-lens for optical MEMS, a highly smooth rounded column surface is required. Basically commonly used wet etchants (e.g. pure KOH and TMAH) provide very high etch rates at a rounded profile and exhibit a quite rough surface as the many faceted orientations appear on such kinds of surfaces. Hence, the removal of scalloping on rounded column surfaces is a challenging issue.

Due to the selective adsorption of surfactant molecules, the etch rate of {110} in TMAH + Triton solutions is significantly reduced. The etch rates of the planes lying between {111} and {110} at 61°C are almost same with values less than 0.05 μm. This property has been studied to remove the scalloping at the sidewalls of DRIE etched patterns on {110} silicon wafers. The study of scalloping removal is performed by short time dipping of DRIE etched patterns in TMAH + Triton. In order to investigate the effect of crystallographic directions on the surface roughness and the final shape of etched profile, a ring shape structure with internal and external diameters 200 μm and 300 μm respectively is fabricated by DRIE of about 40 μm deep. Thereafter, wet anisotropic etching is employed for 10 min in 25 wt% TMAH + 0.1 vol% Triton at 61°C. This time is long enough to clearly observe the change in profile. The SEM pictures of resultant profile and the surface morphologies of its sidewalls at different locations are shown in Fig. 13. The qualitative analysis (SEM image) reveals that the portion between two {111} planes centered by <110> orientation (marked by F), where corrugation patterned is disappeared, is sufficiently smooth for optical applications. This is because that the planes appearing in this portion exhibit very slow and almost same etch rates for 61 °C. Due to symmetry, the portion Q also has the smooth surface. This property is very useful for the fabrication of cylindrical lens, which can be easily integrated with other optical elements on a silicon substrate.

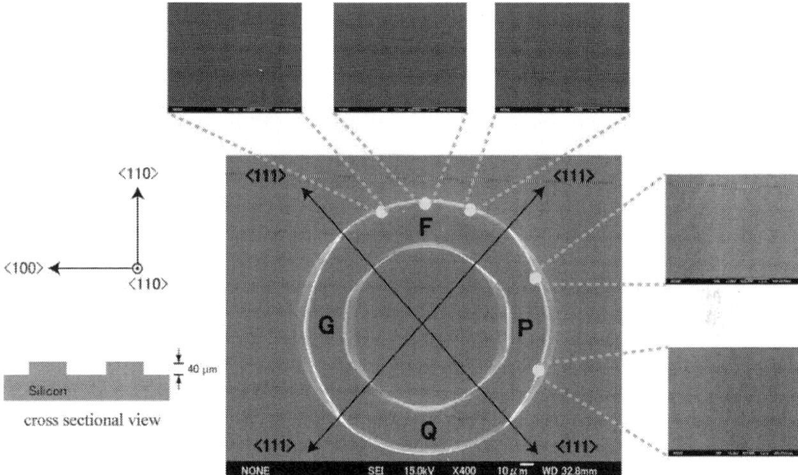

Fig. 13. SEM pictures of the ring structure and the morphologies on its different side walls after wet anisotropic etching at 61 °C.

The details of cylindrical lens portion are schematically illustrated in Fig. 14(a). The total range of the portion within two {111} planes on {110} surface is 70.54°. Since the {111} planes are the slowest etch rate planes in wet anisotropic etchants, in order to maintain the roundness of the curved profile, these planes should not be included in the design. Hence, the curved portion making an angle of 70°, is only useful portion for achieving smooth rounded vertical walls using DRIE followed by TMAH + Triton treatment. In this work, in order to demonstrate the application of surfactant-added TMAH for the removal micro-corrugation at rounded surface and the fabrication of cylindrical lens with smooth surface, the curved portion with hatched lines is selected.

The etching time for scalloping removal should be controlled accurately in order to achieve highly smooth etched surface finish while maintaining desired shape profile. In this work, the mechanism of scalloping removal is that the top area of the corrugation is etched with highest etch rate. Firstly, the left equation corresponds to the calculation of wet etching time in order to remove micro-corrugation at {110} side wall (Fig. 14(b) A-A'), where T is the etching time, $R_{\{170\}}$ is the highest etching rate at 61 °C, α is the angle between {110} and {170}, and H_1 is the height of corrugation. At 61 °C, the etching rates for {170} and {110} are the largest and smallest respectively. {170} project the corrugation, and the projected region is etched with the highest etching rate. On the other hand, {110} appears at the bottom of the corrugation, minimizing the over-etching of the sidewall. The projected region of the corrugation is becomes small as the etching process, and it finally flattens out. In our case, α is measured as 37° and H_1 is 69.428 nm. The etching time of 42 seconds is obtained. Secondly, removing of micro-corrugation on other planes {ijk} is shown in Fig. 14(b) B-B'. Here micro-corrugation caused by DRIE is considered as the same scalloping shape. Therefore, the angle β between {ijk} and {xyz}, where {xyz} has the highest etch rate between {110} and {ijk}, is $\beta = \alpha = 37°$. Also, the height $H_2 = H_1 = 69.428$ nm. $R_{\{ijk\}}$ is the etching rate of those planes located between {110} and {111}. Now the etching formula is described as the right equation. As noted before, etching rate of planes lying between {110} and {111} at 61 °C exhibit little variation. For this reason, we consider $R_{\{ijk\}} = R_{\{110\}}$. Another important parameter is the etch rate of highest etch rate planes in corrugated structures. Luckily, the distributions of etching rates on the {100} vicinal planes exhibit nearly identical property, which implies $R_{\{xyz\}} = R_{\{170\}}$. Thanks to the wet etching characterization in this intriguing etchants. It is feasible for the fabrication of cylindrical lens.

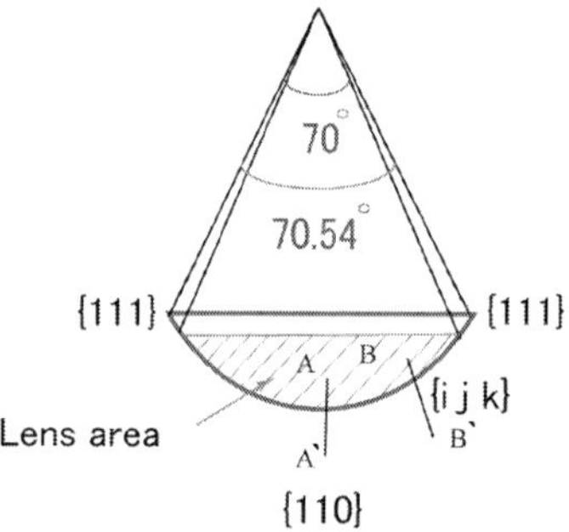

(a) Top view of cylindrical lens

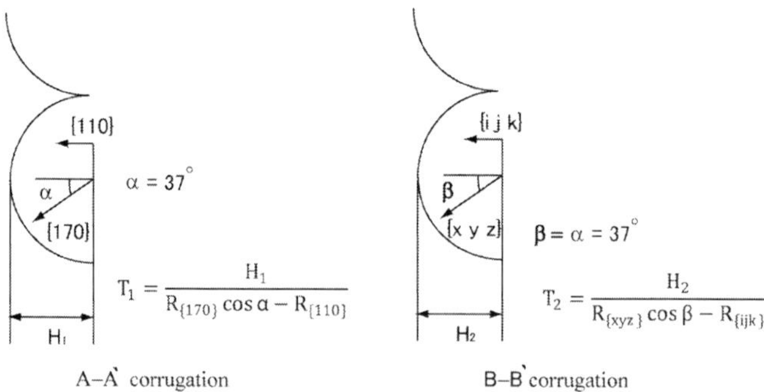

(b) Mechanism of corrugation removing

Fig. 14. (a) Schematic top view of cylindrical lens and (b) Mechanism of micro-corrugation removal by wet anisotropic etching at the sidewalls of cylindrical lens fabricated by DRIE.

Fig. 15 shows the SEM pictures and AFM measurement results of silicon surface roughness before and after wet etching. The roughness of sidewalls has been improved to about 1 nm.

This surface roughness is sufficiently enough for optical MEMS (MOEMS) applications. Thus, the small time etching in 25 wt% TMAH + 0.1 vol% Triton at 61 °C successfully remove the micro-corrugation without altering the desired shape of the DRIE etched profile and provides almost homogeneous surface finish.

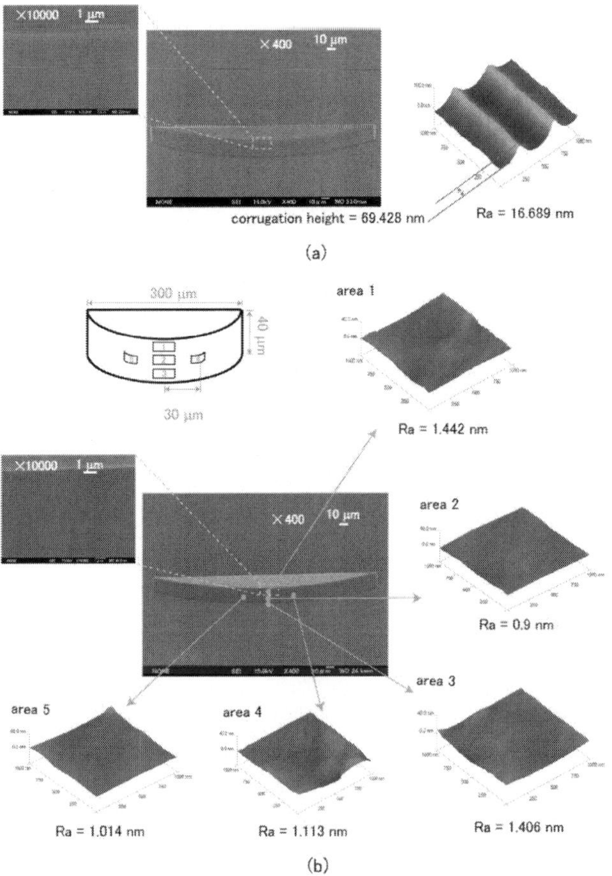

Fig. 15. SEM pictures of cylindrical lens after (a) DRIE and (b) DRIE followed by short time etching treatment in 25 wt% TMAH + 0.1 vol% Triton at 61 °C. The AFM results reveal surface roughness at different locations of vertical sidewalls. Dimension of each zoomed figure is 9 μm x 9 μm (Reproduced with permission from IOP).

4.3 Sharp tips with high aspect ratio

(111) planes are often employed as self-stop sides in the development of MEMS structures. However, on (111) silicon wafers, if certain designated planes are exposed by dry etching, novel structures dominated by fast planes might be formed in the subsequent wet anisotropic etching. In the etchant of TMAH + Triton X-100, although (110) and its vicinal

planes are strongly affected by the adsorption of surfactant molecules, having lower etch rates than other planes, there is still a local maximum plane (221) located between (110) and (111) planes at 80°C, as shown in Fig. 16. That means (221) planes are capable of being applied to tips as the undercutting sides. The data are obtained from hemispherical samples. As we mentioned previously, there is some difference of etch rates on stressed and flat surfaces, but influence from hemispheres is limited and the relative value among the planes is sufficiently small to be ignored.

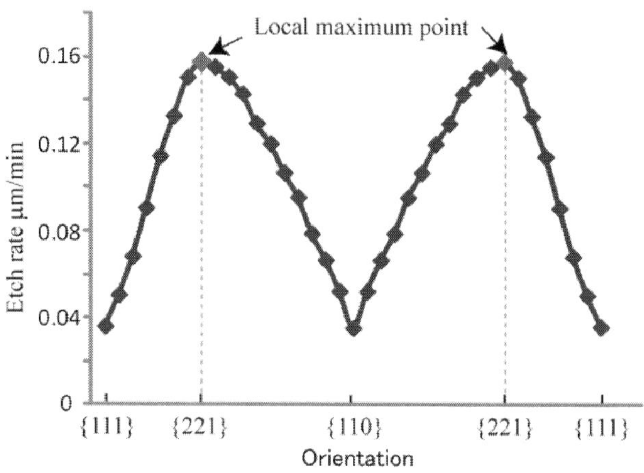

Fig. 16. Etch rate distribution between (111) and (111) in 25 wt% TMAH + 0.1 vol% Triton X-100 at 80 °C.

Here, we use a mask of equilateral triangle on (111) silicon with each side length of 60 μ min the formation of silicon tips. Wafers are firstly deep dry etched of 50 μ m and then dipped into TMAH + Triton X-100 solutions. Fig. 17 shows a circularly graphic net of (111)-centered crystal orientations analyzing the wet etched shapes. Once dry etched, six fast etching (221) planes become dominant until they meet together. The angle from periphery to (221) is about 10°, leading to a much more oblique slope than conventional ones surrounded by (311) or (411) on (100) silicon wafers etched in pure KOH or TMAH. Therefore, it is allowed to fabricate silicon tips with very high aspect ratio in surfactant-modified etchants. The picture of etched sample, not over-etched, permits the examination of etched planes. The top view of one etched tip after 18 min in the TMAH + Triton X-100 at 80 °C, as shown in Fig. 18, exhibits that the etched corner is mainly composed of two planes which have fast and same etch rates. The included angle between two adjacent lines is measured as 150±2°. Note that it is on (111) silicon and the vertex angles are directed at <112>. Making the crystalline projection as a reference, this angle indicates those orientations are located at the red line. Moreover, as mentioned above, planes with the local maximum etch rates would control the control the final. Here, in the solution of TMAH + Triton, this point lies in between (111) and (110), i.e., in blue line. The junction near (221) means those planes constituting the structure which also are in the same crystal class. This result is in good persistence with previous design, indicating the truth of a new tip with very high aspect ratio.

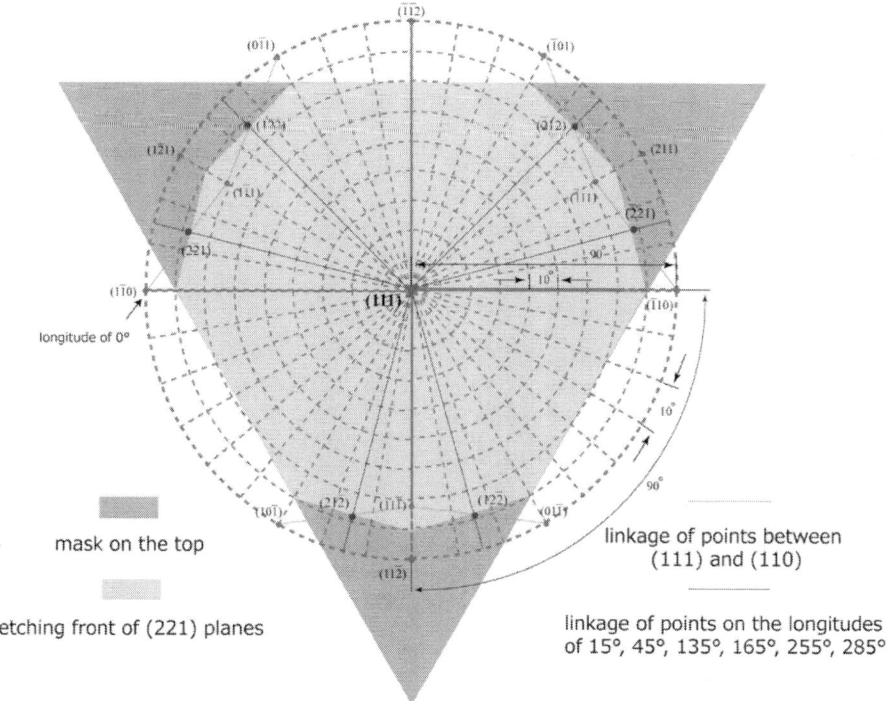

Fig. 17. A circularly graphic net of (111)-centered crystal orientations, illustrating the etched plane of (221).

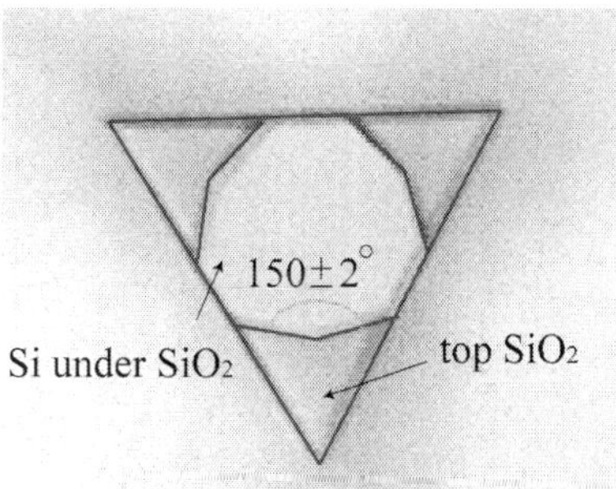

Fig. 18. One photo from optical-microscope reflecting the intermediate status during wet etching with a triangular mask on Si (111) in 25 wt% TMAH + 0.1 vol% Triton at 80 °C.

Fig. 19 shows an SEM image of the etched tip after 33 min, on which there is still a mask cap left. Furthermore, the completed tip is exhibited in Fig. 20. It is clear that the tip has a height of 30 μm, leading to the high aspect ratio of 6:1, and the front angle is measured as about

18°, which conforms with the included angle of two (221) planes. Moreover, the curvature radius of less than 10 nm is achieved without another oxidation-sharpening treatment.

Fig. 19. An SEM image of the etched tip with a cap.

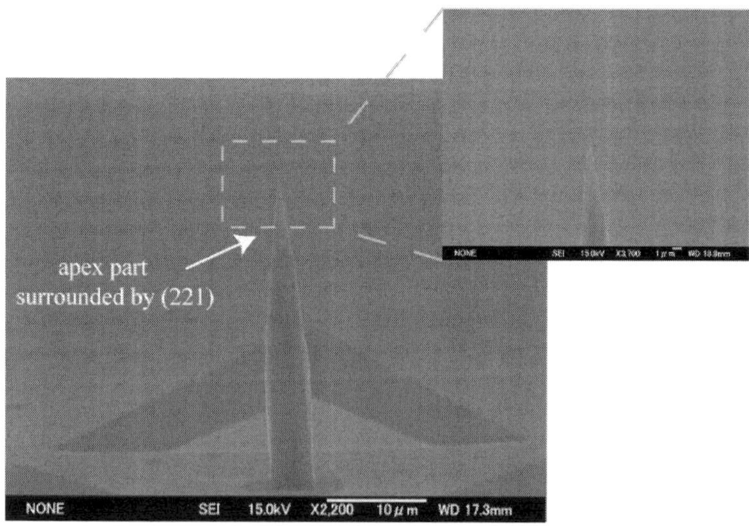

Fig. 20. A completely etched sharp tip of high aspect ratio.

5. Conclusion

In this chapter, firstly the etch rate anisotropy in surfactant-modified etch solution is investigated, showing intriguing properties that are different from that of pure alkaline solutions. The etch rates of exact and vicinal {100} planes are almost unaffected when the surfactant is added, while the etch rates of exact and vicinal {110} planes are reduced significantly. The improved anisotropy ({mnl}/{100}) at high temperature provides better conformity to the mask profile for the formation of a micro cavity. The activation energy of TMAH + Triton (0.1-0.2 eV) is lower than that of pure TMAH (or KOH) solution (0.5-0.7 eV), showing to some extent diffusion-controlled etching process.

Secondly, the underlying effect of the surfactant in etching is understood microscopically and proved macroscopically that enables manufacturing of advanced and exciting structures for MEMS. Thicknesses of surfactant layers are investigated depending on the variation of orientation, temperature et al. The pre-adsorbed surfactant layers are formed and their effects on etch rate, surface roughness and corner undercutting indicate that the dissolved surfactant is adsorbed on the surface during etching.

Finally three applications by using surfactant-modified etching process are exhibited, involving the fabrication of conformal structures, scalloping removing for vertical micro-lens, and sharp tips with high aspect ratio. Much effort will be dedicated on other potential aspects in MEMS, and more advanced devices made by this etching technique could be anticipated in future.

6. Acknowledgment

We acknowledge support by MEXT (Micro/Nano Mechatronics G-COE and grant-in-aid for scientific research (A) 19201026 and 70008053) and the Chinese High-level University Program.

7. References

Bütefisch, S.; Schoft, A. & Büttgenbach, S. Three-Axes monolithic silicon low-g accelerometer. *Journal of Microelectromechanical Systems,* Vol.9, No.4, (December 2000), pp. 551-556, ISSN 1057-7157

Gennissen, P, T, J. & French, P, J. Sacrificial oxide etching compatible with aluminium metallization, *Proceedings of IEEE 1997 9th International Conference on Solid-State Sensors and Actuators,* pp. 225–228, ISBN 0-7803-3829-4, Chicago, Illinois, USA, June 16-19, 1997

Gosalvez, M, A.; Pal, P.; Tang, B. & Sato, K. Atomistic mechanism for the macroscopic effects induced by small additions of surfactants to alkaline etching solutions. *Sensors and Actuators A*, Vol.157, No.1, (January 2010), pp. 91-95, ISSN 0924-4247

Gosalvez, M, A.; Tang, B.; Pal, P.; Sato, K.; Kimura Y. & Ishibashi, K. Orientation and concentration dependent surfactant adsorption on silicon in aqueous alkaline solutions: explaining the changes in the etch rate, roughness and undercutting for MEMS applications. *Journal of Micromechanics and Microengineering,* Vol.19, No.12, (December 2009), pp. 125011, ISSN 1361-6439

Hoffmann, M. & Voges, E. Bulk silicon micromachining for MEMS in optical communication systems. *Journal of Micromechanics and Microengineering,* Vol.12, No.4, (July 2002), pp. 349-360, ISSN 1361-6439

Inagaki, N. ; Sasaki, H. ; Shikida, M. & Sato, K. Selective removal of micro-corrugation by anisotropic wet etching, *Proceedings of IEEE 2009 15th International Conference on Solid State Sensors, Actuators and Microsystems,* pp. 751-754, ISBN 978-1-4244-4193-8, Denver, Colorado, USA, June 21-25, 2009

Lee, K, N.; Lee, D, S.; Jung, S, W.; Jang, Y, H.; Kim Y, K. & Seong, W, K. A high-temperature MEMS heater using suspended silicon structures. *Journal of Micromechanics and Microengineering,* Vol.19, No.11, (November 2009), pp. 115011, ISSN 1361-6439

Ohara, J. ; Kano, K. & Takeuchi, Y. Development of fabrication process for integrated micro-optical elements on Si substrate. *Sensors and Actuators A,* Vol.143, No.1, (May 2008), pp. 77-83, ISSN 0924-4247

Pal, P. & Sato, K. Various shapes of silicon freestanding microfluidic channels and microstructures in one step lithography. *Journal of Micromechanics and Microengineering,* Vol.19, No.5, (May 2009), pp. 055003, ISSN 1361-6439

Pal, P.; Sato, K.; Gosalvez, M, A.; Kimura, Y.; Ishibashi, K.; Niwano, M.; Hida, H.; Tang, B. & Itoh, S. Surfactant adsorption on single crystal silicon surfaces in TMAH solution: Orientation-dependent adsorption detected by in-situ infra-red spectroscopy, *Journal of Microelectromechanical Systems,* Vol.18, No.6, (December 2009), pp. 1345-1356, ISSN 1057-7157

Pal, P.; Sato, K.; Gosalvez, M, A.; Tang, B. & Hida, H. Advanced MEMS applications using orientation dependent adsorption of surfactant molecules in TMAH solution, *Proceedings of 2010 5th Asia-Pacific Conference on Transducers and Micro-Nano Technology,* Perth, Australia, July 6-9, 2010

Park, J.; Park, K.; Choi, B.; Koo, K.; Paik, S.; Park, S.; Kim, J. & Dan Cho, D. A novel fabrication process for ultra-sharp, high-aspect ratio nano tips using (111) single crystalline silicon, *Proceedings of IEEE 2003 12th International Conference on Solid-State Sensors, Actuators and Microsystems,* pp. 1144-1145, ISBN 0-7803-7731-1, Boston, Massachusetts, USA, June 8-12, 2003

Petersen, K, E. Silicon as a mechanical material. *Proceedings of IEEE,* Vol.70, No.5, (May 1982), pp. 420–457, ISSN 0018-9219

Resnik, D.; Vrtacnik, D.; Aljancic, U.; Mozek, M. & Amon, S. The role of Triton surfactant in anisotropic etching of {110} reflective planes on (1 0 0) silicon. *Journal of Micromechanics and Microengineering,* Vol.15, No.6, (June 2005), pp. 1174-1183, ISSN 1361-6439

Sarro, P, M.; Brida, D.; Vlist, W, V, D. & Brida, S. Effect of surfactant on surface quality of silicon microstructures etched in saturated TMAHW solutions. *Sensors and Actuators A,* Vol.85, No.1-3, (August 2000), pp. 340-345, ISSN 0924-4247

Sato, K.; Shikida, M.; Yamashiro, T.; Asaumi, K.; Iriye, Y. & Yamamoto, M. Anisotropic etching rates of single-crystal silicon for TMAH water solution as a function of crystallographic orientation. *Sensors and Actuators A,* Vol.73, No.1-2, (October 1998), pp. 131-137, ISSN 0924-4247

Sato, K.; Uchikawa, D. & Shikida, M. Change in orientation–dependent etching properties of single–crystal silicon caused by a surfactant added to TMAH solution. *Sensors and Materials*, Vol.13, No.5, (May 2001) pp. 285–291, ISSN 0914- 4935

Seidel, H.; Csepregi, L.; Heuberger, A. & Baumgartel, H. Anisotropic etching of crystalline silicon in alkaline solutions: I. Orientation dependence and behavior of passivation layers. *Journal of The Electrochemical Society*, Vol.137, No.11, (November 1990), pp. 3612–3626, ISSN 1945-7111

Shikida, M. ; Sato, K. ; Tokoro, K. & Uchikawa, D. Difference in anisotropic etching properties of KOH and TMAH solutions. *Sensors and Actuators A*, Vol.80, No.2, (March 2000), pp. 179–188, ISSN 0924-4247

Shikida, M.; Hasada, T. & Sato, K. Fabrication of a hollow needle structure by dicing, wet etching and metal deposition. *Journal of Micromechanics and Microengineering*, Vol.16, No.10, (October 2006), pp. 2230-2239, ISSN 1361-6439

Tabata, O.; Asahi, R.; Funabashi, H.; Shimaoka, K. & Sugiyama, S. Anisotroplc etching of silicon in TMAH solutions. *Sensors and Actuators A*, Vol.34, No.1, (July 1992), pp. 51-57, ISSN 0924-4247

Tanaka, H.; Abe, Y.; Inoue, K.; Shikida, M. & Sato, K. Effects of ppb–level metal impurities in aqueous potassium hydroxide solution on the etching of Si{110} and {100}. *Sensors and Materials*, Vol.15, No.1, (January 2003) pp. 43–51, ISSN 0914- 4935

Tang, B.; Amakawa, H.; Shikida, M.; Hida, H; Inagaki, N.; Pal, P. & Sato, K. Characterization of etching anisotropy in a surfactant-added TMAH solution, and its application to scalloping reduction, *Proceedings of 2010 5th Asia-Pacific Conference on Transducers and Micro-Nano Technology*, Perth, Australia, July 6-9, 2010

Tang, B.; Gosalvez, M, A.; Pal, P.; Itoh, S.; Hida, H.; Shikida, M. & Sato, K. Adsorbed surfactant thickness on a Si wafer dominating etching properties of TMAH solution, *Proceedings of IEEE 2009 20th International Symposium on Micro-Nano Mechatronics and Human Science*, pp. 48-52, ISBN 978-1-4244-5094-7, Nagoya, Japan, Nov 9-11, 2009

Tang, B.; Pal, P.; Gosalvez, M, A.; Shikida, M.; Sato, K.; Amakawa, H. & Itoh, S. Ellipsometry study of the adsorbed surfactant thickness on Si{110} and Si{100} and the effect of pre-adsorbed surfactant layer on etching characteristics in TMAH. *Sensors and Actuators A*, Vol.156, No.2, (December 2009), pp. 334-341, ISSN 0924-4247

Tang, B.; Sato, K.; Tanaka H. & Gosalvez, M, A. Fabrication of sharp tips with high aspect ratio by surfactant-modified wet etching for the AFM probe, *Proceedings of IEEE 2011 24th International Conference on Micro Electro Mechanical Systems*, pp. 328-331, ISBN 978-1-4244-9632-7, Cancun, Mexico, January 23-27, 2011

Tang, B.; Shikida, M.; Sato, K.; Pal, P.; Amakawa, H.; Hida, H. & Fukuzawa, K. Study of surfactant-added TMAH for applications in DRIE and wet etching-based micromachining. *Journal of Micromechanics and Microengineering*, Vol.20, No.6, (June 2010), pp. 065008, ISSN 1361-6439

Yang, C, R.; Chen, P, Y.; Yang, C, H.; Chiou, Y, C. & Lee, R, T. Effects of various ion–typed surfactants on silicon anisotropic etching properties in KOH and TMAH solutions. *Sensors and Actuators A*, Vol.119, No.1, (March 2005), pp. 271-281, ISSN 0924-4247

Yang, C, R.; Yang, C, H. & Chen, P, Y. Study on anisotropic silicon etching characteristics in various surfactant-added tetramethyl ammonium hydroxide water solutions. *Journal of Micromechanics and Microengineering,* Vol.15, No.11, (November 2005), pp. 2028-2037, ISSN 1361-6439

Zubel, I. & Kramkowska, M. The effect of isopropyl alcohol on etching rate and roughness of (100) Si surface etched in KOH and TMAH solutions. *Sensors and Actuators A,* Vol.93, No.2, (September 2001), pp. 138-147, ISSN 0924-4247

8

Macromodels of Micro-Electro-Mechanical Systems (MEMS)

Anatoly Petrenko
Systems Design Department, National Technical University of Ukraine
Ukraine

1. Introduction

Micro-electro-mechanical Systems (MEMS) are components with micron-scale moving parts based on materials and processes of microelectronics fabrication. This is a good example of on-chip integration of electronics, microstructures, microsensors and microactuators. Accurate simulation of MEMS requires precise modeling of all effects of mechanical and damping forces, electrostatic forces and inner stresses, heat transfer, thermal expansion, piezoelectric stresses etc.

Modern methodology of MEMS design implies that the entire MEMS can be investigated only at higher abstraction levels such as **schematic and system** ones, where accurate macromodels can be used [1]. On the other hand, at component or device levels the physical behavior of three-dimensional continuums is described by **partial differential equations** (PDE) easily solvable by Finite Element or Finite Difference Element Methods (FEM or FDM) [2,3], available in ANSYS -like software. Component level simulations are classified in single - domain and coupled - domain simulations, both being very computer time- consuming.

The goal of this chapter is to consider methods of automatically obtaining macromodels of MEMS and their mechanical or non-electric components from ANSYS models as equivalent electric circuits or low order differential ordinary equations for further use in circuit design software. This can be done by using different model order reduction techniques developed in recent years.

When dealing with the modern MEMS, the possibility for using a single environment to simulate objects, where different physical processes such as electrical, mechanical, optical, thermal etc. take place, plays an important role. Here we have to represent different subsystems of the initial MEMS as equivalent models of the same physical nature permitting to combine them for solution in a single computational process. After that, the complete behavioral model of the entire MEMS and its subsystems can be compiled either in VHDL-AMS language (as sets of ODE) or in SPICE-like language (as equivalent electric circuits).

The Microsystems design exploits various analytical and numerical methods for virtual prototyping of MEMS. It also demands for libraries of electromechanical, optical models and microfluid components, including springs, bulks, buffers, capacitors, inductances, operational amplifiers, transistors etc. Three basic possible approaches of MEMS design procedure are illustrated below: FEM/FDM Model, Reduced Order Model (ROM), Coupled system-level model.

2. FEM/FDM model

MEMS typically involve multiple energy domains such as kinetic energy, elastic deformation, electrostatic or magneto static stored energy and fluidic interactions. The difficulty in the modeling of MEMS devices is mainly caused by the tight coupling between the multiple energy domains. Individual physical effects are governed by partial differential equations (PDE), typically nonlinear. When these equations become coupled, the computational challenges of highly meshed numerical simulation become formidable.

FEM relies on highly localized interpolation functions (or *mesh element functions*) for approximation of the solution of PDE. These mesh element functions are generated by meshing the domain of interest and parameterize the desired solution locally on each mesh element. This parameterized solution converts a continuous (PDE) problem to a coupled system of ordinary differential equations (ODE) that can be integrated in time. The resulting ODE system usually has many degrees of freedom (perhaps **several variables per mesh element**). If a fine mesh is required, the problem size grows rapidly, with a corresponding rapid growth in computational cost for explicit dynamic simulation. Consequently, it is very expensive to use FEM model in system-level simulations during MEMS iterative design. As a result, FEM models are mostly used to analyze the performance of MEMS components and to couple their multiphysics effects.

By reading ANSYS binary FULL file it is possible to assemble a MEMS component state-space model in the form of first order systems or second order ordinary differential equations (ODE)

$$E_r z' + A_r z = B_r f \,, \qquad Y = C_r z \tag{1}$$

$$Mx'' + Dx' + Kx = Bf \,, \; Y = Q^T x + R^T x' \,, \tag{2}$$

where A_r, E_r, C_r, B_r, M, D, K, B, C- are the system matrices, B_r, B are the input and the C_r, C - output matrices, f is input force. In mechanics matrices M, D and K are known as the *mass*, *damping* and *stiffness* matrices correspondingly. Usually damping is included in the model as Rayleigh damping. The damping matrix D is computed as a linear combination of the stiffness K and the mass M matrices:

$$D = \alpha M + \beta K,$$

where α, β are constant coefficients.

In (1) the state space vector z is defined through the unknowns deflections $u(x,t)$ and pressures $p(x,y,t)$ into the node points being automatically generated in MEMS structure:

$$z = [u_1 \ldots u_N \frac{\partial u_1}{\partial t} \ldots \frac{\partial u_N}{\partial t} p_{11} \ldots p_{MN}]^T \tag{3}$$

By defining

$$E_r = \begin{bmatrix} D & M \\ M & 0 \end{bmatrix} \quad A_r = \begin{bmatrix} K & 0 \\ 0 & -M \end{bmatrix} \quad B_r = \begin{vmatrix} B \\ 0 \end{vmatrix} \quad C_r = \begin{vmatrix} Q \\ R \end{vmatrix} \quad z = \begin{vmatrix} x' \\ x'' \end{vmatrix} \tag{4}$$

second equations (2) can be transfer to the first (1).

The FULL file contains all the information about the system: the system element matrices, Dirichlet boundary conditions, equation constrain and the load vector. It is generated using ANSYS partial solver, which enables to assemble system element matrices for the desired analysis without solving them and it therefore computationally fast. The speed of the reading operation has been optimized taking into account that the element matrices are sparse. The load vector directly gives the matrix-vector product *Bf* and thus describes the distribution of all loads being applied. In order to obtain the *B* matrix, and thus being able to modify the inputs singularly, it is necessary to repeat the partial solution for each input of interest.

3. ROM (Reduced Order Model)

It would be easier and more intuitive for the designer to explore the design space if the MEMS model had only a few variables with a clear relationship between them and the overall device performance. *Reduced-order models (ROM),* also called *macromodels,* lend themselves very well to these purposes. The main idea behind the reduced order model is that the number of ordinary differential equations (ODE) needed to simulate the system has been reduced from perhaps many thousands in the case of the full FEM simulation, to just a few basis function coordinates (fig.1).

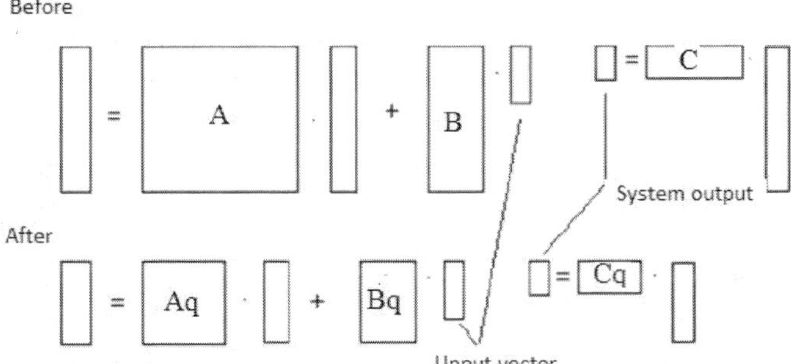

Fig. 1. Reduced order model illustration

Such the macromodel simulation can be very efficient computationally compared to the FEM model. A designer can use the FEM model for different component geometry and materials trying and the ROM model for investigation of different input forces effect (fig.2).

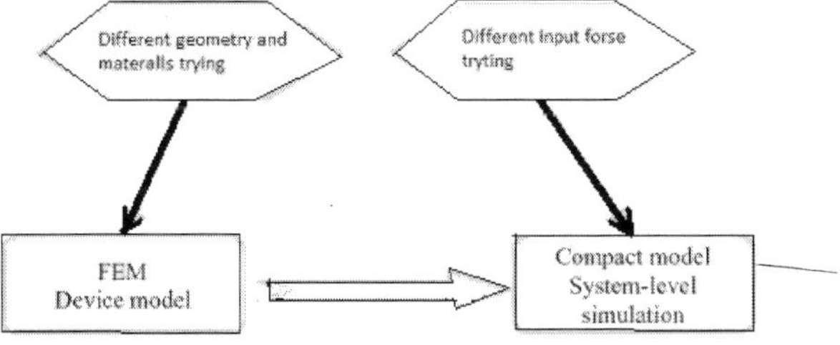

Fig. 2. Compact reduced order model in MEMS design [33]

3.1 Modal decomposition ROM

Modes (or resonances) are inherent properties of a structure. Resonances are determined by the material properties (mass, stiffness, and damping properties) and boundary conditions of the structure. Each mode is defined by a *natural (modal or resonant) frequency, modal damping*, and a *mode shape*.

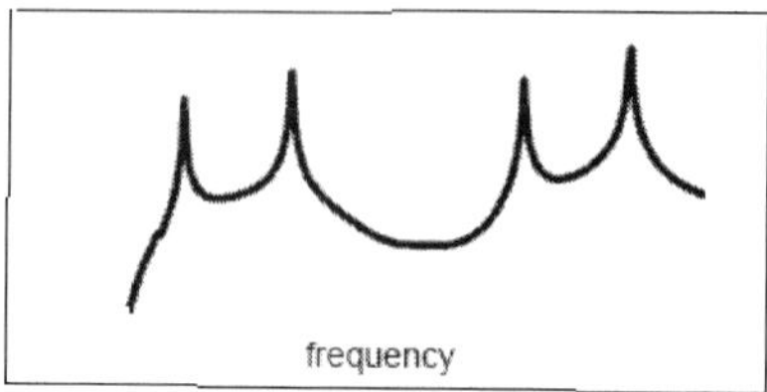

Fig. 3. Simple Plate Frequency Response Function

If either the material properties or the boundary conditions of a structure change, its modes will change as well. The overall response of a structure at any frequency is a *summation of responses due to each of its modes*. It is also evident that close to the frequency of one of the resonance peaks, the response of *one mode will dominate the frequency response* [6], fig.3.

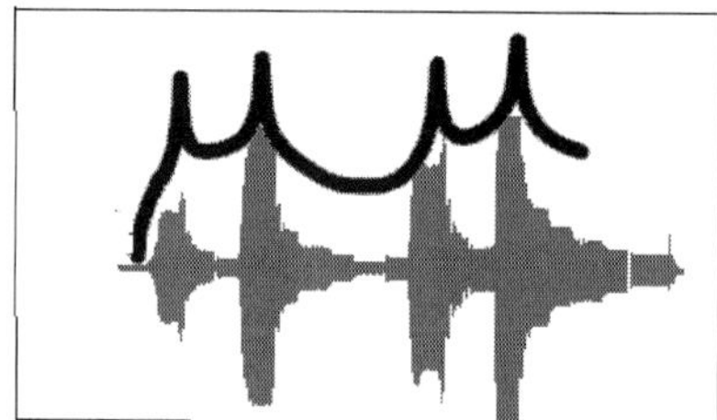

Fig. 4. Overlay of Frequency and Time Response Functions

Now if we overlay the time trace with the frequency trace it is possible to notice that maximum oscillations at the time are corresponding to the frequencies of maximum frequency response function (fig.4).

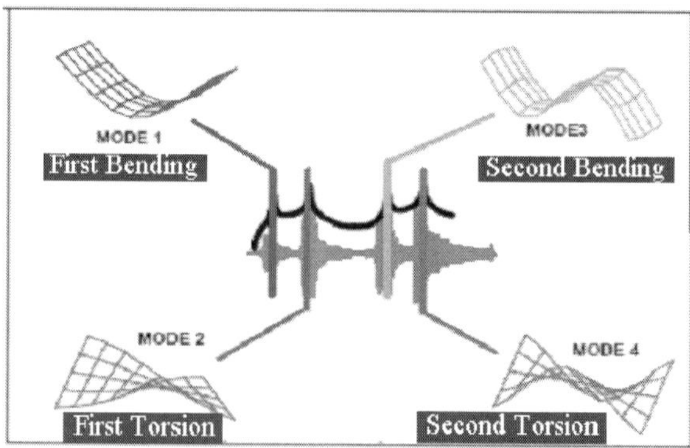

Fig. 5. Flexible Body Modes [6].

Let's see what happens to the deformation pattern on the structure at each one of these natural frequencies. Modes are further characterized as either *rigid body* or *flexible body* modes. All structures can have up to *six rigid body* modes, three translational modes and three rotational modes. Many deformation problems are caused, or at least amplified by the excitation of one or more flexible body modes. Fig.5 shows some of the common fundamental (low frequency) modes of a plate. The fundamental modes are given names like those shown in Fig.5. The higher frequency mode shapes are usually more complex in appearance, and therefore don't have common names.
Modal decomposition method uses a weighted sum of n mode shapes (modal amplitudes or eigenvalues) q_i, basis function (an eigenvector) $\varphi_i\ (x,\ y,\ z))$ of the mechanical structure to represent its deflection u:

$$u(x,y,z,t) = u_{eq} + \sum_{i=1}^{n} q_i(t)\varphi_i(x,y,z) \ , \tag{5}$$

where u_{eq} is the initial displacement produced by the initial load. For MEMS it is usually sufficient that few modes accurately describe dynamical response of the system. This approach is equivalent to the projection of the original PDE, describing the MEMS behavior, on the subspace defined by the basis functions.
By inserting (5) into the equation of motion (2), taking $\mathbf{x} = \varphi\ e^{\lambda t}$ and multiplying by $\varphi^T{}_i$ from the left, and using the orthogonality of eigenvectors, the equation (2) can be reduced to

$$\ddot{q}_i - 2\sigma_i q_i + (\omega_i^2 + \sigma_i^2) = \varphi_i^T f \ , \quad i=1,2,\ldots n \tag{6}$$

The eigenvectors satisfy the orthogonality conditions $\varphi^T{}_k\, M\varphi_i = \varphi^T{}_k\, D\varphi_i = \varphi^T{}_k\, K\varphi_i = 0$ for $\lambda_k _\neq \lambda_i$. With the normalization of the eigenvectors $\varphi^T{}_k\, M\varphi_k \equiv 1$, it can be shown that

$$\varphi^T{}_k\, D\varphi_k = -2\sigma_k \text{ and } \varphi^T{}_k\, K\varphi_k = \omega^2{}_k + \sigma^2{}_k \text{ if } \lambda_k = \sigma_k + j\omega_k.$$

Note that the orthogonality condition does not apply in the case of multiple eigenvalues. In such case, special modal analysis techniques are needed for decoupling of modes [7].
The decoupling of modes yields that transfer functions can be written as a sum of *modal transfer functions*. Using Fourier transforming expansion and equations (6), the transfer matrix can be derived as

$$H(\omega) = \sum_{i=1}^{n} H_k(\omega) = \sum_{i=1}^{n} \frac{\varphi_i \varphi_i^T}{(j\omega - \sigma_i - j\omega_i)(j\omega - \sigma_i + j\omega_i)} \tag{7}$$

This relation is the basis of modal analysis. It relates the *measurable* transfer functions to the modal properties ω_i, σ_i, and φ_i. Each i- th mode contributes with a modal transfer matrix H_i to the complete transfer matrix. Fig.6 shows an example of a theoretical transfer function (7) with three *modal peaks* corresponding to modes at 1, 2, and 4 Hz which are all damped with a logarithmic decrement of 1% ($-\sigma_i/f_i$ = 0.01). Also the individual modal transfer functions (7) are plotted from which the complete transfer function is computed.
By the way the eigenvalues q_k and the corresponding eigenvectors φ_k for k = 1, 2, . . ., n can be found as an solution of the modified eigenvalue problem

$$(M\lambda^2 + D\lambda + K)\varphi = 0\ . \tag{8}$$

The relationships between natural frequencies f_k, logarithmic decrements δ_k and the eigenvalues are $f_k = 2\pi\omega_k$ and $\delta_k = -\sigma_k/f_k$.

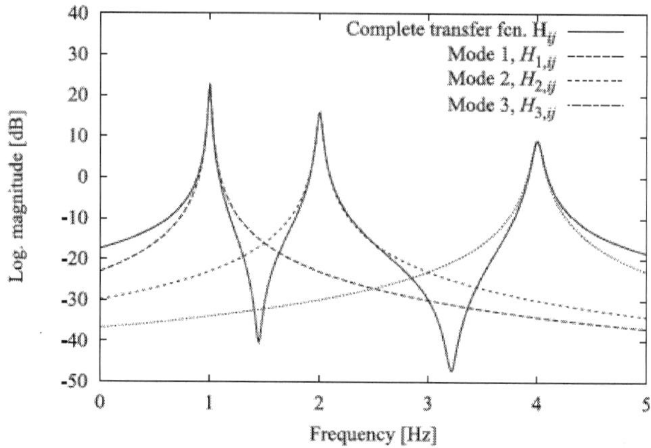

Fig. 6. Example of the modal decoupling in a transfer function [7].

3.1.1 Modal ROM based on proper orthogonal decomposition methods

Reduced order modeling using modal basis functions was originally developed by [11-13] and has been continuously improved by several authors. The basis functions $\varphi_i(x,y,z)$ in (5) may be chosen in two ways by:

- Using the undamped linear mode shapes of the undeflected microstructure as basis functions. For simple structures with simple boundary conditions, the mode shapes can be found analytically. For complex structures or complex boundary conditions, the linear mode shapes are obtained numerically using the finite element method. However, it is usually difficult to determine, *a priori*, an optimum set of eigenvectors φ_k, particularly when irregular geometries are involved.
- Using snapshots obtained from experiments under a training signal or full FEM model runs. Then the proper orthogonal decomposition method (Singular Value Decomposition –*SVD* , Karhunen -Loève decomposition -*KL* and neural networks-based generalized Hebbian algorithm -*GHA*) is applied to the time series for extracting the mode shapes of the device structural elements.

The choice of orthogonal basis functions φ_k can be done by the following way [8]. First the MEMS dynamics are simulated using a slow but accurate technique such as FEM or FDM. Sets of runs may be used to suitably characterize the operating range of the device. The spatial distributions of each state variable $u(x, y, z, t)$ are then sampled at a series of N_s different times during these simulations, and the sampled distributions are stored as a series of vectors, $\{u_i\}$, where each of them corresponds to a particular "snapshot" in time. Now suppose we would like to pick orthogonal basic n functions $\{\varphi_1, \varphi_2,..., \varphi_n\}$ in order to represent the observed state distributions as closely as possible. One way to do this is to attempt to minimize a least squares measure of the "error" distances between the observed states and the basis function representation of those states:

$$\sum_{i=1}^{n}[u_i - proj(u_i, span\{\varphi_1,...,\varphi_n\})]^2 = \sum_{i=1}^{n}\sum_{j=1}^{n}\varphi_i\varphi_j^T u_i , \quad (9)$$

where $proj(v,S)$ is the projection of the vector v into subspace S. In other words, we minimize a least squares measure of the "error" distances between the observed states and the basis function representation of those states.

Singular value decomposition (SVD) takes a rectangular matrix of modal experimental data (defined as U, where U is a N x n matrix) in which N is a number spatial points of each snapshot, n is a number of snapshots. The SVD theorem states:

$$U_{Nxn} = W_{NxN} S_{Nxn} V^T_{nxn} ,$$

where the columns of W are the left singular vectors (*spatial points vectors*); S (the same dimensions as U) has singular values and is diagonal (*mode amplitudes*); and V^T has rows that are the right singular vectors (*snapshots numbers vectors*).

The SVD represents an expansion of the original data in a coordinate system where the covariance matrix is diagonal. Calculating the SVD consists of finding the eigenvalues and eigenvectors of matrices UU^T [NxN] and U^TU [n x n]. The eigenvectors of U^TU make up the columns of V, the eigenvectors of UU^T make up the columns of W. Also, the singular values in S are square roots of eigenvalues from UU^T or U^TU. The singular values are the diagonal entries of the S matrix and are arranged in descending order: $\sigma_1 \geq \sigma_2 \geq ... \geq \sigma_n \geq 0$ .The singular values are always real numbers. If the matrix U is a real matrix, then W and V are also real.

It was shown [11] that the proper orthogonal basis functions $\{\varphi_1, \varphi_2,..., \varphi_n\}$ minimizing (9) can be chosen by setting

$$\phi_i = v_i , \; i = \{1,2,...,n\}$$

where v_i is the columns of V.

After finding basis functions φ_k the linear equations for modal amplitudes q_k estimation are formed for each temporal snapshot on the base of selected before "snapshot" spatial points and equation (5).

Karhunen-Loève decomposition (KL) can be viewed as a statistical procedure [9] .One initially supposes that the observed system dynamics can be modeled as a second-order ergodic stochastic process. The method consists then in constructing a spatial autocorrelation tensor from data obtained through numerical or physical experiments and performing its spectral decomposition.

KL differs from SVD in way of finding basis functions φ_i through the temporal snapshot correlation matrix entries $a_{ij} = \frac{1}{n}\sum_{\substack{k=1 \\ m=1}}^{n}[u_i(x,t_m)u_j(x,t_k)]$

as

$$\varphi_i(x) = \sum_{k=1}^{n} b_{k,i} u_k(x) ,$$

where n is number of temporal snapshots and $b_{k,i}$ are eigenvectors of the matrix A, or the solutions of the equation $Ab=\lambda b$.

Generalized Hebbian algorithm (GHA) is an Artificial Neural Network (ANN) approach of performing Principal Component Analysis (PCA) on a set of data and can be also used as a learning procedure for the approximation of *PDE* solutions by the expression (5) [10] .
The modal approach to MEMS macromodeling is illustrated by sensor device (fig.7), which is described by coupling a 1-D elastic beam equation with electrostatic force and 2-D compressible isothermal squeeze-film Reynold's equation [11]. This device consists of a deformable elastic beam microstructure that is electrostatically pulled in by an applied voltage waveform. The dynamics of the beam are first simulated using a finite-difference analysis. A quarter of the beam is initially meshed using a 20 x10 node 2-D grid.

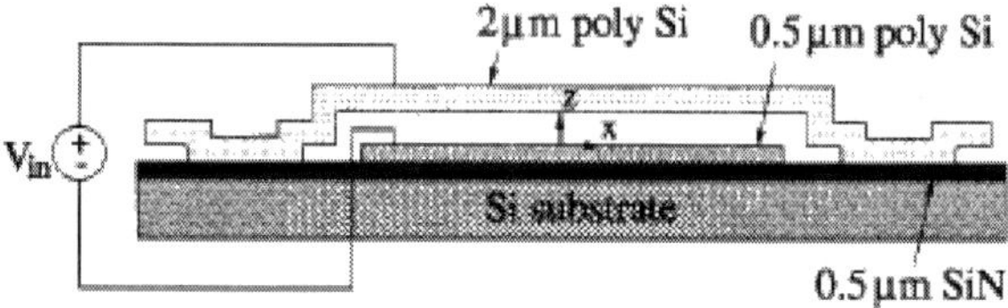

Fig. 7. Fixed-fixed beam pull-in time pressure sensor device [11]

The state at each node consists of three quantities: *z, dz/dt* and *p*. Since *z* and *dz/dt* are simulated in 1-D, this results in coupled nonlinear ODE which must be integrated in time. Basis functions are generated for pressure *p* and displacement *z* based on runs of the finite-difference code for an ensemble of four different step voltages: 9V, 10V, 12V, 16V. One hundred samples of pressure and displacement are taken during these four runs at fixed time intervals. These samples are used to generate the basic functions. The resulting basis functions for displacement and pressure are shown in Fig.8 [11].

3.1.2 Nonlinear modal ROM

The device governing equations are generally derived from Lagrange equations, after expressing the internal (elastic and kinetic) and external (electrostatic) energy of the system in terms of modal amplitudes and symbolical calculation of the gradients. Assuming that the device undergoes small displacement, the basis chosen results in diagonal mass and stiffness matrices, which can be pre-computed. Linear elastic undamped normal modes of the undeflected device have been often chosen as basic functions to approximate the solution of an electromechanical problem discretized using finite element methods. Modal representation is very efficient since it *requires only one equation per mode* and involved conductor to describe the entire system.
The nonlinear energy terms, instead, are generally expressed as analytical functions of the modal coordinate. In [13] a single static full finite element simulation is used to determine the number of modal functions needed to capture the device behavior and their expected amplitude. Then this information is used to construct the electrostatic energy term. A 3D full model electrostatic simulation is run for values of the modal amplitude that span the operating range of the device, in order to compute the capacitance/deformation curve by using ANSYS' transducer element TRANS126. The results are fitted with a rational fraction of multivariate polynomials using a nonlinear function fitting scheme. In order to model large-displacement behavior of the device, the strain energy is derived by fitting data from a

set of full finite element simulations [13]. A similar procedure is proposed in [17]. In this case, the force-displacement function and the modal strain energy are still derived from a series of FEM simulations, but a polynomial multi-variable fit is used. In order to reduce the complexity of the fitting step, dominant and relevant modes are first characterized. Both the procedures in [13] and [17] can be partially automated and extended to include other conservative energy domains. Other algorithms have also been proposed for the approximation of dissipative energy terms [18, 19]. Dedicated methods have been demonstrated for actuated microbeams, still using the linear undamped mode shapes of the device as basic functions in the Galerkin procedure [20], where two expressions of the nonlinear electrostatic term were proposed as a function of modal coordinates, each including all nonlinearities up to the fifth order, obtained via mathematical manipulation [21]. A new fuzzy-logic model (FLM) for MEMS is presented in [22] in which for reducing the number of data needed for macromodel identification, cluster estimation of a model structure and back propagation method of structure parameters adaptation are chosen to fit the data. As a result the dynamic coupled simulation of a magnetic microactuator takes only several minutes and the force macromodel yielded errors is less than 1.5% for a 5-μm displacement. These FLMs combine fuzzy sets with fuzzy rules that have the capability to model the complex nonlinear behavior.

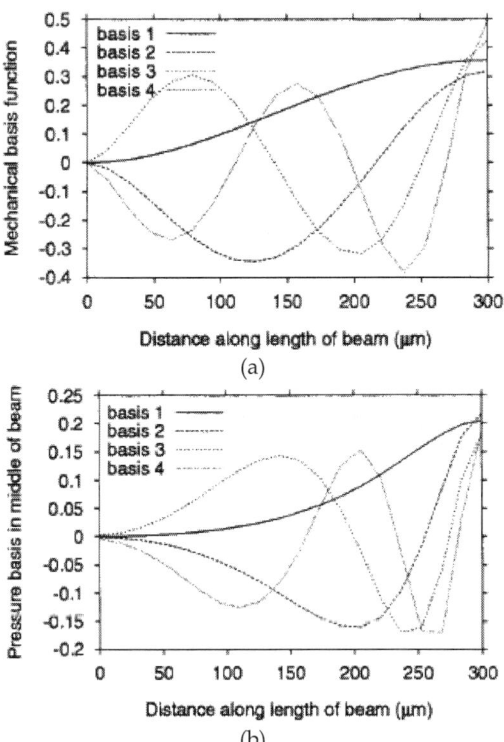

Fig. 8. Basis functions for (a) displacement $z(x, t)$ and (b) pressure $p(x, y, t)$ [11]

Macromodels obtained via modal basis functions methods have been demonstrated to reproduce results obtained with full physical level simulation with an accuracy of some percentage points and a reduction of the computational complexity simulation. It was shown that even when the problem is mechanically nonlinear, the linear normal modes can serve as basic functions. The approach can be outlined as follows [13]:

1. Compute the linear modes φ_i of the elastic problem.
2. Substitute $u\ (x\ ,y,\ z\ ,t)$ in the governing equation for the deflection (e.g., the Euler-Bernoulli or Timoshenko equations for beams).
3. Obtain a system of n-coupled second-order ordinary differential equations for the $q_i(t)$.
4. Solve the equations to compute the dynamic response either numerically or analytically.

3.1.3 VHDL-AMS export of modal ROM

In general, the equation (5) describes a coordinate transformation of finite element displacement coordinates (mesh element coordinate) to modal coordinates of the macromodel (basic functions or a degree of freedom-d.o.f.):

$$u(x,y,z,t)=u_{eq}+\sum_{i=1}^{n}q_i(t)\varphi_i(x,y,z)\ .$$

The deformation state of the structure given by N nodal displacements u_i $(i=1,2,\ldots,N\ )$ is now represented by a linear combination of n modes weighted by their amplitudes q_j $(j=1,2,\ldots,n\ )$ where $n << N$.

The governing equation of motion describing the ROM of electrostatic actuated MEMS structures in modal coordinates:

$$m_j\ddot{q}_j+2\xi_j\omega_j m_j\dot{q}_j+\frac{\partial}{\partial q_j}W_{st}(q_1,\ldots,q_n)=\frac{1}{2}\sum_r\frac{\partial}{\partial q_j}C_{ks}(q_1,\ldots,q_n)\cdot(V_k-V_s)^2+\sum_{i=1}^{n}\varphi_i f_i\ , \tag{10}$$

where m_j is the modal mass, ω_j is the eigenfrequency, ξ_j the linear modal damping ratio, W_{st} is the modal strain energy function, C_{ks} is the modal capacity-stroke function, r is the number of capacities involved for microsystems with multiple electrodes, V is the electrode voltage applied and f_i is a local force acting at the i-th node. The current I_k at each electrode k is defined by:

$$I_k=\frac{\partial Q_k}{\partial t}=\sum_r(C_{ks}\cdot(\frac{\partial V_k}{\partial t}-\frac{\partial V_s}{\partial t})+\frac{\partial C_{ks}}{\partial t}\cdot(V_k-V_s)) \tag{11}$$

An essential prerequisite to establish (10) and (11) are proper modal strain energy and capacity-stroke functions. Both are derived from a series of FEM runs at various deflection states in the operating range. The received data are used for polynomial functions fitting in order to compute the local derivatives, which describe force and stiffness terms. As a matter of fact, shape function methods can be applied to nonlinear systems, too [13]. Geometric nonlinearities, as, for instance, stress stiffening, can be regarded if the modal stiffness is computed from the first derivative of the strain energy function with respect to the modal amplitudes. Capacitance-stroke functions provide non-linear coupling between each eigenmode and the electrical quantities (i.e. electrostatic modal forces, electrical current) if stroke is understood as modal amplitude. Damping parameters are assigned to each eigenmode.

The first step of the ROM generation is to determine which modes are really significant, and to estimate a proper amplitude range for each mode. Several criteria can be applied, for instance, the lowest eigenmodes of a modal analysis, modes in operating direction, or modes, which contribute to the deflection state at a typical test load. Next the dependencies of the strain energy W_{st} and capacities are described by polynomial functions being fitted. The necessary data points are obtained by imposing each eigenmode with varying amplitude on the mechanical modell for the non-linear strain energy and on an electrostatic space model for capacitance. This process is computationally expensive but has to be done just once. The result is a black-box model that can be applied to any load situation. In the concept of the modal superposition method, each eigenmode represents a single independent resonator with modal mass m_i and modal damping ξ_i.

The export of the ROM to VHDL-AMS is performed in two steps [16]. At first, an initialization file containing all necessary information of the macromodel, such as the fitted polynomial coefficients and orders, is generated. Then, the source code in VHDL-AMS is automatically generated. The main problem of exporting the ROM in VHDL-AMS is to express the fitted functions of the non-linear strain energy and of the capacities which are part of coefficients of the differential algebraic equations (DAE), which can be mapped to the simultaneous statements of VHDL-AMS. If simulators support description in matrix notation properly (as MATLAB), the exported VHDL-Models will become more compact and clear.

The Modal ROM approach was implemented as the available ROM-Tool in ANSYS/Multiphysics since Release 7 **(ROM144).** It contains some terminals (fig. 9) and provides necessary functions:

- the master node terminals which describe the displacement u_i and the inserted forces $F_{N,i}$ at these nodes;
- the modal terminals with the modal amplitude q_i and modal force $F_{M,i}$ for the chosen modes;
- the electrical terminals which provide the voltages V_i and currents I_i for the electrodes of the system.

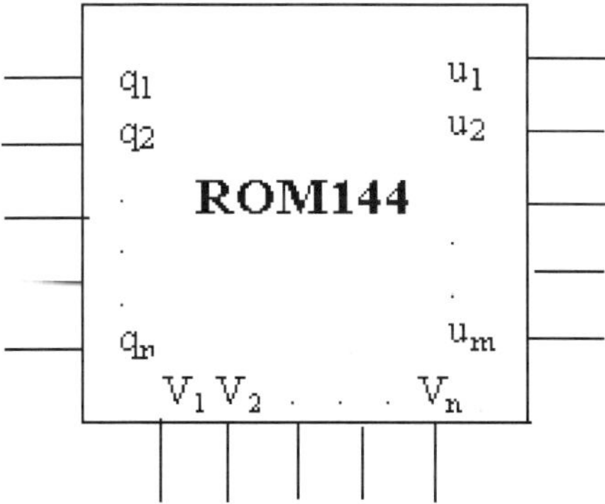

Fig. 9. ROM144 functional block

The ROM144 takes input equations (5),(10),(11) with characteristic constants (modal masses, modal damping ratios and eigenvectors of the master nodes) and provides the special functions of calculating the strain energy W_{mech} (q_i, q_j, q_k) and capacitances $C_{op}(q_i, q_j, q_k)$as well as their first derivatives with respect to the modal amplitudes q_i, q_j and q_k using the information of the polynomials degrees defined in other packages.

Modal ROM144 approach speeds up computations in 40 times in comparison with the FEM model while pull-in time errors is less than 2%.

Modal ROM approach was implemented also in *INTEGRATOR system of CoventorWare* (http:// www.coventor.com), *MEMS Pro* (http://www.memspo.com) and *MEMSCAP* (http:// www.memsscap.com).

3.1.4 Galerkin's approximated ROM

As in (5) the desired PDE solution $u(x,y,z,t)$ can be approximated by spatially varying arbitral basis functions $\beta_i(x,y,z)$ with time varying coefficients $\alpha_i(t)$:

$$u(x,y,z,t) = \sum_{i=1}^{n} \alpha_i(t)\beta_i(x,y,z) \tag{12}$$

For the Galerkin's method the PDE residual $(L(u)-f)$ is orthogonal to each a_i of the basic functions in the operating range H:

$$(\alpha_i, L(u)-f) = \int \alpha_i^T (L(u)-f)dt = 0, \quad i = 1,n. \tag{13}$$

where L is a differential operator (possibly nonlinear), and f is an input vector.

The basic functions $\beta_i(x,y,z)$ can be chosen arbitrarily, as long as their elements satisfy all of the boundary conditions and are sufficiently differentiable. It means that they can be not only eigenmodes as it was shown before, but they may be Tchebychev, Legendre, Hermit polynomials or even wavelets functions, which were introduced in the past two decades and are gaining increasing popularity. Indeed wavelets have many excellent properties: such as orthogonality, compact support, exact representation of polynomials to a certain degree, and flexibility to represent functions at different resolution levels. The wavelet-Galerkin method is a Galerkin scheme using wavelet functions as the basic functions. However, wavelet functions do not satisfy the boundary conditions. Thus the treatment of general boundary conditions is a major difficulty for the application of the wavelet- Galerkin method. The wavelet interpolation Galerkin method is a Galerkin scheme where basic functions are a class of interpolating functions generated by autocorrelation of the usual compactly supported Daubechies scaling functions [23]. Daubechies' functions are easy to construct. For an even integer L, we have the Daubechies' scaling function $\varphi(x)$ and wavelet $\psi(x)$ satisfying

$$\begin{aligned} \varphi(x) &= \sum_{i=1}^{L-1} p_i \varphi(2x-i) \\ \psi(x) &= \sum_{i=2-L}^{1} (-1)^i p_{1-i} \varphi(2x-i) \end{aligned} \tag{14}$$

The scaling function $\varphi(x)$ is supported in the interval $[0, L - 1]$ while the corresponding wavelet $\psi(x)$ is supported in the interval $[1 - L/2, L/2]$. The parameter L will be referred to as the degree of the scaling function $\varphi(x)$. The coefficients p_i are called the wavelet filter coefficients and they satisfy the same conditions. The constructed scaling function $\varphi(x)$ and wavelet $\psi(x)$ have also the prescribed properties.
The autocorrelation functions $\theta(x)$, which are used for generating basic functions, can be defined as follows:

$$\theta(x) = \int_{-\infty}^{\infty} \varphi(\tau)\varphi(\tau - x)d\tau \tag{15}$$

and act as the scaling function

$$\theta_{J,k}(x) = \theta(2^J x - k)\,. \tag{16}$$

Wavelets have proven to be an efficient tool of analysis in many fields including the solution of PDE. In [23] a new wavelet interpolation Galerkin method is used for the numerical simulation of MEMS devices under the effect of squeeze film damping. The air film pressure is expressed as a linear combination of a class of basic functions generated by autocorrelation of the usual compactly supported Daubechies scaling functions, which are the first- generation wavelets. The wavelet interpolation Galerkin method was used to predict the frequency response of the accelerometer with a spring mass-damper model with a parallel-plate electrostatic force. Various numerical tests have been conducted by changing the degree of the Daubechies wavelet L and the number J of the scale. Better accuracy can be achieved by increasing L and J. The higher L is, the smoother the scaling function becomes. The price for the high smoothness is that its supporting domain gets larger. The higher J is, the more accurate the solution becomes. The number of differential equations and the CPU time increase significantly as J increases. The solutions for $L = 6$ and $J = 4$ have results higher than the finite difference method.
The present wavelet interpolation Galerkin method is not suitable to solve problems defined on nonrectangular domains, since higher-dimensional wavelets are constructed by employing the tensor product of the one-dimensional wavelets, so their application is restricted to rectangular domains. But usage of the second-generation wavelets which are constructed in the spatial domain can expend in future PDE solutions to complex domains.

3.2 Moment matching based ROM

Moment matching model order reduction is based on the approximation of the original n-dimensional system transfer function $F(s)$ of the original n-dimensional system with a rational function with a lower degree $q<<n$ [24]. This is done by matching some terms of the Taylor expansion of $F(s)$ around a certain expansion point.
For a state space equation

$$\begin{aligned} \dot{X}(t) &= AX(t) + Bv(t) \\ Y(t) &= C^T X(t) \end{aligned} \tag{17}$$

let's perform a Laplace transform to obtain its frequency domain transfer function

$$F(s) = C^T (sI - A)^{-1} B \tag{18}$$

and expand it into Taylor series as

$$F(s) = -\sum_{k=0}^{\infty} m_k s^k \tag{19}$$

where $m_k = C^T A^{-k}(A^{-1}B)$.

Moment matching is directly connected to the Krylov subspace formed by the pair of matrices (A^{-1}, $A^{-1}B$) [24]. The Krylov subspace is spanned by the column vectors in the following collection of matrices:

$$\{A^{-1}B, (A)^{-1}A^{-1}B, \ldots, (A)^{-i} A^{-1}B, \ldots\} \tag{20}$$

where the column vectors are called the Krylov vectors. The q-th order Krylov subspace is denoted by

$$K_q(A^{-1}, A^{-1}B) \tag{21}$$

which is spanned by the leading q linearly independent Krylov vectors in (20). Let $V \in R^{n \times q}$ be any matrix which columns span the Krylov subspace $K_q(A^{-1}, A^{-1}B)$

$$\text{Spancolumn}\{V\} = \text{span}\{A^{-1}B; A^{-2}B; \ldots; A^{-q}B\}.$$

Here V is the orthogonal projection matrix that maps the n-dimensional state-space into a q dimensional state-space and satisfies $V^T V = I$. If the columns of V are orthogonal and B is a column vector, it can be shown that the following identities hold [24]:

$$(A)^i A^{-1}B = V(A_q)^i A_q^{-1} B_q \tag{22}$$

for $i = 0, 1, \ldots, q - 1$. These identities can be used to verify that at least the q leading moments of the full-order and reduced-order transfer functions are matched. Finally, we get the reduced order system of much smaller order (or state-space dimension) by performing variable change $x(t) = V\tilde{x}(t)$ and multiplying on V^T both sides of the equations (17):

$$\begin{aligned} \tilde{x}'(t) &= A_q \tilde{x}(t) + B_q v(t) \\ y(t) &= C_q{}^T \tilde{x}(t) \end{aligned} \tag{23}$$

where $A_q = (V^T A V)$, $B_q = V^T B$, $C_q = C^T V$.

As a common practice, the block vectors forming the Krylov subspace are orthogonalized by using the Arnoldi algorithm for a numerical stability. Lets describe the block Arnoldi algorithm for a single column input matrix, i.e., $B = b$, where $b \in R^n$, so that the algebraic operations involved can be seen.

3.2.1 Arnoldi Algorithm

i. LU factorize matrix A: $A = LU$.

ii. Solve $\tilde{v}_1$ from: $A\tilde{v}_1 = b$.

iii. Compute $h_{11} = \|\tilde{v}_1\|$ and $v_1 = \tilde{v}_1 / h_{11}$.

iv. For j = 2, . . . , q:
Solve $\tilde{v}_j$ from: $A\tilde{v}_j = Av_{j-1}$.
For $i = 1, \ldots, j-1$: $h_{ij} = v_i^T \tilde{v}_j$.

$$w_j = \tilde{v}_j - \sum_{i=1}^{j-1} v_i h_{ij}$$

$h_{jj} = \|w_j\|$, $v_j = w_j / h_{jj}$.

Note that the Arnoldi algorithm terminates when $h_{jj} = 0$, which means that the subsequent vectors belong to the subspace already generated. The Arnoldi algorithm is basically a Gram-Schmidt procedure for orthogonalizing the Krylov vectors. The variant of the Arnoldi algorithm (also known as *PRIMA*) with some extra computational effort preserves the passivity of the original system [27].
The Moment matching method of model order reduction can be extended on nonlinear systems. In these cases the original nonlinear systems has to be changed previously (linearized or piecewise-linearized, approximated by a quadratic systems, divided into several linear systems, etc.) [28].

3.2.2 Second order systems

A size of a second order equations system (2) can be also reduced by transforming it to the first order system (1), and then applying the methods described before. However, the reduction of second order systems by such transformation ignores the physical meaning of the original matrices and gives a reduced order model in a first order form. It is desirable for the reduced system to preserve the form of the original system (2). Approaches, that deal directly with the system (2) reduction have been proposed in the framework of Krylov subspaces methods [29, 30].
The transfer function for the system (2), with zero initial conditions, is given by:

$$H(s) = C^T (s^2 M + sD + K)^{-1} B \tag{24}$$

If the system is **undamped**, i.e. $D = 0$, the Arnoldi process can be applied for the computation of a basis for the Krylov subspace K_q $(K^{-1}M, K^{-1}B)$ which is used for the projection matrix V building. The transfer function can be expanded into Taylor series as

$$H(s) = -\sum_{k=0}^{\infty} m_k s^{2k} ,$$

where $m_k = C^T (K^{-1}M)^{-k} (A^{-1}B)$.
Then the reduced system matrices with variable change $x(t) = V\tilde{x}(t)$ are obtained by the orthogonal projection:

$$M_q = V^T M V,\; K_q = V^T K V,\; B_q = V^T B,\; C_q = C^T V \tag{25}$$

and the matching properties of the method are conserved [31].
It is worth noting that, starting from a second order system in the form (2), Krylov subspace methods require the knowledge of K^{-1}. For high dimensional systems, the explicit calculation of the inverse of K is computationally not affordable. Its computation is therefore

replaced by the solution of linear systems of equations through a *LU*- decomposition and defying $K^{-1} = U^{-1} L^{-1}$.

If the system is **damped**, i.e. $D \neq 0$, in case of Rayleigh damping, it demonstrates that the damping matrix can be neglected during the reduction process and it can be computed afterwards, as a linear combination of the reduced stiffness and mass matrices [29]. That is why it is possible to recalculate the matrix *D* also similar to matrices *M* and *K* as in (25):

$$D_q = V^T D V.$$

The reduction of a nonlinear system (2) can be done by using linear model order reduction techniques and considering nonlinearities as inputs. In the general case, the stiffness matrix is nonlinear since its entries dependent on the nodal displacements x. So damping matrix entries also are nonlinear functions of the nodal displacements x and the applied voltages V. If the nonlinearities are confined in the input function f, the system can be reduced using linear model order reduction technique. The only complication is that, after the reduction, the argument of x has to be recovered by the projection $x = Vx_q$ [31,32].

As a typical example Fig. 10 displays the transient simulation and frequency response of the original and reduced models of the microgyroscope being received via usage of the Arnoldi process [33].

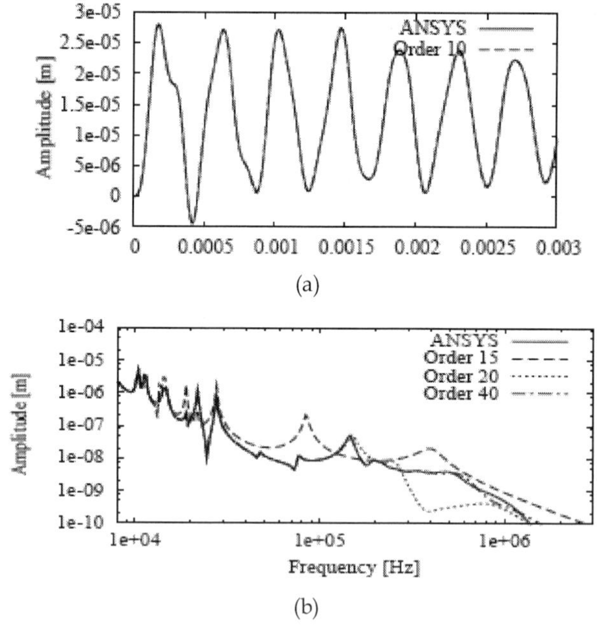

Fig. 10. Comparison of transient behavior (a) and transfer functions (b) for the full and reduced microgyroscope models [33].

We see that the solution obtained by reduced model of order 10 is already very close to the true ANSYS excitation (fig.10,a) while the reduced models of order 15 up to 20 show

considerable deviations at the high frequency range (fig.10,b). The model with order of 40 shows a perfect match for the lower eigenfrequencies and it is quite closer for higher frequencies, though this is not so important for the gyroscope.
Using Krylov/Arnoldi approach, only a postprocessor to ANSYS is necessary to generate a macromodel in one of the well established model description languages: pure C code, HDL-A, MAST, Modelica and the new standardized VHDL-AMS which are supported by powerful system simulators. Such approach was implemented in *mor4ansys* (pronounced "more for ANSYS") that was developed by IMTEK [33] and which provides the reduced model of order 20 - 30 with an accuracy of a few percents when the dimension of an original FEM is up to 100 000.

3.3 Equivalent-circuit ROM

Taking in to account relations between displacements x, velocities v and accelerations a: $a = dv/dt, \quad x = \int v dt$, it is possible to present the equation (2) in the form [34,35]:

$$\frac{d}{dt}(Mv) + Dv + \int Kv dt = F(t) \quad \text{or} \quad \tilde{C}\dot{v} + \tilde{G}v + \tilde{L}v = F(t)\ , \tag{26}$$

where $\tilde{C} = M, \tilde{G} = D, \tilde{L} = K$ are equivalent matrices of capacitances, conductance and inductances.

$$\tilde{C}, \tilde{G}, \tilde{L}$$

The elements of matrices $\tilde{C}, \tilde{G}, \tilde{L}$ are formed from the elements of the mass, damping and stiffness matrices in the following ways:

$$C_{ij} = -m_{ij}, \quad i,j = 1(1)N, \quad i \neq j; \quad C_{ii} = \sum_{j=1}^{N} m_{ij}, \quad i = 1(1)N; \quad L_{ij} = -1/k_{ij}, \quad i,j = 1(1)N, \quad i \neq j;$$

$$L_{ii} = 1/\sum_{j=1}^{N} k_{ij}, \quad i = 1(1)N; \quad G_{ij} = -d_{ij}, \quad i,j = 1(1)N, \quad i \neq j; \quad G_{ii} = \sum_{j=1}^{N} d_{ij}, i = 1(1)N. \tag{27}$$

where N is a number of equations or nodes of the MEMS structure.
In this approach a capacitive-inductive-resistive model of the circuit is built which correctly reflects mass, damping and stiffness matrices. Nodal potentials in this model correspond to the displacement velocities v.
Connected inductors, capacitors and conductors are placed in parallel between each two nodes and each node and ground (for the case i = j), fig.11. Values of these elements are defined by equations (27). Similar approach was used for solving thermo-structural analysis [35].
Displacements x are defined by the current of inductances L_{ii}, which are connected between a node and ground. Such approach suggests significant advantage if the coefficients of the mass and stiffness matrices become time-dependant.
The task of reduction of MEMS model order turns now into reduction of the equivalent RLC circuit size. There are two ways of performing such circuit size decreasing with the minimal accuracy loss:

- sequential excluding of the internal circuit nodes by application of Y-Δ (star-triangle) transformation [36, 39];
- building a macromodel as a four- terminal [40].

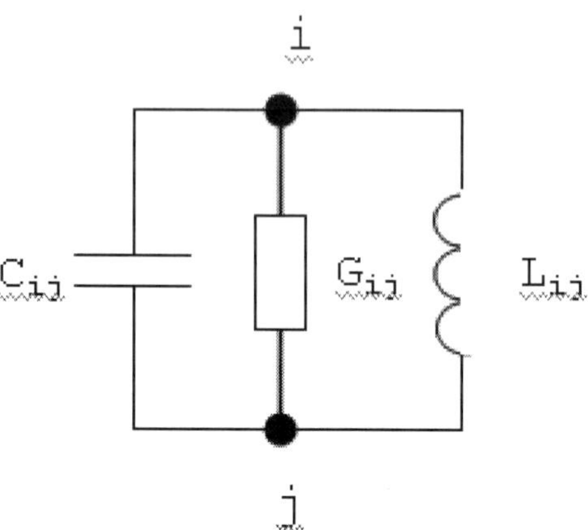

Fig. 11. Circuit element for composing the equivalent-circuit model

3.3.1 Y-Δ transformation based circuit size reduction methods

The essence of the methods based on the Y-Δ transformation consists in the following. Let's *i*-th node and *k* its neighbors are located as shown on fig.11. Then the component equation of *i*-th row will look like

$$Y_i V_i - y_1 V_1 - y_2 V_2 - \ldots - y_n V_n = 0 \ , \tag{28}$$

where $Y_i = \sum_{j=1}^{k} y_j$.

Let's define V_i as

$$V_i = \left(\sum_{j=1}^{k} y_j V_j \right) \Big/ Y_i \tag{29}$$

and replace V_i in other *k* equations , that is equivalent to excluding *i*- th node from a circuit. Then, for example, the equation of the first node transfers to:

$$(\overline{Y}_1 + y_1 - y_1^2 / Y_i) V_1 - \left(\sum_{j=2}^{k} y_1 y_j V_j \right) / Y_i - \sum_{\substack{r=1 \\ r \neq i}}^{k1} y_r V_r = 0 \tag{30}$$

where $\overline{Y}_1 = \sum_{\substack{r=1 \\ r \neq i}}^{k1} y_r$ is a sum of all node conductance excluding *i*- th node, *k1* is a number of nodes being connected to node 1.

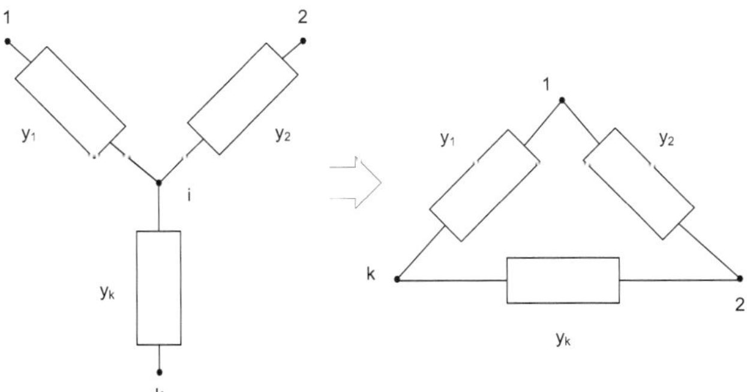

Fig. 12. A working node of the RLC-circuit: conductance are added between node 1 and all neighbors nodes

The equation (30) can be simplified:

$$(\bar{Y}_1 + \left(\sum_{j=2}^{k} y_1 y_j \right) / Y_i) V_1 - \left(\sum_{j=2}^{k} y_1 y_j V_j \right) / Y_i - \sum_{\substack{r=1 \\ r \neq i}}^{k1} y_r V_r = 0 \tag{31}$$

Note that it is equivalent to adding k-*1* new elements between first node and *k-1* former neighbors of *i*-th node on Fig. 12.
For any two *a*-th and *b*-th nodes, which are neighbors to *i*- th node, elimination of *i*- th node will add a new element between these two nodes, which is equal to:

$$y_{ab} = (y_a y_b) / Y_i \tag{32}$$

or in the p-polynomial form, taking into account existence of R, L and C elements, as shown on fig. 11:

$$y_{ab} = \left(g_a + \frac{b_a}{p} + pc_a \right) \left(g_b + \frac{b_b}{p} + pc_b \right) \Big/ \left(G_i + \frac{B_i}{p} + pC_i \right) \tag{33}$$

where $C_i = \sum_{j=1}^{k} C_j$, $B_i = \sum_{j=1}^{k} B_j$, $G_i = \sum_{j=1}^{k} G_j$ are sums of all the capacitances, reciprocal inductances and conductance connected to *i*-th node.
In order to simplify (33) two constant time values $\tau_{RCi} = C_i / G_i$ and $\tau_{LCi} = \sqrt{C_i / B_i}$ are introduced for any node in the circuit. The time constant of *i*-th node is defined as $\tau_i = \max(\tau_{RC}, \tau_{LC})$ and it is considered to be fast if

1. $\tau_i < \tau_{\min} = 2\pi / \omega_{\max}$,

where $\tau_{\min}$ - a time constant which depends on maximal circuit frequency $\omega_{\max}$ being defined by user. So, a fast node will satisfy the following conditions:

$$\omega_{\max} C_i < G_i,\ G_i < B_i / \omega_{\max} \text{ i } \omega_{\max} C_i < B_i / \omega_{\max}.$$

In order to eliminate a fast node from *RLC* circuit, let us consider the following two cases.

2. If $\tau_{RCi} >> \tau_{LCi}$, the equation (33) can be transformed into:

$$y_{ab} = (g_a + pc_a)(g_b + pc_b) / (G_I + pC_i) \tag{34}$$

and its decomposition into Taylor series will look like:

$$y_{ab} = \frac{g_a g_b}{G_i} + p\frac{c_a g_b + c_b g_a}{G_i} + p^2 \frac{4c_a c_b}{G_i} . \tag{35}$$

For typical R,L,C values (R=0,1÷1000 ohm, L=0,01÷10 nH, C=0,001÷10 pF, ω=0,1÷10 GHz) the contribution of the last term in (35) will be significantly smaller that previous ones. The constant term in (35) gives the value of the conductance, which appears between a- th and b-th nodes during the elimination of i- th node, and the second term in (35) defines the value of capacitance.

3. If $\tau_{LCi} >> \tau_{RCi}$, the equation (33) is transformed into:

$$y_{ab} = \frac{1}{p}\frac{b_a b_b}{B_i} + p\frac{c_a b_b + c_a b_b}{B_i} . \tag{36}$$

The first term with $1/p$ in (36) defines the value of the reactive conductance (the inductances' reciprocal value), which appears between a- th and b- th nodes during the elimination of i- th node, and the second term in (36) gives the value of the capacitance.
Final formulas for additional elements between a-th and b-th nodes during the elimination of i- th node for all possible cases are given in tables 1 and 2. All these formulas can be deducted from (35) and (36). The only exception is the case when a- th and b- th nodes are connected to i- th node through a capacitor when $C_{ab} = C_a C_b / C_i$.
In order to eliminate a node from equivalent MEMS circuit with the smallest inaccuracy, the following two criteria should be taken into account:

- a node should be fast;
- time constants τ_{LCi}, τ_{RCi} should not be compared (have almost the same values).

In practice, usually lower eigenfrequencies for mechanical systems are most interesting. Therefore, a compromise between accuracy and size of the received circuit models can be reached by proper selection of τ_{min} value. So the algorithm of equivalent-circuit ROM can be described as following:

1. Time constants should be calculated for all the nodes of the circuit.
2. All the nodes should be put into a priority queue, sorted by time constants, apart from input, output and ground nodes.
3. Take the first i- th node in the queue (the one with the smallest time constant $\tau_i < \tau_{min}$).
4. Find the neighbors k of i- th node.
5. Eliminated i- th node according to the rules described in tables 1, 2 (depending on τ_{LCi}, τ_{RCi} values).
6. Time constants in the set k (which consists of the neighbor of i-th node) should be recalculated
7. Take i- th node from the head of the priority queue and jump to step 3.

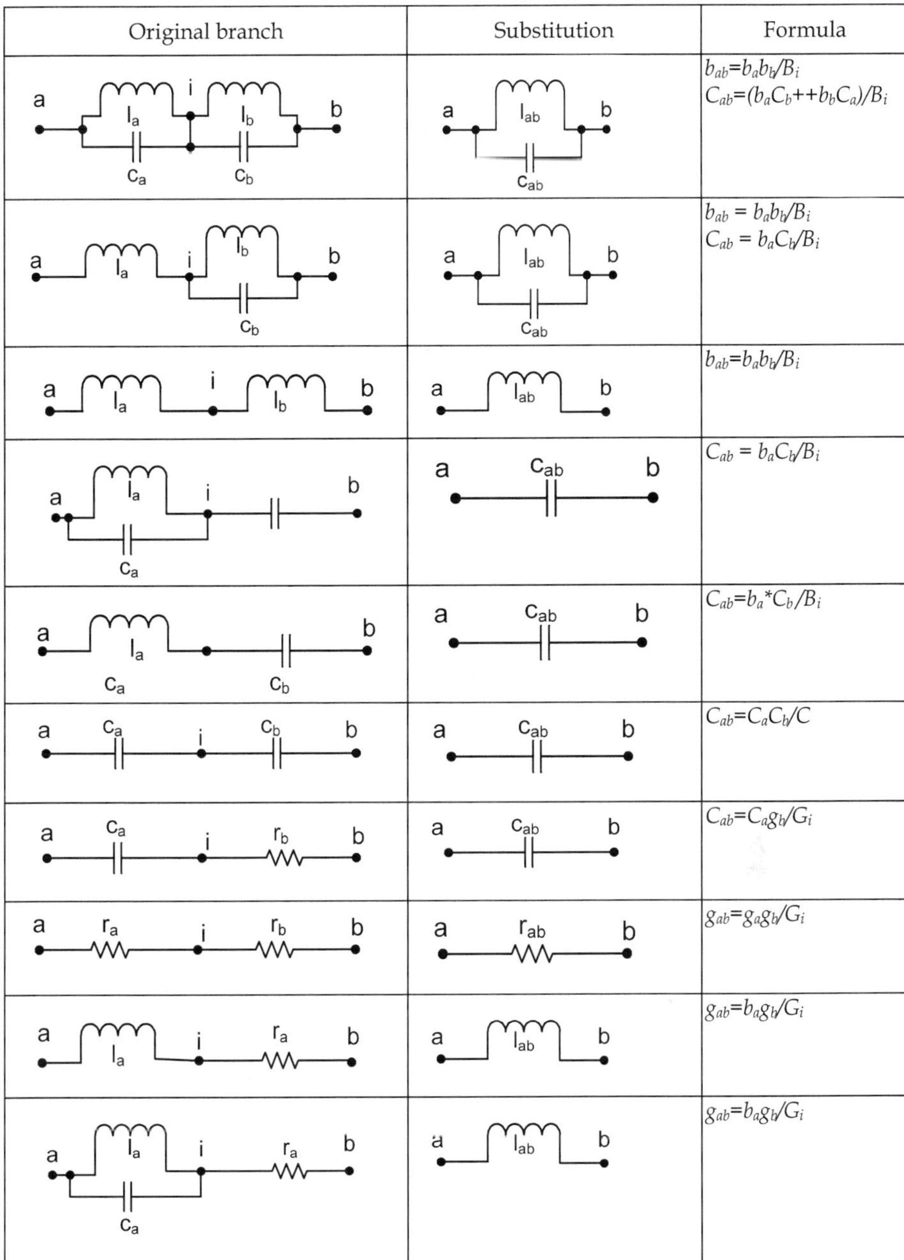

Original branch	Substitution	Formula
		$b_{ab}=b_ab_b/B_i$ $C_{ab}=(b_aC_b++b_bC_a)/B_i$
		$b_{ab} = b_ab_b/B_i$ $C_{ab} = b_aC_b/B_i$
		$b_{ab}=b_ab_b/B_i$
		$C_{ab} = b_aC_b/B_i$
		$C_{ab}=b_a{}^*C_b/B_i$
		$C_{ab}=C_aC_b/C$
		$C_{ab}=C_ag_b/G_i$
		$g_{ab}=g_ag_b/G_i$
		$g_{ab}=b_ag_b/G_i$
		$g_{ab}=b_ag_b/G_i$

Table 1. Transformations for $\tau_{LCi} >> \tau_{RCi}$ case

Original branch	Substitution	Formula
a r_a i r_b b C_a C_b	a b C_{ab}	$g_{ab}=g_ag_b/G_i$ $C_{ab}=(g_aC_b+{}_bC_a)/G_i$
a r_a r_b b C_b	a b C_{ab}	$g_{ab}=g_ag_b/G_i$ $C_{ab}=g_aC_b/G_i$
a C_a r_b b C_b	a C_{ab} b	$C_{ab}=g_bC_a/G_i$
a C_a i r_b b	a C_{ab} b	$C_{ab}=g_bC_a/G_i$
a C_a i C_b b	a C_{ab} b	$C_{ab}=C_aC_b/C$
r_a i r_b b	a r_{ab} b	$g_{ab}=g_ag_b/G_i$
a l_a r_b b C_b	a r_{ab} b	$g_{ab}=b_ag_b/B_i$
a l_a i r_b b	a r_{ab} b	$g_{ab}=b_ag_b/B_i$
a l_a C_b	a C_{ab} b	$g_{ab}=b_aC_b/B_i$
a l_a i l_b b	a l_{ab} b	$g_{ab}=b_ab_b/B_i$

Table 2. Transformations for $\tau_{RCi} >> \tau_{LCi}$ case

3.3.2 Building MEMS macromodel as a four- terminal

The main idea of this method is to develop a MEMS macromodel as a four-terminal circuit (n-terminal in a general case) as:

$$\begin{bmatrix} I_a \\ I_b \end{bmatrix} = \begin{bmatrix} Y_{aa} & Y_{ab} \\ Y_{ba} & Y_{bb} \end{bmatrix} \begin{bmatrix} U_a \\ U_b \end{bmatrix}, \tag{37}$$

where I_a, U_a are current and voltage at the macromodel input, I_b, U_b are current and voltage at the output. It is suggested to obtain equation (37) directly from the general matrix of the circuit (e.g. admittance matrix Y) according to the expression [40]:

$$Y_{tp} = \frac{1}{\Delta_{aa,bb}} \begin{bmatrix} \Delta_{bb} & -\Delta_{ba} \\ \Delta_{ab} & \Delta_{aa} \end{bmatrix} \tag{38}$$

where Δ_{ij} – an algebraic complement to a_{ij} element of the initial circuit admittance matrix Y, which is equal to a determinant of the matrix being obtained from the matrix Y after eliminating i – th row and j- th column on the crossing of which the given element a_{ij} is situated. In addition the sign of Δ_{ij} is defined by the factor $(-1)^{i+j}$;

$\Delta_{aa,bb}$ – redoubled algebraic complement which equals to the determinant of the matrix obtained by eliminating a-th,b-th rows and a-th ,b-th columns and its sign is defined by the factor $(-1)^{a+a+b+b}$.

The necessary algebraic complements can be calculated by the initial matrix inversion procedure, if matrix elements are defined for the selected frequency ω_0, which is fixed by an user:

$$Y^{-1} = \frac{1}{\Delta} \begin{bmatrix} \Delta_{11} \; \Delta_{21} \cdots \; \Delta_{n1} \\ \Delta_{12} \; \Delta_{22} \cdots \; \Delta_{n2} \\ \\ \Delta_{1n} \; \Delta_{2n} \cdots \; \Delta_{nn} \end{bmatrix},$$

where Δ - is the determinant of the initial matrix Y and $\Delta_{aa,bb} = \dfrac{\Delta_{bb}\Delta_{aa} - \Delta_{ba}\Delta_{ab}}{\Delta}$.

So if numerical values of the inverted Y^{-1} matrix elements are computed:

$$Y^{-1} = \begin{bmatrix} g_{11} \; g_{21} \cdots \; g_{n1} \\ g_{12} \; g_{22} \cdots \; g_{n2} \\ \\ g_{1n} \; g_{2n} \cdots \; g_{nn} \end{bmatrix} \tag{39}$$

it is possible to find parameters of the equivalent four- terminal (38). By choosing elements $g_{aa}, g_{bb}, g_{ab}, g_{ba}$ of the inverted matrix (39) and computing $g_{aa,bb} = g_{aa}g_{bb} - g_{ab}g_{ba}$ it is possible to find parameters of the reduced model (37) from the relations:

$$\begin{aligned} Y_{aa} &= g_{bb}/g_{aa,bb}; \\ Y_{ab} &= -Y_{ba} = g_{ba}/g_{aa,bb}; \\ Y_{bb} &= g_{aa}/g_{aa,bb}. \end{aligned} \tag{40}$$

Since macromodel parameters are determined by sum of the real and imaginary parts:

$$Y_{ij} = a_0 + i\, a_1\,, \tag{41}$$

it is convenient to represent the circuit macromodel also as a sum of real and imaginary terms:

$$Y_{tp} = \begin{bmatrix} Y_{aa0} Y_{ab0} \\ Y_{ba0} Y_{bb0} \end{bmatrix} + i \begin{bmatrix} Y_{aai} Y_{abi} \\ Y_{bai} Y_{bbi} \end{bmatrix}. \tag{42}$$

Investigation the expression (42) for different frequencies confirms that for linear circuits at least for frequency range $\omega \le \omega_0$, where ω_0 – frequency at which elements of the inverted matrix (40) are initially defined, real terms of the computed parameters (42) preserve their values and imaginary terms change proportionally to the selected frequency. This confirms the possibility to use macromodel parameters being computed at once frequency in a broad frequency range.

As it was noted earlier, the macromodel form (42) is inconvenient for direct implementation in the circuit simulation software. A real part of Y_{ij} parameters for the stable passive macromodels ($a_0>0$) has to be positive while an imaginary part could be negative. If in (41) $a_o > 0,\ a_1 > 0$ the Y_{ij} component of the four- terminal may be presented by parallel connection of conductance, capacitance and inductance [41]:

$$G_{ij} = \mathrm{Re}\,(Y_{i,j}),\ C_{ij} = k\,\mathrm{Im}\,(Y_{ij})/\,\omega_0,\ L_{ij} = 1/[(k-1)\mathrm{Im}\,(Y_{ij})\,\omega_0]. \tag{43}$$

If $a_o > 0,\ a_1 < 0$ the respective component Y_{ij} may be presented in the same way as parallel connection of conductance, capacitance and inductance but:

$$G_{ij} = \mathrm{Re}\,(Y_{i,j})\,,\ C_{ij} = (k-1)\,\mathrm{Im}\,(Y_{ij})/\,\omega_0 \ \text{and}\ \ L_{ij} = 1/(k\,\mathrm{Im}\,(Y_{ij})\,\omega_0), \tag{44}$$

where k is defined by reactive components values ratio.

So it is possible to select priori the equivalent-circuit macromodel, for example, shown on Fig.13, and to define parameters of its components using expressions (43) and (44).

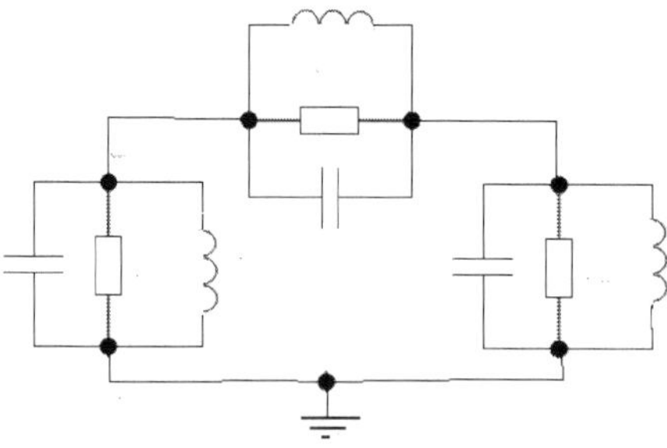

Fig. 13. Equivalent-circuit macromodel of a four-terminal

The important feature of the equivalent circuits MEMS macromodel is a possibility to use optimization procedures for accurate adjustment of the macromodel components values to meet the device characteristics being obtained by FEM model. If a whole MEMS equivalent circuit is rather large it is possible to divide it into some subcircuits and apply transformation (37) to each subcircuit.

Then to combine individual four-terminal circuits into one equivalent MEMS four-terminal circuit taking in account ways of connectivity of different subcircuits (parallel, sequence or mixed) and proper recalculating four-terminal circuits parameters systems (*y, z, h, a*) [40].

3.3.3 NetALLTED equivalent-circuit ROM subsystem

Equivalent-circuit ROM for MEMS was implemented in the circuit simulation package NetALLTED (**ALL TE**chnologies **D**esinger) which was developed not only for simulation and analysis, but for processing project procedures such as parametric optimization tasks; optimal tolerance assignments; centering availability regions; yield maximization [42]. NetALLTED is widely used for design of Nonlinear Dynamic Systems composed of either/and electronic, hydraulic, pneumatic, mechanical, electromagnetic, and other elements and it is available through the Internet (http://allted.kpi.ua/). ROM developing approach in hand provides more than 99% reduction of elements and node numbers of equivalent- circuit ROM. For example, for the accelerometer only 3 nodes and 6 elements are left from initial 1,883 nodes and 62,826 elements.

Let's consider the example of the equivalent- circuit ROM for the beam working on bending and find its eigenfrequencies [34]. The left end of the beam is fixed motionlessly, right one is free. Force *f* is applied to the right end perpendicularly to the beam axe.

The initial FEM model was constructed using ANSYS Multiphysics v.10.0 and the beam ANSYS library's BEAM3 finite element with the length of 0.5 um for following beam parameters: L = 25 um; beam cross-section is a square one with height of 3 um and width of 2 um. Beam material properties: coefficient of elasticity $E=2\cdot10^{11}$, Pa = 0.2 N/mkm^2; Poisson coefficient μ = 0.3; material density ρ = $6\cdot10^3$ kg/m^3 = $6\cdot10^{-9}$ mg/mkm^3. The initial equivalent beam circuit contains 101 nodes and 314 elements.

The developed equivalent-circuit ROM with 5 nodes and 14 elements is presented on fig.14.

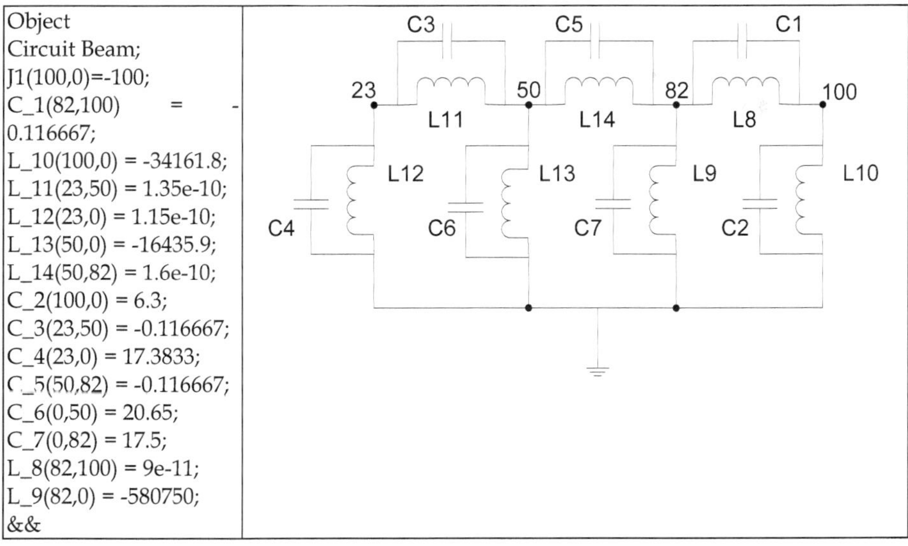

```
Object
Circuit Beam;
J1(100,0)=-100;
C_1(82,100) = -0.116667;
L_10(100,0) = -34161.8;
L_11(23,50) = 1.35e-10;
L_12(23,0) = 1.15e-10;
L_13(50,0) = -16435.9;
L_14(50,82) = 1.6e-10;
C_2(100,0) = 6.3;
C_3(23,50) = -0.116667;
C_4(23,0) = 17.3833;
C_5(50,82) = -0.116667;
C_6(0,50) = 20.65;
C_7(0,82) = 17.5;
L_8(82,100) = 9e-11;
L_9(82,0) = -580750;
&&
```


Fig. 14. Equivalent beam circuit

The two lower eigenfrequencies of the beam are defined by computation. The results of ANSYS Multiphysics frequency analysis as well as the results of the equivalent- circuit ROM simulation for different τ_{min} by NetALLTED [36,43] are given in table 3.
It is obvious that the accuracy of a macromodel obtained with $\tau_{min} = 3*10^{-5}$ is rather high and there are only 5 nodes and 14 elements in the reduced circuit. For more accurate simulation it is possible either to use a reduced circuit obtained with smaller values of τ_{min} (when a size of equivalent circuit ROM increases), or to adapt the received ROM with help of the optimization methods.

	ANSYS results	ALLTED results				
		Source circuit	Reduced circuit			Optim. circuit
τ_{min} , s	-	-	$5*10^{-6}$	10^{-5}	$3*10^{-5}$	$3*10^{-5}$
Node number	-	101	24	12	5	5
Element number	-	314	76	38	14	14
Reduction by nodes, %	-	-	76,2376	88,1188	95,0495	95,0495
Reduction by elements, %	-	-	75,7962	87,8981	95,5414	95,5414
1st peak, Hz	1336,2	1336,3	1336,1	1334,9	1327	1336,2
2nd peak, Hz	4009,3	4009,3	4009,4	3993,3	3612,1	4009,4
Maximal error, %	-	-	0,01	0,3	9,9	0,003

Table 3. Frequency beam analysis results

The macromodel of the four-terminal for the same beam is being developed in according with (38)-(44) with only 3 nodes and 6 elements that presented on fig.15.

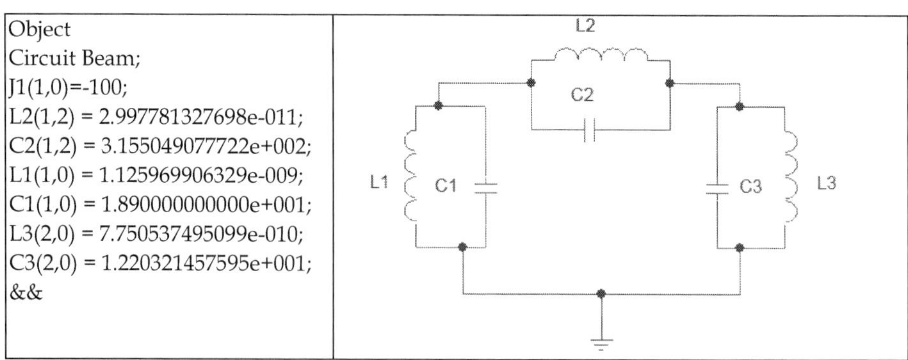

Fig. 15. Equivalent beam circuit being considered as a four- terminal

The results of simulations of the four- terminal equivalent circuit by NetALLTED [36,43] are given in table 4.

	Source circuit	Reduced circuit (as n-ports)
Node number	101	3
Element number	314	6
1st peak, Hz	1336.3	1337.0(0.05%)
2nd peak, Hz	4009.3	4008.9(0.01%)

Table 4. Results of frequency analysis of the four- terminal equivalent circuit for a beam

The advantage of the equivalent circuit approach is obtaining small size of the equivalent reduced circuit as well as the possibility to get required frequencies with a high accuracy, using NetALLTED optimization possibilities (fig.16). The disadvantage is a necessity to make the additional analysis of reduced circuit in order to find the most sensitive elements and take them as variable parameters during optimization procedure.

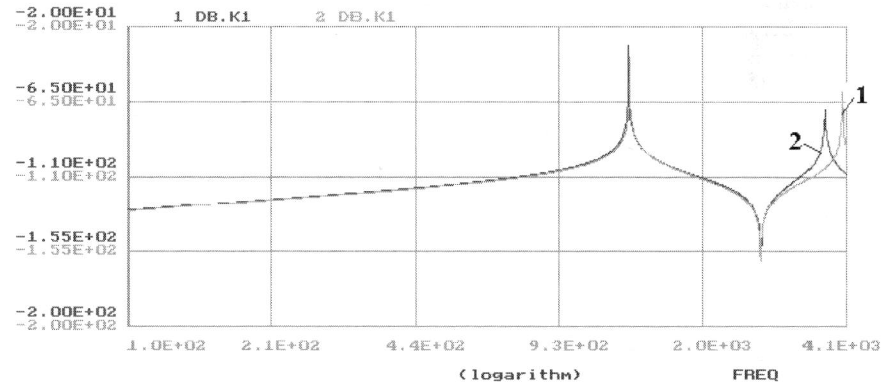

Fig. 16. ROM transfer function before (1) and after (2) optimization.

For example, for reduced circuit, shown at fig.14, four variable elements have been chosen (*L8, L11, L12, L14*) and the objective function *OF ERROR1= F8(1336.3,4009.3/T1,T2)* was constructed which contains the requirement to obtain resonance peaks at frequencies *T1=1336.3* Hz and *T2=4009.3* Hz in according to the ANSYS analysis results (table 3). The objective function contains also current values of frequency response extremes *T1=MAXA(db.K1,100,1600)* and *T2=MAXA(db.K1,1700,4100),* which are calculated with help of MAXA directive for defining time or frequency when the output (*db.K1* in our case) reaches its maximum value in the specified time or frequency range (*100-1600* Hz). Among available 12 optimization methods being incorporated in NetALLED the Random Search Method (METHOD=40) with search interval reducing has been used to optimize the beam macromodel parameters.

4. MEMS coupled system-level model

The modeling of MEMS provides a very challenging task in modern engineering. This field of research is inherently multiphysics of nature, since different physical phenomena are tightly intertwined at microscale. Typically, up to four different physical domains are usually considered in the analysis of microsystems: *mechanical, electrical, thermal* and *fluidic.* For each of these separate domains, well-established reduced order modeling and analysis techniques are available. However, one of the main challenges in the field of microsystems engineering is to connect models for the behavior of the device in each of these domains to equivalent lumped or reduced-order models without making unacceptably inaccurate assumptions and simplifications and to couple these domains correctly and efficiently.

Micromechanical membrane devices (capacitive pressure transducers, ultrasonic transducers), surface micro machined devices (RF switches, micro optical devices) as well as bulk micro machined devices (accelerometers, inclinometers, laser scanning mirrors) are driven or sensed by nearly parallel electrode pairs in many cases. The motion of these electrodes is strictly normal to the surfaces.

It means that MEMS electrical parts of these MEMS had to be combined with mechanical ones (fig.16). Typically, block-diagram descriptions and lumped-element circuit models for components are connected into a full system. Mostly, this description is used for functional analysis of a design concept.

The coefficients and electrostatic nodal force are obtained from the capacitance-displacement function *C(w)* of the associated electrode portions and gap space. The function can be input by one of three means [44]:

- as analytical function if the electrode portions make up a plate capacitor geometry with homogeneous intermediate field;
- as polynomial approximation of a function given by data points;
- as data table wherein the element subroutine interpolates values during solution.

The electrostatic forces acting on the movable conductor of the device are included in the model as nonlinear input forces, which are applied on nodes distributed over the conductor surface. Nodes divide the surface into N smaller portions. A lumped force is applied to k- th node at the center of each portion, in its preferential direction of movement x_i. The capacitance C_k between the k-th portion and the fixed electrode of the device is computed, for its undeformed configuration, using an electrostatic analysis

The entity of each force f_k is then approximated as:

$$f_k = \frac{1}{2}\frac{\varepsilon A_k}{(d_0 + w_i^k)^2}(V_p - V_n)^2 \tag{45}$$

where d_0 is the initial distance between the conductor and the electrode, ε is the relative dielectric permittivity, w_i^k is the displacement of the k- th node along the direction x_i and $(V_p\text{-}V_n)$ is the voltage difference between the structure and the fixed electrode.

The capacitance C_k can be calculated from

$$C_k = \varepsilon A_k/(d - w_i^k)\,. \tag{46}$$

If C_k varies considerably with the deformation of the structure, then a series of electrostatic computations for different device deflections in its operation range can be performed. The results can be used for extracting the dependency $C_k\,(w_i^k)$ and calculating the electrostatic force.

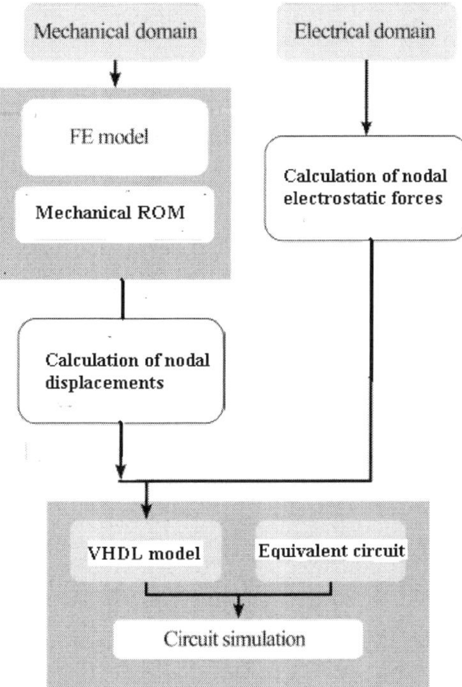

Fig. 16. MEMS system-level simulation approach [33]

According to the assumption that conductors are equipotential, all the nodes connected to a certain conductor are subjected to the same voltage boundary conditions. The total current flowing in the conductor is simply given by the sum of the currents at those nodes:

$$i_k = \frac{d}{dt}[C_k(V_k - V_n)].$$

The nodes used for electrostatic force application can be also used for monitoring the distance between the movable structure and the electrode. When this is equal to the transduction gap, the contact condition is reached and contact forces with the stiffness of the contact K_n, given by equation

$$F_{cont} = K_n(d - gap_{min}), \tag{47}$$

can be applied to nodes.

Computation of the electrostatic forces adds some complexity to the development of the MEMS system-level model, but this is largely compensated by the speed-up simulation of the full model.

There is the special ANSYS' transducer element *TRANS126* which has the possibility to calculate the capacitance of a parallel plate capacitor model (at particular nodes or as a whole) as [47]

$$C(w)=\frac{C_0 d_0}{d_0-w}=\frac{C_0}{d_0}\left(1+\frac{w}{d_0}+\frac{w^2}{d_0^2}+\frac{w^3}{d_0^3}+\frac{w^4}{d_0^4}+...\right) \tag{48}$$

where d_o and w are the initial distance and the displacement between the plate.
The element has two nodes; the gap distance is calculated as the sum of the initial displacement and the difference of the nodal displacements in the direction of the element. The force is calculated by equation being similar to (44) so that in the constant voltage case

$$F=\frac{1}{2}\frac{\partial C(w)}{\partial d}V^2=\frac{C_0 V^2}{2}\left(\frac{1}{d_0}+2\frac{w}{d_0^2}+3\frac{w^2}{d_0^3}+4\frac{w^3}{d_0^4}+...\right). \tag{49}$$

The element has also contact capabilities (47). It is possible to specify a minimal gap and a spring stiffness K_n for the repelling force.
The drawback of this kind of element is that it is limited to the case where the electrodes are (almost) parallel plates, so that the stroke/capacitance function can be evaluated from a single degree of freedom. But the extensions for rotation plates and for 2D cases were developed. The last one with a triangular shape (element *TRANS109*) is useful for simulating structures such as comb drivers and optical MEMS, in which capacitance between the device parts is generally a function of a two-directional displacement. *TRANS126* and *TRANS109* elements enable a huge reduction of the complexity of the system-level simulation.
Let's consider for example the system-level macromodel of an ultrasonic transducer which has two plates with the bottom electrode area A_c and the plate dimension L and which can be presented by nonlinear capacitance:

$$C_{eq}=C_o+(C_{L/2}-C_0)(1-e^{-\tau t}), \tag{50}$$

where C_0 is the smallest capacitance in the absent of voltage V: $C_0=\frac{\varepsilon_0 A_c}{d_e}$

$C_{L/2}$ is the largest capacitance when a plate center displacement $w(\frac{L}{2},t)$ is calculated from the ROM macromodel equations (26):

$$C_{L/2}=\frac{\varepsilon_0 A_c}{d_e-w\left(\frac{L}{2},t\right)},$$

where d_e is an equivalent gap ($d_e=d_0+\frac{d_1}{\varepsilon_1}+\frac{d_{ins}}{\varepsilon_{ins}}$); ε_0 is the absolute dielectric permittivity of the vacuum, ε_1 is the relative dielectric permittivity of the poly silicon , ε_{ins} is the relative dielectric permittivity of the insulator; w is a plate deflection .
The largest value of this capacitance corresponds to the value $w_{\max}=d_0$. The electrostatic force acting on the capacitor surfaces is the Coulomb force:

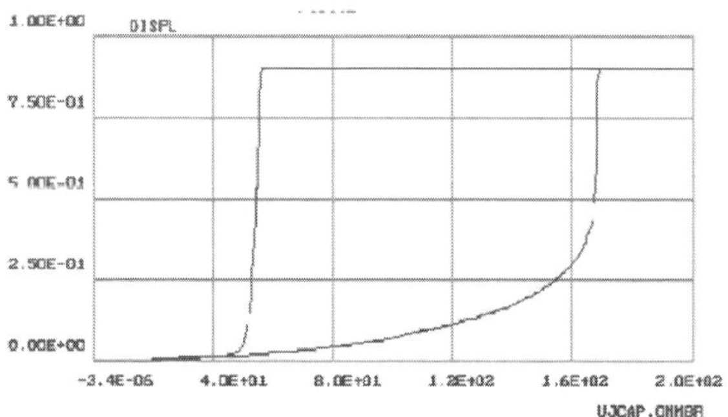

Fig. 17. Capacitive –voltage characteristic an ultrasonic transducer

$$\tilde{F}_{elec} = -\frac{\partial E}{\partial w} = \frac{V_{in}^2}{2}\frac{\partial C_{eq}}{\partial d} = \frac{\varepsilon_0 A_C V_{in}^2}{2\left(d_e - w(\frac{L}{2},t)\right)^2}.$$

As the displacement $w(\frac{L}{2},t)$ reaches the value $w_{\max} = d_0$, the hard stop will restrict its further increase, but input voltage V_n and $\tilde{F}_{elec}$ can be further increased. Assume that the input voltage V_n is a superposition of a constant voltage V_{DC} and a time dependent signal $V(t)$ and that $V_{DC} >> V(t)$. The elastic-plastic properties of the points of contact are simulated by a spring with rigidity K_n and a damper of damping factor b_0. When the beam center moves past w_{max}, it starts interacting with the spring that represents the contact. The damper is introduced to take into account the energy dissipation at the contact. The following equations are used to represent this model:

$$R = \begin{cases} 0 & for\ w \le w_{\max} \\ K_n(w_{\max} - w) - b_o w' & for\ w > w_{\max} \end{cases},$$

where R is the interaction force.

The process of interaction simulated by the force R sometimes can be highly sensitive to the values of the model parameters (rigidity and damping factor), especially if $w_{\max} \approx d_e$. The interaction can induce high-frequency motions and slow down the rate of convergence considerably. Therefore, a good deal of attention must be paid to accurately modeling and representing this process using experimental data.

If the input voltage is increased more, there will be no equilibrium and the plate collapse takes place. In this case, a hard stop or some other arrangement must be introduced to limit the plate motion. During the plate collapse, the difference between the electrostatic force and the elastic force of the spring will continue to increase. Therefore, when the plate drops down into the hard stop, it is not enough to reduce the input voltage below to release the plate. The input voltage should be reduced more to make the electrostatic force at least equal to the elastic force. Hence, the plate capacitance exhibits the *hysteretic* behavior with respect to voltage change.

But if the insulation layer is rather thick its restrictive effect should be taken into account. If the maximal center displacement w_{max} is equal to initial thickness air gap the moving plate touches the insulator top surface of the electrode when the input V_n voltage reaches the value of V_{max} (V_{max} may be calculated from simulation). But as soon as the input voltage will be decreased under this value the plate will leave the hard stop. So, its capacitance does not demonstrate the hysteretic behavior (fig.17).
Parameter τ in (50) defines an ultrasonic transducer frequency band and can be calculated through the plate displacement $w(\frac{L}{2},t)$ and its velocity $v(\frac{L}{2},t) = w'(\frac{L}{2},t)$, which are defied from ROM equations in the following way:

$$\tau = \frac{1}{3} w(\frac{L}{2},t) / v(\frac{L}{2},t) \tag{51}$$

The coefficient 3 appears in (51) due to the fact that a capacitance recharge to 98% for a time value which is equal approximately 3τ.
It is possible to see two included procedures in according to fig.18: one for development of ROM for an ultrasonic transducer plate, where a deflection and speed of central point`s deflection of transducer plate is calculated for the value V_n, and second - for determine of MEMS system- equivalent capacity value (SLM), using values of plate central point co-ordinates .Then the cycle of calculations recurs whereupon.

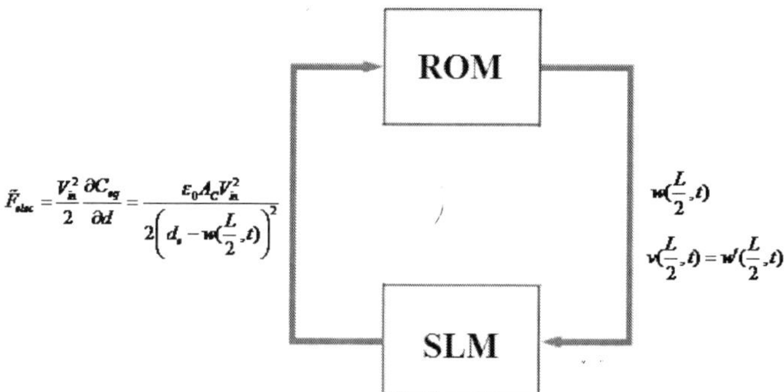

Fig. 18. ROM- system-level model coupled simulations

Instead of using two sequence procedures mentioned above it is possible using functional possibilities of the circuit simulator NetALLTED to built a single system-level equivalent circuit model for an ultrasonic transducer by introducing directly into the equivalent- circuit ROM of mechanical MEMS part the additional arbitrarily connected element (a Depended Source) with an informative function which is determined by equation (50) [42]. Optimization procedures of NetALLTED allow getting the desirable values of this transducer capacity and through it to get a desirable value of output signal of an ultrasonic transducer system-level model by the changing ROM parameters, which, in turn, are depended upon an ultrasonic transducer construction sizes and used material properties.

Providing a single representation of a MEMS operating in multiple physical domains, the electrical circuit approach is very convenient. Moreover, powerful mathematical techniques and circuit simulation programs are available for solving design tasks. It is possible to develop a library of schematic model for different MEMS elements and then use their combinations to build a system-level macromodels for entire rather complicated MEMS constructions.

5. Conclusion

In this chapter, the methods and issues encountered in the development of MEMS macromodels at the system level have been presented. System level modeling is the highest and most abstract level of modeling. This level requires various devices` linking of MEMS *component level models* - both electronic and micromechanical - into a micro-electro-mechanical system. *System-level models* of MEMS components are needed to allow a fast and sufficiently exact investigation of their behavior to simulate entire MEMS.

Starting point for the extraction of a reduced order model (ROM or a macromodel) is already its description with a large ODE system, which is typically derived using physical modeling techniques based on Finite Element Method (FEM) which is rather time consuming. Macromodels application allows the extraction of lower order ODE system that reproduces the input/output behavior with good accuracy. Particular attention has been posed in the chapter on the possibility to get a macromodel circuit presentation.

There are special methods for generating ROM for MEMS components and entire MEMS based on FEM descriptions. To derive macromodels of smaller sizes different approaches (*Modal decomposition, Moment matching, Equivalent circuit presentation)* were developed. Usage of the reduced MEMS components models allows applying successfully modern circuit simulators in workflow for MEMS design on system level.

Three automatic procedures to generate device reduced order macromodels, being based on full FEM/FDM models, were demonstrated in this chapter. Two of them are suitable for simulators with possibilities to get input information in the equation forms (ODE or OAE). The third one in opposite produces macromodels in circuit presentation and so it is more suitable for circuit simulators. The Modal ROM approach is based on using natural (modal or resonant) frequencies of MEMS structure and it is spread mostly in the USA and Asia. The Moment matching ROM approach is based on using the Krylov subspace for transfer functions and it is popular in Western Europe and Asia. The Equivalent circuit ROM approach is based on using a capacitive-inductive-resistive circuit model for mass, damping and stiffness matrices and it is used mostly in Eastern Europe. It is worth to notice that the Modal ROM approach requires some full ANSYS runs to perform a proper orthogonal decomposition during basic functions determination in opposite to the Moment matching and the Equivalent circuit ROM approaches for which it is enough to use ANSYS only for FEM model matrices building.

It seems to be interesting and perspective trying to combine mentioned approaches, for example, to start with Krylov/Arnoldi reduction of ODE dimension, then to build the proper equivalent circuit for obtained ODE systems and finally to apply Y/Δ transformation or n-port transformation for further reducing macromodel order.

6. References

[1] S. D. Senturia, "CAD challenges for microsensors, microactuators, and microsystems," *Proc. IEEE*, vol. 86, pp. 1611–1626, 1998.

[2] E. B. Rudnyi and J. G. Korvink, "Model order reduction for large scale engineering models developed in Ansys." *Lect. Notes Comput. Sc.*, vol. 3732, pp. 349–356, 2006.

[3] M. G.Mand G. Ostergaard, "Electro-mechanical trasducer for MEMS analysis in Ansys," in *Proc. Int. Conf. on Modeling and Simulation of Mycrosystems (MSM)'99*, pp. 270–273, 1999.

[4] W. Z. Lin, K. H. Lee, S. P. Lim, and Y. C. Liang, "Proper orthogonal decomposition and component mode synthesis in macromodel generation for the dynamic simulation of a complex MEMS device," *Journal of Micromechanics and Microengineering*, vol. 13, no. 5, pp. 646–654, 2003.

[5] A.I.Petrenko," Design Methology and Workflow for MEMS Design ", *Proc. MEMSTECH'2011, 11-14 May, Polyana-Svalyava (Zakarpattya), UKRAINE,* pp.12-15, 2011.

[6] Peter Avitabile. "Experimental Modal Analysis- A Simple Non-Mathematical Presentation", *Modal Analysis and Control Lab., University of Massachusetts Lowell,* 16 p.,2001.

[7] Gunner C. Larsen, Morten H. Hansen, Andreas Baumgart, Ingemar Carl. " Modal Analysis of Wind Turbine Blades", *Riso–R–1181(EN), ISBN 87–550–2697–4 (Internet), Riso National Laboratory, Roskilde,* Denmark, 72 p., February 2002.

[8] W. Z. Lin, S. P. Lim, and Y. C. Liang, "Proper orthogonal decomposition and component mode synthesis in macromodel generation for the dynamic simulation of a complex MEMS device," *J. Micromech. Microeng.*, vol. 13, pp. 646–654, 2003.

[9] H. M. Park and D. H. Cho, "The use of the Karhunen-Lo`eve decomposition for the modeling of distributed parameter systems," *Chem. Eng. Sci.*, vol. 51, no. 1, pp. 81–98, 1996.

[10] Y. C. Liang, W. Z. Lin, H. P. Lee, S. P. Lim, K. H. Lee, and D. P. Feng, "A neural-network-based method of model reduction for the dynamic simulation of MEMS," *Journal of Micromechanics and Microengineering*, vol. 11, no. 3, pp. 226–233, 2001.

[11] Elmer S. Hung and Stephen D. Senturia." Generating Efficient Dynamical Models for Micro-electro-mechanical Systems from a Few Finite-Element Simulation Runs", *Journal of Microelectromechanical Systems*, vol. 8, no. 3, September 1999.- pp.280-289.

[12] Lynn D. Gabbay, Jan E. Mehner, and Stephen D. Senturia. "Computer-Aided Generation of Nonlinear Reduced-Order Dynamic Macromodels—I: Non-Stress-Stiffened Case", *Journal of Microelectromechanical Systems,,* vol. 9, no. 2, June 2000.- pp.262-269.

[13] Jan E. Mehner, Lynn D. Gabbay, and Stephen D. Senturia. "Computer-Aided Generation of Nonlinear Reduced-Order Dynamic Macromodels—II: Stress-Stiffened Case", *Journal of Microelectromechanical Systems*, vol. 9, no. 2, June 2000.- pp.270-278.

[14] Schlegel, M.; Bennini, F.; Mehner, J.; Herrmann, G.; Muller, D.; Dozel, W.: "Analyzing and Simulation of MEMS in VHDL-AMS Based on Reduced Order FE-Models", *IEEE Sensors 2003*, Second IEEE International Conference on Sensors, Toronto, Canada, 2003.

[15] Bennini, F.; Mehner, J.; Dotzel, W. " Computational Methods for Reduced Order Modeling of Coupled Domain Simulations", *Proc. of 11 International Conference on Solid State Sensors and Actuators (Transducers 01),* Germany, 2001.

[16] M. Schlegel, F. Bennini, J. E. Mehner, G. Herrmann, D. Mueller, and W. Doetzel, "Analyzing and simulation of MEMS in VHDL-AMS based on reduced-order FE models" , *IEEE Sensors J.*, vol. 5, no. 5, 2005.

[17] J. Chen and S. M. Kang, "An algorithm for automatic model reduction of nonlinear MEMS devices," in *Proc. IEEE Int. Symp. Circuits and Syst.*, pp. 445–448, 2000.

[18] E. S. Hung, Y.-J. Yang, and S. D. Senturia, "Low-order models for fast dynamical simulation of MEMS microstructures" , *Digest of Technical Papers, IEEE Int. Conf. on Solid-State Sensors Actuators and Microsystems (Transducers'97)*, pp. 1101–1104, 1997.

[19] L. H. Feng, "Review of model order reduction methods for numerical simulation of nonlinear circuits," *Appl. Math. Comput.*, vol. 167, no. 1, pp. 576–591, 2005.

[20] M. I. Younis, E. M. Abdel-Rahman, and A. Nayfeh, "A reduced order model for electrically actuated microbeam-based MEMS" , *IEEE J. Microelectromech. S.*, vol. 12, no. 5, pp. 672–680, 2003.

[21] A. H. Nayfeh, M. I. Younis, and E. M. Abdel-Rahman, "Reduced-order models for MEMS applications," *Nonlinear Dynamics*, vol. 41, no. 1–3, pp. 211–236, 2005.

[22] Chun-Hsu Ko ,Jin-Chern Chiou." Fuzzy Macromodel for Dynamic Simulation of Microelectromechanical Systems" , *IEEE Trans.on Systems, Man and Cybernetics – Part A: Systems and Humans*, vol. 36, no. 4, pp.823-830, JULY 2006

[23] Pu Li, Yuming Fang. "A Wavelet Interpolation Galerkin Method for the Simulation of MEMS Devices under the Effect of Squeeze Film Damping", Hindawi Publishing Corporation, *Mathematical Problems in Engineering*, Article ID 586718, pp.1-25,2010.

[24] Z. Bai, "Krylov subspace techniques for reduced-order modeling of large-scale dynamical systems," *Applied Numerical Mathematics*, vol. 43, pp. 9–44, 2002.

[25] Zhaojun Bai, Daniel Skoogh," Krylov Subspace Techniques for Reduced-Order Modeling of Nonlinear Dynamical Systems", 2002

[26] Jan Lienemann. "Complexity reduction techniques for advanced MEMS actuators simulation," *Ph.D. dissertation, Albert-Ludwigs Universit at Freiburg im Breisgau*, 298 p.,2006.

[27] A. Odabasioglu, M. Celik, L. T. Pileggi, PRIMA: Passive reduced-order interconnect macromodeling algorithm, *IEEE Trans Comput Aid D*, 17 (8) pp. 645–654,1998.

[28] Lihong Feng. "Review of model order reduction methods for numerical simulation of nonlinear circuits", *Applied Mathematics and Computation, vol. 167* , pp. 576–591,2005.

[29] Z. J. Bai and Y. Su, "SOAR: A second-order Arnoldi method for the solution of the quadratic eigenvalue problem," *SIAM J. Matrix Anal. A*, vol. 26, no. 3, p. 640659, 2005.

[30] Z. Bai, P. M. Dewilde, and R. W. Freund, "Reduced-order modeling," *Numerical Analysis*, vol. 02, pp. 1–59, 2002.

[31] M. Rewienski and J.White, "A trajectory piecewise-linear approach to model order reduction and fast simulation of nonlinear circuits and micromachined devices," *IEEE Trans. Comput.Aid. D.*, vol. 22, p. 155170, 2003.

[32] J. Chen, Sung-Mo Kang, Jun Zou,Chang Liu and José E. Schutt-Ainé. "Reduced-Order Modeling of Weakly Nonlinear MEMS Devices with Taylor-Series Expansion and Arnoldi Approach", *IEEE J. Microelectromech. S*, vol. 13, no. 3, pp.441-446, JUNE 2004.

[33] L. Del Tin, J. Iannacci, R. Gaddi, A. Gnudi, E. B. Rudnyi, A. Greiner and J. G. Korvink. "Non Linear Compact Modeling of RF-MEMS Switches by means of Model order rerduction", *Solid-State Sensors, Actuators and Microsystems International Conference, 2007. TRANSDUCERS 2007. Lyon*, 10-14 June 2007, pp.635 - 638.

[34] A.Petrenko, V.Ladogubets, O.Beznosyk, O.Finogenov, "Using Optimization Procedures to Calculate Parameters of MEMS Macromodels", *Proc. CADSM'2009, 24-28 February, Polyana-Svalyava (Zakarpattya), UKRAINE*, pp.511-514, 2009.

[35] Hsu J.L., Vu-Quoc . A rational formulation of termal circuit models for electrothermal simulation. -part 1: finite element method. *IEEE Trans. Circuits & Systems – I:Fund.Theory &Appl.*, 43, 9, pp.721-732, 1996.

[36] Beznosyk O., Ladogubets V., Finogenov O., Tchkalov O. "Using circuit design software to simulate microelectromechanical components // MEMSTECH 2008,

IV-th International Conference on Perspective Technologies and Methods in MEMS Design, Polyana,Ukraine, pp.130-133, 2008.

[37] H. Tilmans, "Equivalent circuit representation of electromechanical transducers: I. lumped-parameter systems," *IEEE J. Electromicromech. Syst.*, vol. 6, no. 1, pp. 157–176, Mar. 1996.

[38] T. Veijola, H. Kuisma, and J. Lahdenper.a, "Dynamic modelling and simulation of microelectromechanical devices with a circuit simulation program," in *Proc. Int. Conf. on Modeling and Simulation of Mycrosystems (MSM)'98*, pp. 245–50, 1998.38. Sheehan JB.N.

[39] TICER: Realizable Reduction of Extracted RC Circuits // *Digest of Technical Papers – IEEE/ACM Proc. of ICCAD*, pp. 200-203, 1999.

[40] Petrenko A., Sigorsky V., "Algorithmic analysis of electronic circuits", *Western Periodical Corp., San Francisco*, 618 p, 1975.

[41] Petrenko A.I.," RLC – circuits models size reduction ", *Proc. CADSM'05, Lviv-Polena*, p.3-8, 2005.

[42] Petrenko A.I., Ladogubets V.V., Tchkalov V.V., Pudlowsky Z.J. ALLTED - a Computer - Aided System for of Electronic Circuit Design, *UICEE.(UNESCO), Melbourne*, 205 p, 1997.

[43] Beznosyk O., Finogenov O., Ladogubets V. Presentation of a System of Ordinary Differential Equations as an Equivalent Electrical Circuit.- // *Perspective Technologies and Methods in MEMS Design : VI-th International Conference MEMSTECH'2010, 20-23 April 2010,* Lviv-Polyana, Ukraine: proc. – Lviv : Publishing House Vezha&Co, 2010. – P. 116–120.

[44] J.Wibbeler, J.Mehner, F.Vogel, F. Bennini. " Development of ANSYS/Multiphysics Modules for MEMS by CAD-FEM GmbH", *19-th CAD-FEM Users' Meeting 2001, International Congress on FEM Technology, Berlin, Potsdam, October* , 2001, pp.1-10.

[45] Sven Reitz, Jens Bastian, Joachim Haase, Peter Schneider, Peter Schwarz." System Level Modeling of Microsystems using Order Reduction Methods", *Journal Analog Integrated Circuits and Signal Processing, vol.37, Issue 1,* 2003, pp.7-16.

[46] L. Del Tin, R. Gaddi, E. B. Rudnyi, A. Gnudi, A. Greiner, J. G. Korvink, "Nonlinear compact modeling of RF-MEMS switches by means of model order reduction", *Proc. 14th International Conference on Solid-State Sensors, Actuators and Microsystems (Transducer'07), 10-14 June, Lyon* (France).2007.

[47] M. G.Mand G. Ostergaard, "Electro-mechanical trasducer for MEMS analysis in Ansys," in *Proc. Int. Conf. on Modeling and Simulation of Mycrosystems (MSM)'99*, 1999, pp. 270–273.

[48] J. Iannacci, L. Del Tin, R. Gaddi, A. Gnudi and K. J. Rangra,"Compact Modeling of a MEMS Toggle-Switch based on Modified Nodal Analysis", Symposium on Design, Test, Integration and Packagingof MEMS/MOEMS (DTIP) 2005, Montreux, Switzerland, 0103 June 2005.

[49] Peter Schwarz. "Microsystem CAD: From FEM to System Simulation", *Proc. Intern. Conf. "Simulation of Semiconductor Processes and Devices" (SISPAD98), Leuven, 2-4. Sept. 1998*, pp.141-148, Springer, Wien 1998.

[50] Xuan F. Zha, Ram D. Sriram. " Information and Knowledge modeling for computer supported Micro-electro-mechanical systems design and development", *Proceedings of ASME Design Engineering Technical Conference (DETC;2005) September 24-28, 2005, Long Beach, California,* USA, pp.1-11.

Part 3

RF and Optical MEMS

9

Dynamics of RF Micro-Mechanical Capacitive Shunt Switches in Coplanar Waveguide Configuration

Romolo Marcelli[1], Daniele Comastri[1,2], Andrea Lucibello[1],
Giorgio De Angelis[1], Emanuela Proietti[1] and Giancarlo Bartolucci[1,2]
[1]CNR-IMM Roma, Roma,
[2]University of Roma "Tor Vergata", Electronic Engineering Dept., Roma,
Italy

1. Introduction

Micro-electromechanical switches for Radio Frequency applications (RF MEMS switches)[1]-[4] are movable micro-systems which pass from an ON to an OFF state by means of the collapse of a metalized beam. They can be actuated in several ways but, generally, the electrostatic actuation is preferred because no current is flowing in the device nor power absorption has to be involved in the process.

The bias DC voltage signal is usually separated with respect to the RF signal for application purposes. Anyway, in the simplest mechanical model, a voltage difference V is imposed between the metal bridge, connected to the ground plane of a coplanar waveguide (CPW) structure, and the central conductor of the CPW, which also carries the high frequency signal. Under these circumstances, the switch will experience an electrostatic force which is balanced by its mechanical stiffness, measured in terms of a spring constant k. The balance is theoretically obtained until the bridge is going down approximately (1/3) of its initial height. After that, the bridge is fully actuated, and it needs a value of V less than the initial one to remain in the OFF (actuated) position, because contact forces and induced charging effects help in maintaining it in the down position. A general layout of the switch is diagrammed in Fig. 1a, with its simplified equivalent lumped electrical circuit. In Fig. 1b the cross-section of the device is shown, with the quantities to be used for the definition of the geometry and of the physical properties of the structure.

The actuation as well as the de-actuation are affected also by the presence of a medium (typically air, or preferably nitrogen for eliminating humidity residual contributions in a packaged device) which introduces its own friction, causing a damping, and altering the speed of the switch [5]-[7]. Several models are currently available to account for a detailed treatment of the damping, including also the presence of holes in the metal beam [8]-[11]. Moreover, the damping modifies the natural frequency of oscillation for the bridge. In particular, the actuation and de-actuation mechanisms will be consequently affected, leading to *simple oscillations* (no fluid damping contribution) or *damped oscillations* (fluid contribution) up to *over-damping* for particular values of the bridge dimensions or material properties. Experimental problems related to the dynamic characterization of electrostatically actuated switches have been also considered elsewhere to predict and interpret experimental results [12]. Vibrations and damping are key issues in specific applications, and modeling as well as experimental determination are needed [13]. Optical analysis is one of the most advanced techniques for the characterization of movable Microsystems [14][15]. Other contributions to the motion of the switch are given by the repulsive "contact forces" of the bridge with respect to the plane, when it is already actuated or very close to the plane of the CPW, i.e. close to the substrate. They are due to the interaction between the two surfaces and to the local re-arrangement of the charges. The van der Waals force having an attractive effect has to be also included [5][7]. Both last contributions are, of course, very important when the bridge is close to the bottom RF electrode, in contact with the dielectric used for the capacitance needed to get the best RF isolation for the switch, and close to the actuation pad surface. It is difficult to manage the theory involving all of these contributions, and usually a phenomenological approach is followed, trying to individuate the most important parameters useful to describe the required mechanical and electrical response. For instance, higher is the ratio between the bridge thickness and the bridge length, higher will be the spring constant value and, consequently, the robustness of the switch.

In this chapter it will be demonstrated the usefulness of a fully analytical approach, as compared to the commercial software predictions, thanks to the implementation of all the contributions needed for describing the mechanical response of the double beam structure. In particular, further to predict the relevant quantities useful for the dynamical and electrical characteristics of the switch (actuation and release times, capacitance dynamics, ...) a possible optimization of the structure will be proposed for no-contact actuation of the device, in order to minimize the surface and charging effects. One more contribution of this chapter is in the analytical derivation of the actuation voltage, depending on the strength of the applied voltage and the biasing area. This is very important when non-centered actuation voltage is applied, and simplified approaches are no more valid. In fact, this is the real case for devices implemented in RF configurations using the switch as a building block because, for application purposes, the RF and the DC paths have to be distinguished between them. Currently, many papers about the linear and non-linear dynamics of the switch are available in literature [3],[16]-[18] , including also possible collateral effects due to the Casimir force [19][20] or self-actuation mechanisms due to the level of the RF power [21]-[23]. Actually, power is a quantity to be carefully considered for potential applications in bolometers, and for this reason self-actuation was also proposed in nano- and micro-systems to get the power value from the actuation onset [24]. Nonlinear response of a double clamped beam under vibrations induced by RF fields have been recently studied, leading to the onset of chaos under specific solicitations and boundary conditions [25].

Analytical approaches have been also used for the circuital modeling of the switches, by fitting data [3][25][26] or deriving in closed form the expressions for the capacitance and for the resistors and inductors involved in the equivalent circuit [27][28]. Methods with EM considerations in circuit derivation have been discussed [30][31]. Inter-modulation products in the response of RF MEMS capacitors have been also investigated [32]. Thermal effects due to the power handling have to be considered as an additional issue.

A comprehensive study involving analytical and numerical predictions has been performed to obtain a full modeling of the electro-mechanical response of shunt capacitive micro-electro-mechanical (MEMS) switches for radio-frequency (RF) signal processing. The analytical approach was based on uni-dimensional equations, and it has been settled up for

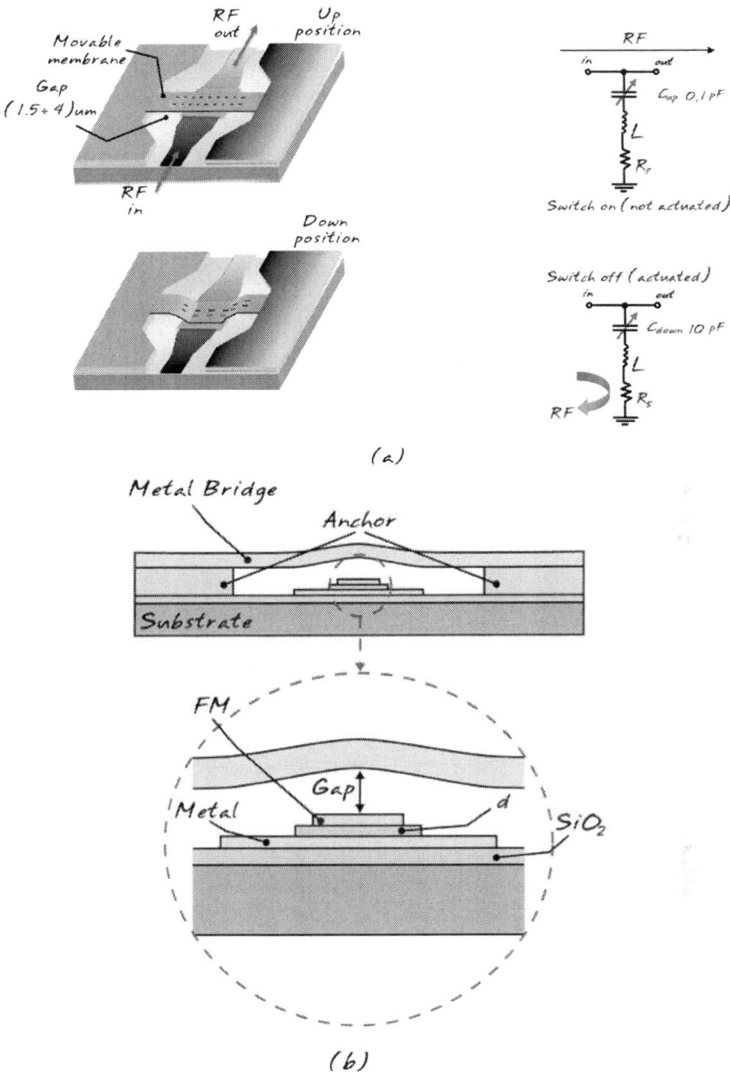

Fig. 1. (a) Schematic diagram of a RF MEMS capacitive shunt switch in coplanar configuration (left) and its equivalent circuit (right); (b) cross section of the switch structure, where the metal bridge is suspended by means of dielectric anchors on a multilayer composed by: (i) the air gap *g* with respect to+ (ii) a metal thin layer at a floating potential (FM) to be used for improving the capacitance definition in the down position, (iii) a dielectric layer with thickness *d* deposited onto (iv) the metal M of the central conductor of the CPW, and finally (v) the SiO_2 thermally grown layer onto the high resistivity silicon wafer.

studying the dynamics of RF MEMS shunt capacitive switches in a coplanar waveguide (CPW) configuration. The capacitance change during the electrostatic actuation and the de-actuation mechanism have been modeled, including the actuation speed of the device and its dependence on the gas damping and geometry. Charging and surface effects have been discussed, in order to select the physical and geometrical parameters useful for obtaining a reliable double clamped structure, eventually proposing simple configurations having no contact, thus minimizing surface and charging effects on the pads used for the electrostatic actuation. The model has been implemented by means of a MATHCAD program. Energy considerations have been included to account for the voltage necessary to maintain the switch actuated, by using a DC bias lower than the actuation voltage. The contribution of the charging mechanism has been discussed to describe both the maintenance voltage and the effects on the device reliability. Moreover, two- and three-dimensional models have been developed by using the commercial software package COMSOL Multi-physics, and they have been compared with the analytical and numerical results.
The above approach has been done for predicting the dynamical response of a capacitive shunt switch, but it is generally valid for resistive series as well as for capacitive shunt devices, because it has been implemented for a double clamped beam. Experimental findings confirm the validity of the proposed analytical approach, which is especially useful when a fast preliminary design of the device is needed.

2. One-dimensional model of the mechanical response of shunt capacitive RF MEMS switch

The equation which accounts for the most part of the above introduced physical mechanisms can be written as:

$$m\ddot{z} = F_e + F_s + F_p + F_d + F_c \tag{1}$$

where: $m = \rho At$ is the bridge mass, computed by means of the material density ρ, the area A and the thickness of the bridge t; $F_e = \frac{1}{2}\frac{\partial C}{\partial z}V^2$ is the electric force due to the applied voltage V and to the change of the capacitance along the direction of the motion z; $F_p = -k[z-(d+g)]$ is the force due to the equivalent spring characterized by its constant k, acting against the electrical force to carry back the bridge to the equilibrium position; $F_s = -k_s[z-(d+g)]^3$ is the nonlinear stretching component of the spring constant [3]; $F_d = -\alpha\dot{z}$ is the damping force due to the fluid, dependent on the bridge velocity $\dot{z}$ and on the damping parameter α, which, in turns, is related to the geometry of the bridge and to the viscosity of the medium; F_c is the contact contribution, which can be divided in the van der Waals and surface forces, the first acting as attractive and the second one as repulsive, with a possible equilibrium position at a given distance from the bottom electrode of the switch [5].
The total capacitance of a shunt capacitive MEMS switch can be described in terms of two series capacitors, each of them having its own dielectric constant. This is only a formal way to approach the problem, because the intermediate plate is a dielectric interface and not a metal one. From the above considerations, and with reference to Fig. 1b, the total capacitance will be:

$$C(z) = \frac{\varepsilon_0 \varepsilon_r A}{d + \varepsilon_r (z - d)}; \qquad z \in [d, \quad d + g] \tag{2}$$

Where ε_0=8.85×10^{-12} is the vacuum dielectric constant in MKS units and ε_r is the relative dielectric constant of the material covering the bottom electrode. The derivative of $C(z)$, to be used in the definition of the electric force $F_e = \frac{1}{2}\frac{\partial C}{\partial z}V^2$ is given by:

$$\frac{\partial C}{\partial z} = -\frac{\varepsilon_0 \varepsilon_r^2}{[d + \varepsilon_r (z - d)]^2} A \tag{3}$$

and Eq. (1) can be re-written as:

$$m\ddot{z} + k[z - (d + g)] + k_s[z - (d + g)]^3 + \alpha \dot{z} = -\frac{1}{2}\frac{\varepsilon_0 \varepsilon_r^2 A}{[d + \varepsilon_r (z - d)]^2} V^2 \tag{4}$$

which can be transformed in:

$$\ddot{\zeta} + \beta \dot{\zeta} + \omega^2 \zeta + \frac{k_s}{m}\zeta^3 = B(\zeta) V^2 \tag{5}$$

where:

$$\begin{aligned} \zeta &= z - (d + g) \\ \beta &= \frac{\alpha}{m} \\ \omega &= \sqrt{\frac{k}{m}} \\ B(\zeta) &= -\frac{1}{2m}\frac{\varepsilon_0 \varepsilon_r^2}{[d + \varepsilon_r (\zeta + g)]^2} A \end{aligned} \tag{6}$$

The spring constant k is a measure of the potential energy of the bridge accumulated as a consequence of its mechanical response to the electrical force due to the applied voltage V. An approximated definition of it for central actuation can be given by [16]:

$$k = K_1\left(32Ewr^3\right) + K_2\left[8\sigma(1 - \nu)wr\right] \tag{7}$$

where:

$$\begin{aligned} K_1 &= \frac{1}{2 - \left(2 - \frac{L_c}{L}\right)\left(\frac{L_c}{L}\right)^2}; K_2 = \frac{1}{2 - \frac{L_c}{L}} \\ r &= \frac{t}{L} \end{aligned} \tag{8}$$

L is the bridge total length, L_c is the switch length in the RF contact region (width of the central conductor of the CPW), w is the bridge width, before the contact region, which can be also an averaged value as an approximation of the real cases if a tapering is present, t is the Au thickness of the bridge. The other parameters are the Young modulus E, the residual stress σ and the Poisson coefficient ν. As well established, the Young modulus is an intrinsic property of the material, and specifically it is a measure of its stiffness [33]
The Poisson coefficient ν is a measure related to the response of a material when it is stretched in one direction, and it tends to get thinner in the other two directions.
σ is mainly related to the process for obtaining the mechanical structure, and it must be measured in the real case. Actually, σ is a measure of the stress which remains after the original cause of the stresses (external forces, heat gradient) has been removed. They remain along a cross section of the component, even without the external cause.
A new, and more accurate definition of k has been recently given [35] for the treatment of miniature RF MEMS switches. Actually, it turns out that [35]:

$$\begin{aligned} L_{nc} &= L - L_c \\ L_d &= L_{nc}/2 \\ k_{new} &= \frac{2Ewt^3}{L_d^3} + \frac{Ewtg^2}{L_d^3} + \frac{2\sigma(1-\nu)tw}{L_d} \end{aligned} \quad (9)$$

where k_{new} is the new definition of the spring constant, to be compared with that given in Eq. (8). Generally, $k_{new} > k$ and bridges shorter than 500 μm ca. are better approximated by the k_{new} definition, exhibiting higher actuation voltages.
A recent experimental approach was also adopted for evaluating the contribution of the spring constant and for modeling it on the base of nano-indentation techniques [35].
All the quantities previously introduced have to be re-defined because of the presence of holes in the released beam. The holes need to be used for an easier removal of the sacrificial layer under the beam, and for mitigating the stiffness of the gold metal bridge, i.e. for better controlling the applied voltage necessary for collapsing it, to have not values too high because of the residual stress.
In this framework, we have re-calibrated the material properties accounting for the holes distribution on the metal beam. Literature definitions [3][37] are generally accepted for analytically describing the effect of the holes by means of the pitch, i.e. the center-to-center distance p between the holes and the edge-to-edge distance l. The situation is explained in the Fig. 2. In this way, the *ligament efficiency* will be given by the term $(1-(l/p))$ and such a term will be used in this paper for evaluating the effective quantities which are decreased with respect to the original one. Following this approach, $\sigma_{eff}=\sigma(1-(l/p))$, while $E_{eff}=E(1-(l/p))$, $\nu_{eff}=\nu(1-(l/p))$.
For the effective mass, we preferred to use a definition based on the ratio between the area with and without the holes, thus obtaining $m_{eff}=m(A/A_0)$, where A_0 is the geometrical area of the beam and A is the effective one considering the presence of the holes. All the evaluations which will be shown in this chapter are based on the previously defined quantities, calculated accounting for their effective contribution, but they have been written without the *eff*-pedex, for the sake of simplicity.

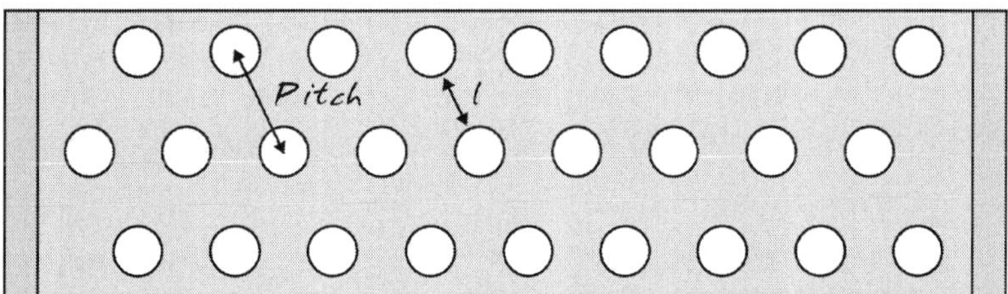

Fig. 2. Typical shape of a perforated beam used for RF MEMS double clamped switches. Holes are realized for facilitating the sacrificial layer removal and their position, number and dimensions are properly tailored depending on the application.

We included in this contribution a generalization of the mechanical evaluation for *k*. Actually, in the previous definitions, the spring constant is approximated accounting for a load in the center of the beam, but a constant load can be applied in different places along the beam, and the *k* value has to be calculated depending on the real position and extension of the applied force with respect to the bridge area. Two contributions are expected, the first one coming from the inertia moment of the beam and its Young modulus, and the second one having a technological origin due to the residual stress of the released bridge. Generalizing the equations introduced in literature about this topic [3] we can write:

$$k' = -\frac{LEI}{2}\frac{1}{\int_{x1}^{x2} f(a)da} \quad ; \quad f(a) = \frac{1}{48}\left(L^3 - 6L^2a + 9La^2 - 4a^3\right)$$
$$k'' = \frac{L}{2}\frac{1}{\int_{x1}^{x2} g(a)da} \quad ; \quad g(a) = \frac{L-a}{2\sigma(1-\nu)wt} \tag{10}$$
$$k = k' + k''$$

Where $I = \frac{wt^3}{12}$ is the inertia moment classical calculation. Following the above definitions, we can effectively derive a *k*-value depending on the actuation position and on the size of the pads used to impose the voltage between the beam and the bottom electrode necessary for such a collapse, without approximations.

The threshold voltage, obtained by assuming that the mechanical structure collapses after trespassing the critical distance $(1/3)g$, is given by [3][17]:

$$V_{threshold} = \sqrt{8\frac{kg^3}{27\varepsilon_0 A}} \tag{11}$$

For the results coming from all the next simulations, we assume the following parameters for the bridge: L=600 µm as the bridge total length, L_c=300 µm as the switch length in the RF contact region (width of the central conductor of the CPW), w=100 µm as the bridge width, before the contact region, to be considered as an averaged value to fit the real cases if a tapering is present, w_S=100 µm for the switch width (transversal dimension of the switch, parallel with respect to the CPW direction), d=thickness of the dielectric material=0.2 µm, with dielectric constant ε=3.94 (SiO_2), t=1.5 µm for the gold bridge, ρ=19320 kg/m^3 for the

gold density, E=Young modulus=80×10^9 Pa, ν=0.42 for the metal Poisson coefficient and σ=18 MPa as the residual stress of the metal (measured on specific micromechanical test structures). The above defined physical quantities, as already discussed in the previous section, have been re-calculated because of the ligament efficiency.
An evaluation of k and of the actuation voltage has been given, according to the general geometry introduced in Fig. 3. In particular, holes with 10 μm diameter, side-to-side

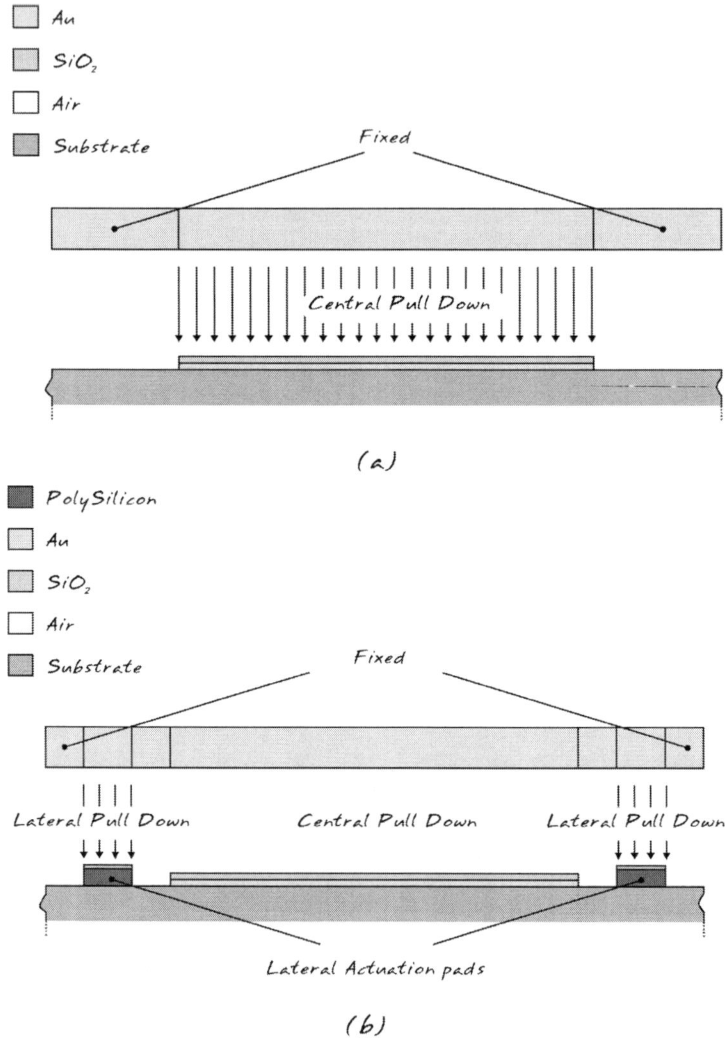

Fig. 3. Cross section of the geometry used for the simulation of the pull down of the metal beam: (a) by using the central actuation, and (b) by means of the lateral pads.

distance l=10 μm and center-to-center distance *pitch*=20 μm have been imposed. For the position of the actuation pads, we have chosen two possibilities: (i) central actuation, distributing the force over the entire area defined by the switch width and by the width of the CPW; and (ii) lateral actuation, by means of two poly-silicon pads 50 μm wide and long as the bridge width, centered with respect to the slot of the CPW. With reference to the following Fig. 3 and to Table I, the integration for obtaining the value of k to be used in Eq. (11) has been performed by choosing x_1 and x_2 by means of a coordinate system having the origin on the left side of the bridge. In this way, for the central actuation, x_1=L/2 and x_2=(L/2)+(L_c/2); because of the symmetry of the analyzed structure, the integration is performed only on half of the length where the force is applied, and multiplied by a factor two (see Eq. (10)). For the lateral actuation, the same symmetry is invoked, and we have x_1=(L/2)+(L_c/2)+d_{pad} and x_2=(L/2)+(L_c/2)+d_{pad}+L_{pad}, i.e. from the beginning to the end of the dielectric actuation pad for its entire length L_{pad}, distant d_{pad} from the edge of the central conductor of the CPW.
From the above data and from the k-values and actuation voltages V_{act} calculated for both actuation schemes (central and lateral) we obtained the results shown in Table 1:

	Central Actuation Eq. (7) and Eq. (10)	Lateral Actuation Eq. (10)
k [N m^{-1}]	10, 21	208
V_{act} [V]	13, 18	56

Table 1. Spring constant values and actuation voltages for the exploited geometry. It is worth noting that the voltage needed for the central actuation calculated by Eq. (7) is only an approximation with respect to that calculated by the exact Eq. (10); moreover the lateral actuation can be predicted only by using Eq. (10).

It is evident that the actuation requires a voltage applied in the center of the bridge much lower with respect to that needed at the sides of the double clamped structure. This is immediately understandable because of the higher value of the spring constant when the lateral actuation is used. Actually, in this modeling, the value of k is the parameter which accounts for the geometry used for the actuation of the switch.
Now, the particular and the general cases where voltage and damping are present or absent will be studied.

2.1 Case V=0, β=0

In this elementary case a small voltage can be applied just as a perturbation to the bridge, and after that the source is turned off, leaving the bridge moving in a non-dissipative environment. The cubic term due to the stretching is negligible and Eq. (5) is simplified in the well known equation of the harmonic oscillator:

$$\ddot{\zeta} + \omega^2 \zeta = 0 \tag{12}$$

The simple solution of Eq. (12) for the harmonic oscillator can be found by using a complex formalism, as:

$$\zeta(t) = \zeta_0 \exp[-i(\omega t + \varphi)] \tag{13}$$

and the real part will be:

$$\zeta(t) = \zeta_0 \cos(\omega t + \varphi) \tag{14}$$

with $\zeta_0 = \zeta_{0,\max} = \frac{1}{3}g$, and $\varphi = \frac{\pi}{2}$. The maximum value of the amplitude is given by the maximum allowed oscillation before a possible collapse induced by the applied voltage, and the phase is given by assuming an initial motion towards the bottom electrode. Then, the time dependence of the vertical coordinate to describe the maximum oscillation can be written as:

$$z(t) = (d + g) - \frac{1}{3} g \sin(\omega t) \tag{15}$$

This is a well established solution for the un-damped motion equation, but real cases always need the contribution of damped oscillations, which have been studied and will be presented in detail in the following sections.

2.2 Case V=0, β≠0

The presence of environmental intrinsic damping in the motion of the bridge will cause Eq. (5) to be re-written in the following way:

$$\ddot{\zeta} + \beta\dot{\zeta} + \omega^2\zeta + \frac{k_s}{m}\zeta^3 = 0 \tag{16}$$

As already mentioned, there are literature results concerning the correct analytical modeling of the medium in which a metal membrane is moving. By using a phenomenological approach, the effect of the damping can be modeled by defining a complex radian frequency as:

$$\omega' = \omega - i\Omega \tag{17}$$

where, usually but not necessarily, $\Omega << \omega$. The complex solution for Eq. (16) will be:

$$\zeta(t) = \zeta_0' \exp[-i(\omega' t + \varphi)] \tag{18}$$

from which it can be inferred the relation between the natural radian frequency and that modified by the damping in the following form:

$$\omega' = \omega\sqrt{1 - \left(\frac{\beta}{2\omega}\right)^2} \tag{19}$$

It means that a decrease of the natural frequency is expected when the damping is present. In many physical situations, the condition $\beta/2\omega << 1$ holds. This will depend on the geometry of the bridge and on the intrinsic damping of the medium. The ratio $\beta/2\omega$ is a measure of the influence of the damping, and it will cause a different dynamical response [3][5][6]. The complex solution for the Eq. (16) with damping is:

$$\zeta(t) = \zeta_0' \exp[-i(\omega' t + \varphi)] \exp(-\frac{\beta}{2} t) \tag{20}$$

It could happen that $\beta/2\omega<1$ or $\beta/2\omega>1$ depending on the damping and the resonant frequency values. Consequently, an oscillating response or an over-damping could be obtained. For this reason, the Real Part of the Complex Solution has to be taken:

$$\zeta(t)=\zeta_0'\mathrm{Re}\{\exp(i(\omega't+\varphi))\}\exp\left(-\frac{\beta}{2}t\right) \tag{21}$$

Analogously to the case of non-damped oscillations, and coherently with the maximum allowed oscillation before the possible collapse, it will be:

$$z(t)=(d+g)-\zeta_0'\mathrm{Re}\{\exp(i(\omega't+\varphi))\}\exp\left(-\frac{\beta}{2}t\right) \tag{22}$$

where $\zeta_0'\leq\zeta_{0,\max}'=g/3$ and $\varphi=\pi/2$. Formulations of the damping term for a given geometry can be found in literature [3][5][7] and, as well established, the velocity of the bridge plays a key role together with the geometry and the viscosity of the medium. For such a contribution we can write [16]:

$$\begin{aligned}\alpha&=\frac{3}{2\pi}\mu\frac{A^2}{g^3}\\ \mu&=1.2566\times10^{-6}\sqrt{T}\left(1+\frac{T_0}{T}\right)^{-1}\end{aligned} \tag{23}$$

A temperature dependence has been included, which accounts for the deformations induced by the change in the total length of the beam [3]. μ is the air viscosity, A is the full effective area of the bridge, including the ligament efficiency, T_0=110.33 K and T is the absolute temperature [3]. In our case, T=300 K. Different formulations appear in other papers [7], and it is the result of an approximation introduced in the equations used by us [3], but it does not change the conceptual approach to the problem. Other available results concern with the operation of the switch in harsh environment [38]. The response of the bridge is given in the following Fig. 4a. It is worth noting that by using the values imposed for the geometry of the exploited device, and considering that, by means of the definition of k given in Eq. (10), is valid the condition $\beta/2\omega>1$ for the central actuation, an over-damped flat solution is obtained in this case, without oscillations. On the other hand, for lateral actuation, as it is diagrammed in Fig. 3b, the beam exhibits a higher value for the spring constant, and the condition for the damping is $\beta/2\omega<1$, thus resulting in damped oscillations. See the results shown in Fig. 4b for a beam laterally actuated.

2.3 De-actuation of the bridge

The dielectric used for the RF MEMS switch capacitance is often SiO_2 or Si_3N_4, having ε_r=3.94 or 7.6 respectively, or any other dielectric material assuring a high C_{OFF}/C_{ON} ratio. Current results on high-ε materials are very promising for improving the isolation ratio [38]. The thickness of the dielectric material is usually in the order of tens to a hundred of nm, to provide, with the dimension of the gap g, the proper ratio of the capacitance between the ON (bridge up) and the OFF position (bridge down). By imposing g=2.8 μm and by using a 100 nm thick layer of SiO_2, from Eq. (2), the ratio is defined as:

$$R = \frac{d + \varepsilon_r g}{d} = 1 + \varepsilon_r \frac{g}{d} \approx \varepsilon_r \frac{g}{d} \quad (24)$$

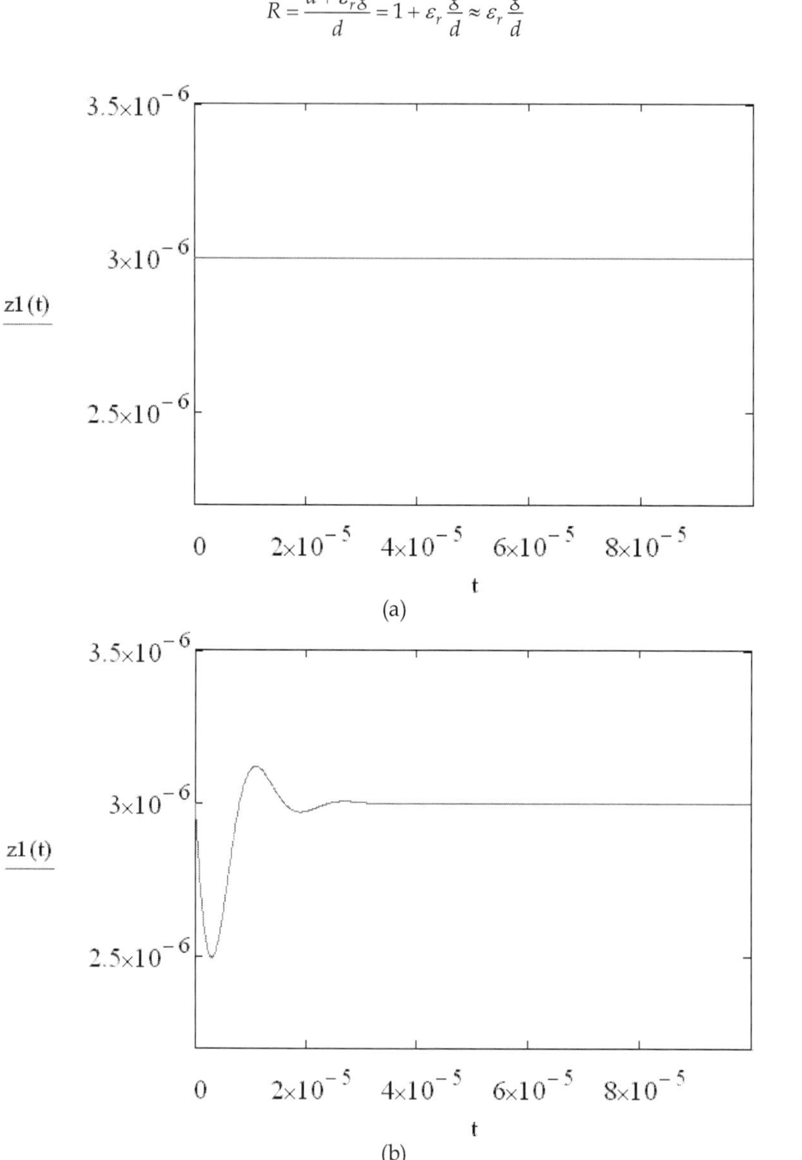

Fig. 4. Damped oscillations around the equilibrium position for (a) a centrally actuated, and (b) a laterally actuated beam. Higher values of k (lateral actuation) correspond to enhanced oscillations before obtaining again the equilibrium position. The time t is in sec and the vertical coordinate is in m.

where the last equation is valid in most practical cases, being $\varepsilon_r(g/d) >> 1$. In our case, the nominal ratio is $R \approx 56$ for the above imposed values of the parameters useful for describing the bridge mechanics by using SiO_2. Actually, as it has been discussed elsewhere[28], the OFF position of the switch is not characterized by a completely flat bridge fully adapted to the bottom electrode, thus contributing to an isolation between the two states lower than expected, unless to use different technological solutions like the floating metal one[39]. In the case of de-actuation, the full solution can be deduced by using the following formulas:

$$z_{deact}(t) = d + g + \zeta_0 \operatorname{Re}\{\exp(i\omega' t)\} \exp\left(-\frac{\beta}{2}t\right)$$
$$\zeta_0 = -g \tag{25}$$

All of the above described analytical solutions of the motion will involve a dynamical response of the capacitance C, which shall change according to the variation of z. Some detailed plots for the change of the capacitance as a function of time will be presented as a novel contribution with respect to the purely mechanical considerations. This is very important in the evaluation of the transient times useful to define the effective response and recovery times of the RF MEMS devices.

In the following Fig. 5 and Fig. 6, an example of the de-actuation response and of the C-response are shown, quite in agreement with similar results obtained elsewhere[29] but in this case we want to stress the extension of the analytical approach to the lateral actuation mechanism.

2.4 Actuation of the bridge

The actuation process is in principle more complicated, because the computation should involve the presence of the external force and of the contact forces when the bridge is very close to the contact area with the substrate. At a first approximation, without including the above contributions, we can write: $\zeta(t) = \zeta_0' \exp[-i(\omega' t + \varphi)] \exp(-\frac{\beta}{2}t)$, which will lead to:

$$z_{act}(t) = d + \zeta_0 \operatorname{Re}\{\exp(i\omega' t)\} \exp\left(-\frac{\beta}{2}t\right)$$
$$\zeta_0 = g \tag{26}$$

This is the solution for the full equation with the initial conditions given by $z(0)=d+g$, i.e. with the bridge in the up position, pushed down by a voltage over the threshold. Actually, only positive values are allowed, because of the constrain due to the presence of the substrate. In particular, the solution for the actuation $z_{act}(t)$ will be:

$$z_{act}(t) = d - g \,|\operatorname{Re}\{\exp(i\omega' t)\}|\, \exp\left(-\frac{\beta}{2}t\right) \tag{27}$$

Intuitively, it means that the beam is expected to bounce before the full, final collapse, and the bridge will attempt to restore its initial position, but the applied voltage will force it to

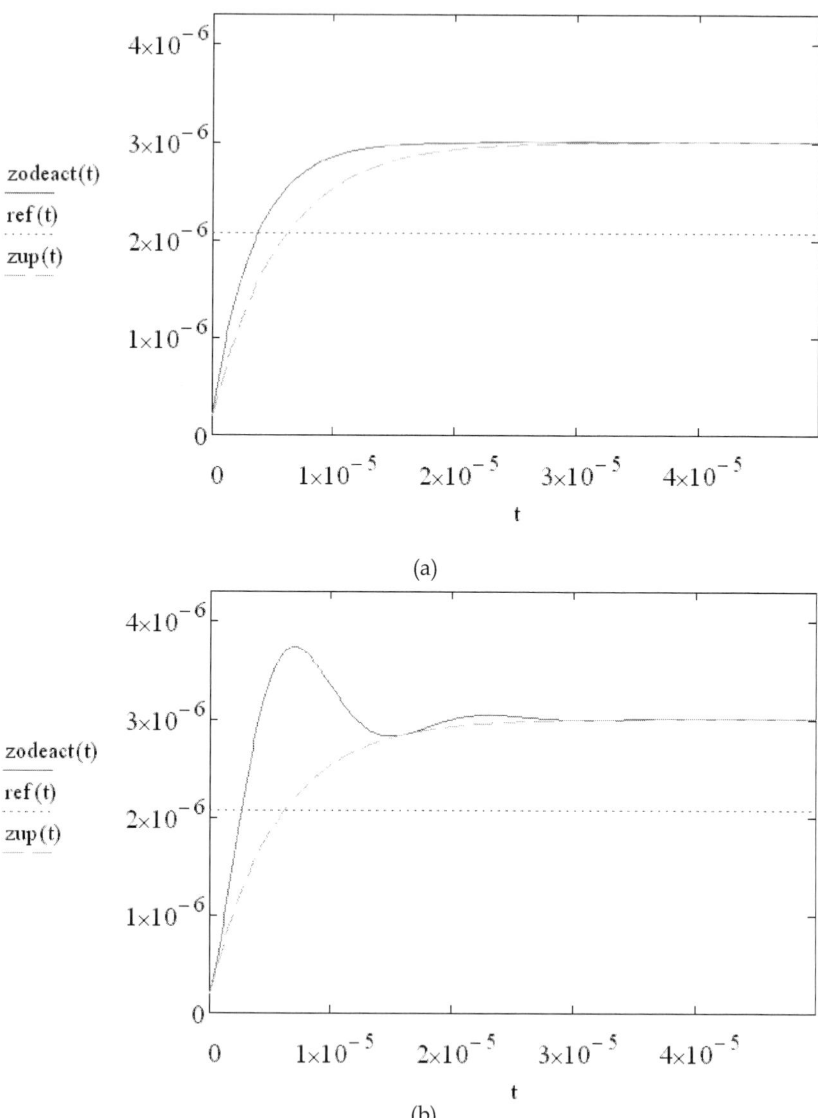

Fig. 5. De-actuation Vs time for the shunt switch used as an example (1.5 μm thick). The dashed curve (zup) accounts only for the exponential restoring mechanism, while the full one (zodeact) accounts also for the air damping. The dotted curve (ref) is the 1/3 limit of the gap for the collapse of the bridge. In (a) the central actuation is simulated, while in (b) it is shown the response for the lateral one.

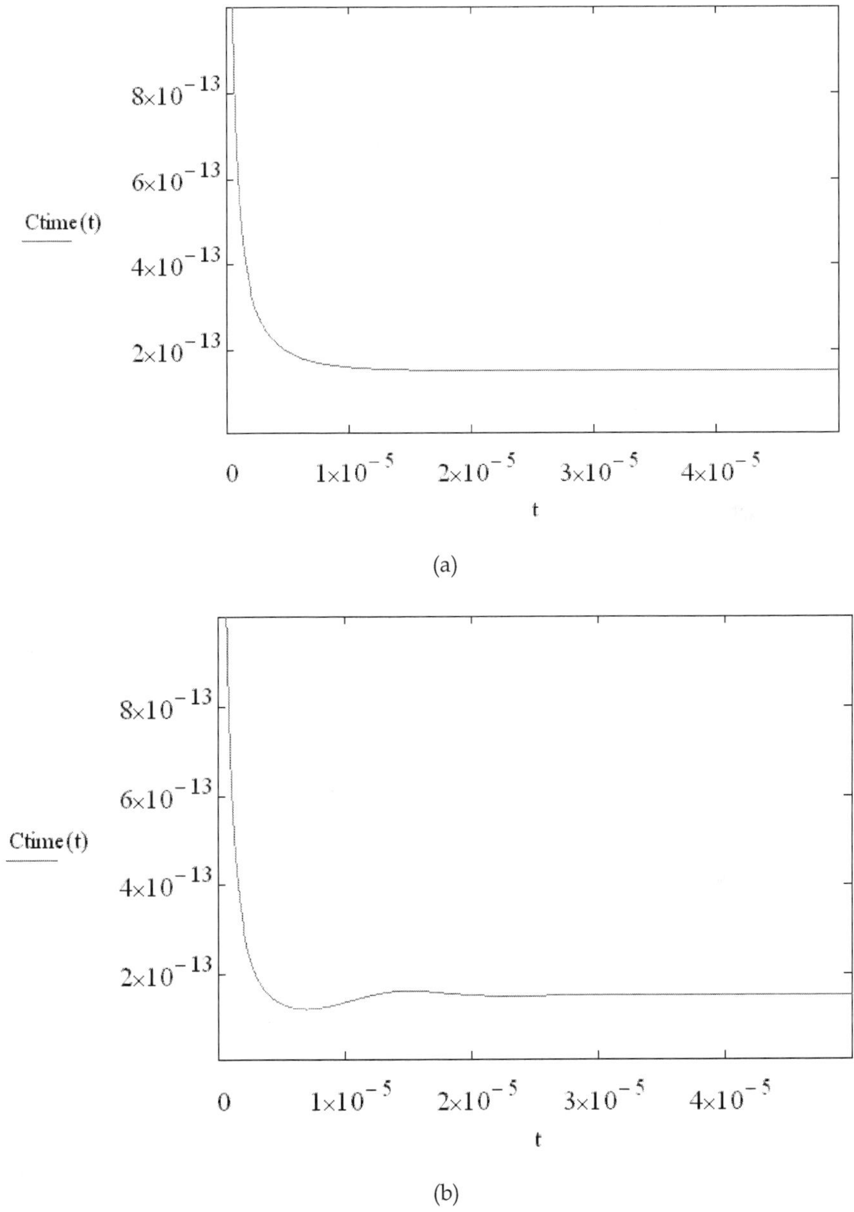

Fig. 6. Change of the capacitance Vs time for the RF MEMS switch (Ctime), in the case of (a) central actuation and (b) the lateral one.

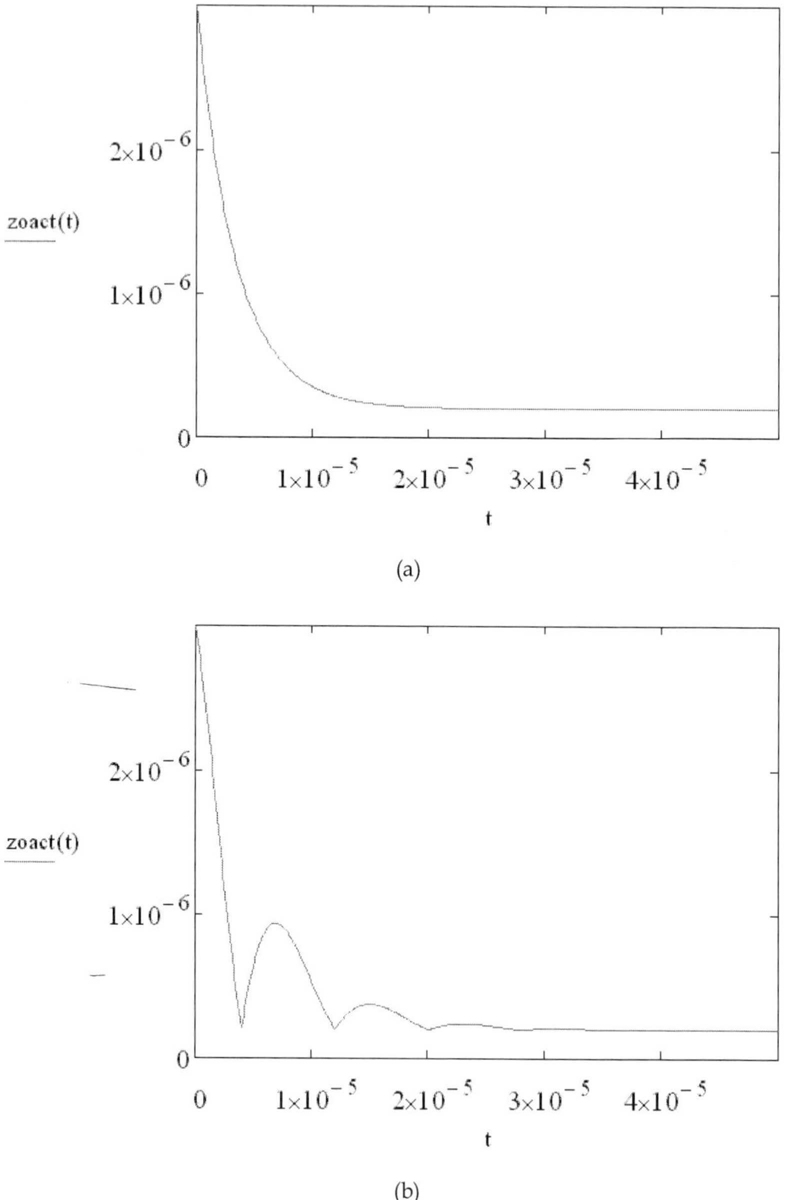

Fig. 7. Actuation of the bridge without accounting for contact forces. The curve (z_{oact}) is used to describe: (a) the central actuation for the 1.5 μm thick beam and (b) the lateral one. It is worth noting the number of bounces before the onset of the full actuation.

stay down. So far, the damped oscillation is due to the analytical solution of the motion equation, corresponding to a possible bounce before the full actuation of the bridge. Following this approximation, we shall have the situation depicted in Fig. 7a and 7b. It is worth noting that only for a high value of k, like it is for the lateral actuation, the bouncing effect is enhanced, while for the central one no bounces are expected for the imposed geometry. In fact, the condition $\beta/2\omega < 1$ is valid only for the lateral actuation, while it is $\beta/2\omega > 1$ for the central one, characterized by an imaginary solution and by over-damping. Of course, a compromise has to be chosen between the required actuation speed of the switch and its robustness and lifetime. It has also to be considered that the beam will be stressed more times when it is thicker, because also in this case we get an increase in the value of the spring constant and of the number of possible bounces, but a thin beam could be less reliable because of the experienced fatigue. A similar situation is obtained when you will have higher values for the residual stress, contributing again in the stiffness of the beam.
On the other hand, at least for a limited time, the van der Waals and contact forces will try to maintain the bridge in the DOWN position, introducing additional corrections. This situation is well described in other papers[18], where a solution similar to that presented in Fig.7 is given by studying an ohmic switch. Because of this effect, it is also claimed that a reliability analysis based on the nominal number of actuations is not correct, because it should include the effective number of bounces before the full actuation, each of them contributing to the fatigue of the beam.

3. Energy considerations and switching times

In this section we shall derive useful formulas for the evaluation of the switching times, based on energy computations. The electrostatic energy given by a voltage generator to a capacitive shunt switch in the ON state (UP position of the bridge), at the threshold for the electrostatic actuation, is given by:

$$E_{initial} = \frac{1}{2} C_{ON} V_{threshold}^2 \tag{28}$$

The threshold value $V_{threshold}$ of the voltage has been used to account for the full actuation of the switch when such a voltage is imposed. The system is non-conservative, and the final energy of the actuated beam (corresponding to the value of the OFF capacitance) will be changed by the contact and dielectric charging contributions, which have to be greater than the restoring mechanical energy for maintaining the beam in the DOWN position:

$$E_{final} = \frac{1}{2} C_{OFF} V_{threshold}^2 + E_c + E_{charge} \geq \frac{1}{2} k g^2 + \frac{1}{4} k_s g^4 \tag{29}$$

Where C_{OFF} is the capacitance when the switch is in the OFF state (bridge DOWN), and $(1/2)kg^2+(1/4)k_s g^4$ is the mechanical spring energy including the stretching contribution. As an implementation of the treatment given in classical literature about this topic [3], we have included additional terms [5],[7] and the contribution coming from the charging of the dielectric [41]. In fact, when the bridge is in the DOWN position, two more effects have to be studied: (i) the contact energy E_c (repulsive and van der Waals), and (ii) the energy due to the charging process of the dielectric E_{charge}. To maintain the bridge in the DOWN position, accounting also for the above two additional terms, a holding down voltage V_{min} less than the threshold one can be applied. In this case the balance is obtained between the mechanical restoring energy and the electrostatic energy for the capacitance in the OFF state, plus the contact and charge terms, as:

$$\frac{1}{2}kg^2 + \frac{1}{4}k_s g^4 = \frac{1}{2}C_{OFF}V_{\min}^2 + E_c + E_{charge} \quad (30)$$

And the maintenance voltage V_{min} can be written as:

$$V_{\min} = \sqrt{\frac{kg^2 + \frac{1}{2}k_s g^4 - 2\left(E_c + E_{charge}\right)}{C_{OFF}}} \quad (31)$$

i.e., V_{min} can be determined by the spring constant k, the gap g, and the value of the capacitance in the OFF state (C_{OFF}). An estimation of the terms like E_c and its dependence on the position of the bridge is available[5],[42]. When successive actuations are performed, and considering E_c as an offset contribution, E_{charge} will increase up to a value where the switch will remain stuck also for V_{min}=0, unless to choose properly materials and geometry for the uni-polar bias scheme. Actually, V_{min}=0 will correspond to the sticking of the bridge, i.e. when $kg^2 + \frac{1}{2}k_s g^4 - 2\left(E_c + E_{charge}\right) = 0$. The threshold voltage, obtained by assuming that the mechanical structure collapses after trespassing the critical distance $(1/3)g$, is given by Eq.(11)[3],[17]. In the cases studied in this paper, and by using the same geometrical and physical data defined for our test structure, the change from the definition of k for central actuation to that for the lateral one can correspond to more than doubling the threshold voltage. From the above Eq (31), simple considerations can be anticipated about the contribution of charging processes and possible sticking of the bridge. It is known that successive applications of the voltage correspond to an accumulation of charge in the dielectric and in an increase of the actuation voltage[43], with an asymptotic trend up to its maximum value[44],[45]. In fact, if it is not given time enough to the charges to be dissipated, they grow up to the maximum contribution allowed by the geometry and by the properties of the dielectric material used for the capacitive response (in the case of central actuation) or by the dielectric used for the actuation pads (lateral actuation). The substrate too, when it is oxidized for improving the isolation, is a source of charges.

The switching times can be evaluated by means of the capacitance dynamics. Actually, exception done for the oscillations superimposed to the exponential decay, the envelope of the z-quote when the threshold voltage is applied will describe the change of the capacitance up to the full actuation of the device. As an approximation, we can say that the switch is *actuated* when the capacitance is 90% of that obtained from the full actuation procedure. In Fig. 8 the curve containing the oscillations of the capacitance, the maximum value when the bridge is in the DOWN position, and the reference value corresponding to $0.9C_{max}$ are shown.

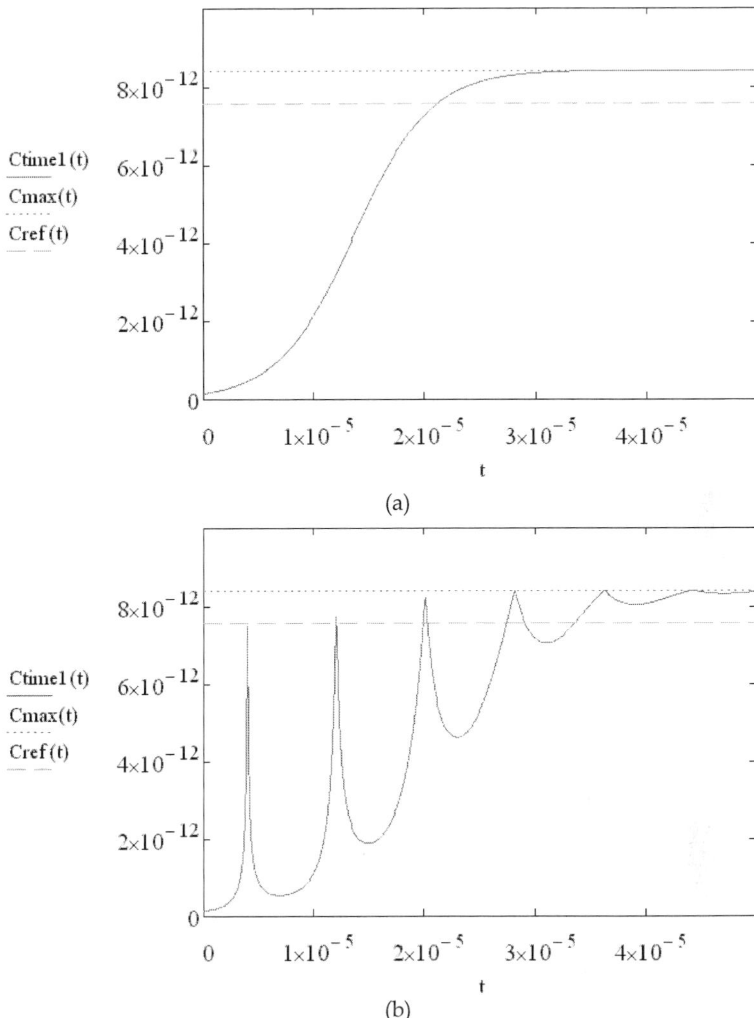

Fig. 8. Simulation of the actuation of the bridge. The curves describing the change of the capacitance (in F) during the actuation time (in s) are: (i) full curve, $C_{time1}(t)$, with the entire response by using k, (ii) flat full reference line C_{max}, i.e. the maximum value obtained when the switch is fully actuated, (iii) C_{ref}–$0.9C_{max}$. Both possible actuations are described: (a) central pull-down, and (b) lateral one.

It is worth noting from Fig. 8 the significant change of the oscillations induced by the different actuation choices. On the other hand, we can simplify the prediction for the actuation time by using the evolution of the envelope describing the actuation, which is written as:

$$C_{act}(t) = \frac{C_{OFF}}{1 + \varepsilon \frac{g}{d} \exp\left(-\frac{\beta}{2}t\right)} \tag{32}$$

The actuation occurs when C_{ref} crosses the envelope described by the previous Eq. (32). In formulae, after simple algebraic passages, it will be:

$$\tau_{act} = -\frac{2}{\beta} \ln\left(\frac{1}{9} \frac{d}{\varepsilon g}\right) \tag{33}$$

By using again the values used for the exploited structure, and assuming that the dielectric is SiO_2, it is ε=3.94 and $\tau_{act}\cong 35$ μs. The same evaluation can be done for the de-actuation, by using the results coming from Fig. 7 and the related analytical treatment. Actually, the evolution of the capacitance can be written, by using the exponential law for the envelope:

$$C_{deact}(t) = C_{ON} \frac{d + \varepsilon g}{d + \varepsilon g \left[1 - \exp\left(-\frac{\beta}{2}t\right)\right]} \tag{34}$$

From which it turns out, if $C_{deact}(\tau_{deact})=0.9C_{ON}$:

$$\tau_{deact} = -\frac{2}{\beta} \ln\left[\frac{1}{9}\left(1 - \frac{d}{\varepsilon g}\right)\right] \tag{35}$$

By using the same values imposed for the actuation, we obtain $\tau_{deact}\cong 12$ μs. It has to be stressed that both evaluations do not include the contact energy neither the charging contributions, so they should be corrected, but the order of magnitude should not change, exception done for some delay in the restoring mechanism when the switch is de-actuated, which should increase the τ_{deact} value.

The above evaluations concern with the response of the switch at the threshold voltage for the actuation. On the other hand, by increasing the applied voltage above such a threshold value, the velocity for the actuation can be increased too, as it can be obtained by considering the energy spent in the actuation process by performing the integral of the motion equation without accounting for contact or charging:

$$\begin{aligned}
&\int_{d+g}^{d} m\ddot{z}dz = m\int_{v(d+g)}^{v(d)} \dot{z}d\dot{z} = \frac{1}{2}m\left[v^2(d) - v^2(d+g)\right] = E_k = \int_{d+g}^{d}(F_e + F_m + F_d)dz \\
&\int_{d+g}^{d} F_e dz = \frac{1}{2}(C_{OFF} - C_{ON})V^2 \\
&\int_{d+g}^{d} F_m dz = -\frac{1}{2}kg^2 \\
&\int_{d+g}^{d} F_s dz = -\frac{1}{4}k_s g^4 \\
&\int_{d+g}^{d} F_d dz = -\frac{\alpha}{\omega}\left[v^2(d) - v^2(d+g)\right]
\end{aligned} \tag{36}$$

Where E_k is the kinetic energy and E_d is the dissipated one, while $\omega = \sqrt{k/m}$. The dissipated energy has been calculated accounting for the dissipated power $P_d = F_d v = \alpha v^2 = \omega E_d$. The effect of k_s is marginal if we are far from the actuation region, but it becomes important close to the full collapse of the beam.

The above modeling is valid under the assumption that we are very close to the full actuation of the switch, but far enough to be obliged in considering the van der Waals and the contact contributions. The initial velocity of the bridge is obtained by $v(d+g)=v_{in}=0$ because it is at rest before the application of the voltage.

From Eq. (36) it turns out:

$$v_{act}(d) = \sqrt{\frac{(C_{OFF} - C_{ON})V^2 - kg^2 - \frac{1}{2}k_s g^4}{m + 2\frac{\alpha}{\omega}}} \tag{37}$$

i.e. the velocity v of the bridge subjected to the force imposed by means of the applied voltage is roughly linearly dependent on the applied voltage V. So far, Eq. (37) describes the velocity an instant before the bridge is collapsed. We can also write:

$$v_{act}(d) = \sqrt{\frac{(C_{OFF} - C_{ON})V^2 - kg^2 - \frac{1}{2}k_s g^4}{m + 2\frac{\alpha}{\omega}}} \rightarrow \\ \rightarrow \left(\sqrt{\frac{(C_{OFF} - C_{ON})V^2 - kg^2 - \frac{1}{2}k_s g^4}{m}}\right)_{\alpha \to 0} \tag{38}$$

Actually, the dissipation causes a decrease in the velocity of the actuation by means of a term depending on a. It is like to substitute the mass of the bridge m with $m'=m+(2a/\omega)$. The dependence of the actuation velocity on the applied voltage V_a is shown in Fig. 9, where the dependence of v_{act} on V_a for the exploited device is presented.

When the bridge is released, no electrostatic force has to be included, and only the change of the potential energy and the contribution of the damping have to be considered for calculating the final velocity of the de-actuated beam, obtaining the result in Eq. (39) for the de-actuation velocity $v_{de\text{-}act}$, with $v(d)=0$ as the initial condition for the velocity:

$$v_{de-act}(d) = \sqrt{\frac{kg^2 + \frac{1}{2}k_s g^4}{m}} \tag{39}$$

Exception done for charging contributions.

From the analysis of the above figures, a significant change is expected comparing the results obtained by using the threshold value $V_{threshold}$ and a voltage higher than the threshold one. The actuation speed is almost doubled by doubling the applied DC voltage. This enhances the dynamical response of the device, but it lowers the lifetime of the switch, tightly related to the mechanical fatigue induced by the electrostatic actuation. It is worth noting that, due also to the increased pull down voltage, the lateral actuation is faster than the central one.

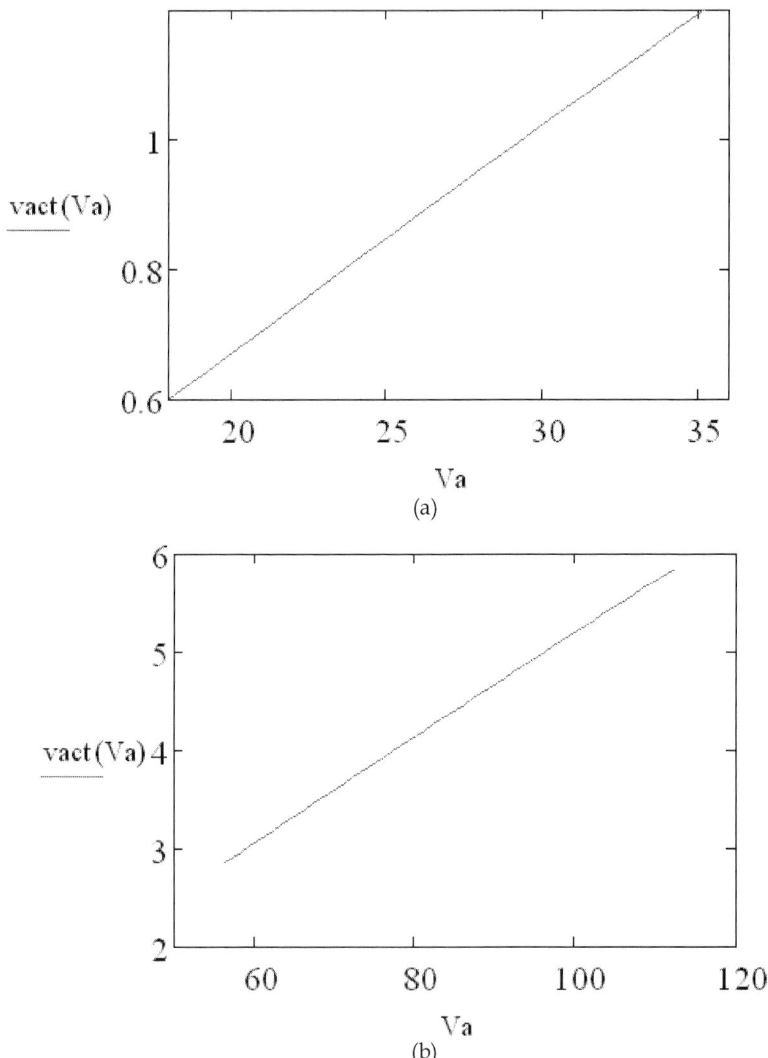

Fig. 9. Actuation speed v_{act} in m/s of the RF MEMS capacitive switch Vs the applied voltage V_a in volt, for the two simulated situations: (a) central actuation, and (b) lateral one.

Using capacitance dynamics considerations at the threshold voltage, we have assumed that the de-actuation and actuation times ($\tau_{de\text{-}act}$ and τ_{act} respectively) are the time values when the capacitance reaches the 90% of its final value. Such a definition is typically used to introduce the electrical response of a lumped circuit, and, by inverting the capacitance equation at these values, we get the phenomenological definitions for the characteristic de-actuation and actuation times. When energy and motion considerations are used, τ_{act} is the time for the full mechanical collapse of the beam and not just a time constant.
Concerning the de-actuation mechanism, by using the approach to lower the applied voltage for the maintenance of the bridge in the down position up to the V_{min} value, no dependence on V_a is expected.
For the actuation, the applied voltage is always present, and, by imposing a generic z-value in the integration for the energy spent in the actuation, it turns out that

$$v_{act}(z) = \sqrt{\frac{\left[C(z) - C_{ON}\right]V^2 - k\left[z-(d+g)\right]^2 - \frac{1}{2}k_s\left[z-(d+g)\right]^4}{m + 2\frac{\alpha}{\omega}}} \tag{40}$$

To obtain the actuation time, the following equation can be used:

$$\tau_{act} = \int_0^{\tau_{act}} dt = \int_{d+g}^{d} \frac{dz}{v(z)} \tag{41}$$

and the applied voltage V can be imposed as a parameter. As a consequence, the actuation time for the exploited configuration is given in Fig. 10 by means of Eq. (41).
A similar result can be obtained for the de-actuation time, but it is independent of the actuation voltage, because the de-actuation occurs when the applied voltage is turned off.
Then, in our case $\tau_{de-act} = \int_0^{\tau_{de-act}} dt = \int_d^{d+g} \frac{dz}{v(z)}$ =18 μs ca.

In the case of the de-actuation the additional force due to the voltage is no more present, and the structure will respond only to the restoring forces dominated by the value of the spring constant.

4. Contact and van der Waals forces

The van der Waals and contact energies introduced in Eq. (29-30) and discussed in literature[5],[42] have attractive and repulsive effects respectively. From a physical standpoint they are due to dipolar contributions induced by atomic interactions. The energy associated with the two effects can be written in the form of the Lennard-Jones potential[42]:

$$E_c = 4\delta\left[\left(\frac{\sigma}{R}\right)^{12} - \left(\frac{\sigma}{R}\right)^{6}\right] = \frac{B}{R^{12}} - \frac{C}{R^6} \tag{42}$$

Where B and C are positive constant values and R is the inter-atomic distance. Sometimes an exponential trend for the repulsive forces is preferred[42]. Because of the analytical formulation of such a potential, a bound state for the atoms is obtained for the minimum of

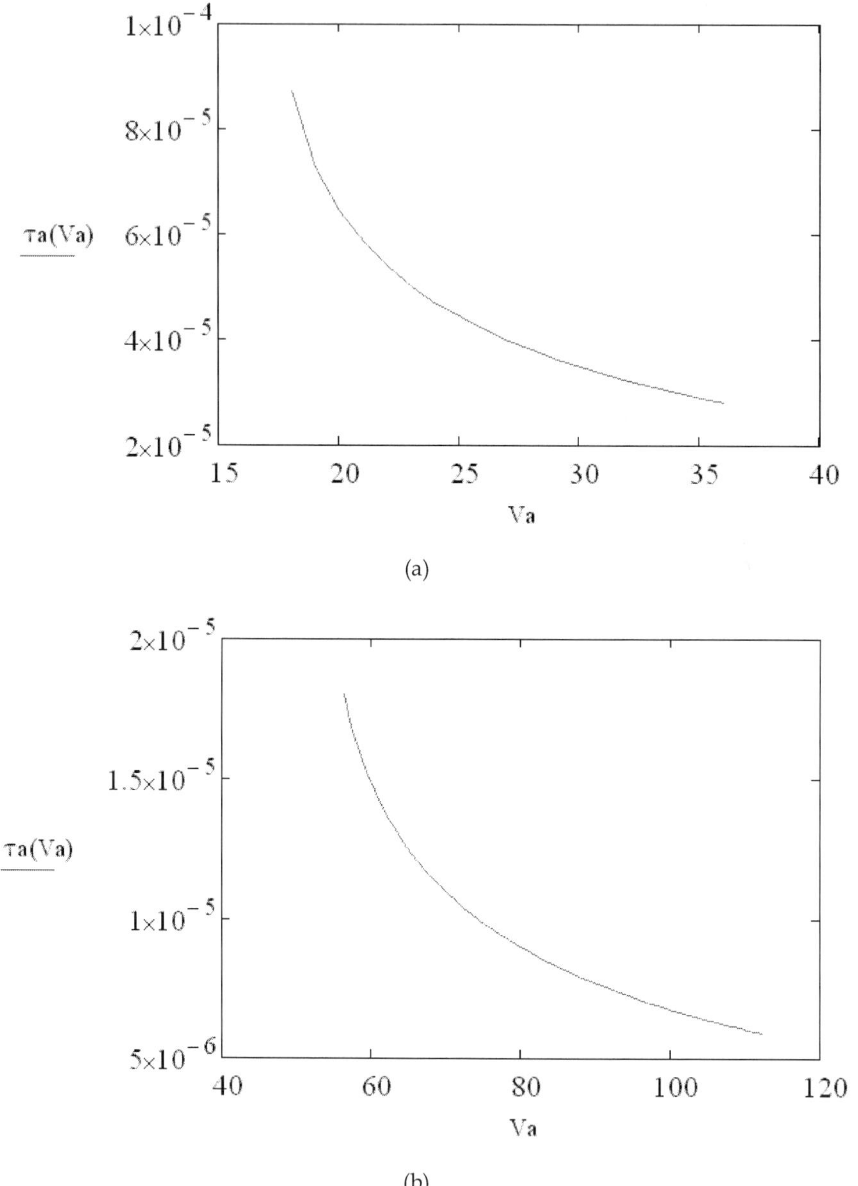

Fig. 10. Central (a) and lateral (b) actuation time τ_a as a function of the applied voltage Va for the studied device.

the energy, i.e. for the value of R vanishing the derivative of E_c. It will happen for $R/\sigma = 2^{1/6}$. Of course, all of the above considerations are free from the ambiguity of possible failures coming from adhesion related mechanisms induced by humidity, which can be present in both the processing and the operating conditions of MEMS devices.
A completely different approach has been followed elsewhere[45], where, further to the problems caused by the processing, the pinning due to inter-solid adhesion is analytically and experimentally treated for typical MEMS materials and structures. Also in this case a potential with a binding energy is derived, where the critical distance is associated with the contact area and the surface energy, and with the restoring force of cantilevers and double clamped bridges. The value for such a critical distance s^* for a cantilever is:

$$s^* = \left(\frac{3}{2}\frac{Et^3h^2}{\gamma_s}\right)^{1/4} \tag{43}$$

It depends on the Young modulus E, the thickness of the beam t, the distance from the substrate h and the surface energy per unit area γ_s. In many cases the surface energy has to be fitted depending on the system, and the order of magnitude should be between 100 and 300 mJ·m^{-2}, but lower values have been also obtained in specific experiments on cantilevers and bridges. It is reasonable to assume a difference depending on the materials used for the contact region (metal-to-metal, metal-to-dielectric, …) and for other surface characteristics, like the roughness. The interfacial adhesion energy is given by[45]:

$$U_S = -\gamma_S w(l-s) \tag{44}$$

Being w the width of the beam, l the length of the cantilever before the contact area and s the length of the contact area with the substrate. The situation is illustrated in Fig. 7 of [45].
A particularly important result is the prediction of the distance at which a cantilever or a beam is detached, by using the definition of a *peeling bound number* N_p, where $N_p>1$ means that the beam can be free again after the contact with the substrate, while it remains in contact with the substrate for $N_p<1$. The results interesting for MEMS materials which can be used for beams to be electro-statically actuated are summarized in the following Table II[45].

Structure	N_p
Cantilever	$\frac{3}{8}\frac{Et^3h^2}{\gamma_S l^4}$
Doubly clamped beam	$\left(\frac{128}{5}\frac{Et^3h^2}{\gamma_S l^4}\right)\left[1+\frac{4\sigma l^2}{21Et^2}+\frac{256}{2205}\left(\frac{h}{t}\right)^2\right]$

Table 2. Peeling bound numbers for cantilevers and doubly clamped beams. σ is the residual tensile stress. l is the full length of the suspended structure, either the cantilever or the doubly clamped beam.

By using the values for the exploited shunt capacitive MEMS device and from [47] γ_S=0.06 J·m^{-2}, in a doubly clamped beam configuration, it turns out that $N_p\approx1.42$ for our perforated 1.5 µm thick and 600 µm long beam. From a quick analysis of the quantities playing a role in the definition of N_p, it is worth noting that it is important to have a thick beam in order to have an effective restoring force. By decreasing the thickness t the value of N_p will decrease too. Moreover, the Young modulus E should be increased, and an additional but less relevant contribution comes from the increase of σ. In conclusion, a mechanically strong structure would be less influenced by the surface adhesion forces.

Some metals like Pt have a good E-value but they are less robust under continuous solicitations. Some others, like Tungsten Carbide, are very hard, but compatibility with standard CMOS processes and integration with microelectronics have to be fully demonstrated. Recent and promising results about molybdenum are available[34]. Materials selection properties are, in general, a hot topic for RF MEMS applications [48]. The threshold value for the actuation voltage is affected by E, which enters the definition of the spring constant k, and it could render unacceptable the voltage to be used for the actuation of the bridge. So far, a reasonable compromise involves the definition of the geometry and of the material to be used. Anyway, just doubling the value of E, i.e. not using Au but Pt, the actuation voltage passes from 29 volt to 30 volt ca. for the central actuation. It means that N_p passes from 1.42 to 1.53 ca. Since Pt is softer than Au, it is a negligible advantage and un-practical solution, because it will also cause problems in terms of repeated actuations of the bridge. An estimation of the van der Waals force has been also given by using a different formulation[49]. Repulsive forces, which can influence by a factor 2 or 3 the full evaluation of the contact contributions, have been neglected. As already discussed, such an interaction is responsible for binding in metallic contacts as well as in crystalline materials. The interaction energy due to van der Waals contribution can be written as:

$$U_{VDW} = -\frac{A}{12\pi d^2} \tag{45}$$

Where A is defined as the Hamaker constant (A=1.6 eV for Si [49], or $A=4.4\times10^{-19}$ J [47]) and d is the separation between the two surfaces. In the case of MEMS, at a microscopic scale, the roughness should avoid the adhesion, because it decreases the effectiveness of the surface contact, and residual gas molecules give a further contribution against a possible stiction [49]. As a further demonstration that it is quite difficult to have non-ambiguous results on the definition of the surface energy constant, other findings [18] have also to be considered, where $\gamma_s=1.37\ \mathrm{J{\cdot}m^{-2}}$ is given in the case of metal-to-metal contact in ohmic switches.

By using another approach [47], the value of γ_s can be obtained accounting that U_{VDW} when the surfaces are in contact between them corresponds to γ_s, i.e. $\gamma_s= U_{VDW}(d=d_0)$, where d_0 is the minimum distance between the two surfaces. Considering that the roughness due to the processing of the device is equivalent to cause a residual air gap in the order of 50-100 nm [51], we can estimate for our purposes

$$\gamma_s= 4.4\times10^{-19}/(12\pi d_0^2) = (1.17 \div 4.67)\times10^{-6}\ \mathrm{J{\cdot}m^{-2}}.$$

If such a value can be considered reasonable for the MEMS technology, no limits on the full length of the bridge should be found. In any case, from all of the above considerations, and from the last ones concerning the roughness, the surface energy should play a minor or negligible role in the sticking of the cantilevers and doubly clamped beams. Humidity only should generate real problems of sticking for un-properly packaged devices

5. Charging effects and sticking

The maintenance voltage, i.e. the voltage V_{min} for holding down the bridge after the actuation, is lower than the threshold one. This will depend on the physics of the electrical contact. Actually, considering the results about surface forces commented in the previous paragraph, they should play no role in sticking, unless introducing other contributions depending on metal-to-metal diffusion and local welding due to aging, heating and power handling. So far, charging of the dielectric material used for the actuation pads should be the main effect in increasing the internal electrical field opposing the actuation one, which will depend on the charging of the device after many actuations [41],[43],[44],[45].
By using Eq. (31), and neglecting the contact contribution previously discussed, we shall have:

$$V_{\min} = \sqrt{\frac{kg^2 + \frac{1}{2}k_s g^4 - 2E_{charge}}{C_{OFF}}} \tag{46}$$

And the sticking will happen for $kg^2+(1/2)k_s g^4=2E_{charge}$. The energy accumulated by means of a charge trapping process will have no effect if a long time is left after the release of the bridge and before the successive actuations, but its contribution on V_{min} is not negligible since the first actuation.
A possible phenomenological approach for explaining the increase in the threshold value under a uni-polar scheme for the actuation voltage of an RF MEMS switch can be given accounting for the change induced in the electric field and, consequently, in the charge accumulated by dielectric layers.
Such an extra-voltage can be written as $\Delta V_{th} = d\,|\vec{E}_{ch}|$, where d is the thickness of the dielectric layer and E_{ch} is the electric field related to the accumulated charge, directed along the normal with respect to the dielectric plane. The charge will decay after the release of the bridge following an exponential trend, in such a way that:

$$\Delta V_{th}(t) = d\,|\vec{E}_{ch}(t)| = d\,|\vec{E}_{ch,0}|\exp\left(-\frac{t}{\tau_{ch}}\right) \tag{47}$$

Where τ_{ch} is the time constant of the decay process during the time t. By using a uni-polar scheme for the voltage to be applied to the MEMS switch (positive pulses only), and by imposing a pulse train with period T and pulse-width τ, successive actuations will be affected by a partial decay of the accumulated charge before the next pulse will actuate the switch again. Such an effect can be formalized by using the following equations:

$$\begin{aligned}
\Delta V_{th}^{(1)} &= d\,|\vec{E}_{ch,0}^{(1)}|\exp\left(-\frac{(T-\tau)}{\tau_{ch}}\right) \\
\Delta V_{th}^{(2)} &= d\left[|\vec{E}_{ch,0}^{(1)}|\exp\left(-2\frac{(T-\tau)}{\tau_{ch}}\right) + |\vec{E}_{ch,0}^{(2)}|\exp\left(-\frac{(T-\tau)}{\tau_{ch}}\right)\right] \\
\Delta V_{th}^{(3)} &= d\left[|\vec{E}_{ch,0}^{(1)}|\exp\left(-3\frac{(T-\tau)}{\tau_{ch}}\right) + |\vec{E}_{ch,0}^{(2)}|\exp\left(-2\frac{(T-\tau)}{\tau_{ch}}\right) + |\vec{E}_{ch,0}^{(3)}|\exp\left(-\frac{(T-\tau)}{\tau_{ch}}\right)\right]
\end{aligned} \tag{48}$$

....

In general, the amount of charge accumulated during successive actuations could be not constant, but we can assume, as a starting point, that $|\vec{E}^{(1)}_{ch,0}| = |\vec{E}^{(2)}_{ch,0}| = \ldots = |\vec{E}_{ch,0}|$. It turns out in:

$$\Delta V_{th}^{(n)} = d\,|\vec{E}_{ch,0}|\exp\left(-n\frac{(T-\tau)}{\tau_{ch}}\right) \quad and$$
$$\Delta V_{th} = d\,|\vec{E}_{ch,0}|\sum_n \exp\left(-n\frac{(T-\tau)}{\tau_{ch}}\right) = d\,|\vec{E}_{ch,0}|\sum_n x^n = d\,|\vec{E}_{ch,0}|\frac{1}{1-x} = \tag{49}$$
$$= \frac{d\,|\vec{E}_{ch,0}|}{1-\exp\left(-\frac{T-\tau}{\tau_{ch}}\right)} \qquad where \quad x = \exp\left(-\frac{T-\tau}{\tau_{ch}}\right) < 1$$

From Eq. (49) it turns out that a limit in the charge accumulation exists also in the simplified case of constant value for the induced electric field after each actuation.
The above approach is valid only in the case of a uni-polar scheme. When a bi-polar voltage is applied, it will result in induced electric fields having opposite polarization, and, independently of the decay time for the accumulated charge, the original situation will be partially restored after each application of the threshold voltage [44]. It is worth noting that the charge accumulated because of this mechanism sometimes needs very long times for the decay, and the dominant Poole-Frenkel effect is difficult to be prevented. Presently, studies are performed for optimizing materials and geometries, eventually using non-contact actuations. The previously defined quantity $d\,|\vec{E}_{ch,0}|$ gives the contribution necessary for evaluating the maintenance voltage after the first collapse of the bridge. Eq. (45) has to be re-written as:

$$V_{\min} = \sqrt{\frac{kg^2 + \frac{1}{2}k_s g^4 - 2E_{charge}}{C_{OFF}}} = \sqrt{\frac{kg^2 + \frac{1}{2}k_s g^4}{C_{OFF}} - \Delta V_{th}^2} = \tag{50}$$
$$= d\sqrt{\frac{kg^2 + \frac{1}{2}k_s g^4}{\varepsilon A d} - |\vec{E}_{ch,0}|^2}$$

Where, from an energetic standpoint, $E_{charge} = (1/2)C_{OFF}\Delta V_{th}^2$
An evaluation of $|\vec{E}_{ch,0}|^2$ can be obtained by the knowledge of the charging mechanisms in the exploited dielectric. Results are available in literature about such a change in the voltage threshold by assuming that the Poole-Frenkel (PF) effect is the dominant one in the charge trapping of MEMS devices [51], taking into account that the current density due to the PF effect is $\vec{J}_{PF} = \sigma|\vec{E}_{ch,0}|$ .and σ is the conductivity of the material. The role of the above contribution in MIM capacitors and the dependence on the applied voltage and temperature has been studied elsewhere [52].
It has to be stressed that C_{OFF} will be quite different from the ideal one if residual air gap contributions have to be included, but in the case of floating metal solutions for the realization of shunt capacitive switches the C_{OFF} value is naturally obtained [28]. As the sticking induced by charging is one of the major problems in the reliability of RF MEMS

configurations, several solutions are currently studied for characterizing or suppressing such an effect [53],[54].

6. Bi-dimensional and three-dimensional mechanical simulations

Starting from the evaluations obtained by means of the uni-dimensional approach described in the previous sections, an extension to 2D and 3D structures has been performed by means of the COMSOL Multi-physics software package [55]. Commercial software begins now to be quite popular for simulating physical processes involving mechanical, thermal, high frequency and many other possible (and contemporary) solicitations for the exploited structure. In fact, only simple geometries can be efficiently simulated by using a uni-dimensional approach, thus estimating actuation times and actuation voltages without using long and complicated simulations with finite element methods. On the other hand, a full simulation is very important especially when the shape of the bridge is tailored in a not simple way. This happens when the cross section has not a constant width, or specific technological solutions, like metal multi-layers for the bridge, and dimples to help the electrical contact in the actuation area are realized. Holes are also present on the beam for improving the sacrificial layer removal and for lowering the spring constant, which is important when the stress induced by the technological process is not acceptable for practical purposes. In all of the above situations, effective quantities can be defined accounting for a re-definition of mass, contact area and beam width. Of course, small changes with respect to the ideal double clamped beam will have a small influence on the response of the entire structure, but more sophisticated geometries and technological solutions need a different evaluation. Moreover, software able to treat combined solicitations of the MEMS device has to be considered if the goal is the definition of a figure of merit for such a technology. For this purpose, 2D and 3D mechanical simulations have been performed to clearly state differences and advantages of such an approach with respect to the uni-dimensional one. An additional consideration is that the deformed shape of the actuated bridge, also in the case of simple geometries, is particularly useful for the prediction of the electrical properties of the device, which could be affected by parasitics for very high frequencies, starting form the millimetre wave range (F > 30 GHz). In the following discussion, parametric and electro-static simulations will be presented, with the aim to compare the central and the lateral actuation, and the expected shape of a simple fixed-fixed beam structure. As the threshold voltage $V_{threshold}$ is not dependent on the width of the bridge, because it is proportional to the ratio k/A between the spring constant k and the area A of the actuation region, the actuation of a bridge with no holes neither tapering along the width can be considered a 2D problem. Some 2D results are presented in Fig. 11-14, where the OFF state of the switch has been obtained by using a central actuation (DC signal along the central conductor of the CPW) or a lateral one by means of symmetrical pads. In both cases the electrostatic package implemented in COMSOL has been used, with a parametric simulation performed by changing the value of the applied voltage. A structure having the same dimensions imposed for the uni-dimensional treatment has been simulated: full length L=600 μm, width W=300 μm for the central conductor of the coplanar structure (corresponding to the bridge length in the actuation region), thickness t=1.5 μm for the bridge. The residual stress is again σ=18 MPa. The central conductor is Au, 0.1 μm thick, covered by SiO_2, 0.2 μm thick. The bridge width is 100 μm, along the direction normal with respect to the 2D view. In the case of Fig. 11, the parametric force needed to get the fully deformed shape of the beam is given by $F = -kg = -46$ μN, corresponding to a pressure of $1.5x10^3$ N/m^2 ca. This is in agreement with the value obtained by using the 2D simulation. On the other hand, the nonlinearity of the mechanical problem allows for a full actuation when the structure begins to be unstable, i.e. when the applied force is sufficient to push the bridge down (1/3) of the gap. This happens at $z = 2(d+g)/3 = 2$ μm, corresponding to a mechanically simulated structure subjected to a force $F = -15$ μN and to a pressure of 500 N/m^2 ca. By calculating the value of $\partial C / \partial z$ when $z = 2(d+g)/3$ and substituting in $V = \sqrt{2F/(\partial C / \partial z)}$ obtained from the definition of the electrostatic force, the voltage V=20 volt is obtained for the threshold, thus demonstrating that the actuation of the bridge can be easily predicted by using the uni-dimensional approach. In this and in the following simulations holes are not included, but from the uni-dimensional simulations, it turns out that V=18 volt is obtained when holes are present, while V=20 volt is the expected threshold voltage for the structure with holes, i.e. only a 10% difference.

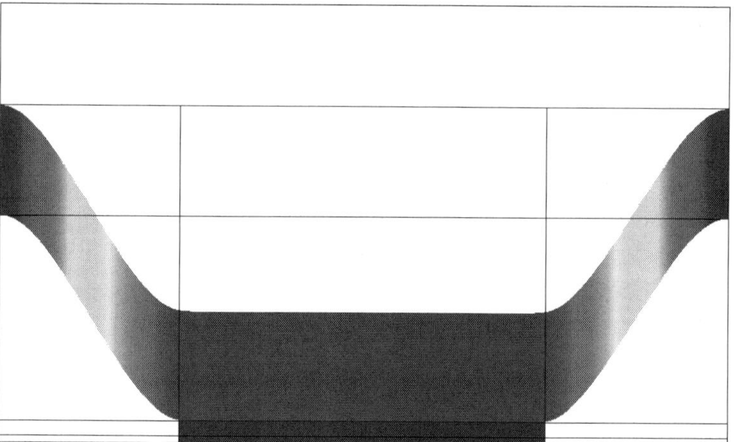

Fig. 11. 2D simulation for the central actuation of the MEMS switch (COMSOL simulation). A mechanical force per unitary area on the central conductor of the CPW has been applied as high as 1500 N/m^2 to obtain the full actuation.

From the result in the next Fig. 12, where the lateral actuation is imposed, it turns out that by properly choosing the shape and the dimensions of the structure, the actuation occurs without having the bridge touching the lateral pads. This could help in decreasing the charging effects for these devices, mainly due to the dielectric used onto the actuation pads, which dramatically affects their reliability in terms of the charge stored. Moreover, the necessity to separate the RF and DC paths is another important reason for preferring a lateral actuation in actual devices.

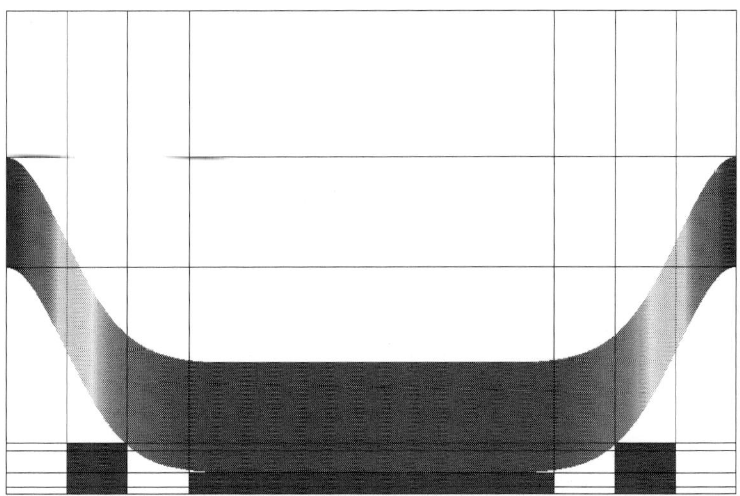

Fig. 12. Lateral actuation of the 2D structure (COMSOL simulation). The same force used for the central actuation was necessary for having a full collapse of the bridge, but applied on smaller lateral pads (50 μm width). A more uniform actuation is obtained, as evidenced by the colour intensity in the central part. Moreover, the beam is contact-less on the pads, which helps in minimizing the charging effects.

An example of the 3D response is given in the following Fig. 13 and Fig. 14, where the actuation has been performed by means of the same force used for the 2D case, with details about the full device. The result is coherent with the prediction performed by using the analytical approach and the 2D actuation. In fact, the mechanical force per unitary area imposed in the simulation for obtaining the full collapse of the bridge corresponds to a pressure of 1500 N/m^2 , i.e. to a force of 45 x 10^{-6} N applied onto an area of 300 x 100 x 10^{-12} m^2. This is the value of the mechanical restoring force applied in the centre of the double clamped beam, and by calculating again the value of $\partial C / \partial z$ when $z = 2(d+g)/3$ and substituting in $V = \sqrt{2F/(\partial C / \partial z)}$ the voltage V=20 volt is obtained also in this case. So, the actuation of the bridge can be easily predicted by using the uni-dimensional approach, also including rough evaluations about the holes contribution, but avoiding long term simulations, exception done for those structures where the distribution of the holes is very complicated or a significant shape tailoring is present and the analytical approach should be forced by the introduction of effective quantities not really matching the actual situation.

By properly choosing the dimensions and the materials, lateral actuation is possible with voltages in the same order of magnitude used for the central one, as from the comparison between the results coming from data in Fig. 13 and Fig. 14, with evidence for contact-less actuation.

Only in simple cases the presence of holes can be approximated by defining an effective stiffness for the metal beam. For the above reason, the 3D simulation is really useful, as already stressed, in the case of configurations which have a very peculiar shape. Also in the case of a moving mesh, i.e. a mechanical solver to be used for the dynamical response of the device, many information can be already obtained from evaluations based on a fully analytical model, without involving cumbersome simulations. A real advantage in having a full 3D modelling of the device is in the combination of mechanical and RF predictions,

Fig. 13. 3D COMSOL simulation of the RF MEMS shunt capacitive switch in the OFF state (bridge in the down position), centrally actuated.

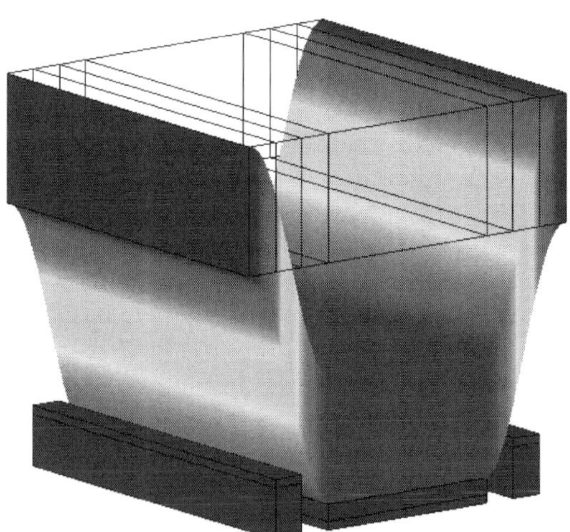

Fig. 14. 3D simulation of the RF MEMS switch in the OFF state (bridge in the down position), laterally actuated when the applied force per unitary area is F=1500 N/m^2. The deformation of the bridge is represented by the change in the colours, from the blue (at rest) to the red (fully actuated).

being based on the construction of the same geometry. Specifically, as it is the case of the COMSOL software package, thermal, power and charging effects could be considered in the same simulation environment. For the above reasons, this will be very useful to get a figure of merit for the RF MEMS technology based on different input conditions.
As a final demonstration of the validity of the proposed theoretical approach for the electro-mechanical analysis, one more simulation has been performed by means of the electrostatic force directly defined within COMSOL as $F_e = \frac{1}{2}\frac{\partial C}{\partial z}V^2 = \frac{1}{2}QE$
where Q is the charge and E is the electric field. The result is shown in Fig. 15, where the structure is pushed down (1/3) of the gap, before the final actuation occurs following the mechanical instability of the structure. It happens by imposing a voltage $V = 22$ volt ca., very close to the calculated threshold value by using the uni-dimensional approach.

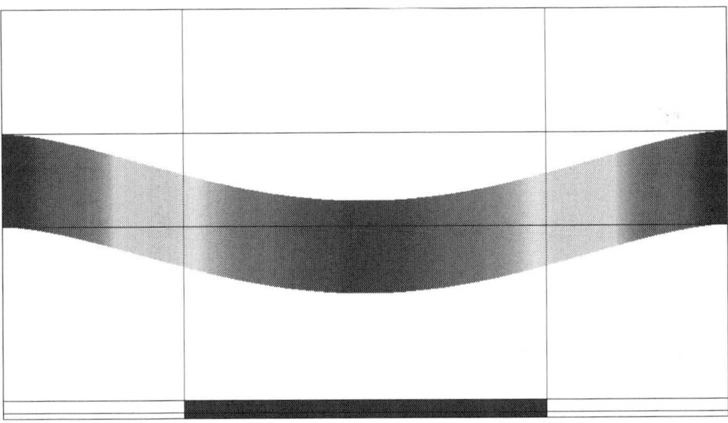

Fig. 15. Electrostatic 2D simulation for the exploited double clamped beam. V = 22 volt was necessary for having a movement of the beam close to (1/3) of the gap.

7. Technology, experimental results and discussion

An actual configuration having the same dimensions described in the previous sections has been realized and preliminary tested. A photo of the structure is given in Fig. 16. SU-8 polymeric sides have been realized by photo-lithography to be used as a support for both the ground planes of the CPW, and the suspended metal bridge. Silicon oxide has been deposited as a dielectric, and the actuation has been performed by means of the central conductor of the CPW.
The realization of double-clamped RF MEMS capacitive shunt switches has been performed by means of negative photo-resist SU-8 for the realization of the ground planes of the coplanar configuration, elevated with respect to the wafer, while positive S1818 photo-resist has been used as a sacrificial layer.
RF MEMS Switches have been manufactured on a 4 inch high-resistivity (ρ > 5000 ohm cm) silicon wafer <100> oriented, having a thickness of 400 μm. For the realization of the devices, a 4 mask sequence has been considered, and the entire fabrication process is sub-divided in five steps:

i. realization of the central conductor of the CPW.
ii. definition of the SiO_2 to be used for obtaining the capacitive configuration, with the aim of a high ratio in the ON/OFF states.
iii. creation of lateral supports made by SU-8 for the double-clamped structure. SU-8 2002 (Microchem Corp., USA) has been used for our purposes. In this configuration, polymer lateral pedestals are obtained, to be metalized for obtaining both the ground planes and the support for the suspended bridge.
iv. reduction of the sacrificial layer has been obtained by means of a purposely designed mask, in order to avoid peaks in the shape coming from the lithography, which act as discontinuities in the next metallization process.
v. as a final step, the switches are obtained by means of the release of the sacrificial layer by using a modified reactive ion etching (RIE) process.

Actuation voltages in the order of 20-25 volt have been obtained in different samples, almost in agreement with the expected value. A possible small under-evaluation of the gold membrane stiffness should be considered in the actual device. In fact, by imposing σ=25 MPa we obtain $V_{threshold}$=21 volt ca. for a uniform beam with the same dimensions and holes. Moreover, because of the stiffness the beam could be also a bit upward with respect to the expected flat geometry and an increased gap could favour and increase in the threshold voltage. Other possible sources of spread with respect to the predicted threshold voltage maybe due to non-uniformity in the Au deposition over the entire 4 inch wafer, and the successive electro-plating process.

An optical characterization of the manufactured device has been also performed, revealing both a good shape of the beam and the resolution of the holes, which have no residuals after the ashing process. Results are given in Fig. 16.

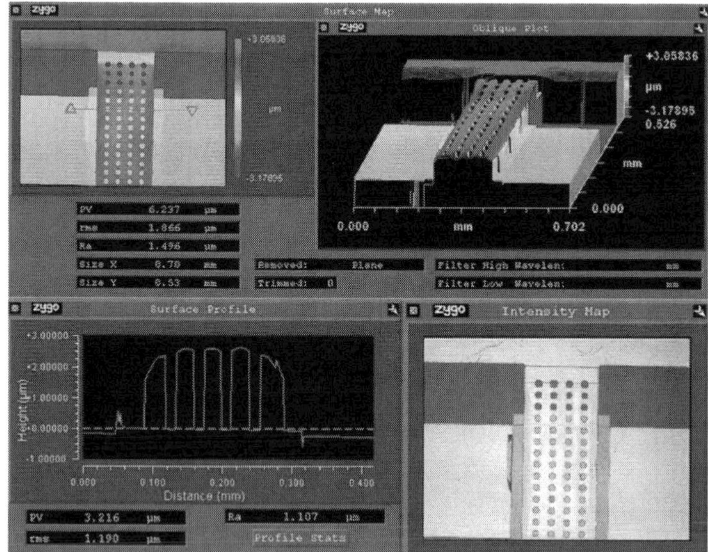

Fig. 16. Optical microscopy characterization of the RF MEMS switch realized by means of SU-8 photo-lithography with evidence for the optimized profile of the beam after the removal of the sacrificial layer.

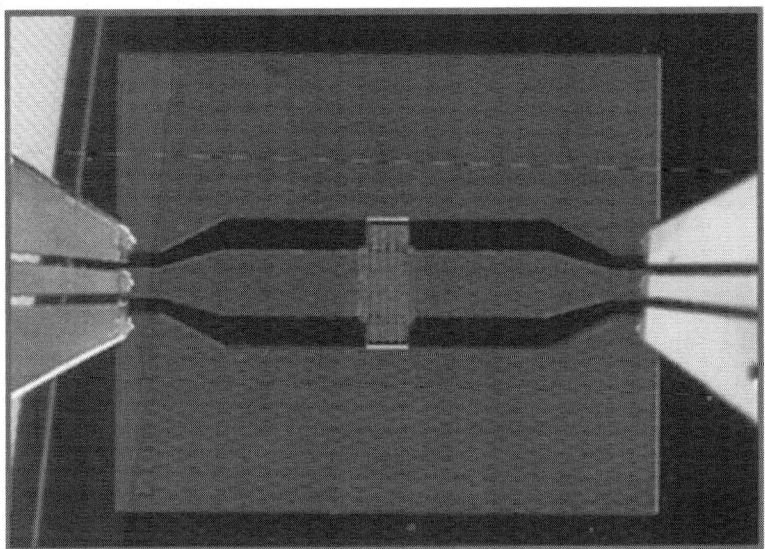

Fig. 17. Test-fixture structure of the RF MEMS switch manufactured by means of SU-8 photo-lithography. The input and output ports are connected to a vector network analyzer by means of coplanar probes for on-wafer characterization.

A further confirmation of the influence of the developed technological processing, and specifically the contribution from the gold stiffness, is evidenced from the mechanical response simulation plotted in Fig. 18 for a laterally actuated beam. In that case, the imposed residual stress is $\sigma = 60$ MPa, leading to two major effects: (i) the increase of the actuation voltage up to values greater than 90 volt, and (ii) a deformation of the bridge, which results in a more rigid shape.

It is worth noting that, looking at the shape of the bridge predicted in Fig. 18, an easier and more uniform actuation in the central part by using the lateral pads could be obtained because of the higher residual stress. On the other hand, the price to be payed in terms of the increase in the actuation voltage is not acceptable for many applications, and a trade-off has to be obtained, usually accounting for actuation voltages not exceeding 50 volt.

The only difference generated by changing the width of the bridge for the studied structure concerns with its RF response. Actually, the central capacitance defined by the bridge, the central conductor of the CPW and the dielectric between them changes the frequency of resonance for the device under test, while no change is recorded for the actuation voltages. Actually, wider beams experience lower frequencies of resonance when the switch is actuated and it works in isolation [56].

Further theoretical and experimental results have been obtained by using another shunt capacitive switch manufactured by using the silicon technology and developed at FBK-irst [57]. Top view and lateral dimensions of the device are shown in the following Fig. 19. The device is a clamped-clamped beam obtained on silicon wafer by means of an eight-mask sequence of technological steps, and the final release of the suspended bridge was obtained removing the sacrificial layer via an ashing process. The configuration is also characterized by the following parameters: gap between bridge and floating metal g=2.85 μm, SiO_2 dielectric thickness on the central conductor of the CPW d=0.1 μm, beam thickness t=4 μm (grown by electroplating). By using the above values and those given in Fig. 19, we get an actuation voltage V=40 V ca. for the lateral actuation. The experimental value was V_{exp}=(41±2) V by using ten devices measured onto the same wafer in different positions

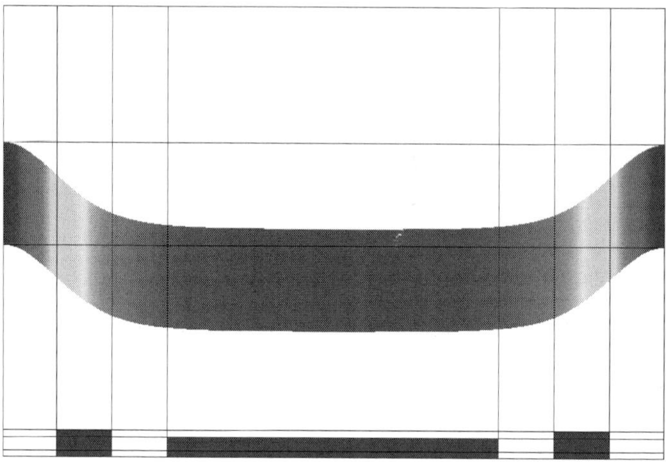

Fig. 18. Simulated deformation for the same bridge experimentally tested with a higher value of the residual stress (σ=60 MPa). An actuation voltage around 95 volt has been predicted.

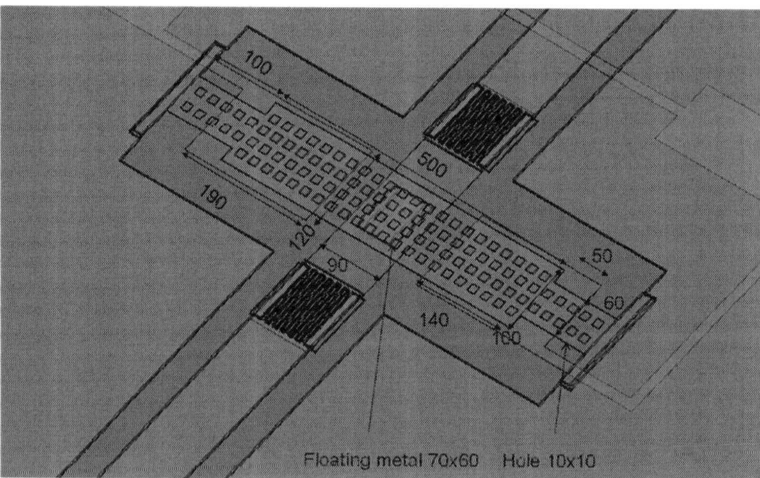

Fig. 19. Shunt capacitive switch manufactured by means of the eigth mask process developed at FBK-irst. Dimensions are in μm.

8. Conclusion

Analytical and numerical modelling for the mechanical response of a shunt capacitive RF MEMS switch have been compared by using an uni-dimensional theory, and 2D and 3D simulations performed by means of a commercial software package (COMSOL Multiphysics). Two actual configurations have been experimentally studied to validate the proposed model.
As a result, it has been demonstrated that RF MEMS mechanics can be predicted in a convenient way by uni-dimensional phenomenological models if evaluations about switching times, threshold voltage and preliminary dynamics have to be studied, without involving cumbersome simulations with a computer. In fact, the most part of the previous quantities depend on the equations to be used for the correct definition of the spring constant, and we demonstrated that the analytical approach based on the knowledge of the materials and of the geometry fulfils the most part of the quantities to be obtained. The actuation velocity and the switching times have been also defined and predicted by means of the analytical approach based on the Mechanical Energy considerations, including the applied DC voltage.
2D and 3D simulations are really useful for configurations having a very peculiar shape, especially for combining mechanical and RF predictions, being based on the same geometry, and this will be very useful to get a figure of merit for the RF MEMS technology (not yet available) based on different input conditions. Concerning the evaluation of the actuation voltage, the main parameter to be defined is the residual stress of the structure, because it dramatically influences the mechanical response of the bridge. On the other hand, with the proper knowledge of the technology used, the evaluation can be easily based on the uni-dimensional approach by defining effective quantities for simple configurations. The relative influence of surface forces and charging contributions has been discussed, to demonstrate that, under proper geometrical and material constraints, only charging effects can be really responsible for un-reliable structures, and a very simple solution to prevent this contribution is to tailor the switch in order to have contact-less actuation pads.

9. Acknowledgment

The activity has been partially funded by the ESA/ESTEC Contract on "High Reliability MEMS Redundancy Switch" ESA ITT AO/1-5288/06/NL/GLC contract No.20847

10. References

[1] G. M. Rebeiz and J. P. Muldavin: "RF MEMS Switches and Switch Circuits", *IEEE Microwave Magazine*, Vol.2, No.4, pp.59-71 (2001).
[2] G. M. Rebeiz, Guan-Leng Tan and J. S. Hayden: "RF MEMS Phase Shifters, Design and Applications", *IEEE Microwave Magazine*, Vol.3, No.2, pp.72-81 (2002)
[3] G. M. Rebeiz, "*RF MEMS, Theory, Design and Technology*", John Wiley and Sons, Hoboken, 2003.
[4] Harrie A.C. Tilmans: "MEMS components for wireless communication", *invited paper at XVI Conference on Solid State Transducers*, Prague, Czech Republic, Sept. 15-18, 2002
[5] E.K. Chan, E.C. Kan and R.W. Dutton: "Nonlinear Dynamic Modeling of Micromachined Switches", *Proceed. Of IEEE MTT-Symposium*, pp.1511-1514 (1997).

[6] E.K. Chan and R.W. Dutton: "Effect of Capacitors, Resistors and Residual Charge on the static and dynamic Performance of Electrostatically Actuated Devices", *SPIE Symposium on Design and Test of MEMS/MOEMS* - March 1999.

[7] D. Mercier, P. Blondy, D. Cros and P. Guillon: "An Electromechanical Model for MEMS Switches", *Proceed. Of IEEE MTT-Symposium*, pp.2123-2126 (2001).

[8] G. De Pasquale, T. Veijola, and A. Somà, "Gas Damping Effects on Thin Vibrating Gold Plates: Experiment and Modeling", *Proceed. of DTIP 2009 Conference*, Roma, 1-3 April 2009, EDA Publishing/DTIP 2009, ISBN:978-2-35500-009-6, pp. 23-28 (2009).

[9] T. Veijola, "Compressible Squeeze-Films in Vibrating MEMS Structures at High Frequencies", *Proceed. of DTIP 2009 Conference*, Roma, 1-3 April 2009, EDA Publishing/DTIP 2009, ISBN:978-2-35500-009-6, pp. 235-238 (2009).

[10] T. Veijola, "3D FEM Simulations of Perforated MEMS Gas Dampers", *Proceed. of DTIP 2009 Conference*, Roma, 1-3 April 2009, EDA Publishing/DTIP 2009, ISBN:978-2-35500-009-6, pp. 243-250 (2009).

[11] P G Steeneken, Th G S M Rijks, J T M van Beek, M J E Ulenaers, J De Coster and R Puers, *Journal of Micromechanics and Microengineering* 15, 176 (2005).

[12] G. De Pasquale, A. Somà, Dynamic identification of electrostatically actuated MEMS in the frequency domain, *Mechanical Systems and Signal Processing, Volume 24, Issue 6, August 2010, Pages 1621-1633.*

[13] R.M. Lin, W.J. Wang, "Structural dynamics of microsystems—current state of research and future directions", *Mechanical Systems and Signal Processing, Volume 20, Issue 5, July 2006, Pages 1015-1043*

[14] P. Castellini, M. Martarelli, E.P. Tomasini, "Laser Doppler Vibrometry: Development of advanced solutions answering to technology's needs", *Mechanical Systems and Signal Processing, Volume 20, Issue 6, August 2006, Pages 1265-1285.*

[15] Matthew S. Allen, Michael W. Sracic, "A new method for processing impact excited continuous-scan laser Doppler vibrometer measurements", *Mechanical Systems and Signal Processing, Volume 24, Issue 3, April 2010, Pages 721-735*

[16] J. B. Muldavin and G. M. Rebeiz, "Nonlinear Electro-Mechanical Modeling of MEMS Switches", *Proceed. of IEEE MTT Symposium*, pp.21119-2122 (2001).

[17] Fuqian Yang, "Electromechanical Instability of Microscale Structures", *J. Appl. Phys*, Vol. 92, No. 2, pp.2789-2794 (2002)

[18] Z. J. Guo, N. E. Mc Gruer and G. G. Adams, "Modeling, simulation and measurement of the dynamic performance of an ohmic contact, electrostatically actuated RF MEMS switch", *J. Micromech Microeng.*, Vol. 17, pp.1899-1909 (2007)

[19] F. Michael Serry, Dirk Walliser and G. Jordan Maclay, "The role of the Casimir effect in the static deflection and stiction of membrane strips in microelectromechanical systems (MEMS)", *J. Appl. Phys.*, Vol. 84, No. 5 pp. 2501-2506 (1998)

[20] Steve K. Lamoreaux, "Casimir Forces: Still surprising after 60 years", *Physics Today*, February 2007, pp.40-45 (2007)

[21] Karl M. Strohm et al., "RF-MEMS Switching Concepts for High Power Applications", *Proceed. of 2001 IMS*, pp.42-46 (2001).

[22] B. Pillans, J. Kleber, C. Goldsmitht, M. Eberly, *Proceedings of the 2002 IEEE MTT-Symposium* 329 (2002).

[23] E.P. McErlean, J.-S. Hong, S.G. Tan, L. Wang, Z. Cui, R.B. Greed and D.C. Voyce, *IEE Proceedings on Microwave Antennas Propagation*, 152, 449 (2005).

[24] D. Dragoman, M. Dragoman, and R. Plana, *J. Appl. Phys.* 105, 014505 (2009).

[25] Brusa, E.; Munteanu, M.G.: "Role of nonlinearity and chaos on RF-MEMS structural dynamics", *Symposium on Design, Test, Integration & Packaging of MEMS/MOEMS, 2009. MEMS/MOEMS.* Volume , Issue , 1-3 April 2009 Page(s):323 - 328

[26] J. B. Muldavin and G. M. Rebeiz, "High-Isolation CPW MEMS Shunt Switches - Part 1: Modeling", *IEEE Trans. Microwave Theory and Tech,* Vol. 48, No. 6, pp.1043-1052 (2000).

[27] Romolo Marcelli, Giancarlo Bartolucci, Gianluca Minucci, Benno Margesin, Flavio Giacomozzi, and Francesco Vitulli: "Lumped Element Modelling of Coplanar Series RF MEMS Switches", *Electronics Letters,* Vol.40, No.20, pp.1272-1273 (2004)

[28] Giancarlo Bartolucci, Romolo Marcelli, Simone Catoni, Benno Margesin, Flavio Giacomozzi, Viviana Mulloni, Paola Farinelli: "An Equivalent Circuital Model for Shunt Connected Coplanar RF MEMS Switches", *Journal of Applied Physics,* Vol. 104, No. 8, pp.84514-1 - 84514-8, ISSN: 0021-8979 print+online (2008).

[29] Dimitrios Peroulis, Sergio P. Pacheco, Kamal Sarabandi, and Linda P. B. Katehi, "Electromechanical Considerations in Developing Low-Voltage RF MEMS Switches", *IEEE Trans. Microwave Theory and Tech,* Vol. 51, No. 1, pp.259-270 (2003).

[30] P. Arcioni et al., "Mastering Parasitics in Complex MEMS Circuits", *Proceed. Of the 35th European Microwave Conference,* pp. 943-946 (2005).

[31] M. Farina and T. Rozzi "A 3-D Integral Equation-Based Approach to the Analysis of Real-Life MMICs-Application to Mcroelectromechanical Systems", *IEEE Trans. On Microwave Theory and Tech.,* Vol. 49, No. 12, pp. 2235-2240 (2001).

[32] D. Girbau, N. Otegi and Lluis Pradell, "Study of Intermodulation in RF MEMS Variable Capacitors", *IEEE Trans. On Microwave Theory Tech.,* Vol. 54, No. 3, pp. 1120-1130 (2006).

[33] http://en.wikipedia.org/wiki/Young's_modulus

[34] C. Goldsmith, D. Forehand, D. Scarbrough, I. Johnston, S. Sampath, A. Datta, Z. Peng, C. Palego, and J.C.M. Hwang, *Proceedings of the IEEE 2009 MTT-Symposium,* 1229 (2009).

[35] Balaji Lakshminarayanan, Denis Mercier, and Gabriel M. Rebeiz "High-Reliability Miniature RF-MEMS Switched Capacitors", *IEEE Trans. on Microwave Theory and Tech.,* Vol.56, No. 4, pp.971-981 (2008).

[36] A. Koszewski, F. Souchon and D. Levy, Procedia Chemistry 1, 626 (2009).

[37] V.L. Rabinov, R.J. Gupta, S.D. Senturia, *Proceedings of the Int. Conf. On Solid - State Sensors and Actuators,* Chicago IL, June, 1997, 1125 (1997).

[38] Anna Persano, Fabio Quaranta, Adriano Cola, Antonietta Taurino, Giorgio De Angelis, Romolo Marcelli, and Pietro Siciliano: "Ta_2O_5 Thin films for Capacitive RF MEMS Switches", *J. of Sensors,* Hindawi Publishing, Volume 2010, Article ID 487061, 5 pages. doi:10.1155/2010/487061

[39] Y. Zhu, H. D. Espinosa, *Int. J. RF and Microwave CAE* 14, 317 (2004).

[40] X. Rottenberg, R. P. Mertens, B. Nauwelaers, W. De Raedt, and H. A. C. Tilmans, *J. Micromech. Microeng.* 15, S97 (2005).

[41] E. Papandreou, M. Lamhamdi, C.M. Skoulikidou, P. Pons, G. Papaioannou, and R. Plana: "Structure dependent charging process in RF MEMS capacitive switches", Microelectronics Reliability Volume 47, Issues 9-11, September-November 2007, Pages 1812-1817, *18th European Symposium on Reliability of Electron Devices, Failure Physics and Analysis*

[42] C. Kittel, *Introduction to Solid State Physics*, John Wiley and Sons Inc., 1996

[43] Lei L. Mercado, Shun-Meen Kuo, Tien-Yu Tom Lee, Lianjun Liu: "A Mechanical Approach to Overcome RF MEMS Switch Stiction Problem", *Proceed. of the 2003 Electronic Components and Technology Conference*, pp.377-384.

[44] Romolo Marcelli, George Papaioannu, Simone Catoni, Giorgio De Angelis, Andrea Lucibello, Emanuela Proietti, Benno Margesin, Flavio Giacomozzi, François Deborgies: *"Dielectric Charging in Microwave Micro-electro-mechanical Ohmic Series and Capacitive Shunt Switches"*; IOP Journal of Applied Physics, Vol. 105, No. 11, pp.114514-1 - 114514-10 (2009).

[45] C. H. Mastrangelo, "Adhesion-Related Failure Mechanisms in Micromechanical Devices", *Tribology Letters*, 1997, http://citeseer.ist.psu.edu/467772.html

[46] Patent http://www.patentstorm.us/patents/5772902.html

[47] E. Buks and M.L. Roukes, "Stiction, Adhesion Energy, and the Casimir Effect in Micromechanical Systems", *Phys. Rev. B*, Vol. 63, pp.33402-1 – 33402-4 (2001).

[48] G. Guisbiers, E. Herth, B. Legrand, N. Rolland, T. Lasri, L. Buchaillot, Materials selection procedure for RF-MEMS, *Microelectronic Engineering, Volume 87, Issue 9, November 2010, Pages 1792-1795*

[49] Brian Stark: "MEMS Reliability Assurance Guidelines For Space Applications", *Jet Propulsion Laboratory Publication 99-1*, Pasadena, California (1999).

[50] Peroulis, D. Pacheco, S.P. Katehi, L.P.B.: "RF MEMS switches with enhanced power-handling capabilities", *IEEE Trans. On Microwave Theory and tech.*, Vol. 52, No.1, pp. 59-68 (2004).

[51] S. Melle, D. De Conto, L. Mazenq, D. Dubuc, B. Poussard, C. Bordas, K. Grenier, L. Bary, O. Vendier, J.L. Muraro, J.L. Cazaux, R. Plana, "Failure Predictive Model of Capacitive RF-MEMS", *Microelectronics Reliability*, Vol. 45 pp.1770–1775 (2005).

[52] J. Franclov´a, Z. Kuˇcerov´a, and V. Burˇs´ıkov´, "Electrical Properties of Plasma Deposited Thin Films", *WDS'05 Proceedings of Contributed Papers, Part II*, pp.353–356 (2005).

[53] J. Iannacci, A. Repchankova, A. Faes, A. Tazzoli, G. Meneghesso, Gian Franco Dalla Betta, Enhancement of RF-MEMS switch reliability through an active anti-stiction heat-based mechanism, *Microelectronics Reliability, Volume 50, Issues 9-11, September-November 2010, Pages 1599-1603*

[54] M. Matmat, K. Koukos, F. Coccetti, T. Idda, A. Marty, C. Escriba, J-Y. Fourniols, D. Esteve, Life expectancy and characterization of capacitive RF MEMS switches, *Microelectronics Reliability, Volume 50, Issues 9-11, September-November 2010, Pages 1692-1696*

[55] http://www.comsol.com/

[56] A. Lucibello, E. Proietti, S. Catoni, L. Frenguelli, R. Marcelli, G. Bartolucci, Proceedings. of the Int. Semiconductor Conference CAS 2007, Sinaia, Romania, October 2007, 259 (2007).

[57] http://www.fbk.eu

10

Characterization and Modeling of Charging Effects in Dielectrics for the Actuation of RF MEMS Ohmic Series and Capacitive Shunt Switches

Romolo Marcelli et al.*
CNR-IMM Roma, Roma, Italy

1. Introduction

Charge accumulation in dielectrics solicited by an applied voltage, and the associated temperature and time dependencies are well known in scientific literature since a number of years [1]. The potential utilization of materials being part of a device useful for space applications is a serious issue because of the harsh environmental conditions and the necessity of long term predictions about aging, out-gassing, charging and other characteristic responses [2], [3]. Micro-mechanical Systems (MEMS) for RF applications have been considered for sensor applications as well as for high frequency signal processing during more than one decade [4], [5], [6], [7], [8], [9]. In this framework, RF MEMS switches are micro-mechanical devices utilizing, preferably, a DC bias voltage for controlling the collapse of metalized beams [8]. Magnetic [10], thermal [11] and piezoelectric [12] actuations have been also evaluated, but the electrostatic one seems to be until now preferred for no current flowing, i.e. a virtual zero power consumption, less complicated manufacturing processes and more promising reliable devices [13]. During the last few years, several research activities started to release the feasibility of RF MEMS switches also for Space Applications [14], [15], [16]. The electrostatic actuation of clamped-clamped bridges or cantilevers determines the ON and OFF states depending on the chosen configuration. As well established, RF MEMS switches are widely investigated for providing low insertion loss [8], no or negligible distortion [17], [18] and somehow power handling capabilities [19], [20], [21], [22] for a huge number of structures already utilizing PIN diodes for high frequency signal processing. Actually, redundancy switches as well as single pole multiple throw (SPMT) configurations, [23], [24], matrices [25] true time delay lines (TTDL) [26], [27] and phase shifters [28], [29] for beam forming networks in antenna systems could benefit from their characteristics. On the

* Andrea Lucibello[1], Giorgio De Angelis[1], Emanuela Proietti[1], George Papaioannou[2], Giancarlo Bartolucci[1,3], Flavio Giacomozzi[4] and Benno Margesin[4]
[1]*CNR-IMM Roma, Roma, Italy*
[2]*University of Athens, Athens, Greece*
[3]*University of Roma "Tor Vergata" – Electronic Eng. Dept., Roma, Italy*
[4]*Bruno Kessler Foundation, Center for Materials and Microsystems, Povo (TN), Italy*

other hand, the reliability of this technology has been not yet fully assessed, because of the limitations introduced by: (i) the mechanical response of the single switches [30], (ii) the necessary optimization of the packaging [31], and (iii) the charging mechanisms. In particular, the charging effect is due to the presence of both the dielectric material used for the realization of lateral actuation pads, deposited to control the collapse of bridges and cantilevers far from the RF path, and the dielectric used for the capacitance in the case of shunt connected microstrip and coplanar configurations. Presently, there is a wide literature about the onset of the mechanism [32], [33], [34] and its control by means of uni-polar and bi-polar actuation voltage schemes [35], [36], [37]. Some results give evidence also for the substrate contribution to charging effects [39] and those related to packaging [38]. Specifically, electromagnetic radiation is a serious issue for space applications [40], [41]. Electrostatic discharge has been discussed in [42], and it is clearly influenced by the deposition process [43]. Besides structural dependence of the charging [44], solutions considering the absence of the dielectric material is also considered, giving evidence for a decrease but not for a complete disappearance for such a contribution [45], [46]. Specific aging schemes based on the temperature are also proposed for long term evaluation of the devices [47]. Advanced studies have been also performed by means of the Kelvin Probe Microscopy, for improving the surface resolution of the charging effect detection [48]. Ohmic contact problems have been evaluated in [49]. Different kind of charging mechanisms can influence the reliability of the MEMS devices, as it has been assessed after the study published in [50].

In this chapter, it will be presented the characterization of two configurations of RF MEMS switches, to demonstrate how the actuation voltage is modified by using a uni-polar bias voltage and how it is under control and stable, at least for a limited number of consecutive actuations, if an inversion in the bias voltage is provided. In particular, the measurements recorded for an ohmic series and for a shunt capacitive configuration will be presented and discussed, considering the main source of charging for both devices. Moreover, experiments performed in both MIM and MEMS reveal that the charging process is strongly affected by the temperature [51]. MIM capacitors have been used to assess the material bulk properties with the aid of Thermally Stimulated Depolarization Current (TSDC) method. The charge storage was found to increase exponentially with temperature in both MIM capacitors and MEMS switches. In particular, in the high temperature range the activation energies in MEMS switches were found to have close values with respect to MIMs, and from TSDC experiments in MIM capacitors they have been found to be rather small. Equivalent circuits accounting for the above charging effects can be used as an effective lumped model, useful for circuital simulations of feeding lines and actuation pads [52].

2. Technology

Suspended bridges have been manufactured in coplanar waveguide (CPW) configuration. The series ohmic switch has been obtained by means of a bridge isolated with respect to the lateral ground planes, closing a capacitive in-plane gap when the proper bias voltage is provided by means of lateral poly-silicon pads. In such a case the bridge is collapsed and the switch passes from the OFF to the ON state. *Vice versa,* the shunt capacitive switch is composed by a metal bridge connecting the lateral ground planes and by a dielectric layer providing a capacitive contribution when the bridge is collapsed. In this case, when the switch is actuated by means of a DC bias voltage, it passes from the ON state to the OFF

one. In order to fabricate micromechanical switches together with integrated resistors and DC blocking capacitors an eight mask process has been developed. Two electroplated gold layers of different thickness are provided for the realization of highly complex moveable bridges and the co-planar waveguides. The substrates are p-type, <100>, 525 μm thick, 5 kΩ·cm high resistivity silicon wafers. A 1000 nm thick thermal oxide is grown as an isolation layer. Next a 630 nm thick un-doped poly-silicon layer is deposited by low pressure chemical vapour deposition (LPCVD), to be used for the resistors and actuation electrodes obtained by selective dry etching of the poly-silicon layer. Then, tetra-ethyl-ortho-silicate (TEOS) is deposited by a LPCVD process to provide the high isolation needed for the actuation electrodes. Contact holes are then defined and etched by a plasma process. After ashing the photoresist mask, a multilayer underpass metal Ti/TiN/Al/TiN is deposited by sputtering. The total thickness of the multilayer has to be the same of the polysilicon, in such a way that metal underpass and actuation electrodes are at the same level. The wafer front side is then covered with 100 nm of low temperature oxide (LTO) to obtain an insulating layer for capacitive shunt switches. The previous step is un-necessary for series ohmic configurations. The vias in the LTO are defined by masking and dry etching. A Cr/Au layer is defined by lithography and wet etched. The main purpose of this layer is to cover with a noble metal the exposed electrical contacts of the series ohmic switches to get low resistive electrical contacts. The sacrificial layer required for obtaining the air gap is formed by a 3 μm thick photoresist, hard baked at 200 °C for 30 minutes to obtain well-rounded edges. As a seed-layer for electrochemical Au deposition a 10/150 nm thick Cr/Au layer is deposited by PVD. The moveable air bridges are defined using a 4 μm thick positive resist. After an exposure to oxygen plasma at 80 °C a 1.8 μm thick gold layer is selectively grown in a gold sulphite bath. The first plating mask is removed with an appropriate solvent and the CPW lines and anchor posts for the moveable air bridges are defined with 5 μm thick positive resist and then a 4 μm thick gold layer is selectively grown. The last plating mask and the seed layer are then wet removed. At this point a sintering in nitrogen at 190 °C for 30 minutes is performed to provide the gold layers with the appropriate tensile stress. Finally the air bridges of the individual switches are released with a modified plasma ashing process (20 minutes oxygen plasma at 200 °C) in order to avoid sticking problems.
The two devices which have been used for the characterization are shown in the photos given in Fig. 1 (series ohmic device, device S1) and in Figure 2 (shunt capacitive switch, device CL).

3. Experimental results

All the measurements have been performed and recorded in a Clean-Room environment, at the temperature T=(23±1) °C, with a relative humidity RH=(35±1) %. A nitrogen flux has been used for providing a dry environment for the devices under test. RF measurements have been used as a validation for the state (ON or OFF) of the switches and for their electrical performances before, during and after the voltage application. In particular, after each cycle used for such a measurement, no changes in the electrical performances of the exploited devices has been recorded. A schematic diagram of the measurement bench is shown in Fig. 3. The reliability of the manufactured devices with respect to the charging effects, and specifically the influence of the pulse shape and of the sign of the voltage (positive or negative) on the actuation mechanism, have been studied by using pulse trains where the rise and fall time, as well as the pulse duration and the separation between pulses

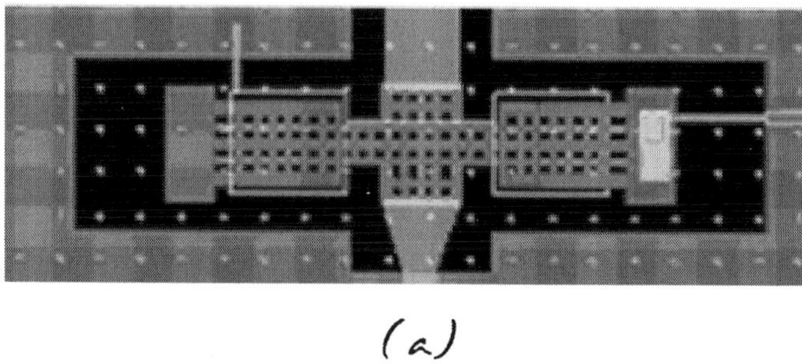

(a)

(b)

Fig. 1. Diagram (a) and photo detail (b) of the implemented ohmic series switch configuration. Lateral wings have been included for improving the electrical contact. A number of switches with different geometrical and physical characteristics have been produced on the base of changes with respect to this one. Actually, the number of dimples as well as the thickness of the bridge and other details of the geometry contribute to the electrical performances. When the switch is actuated, the bridge, isolated with respect to the ground, closes the central conductor of the CPW with a metal-to-metal contact and the device is in the ON state (device S1).

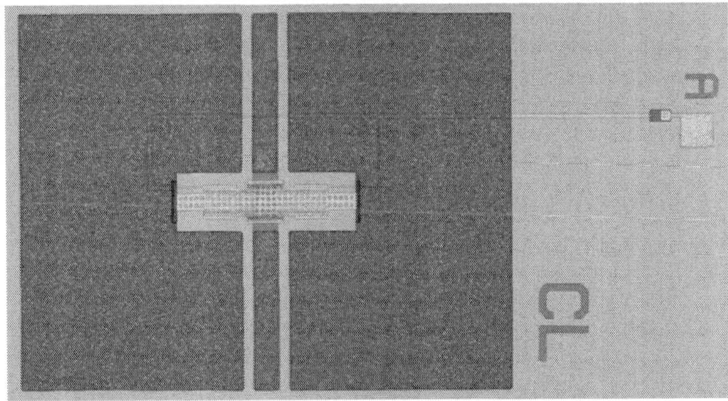

Fig. 2. Coplanar shunt capacitive switch. When the switch is actuated, the bottom side of the suspended bridge is collapsed, touching the dielectric layer placed along the central conductor of the CPW, providing a shunt to ground in a limited frequency range (resonant response), and the device is in the OFF state (device CL).

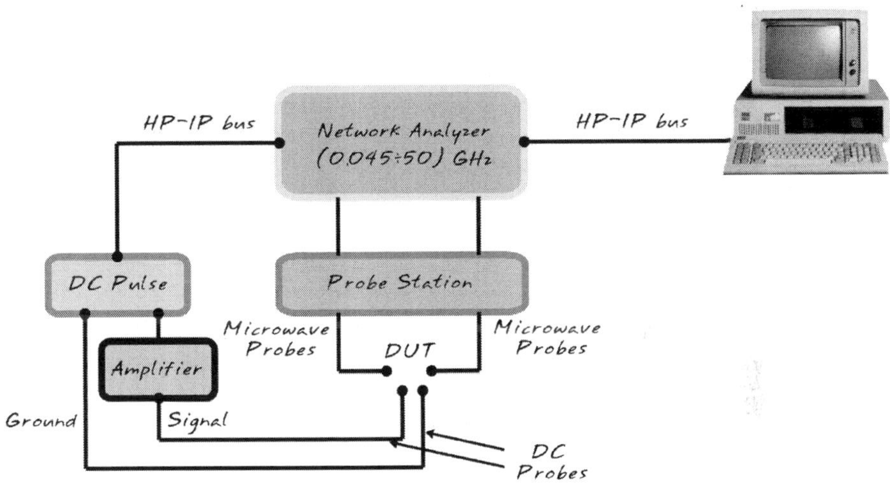

Fig. 3. Schematic diagram of the measurement system used for testing the RF MEMS switches.

have been slowly changed. Moreover, a bi-polar scheme has been applied, with positive voltages followed by negative ones. For the uni-polar experiment as well as for the bi-polar one, the actuation voltage has been recorded when the sudden change in the measured Scattering Parameters due to the bridge collapse was clearly visible on the Vector Analyzer, i.e. by means of an abrupt change in the value of both transmission and return loss. Actually, this occurs during the voltage ramp.

In particular, we paid attention to:

- The pulse-width
- The rise-time and the fall-time of the pulses (ramps)
- The delay between the positive and the negative pulse
- The applied voltage

The first measurements have been performed by using only positive pulses (uni-polar scheme) with a ramp of 1 V/sec and T1=T2=1 min. After that, positive and negative pulses have been used, with the following parameters:

- Ramp=1 V/sec and 2 V/sec
- T1= 1 min and 30 sec
- T2=10 sec.

Both devices given in Fig. 2 and in Fig. 3 have been characterized by using the proposed uni-polar and bi-polar schemes as it is explained in detail in the following text. In the uni-polar scheme, after the actuation, the switch is maintained at the same voltage during the time T1. Then, the voltage has been decreased down to zero, and in the meantime the de-actuation voltage has been measured. The successive ramp was imposed by increasing again the voltage until a new actuation occurs, and also in this case the voltage is maintained constant during the time T1. Every time, the voltage required for the successive actuation was higher than the previous one. The procedure was repeated for recording actuation and de-actuation voltages until a *plateu* value has been obtained. In the bi-polar scheme, the applied DC voltage is composed by positive and negative pulses having a maximum value of ±50 V for the device in Fig. 1 (S1) and ±60 V for the device in Fig. 2 (CL), and in this case the actuation and de-actuation voltages have been measured as absolute values of the imposed pulses. In Fig. 4 the shape of the pulse trains used in the experiments is shown.

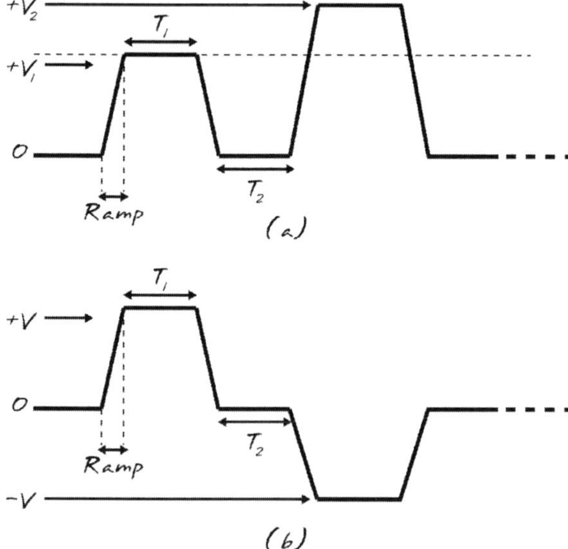

Fig. 4. Shape of the pulse trains used for the experiments on the charging effects. (a) is the uni-polar scheme, while (b) is the bi-polar one.

In the following text and figures, results and comments on the performed measurements are presented. First of all, S1 and CL have been actuated by using a uni-polar, positive voltage scheme. For both of them, the actuation as well as the de-actuation voltages have been measured until a *plateau* voltage has been obtained. The results are shown in Fig. 5 and in Fig. 6 with the values of the voltage and time parameters used for the actuation, and obtained for the corresponding de-actuation. In this case, the applied and measured voltages are always positive.

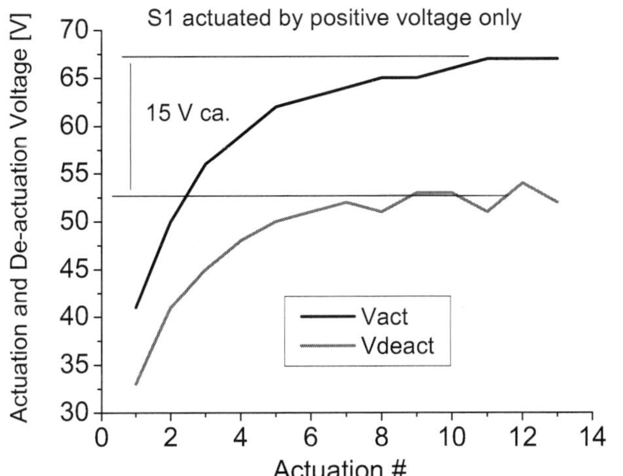

Fig. 5. Response of S1 actuated by using positive voltages only. T1=1 min, T2=10 sec, Ramp=1 V/sec for both trailing and leading edge of the pulse.

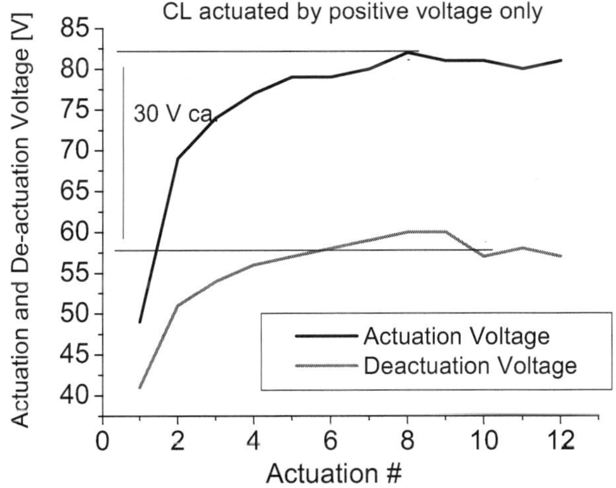

Fig. 6. CL actuated by using positive voltage only. T1=1 min, T2=10 sec, Ramp=1 V/sec, for both trailing and leading edge of the pulse.

Therefore, a bi-polar scheme has been applied by measuring the same devices the day after, when the effect of charging was completely removed, leaving them at rest without voltage nor RF signals applied to the device under test. The results are given in Fig. 7 and in Fig. 8, where the absolute value of the applied voltage is plotted as a function of the performed

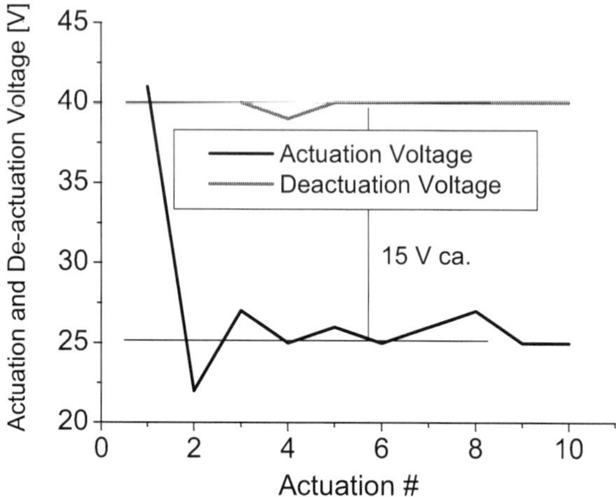

Fig. 7. S1 actuated by using positive and negative voltages. Only the absolute value of the recorded actuation voltage is plotted, but changed from +V to -V after each pulse, with T1=1 min, T2=10 sec, |+V|=|-V|=50 volt and Ramp=1 V/sec.

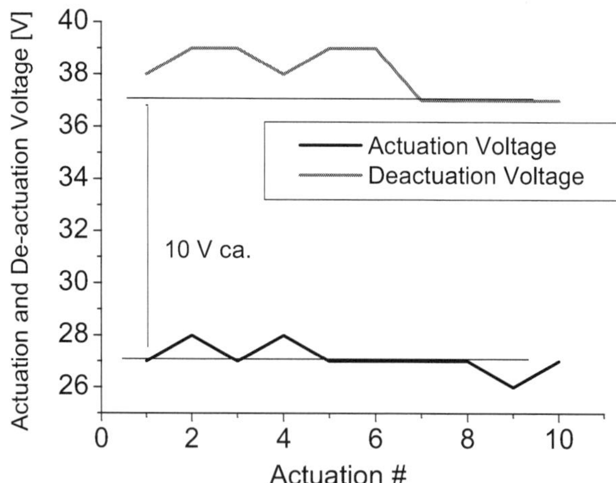

Fig. 8. S1 actuated by using positive and negative voltages. Only the absolute value of the voltage is plotted, but changed from +V to -V after each pulse, with T1=30sec, T2=10 sec, |+V|=|-V|=50 volt and Ramp=1 V/sec. The measurement has been performed 5 min after the one shown in Figure 7. The difference between actuation and de-actuation voltages is a bit decreased, which is an indication of a charging partially re-covered.

actuations. It is worth noting that the measured data have been normalized to positive values, and the reader can have the erroneous feeling that the de-actuation voltage is always higher than the actuation one. This is only a false perspective, and the reason for such a finding is discussed after the presentation of the experimental data.

It is worth noting the difference obtained between the results in Fig. 7, Fig. 8 and Fig. 9. Actually, no dependence on the applied ramps has been obtained, but there is a clear evidence that the process is quite slow, because after times in the order of several minutes, i.e. during the experimental procedure, the charging is still present. From the analysis of the figures where both positive and negative voltages have been applied (Fig. 7, 8 and 9), one could conclude that the actuation voltage is lower than the de-actuation one. In fact, this is due to the kind of plot, because only the absolute value of the applied voltage is given, in order to have a continuous curve, with data not jumping from negative to positive values. The physical reason for that is explained in the discussion at the end of this section.

In the following Fig. 10 and Fig. 11 the same qualitative results are shown for the device CL, where in 30 minutes ca. the charging effect has been almost completely recovered. It turns out from this finding that the same values for the actuation voltages have been recorded, and the same difference between V(actuation) and V(de-actuation) has been obtained.

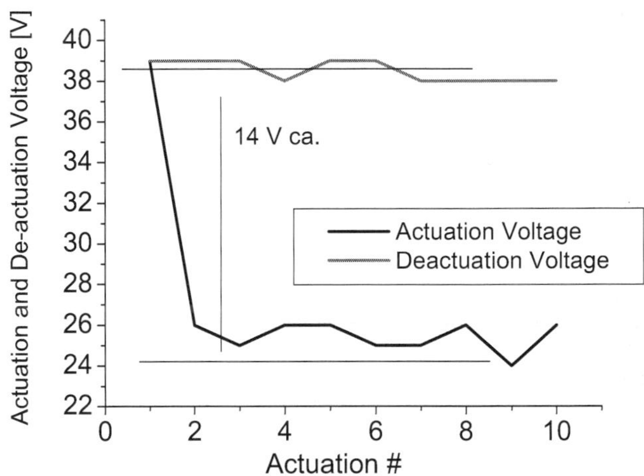

Fig. 9. S1 actuated by using positive and negative voltages. The same parameters used in the previous Fig. 8 have been imposed, i.e.: T1=30 sec, T2=10 sec, |+V|=|-V|=50 volt and Ramp=2 V/sec. The measurement has been performed the day after. The result is quite similar to that shown in Fig. 3, with T1 decreased from 1 min to 30 sec and Ramp passed from 1 V/sec to 2 V/sec. As a consequence, none of the above parameters seems to affect the measures. Moreover, the first actuation is still between 39 and 41 V, but by using positive and negative values it is maintained at a constant value as well as the de-actuation voltage, and it is lower than in the positive case only.

Some of the main findings of the performed measurements are in full agreement with those in [34], and in particular with the conclusion that the devices do not fail if they are subjected to a square wave voltage for the actuation when a C/V curve is taken with a slowly varying

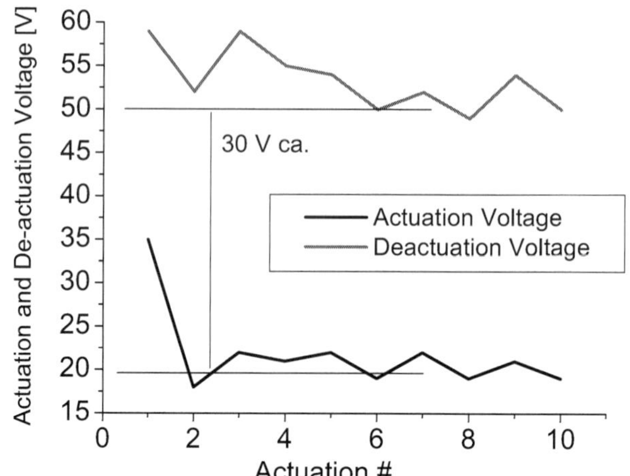

Fig. 10. CL actuated by using positive and negative voltages. T1=1 min, T2=10 sec, |+V|=|-V|=60 V and Ramp=1 V/sec.

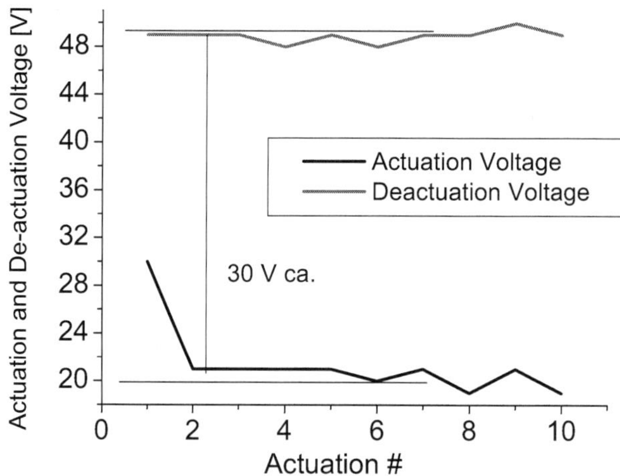

Fig. 11. CL response by using the same parameters as in the case of Fig. 10, but with Ramp=2 V/sec and measurement performed after 30 min. The difference between the two levels has the same value as before.

voltage. Actually, for slow ramps we never experienced a stuck device for both S1 and CL. On the other hand, the reliability tests previously performed on the same devices were never accompanied by a sticking of the series configuration, in spite of the fact that a faster switching was used in that case [47].

It is clear, in the present experimental results, that charging effects are present in both configurations, affecting in a predictable way the performances of the measured devices.

In particular:

- The absolute value of the actuation voltage V_a and of the de-actuation voltage V_d (and the difference between them) is constant when the sign of the pulse is reversed, exception done for the first actuation
- The measured difference in the bi-polar scheme is equal to the difference between the two plateau (i.e. $V_{a,plateau}$-$V_{d,plateau}$) experienced during the charging process when a positive voltage pulse train is applied. This could be used as a measure of the maximum charge which can be accumulated in the device
- The absolute value of the actuation voltage for the series switch S1 is almost half of the first positive value when a train of positive and negative pulses is used
- The absolute value of the actuation voltage for the shunt switch CL subjected to positive and negative pulses is around 20 volt Vs the almost 55 volt used for the positive voltage only.
- The results from the previous two points are "re-normalized" considering that the algebraic sum of the positive actuation voltage (starting from the second actuation), and of the difference between the two plateau values in the case of positive only voltages ($V_{a,plateau}$-$V_{d,plateau}$), gives as a result the first positive actuation voltage. This means that the voltage difference necessary for the actuation is the same. So, when positive and negative voltages are applied to the device S1, for which we have $|V_a|\approx 40$ V, then, after the first de-actuation (occurring in this case again at V=40 V after imposing V=50 V) we always get $|V_a|\approx 25$ V with a difference of 15 V coming from the extra voltage generated by the charging effect due to the previous actuation.
- The same result is obtained for the CL configuration, where $|V_a|\approx 50$ V, and the first de-actuation occurs at $|V_d|\approx 50$ V after imposing a positive voltage V=60 V. After that, the switch is actuated always by applying 20 V. Actually, a difference of 30 V is always observed (in this case this happens independently of the time passed from the previous measurement), in such a way that the sum 20+30=50 V is again the value of the first actuation voltage experienced when positive only pulses are used.

As a consequence of the above discussion, both schemes for actuation (uni-polar and bi-polar) are affected by charging mechanisms, because the dielectric is always present. On the other hand, the bi-polar scheme offers the advantage, with respect to the uni-polar one, in terms of the absolute value of the voltage necessary for actuating the device [37]. This is especially good when a high number of actuations are needed for a frequent re-configuration of architectures based on several RF MEMS, and there is no time for a full de-charging of each individual device. In our devices, we believe that the charging exhibits a saturation value due to the maximum number of charges which can be activated on the surface as well as in the bulk of the dielectric, which slowly goes back to the original situation. In this framework, looking at our experimental results, the utilization of positive and negative pulses allows a faster re-combination process, and the possibility to drive the device always by means of the same absolute value of the voltage, changing the sign of the applied voltage from one actuation to the successive one. A possible interpretation could be that the de-charging process, usually slow, is accelerated when the device is subjected to a gradient of the electric field, passing from positive to negative values and vice-versa. In the following Fig. 12 the bi-polar scheme imposed for S1 and the effect on the actuation voltage is shown.

4. Measurements on test MIM capacitors and discussion

MIM capacitors having the same structure to be used for the actuation pads of the RF MEMS switches have been realized, to study the charging mechanisms related to the materials used

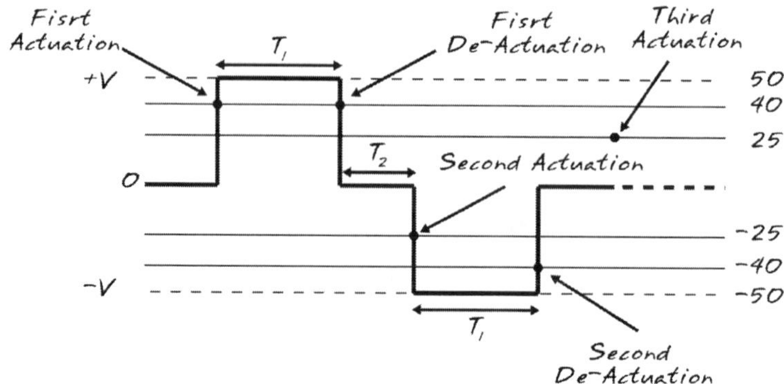

Fig. 12. Bi-polar scheme imposed for the actuation of the switch S1.

for the device actuation. It is worth noting that the MIM is only an approximation of the real actuation, because in this case no residual air gap has to be considered between dielectric and metal bridge. For this reason the MIM should suffer for charging and de-charging effects different with respect to those measured on the real device for both time and kind of processes. On the other hand, it is important to know the properties of the material itself, because it will affect the operation of the device. The scheme and related equivalent circuit of the measurement setup used for characterizing the MIM is shown in Fig. 13. In Fig. 14 the two structures used for the MIM devices are also shown.

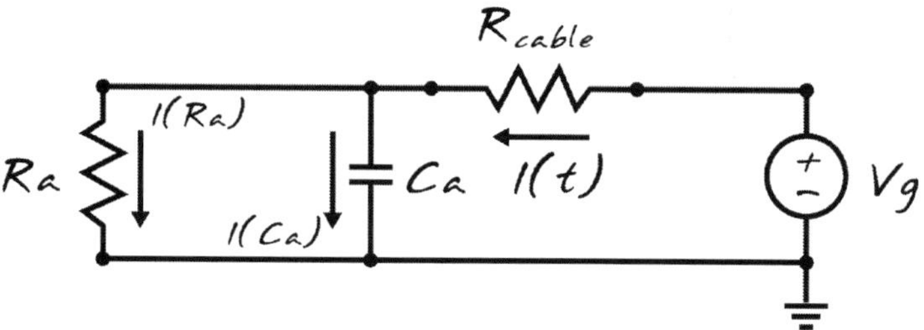

Fig. 13. Equivalent circuit for the measurement setup of the MIM Capacitors. A power supply provides the voltage V_g and the current I, both functions of the time t following a slow ramp. The device under test is a MIM simulating the actuation pad structure, schematized as a capacitor C_a with a high bulk resistance R_a in parallel with respect to C_a.

From the analysis of Fig. 13, the equations governing the voltages and currents on the equivalent lumped components can be written as:

$$
\begin{aligned}
& I(t) = I_{Ca}(t) + I_{Ra}(t) = C_a \frac{dV_{Ca}(t)}{dt} + \frac{V_{Ca}(t)}{R_a} = \frac{-V_{Ca}(t) + V_g(t)}{R_{cable}} \\
& V_{Ca}(t) = V_g(t) - R_{cable} I(t) \\
& \frac{dV_{Ca}(t)}{dt} = \frac{dV_g(t)}{dt} - R_{cable} \frac{dI(t)}{dt}
\end{aligned} \tag{1}
$$

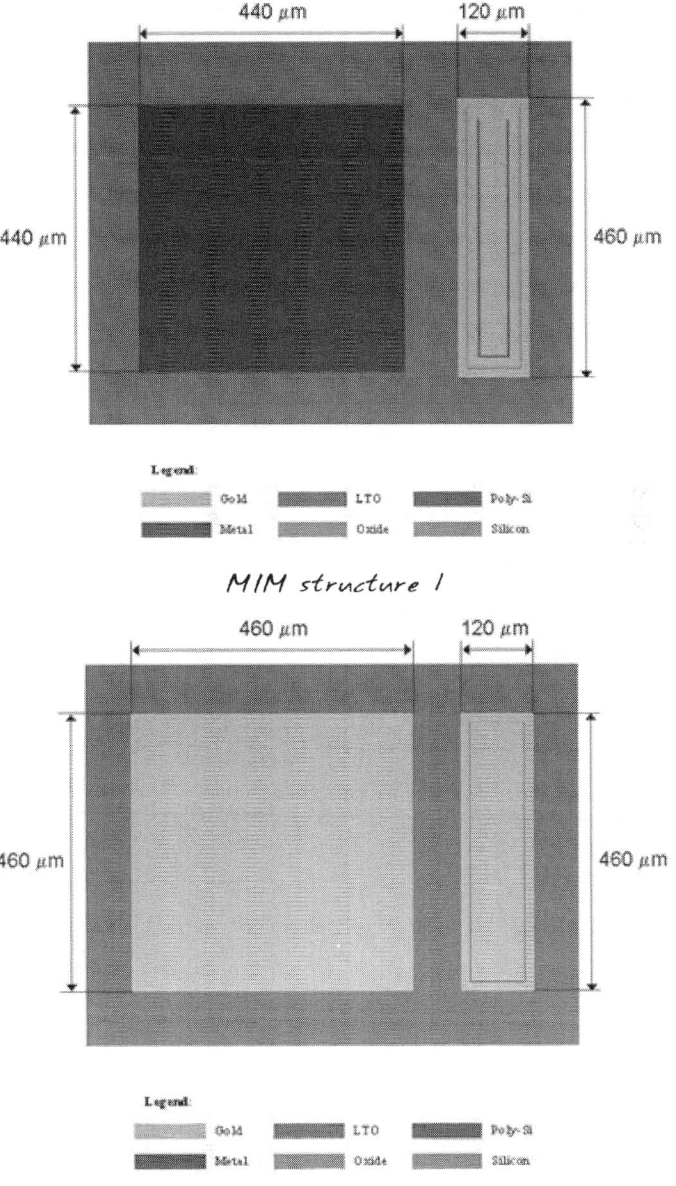

Fig. 14. MIM structures used for the characterization.

From the above equations, it turns out that the measured value of $I(t)$ when imposing $V_g(t)$ will be given by using the following relation:

$$\left(1+\frac{R_{cable}}{R_a}\right)I(t)=C_a\frac{dV_g(t)}{dt}+\frac{1}{R_a}V_g(t)-C_aR_{cable}\frac{dI(t)}{dt}\approx I(t) \quad (2)$$

The last assumption is valid when, as it can be reasonably assumed, $R_{cable}<<R_a$.
From the analysis of the measurements, it will be evident that when the dielectric material of the MIM behaves as an almost ideal dielectric, only the first term on the right hand of the above equation is important, and mainly a capacitive contribution is measured, as expected. Actually, the imposed ramp and the measured current will vary maintaining a constant ratio. On the other hand, because of the non-ideal response of the obtained dielectric material, a parallel resistance has to be included, to account, since the very beginning, for some free charges, and for the asymmetry in the capacitor itself, which has two parallel plates not equal between them for both dimensions and conductivity. So far, a capacitive response is obtained, strictly speaking, only at low voltage levels, and a small charge injection is immediately recorded, evidenced by a linear contribution typical of the second term in Eq. (2). An almost negligible contribution is also given by the last term of the same equation, being the derivative of the current very small and the resistance of the cable small too. The Poole-Frenkel effect, which should dominate the charging processes in the exploited dielectrics, begins to be evident when sufficiently high voltages are imposed after the first ramp, i.e. when V_g is in the order of tens of volt, and the recorded current suddenly increases. In this range, the response of the measured current follows a high-voltage law, while it is almost linear for lower values of V_g.
After some critical value of V_g, the picture given by Eq. (2) has to be strongly modified accounting for the change of both the conductivity and the polarization, which should influence at least the value of R_a.
In particular, the term describing the current flowing in R_a should be better identified by the formula suggested in [61] and [63] for high voltages. Moreover, we can map the time variable t into the applied voltage V_g, as the measurements are performed by using an I Vs V_g plot and, since the voltage is imposed by means of a linear ramp, we can also define $r=dV_g/dt=const$.
In the region of non-linear response for the current Vs voltage, the trend looks like coherent with conclusions in [61], where a dependence on V_g^2 is expected, following also the conclusions in [63]. This will transform Eq. (2) in:

$$I(V_g)=C_a\frac{dV_g}{dt}+f(V_g)-C_aR_{cable}\frac{dI}{dV_g}\frac{dV_g}{dt}=C_a\left(1-R_{cable}\frac{dI}{dV_g}\right)r+f(V_g) \quad (3)$$

Which is valid until the breakdown occurs in the MIM structure, and $f(V_g)$ is a function involving the applied voltage, which is linear like in Eq. (2) until a high voltage dependence is required, as mentioned in [61] and [63]. Because of the charging effect and of the current induced by the Poole-Frenkel effect, Eq. (3) has to be corrected again when a second ramp is applied before the dielectric is naturally de-charged, and the effect will be a current decrease at the same voltage level experienced during the previous ramp. In this case, the Poole-Frenkel current (linked to the current density by $I_{PF}(V_g)=AJ_{PF}(V_g)$) has to be included, having a sign which is opposite with respect to $I(V_g)$.

Wafer #	Dielectric	Thick. [nm] ± 2%	Sample #	V_B [volt]	Charge Injection
1 BE: P TE: Al 1%Si	Nitride	98	C2	> 100	Few volt
			C3	> 100	Few volt
			C3	> 100	Few volt
			C4	> 100	Few volt
			C5	> 100	Few volt
3 BE: P TE: Al 1%Si	TEOS	203	C1	> 100	Few volt
			C2	> 100	Few volt
			C5	> 100	Few volt
4 BE: P TE: Al 1%Si	LTO	114	C6	≈ 100	Few volt
5 BE: P TE: Al 1%Si	PECVD Nitride HF	100	C2	≈ 75	Few volt
7 BE: Al 1%Si+ Ti+TiN TE: Cr / Au	LTO	114	C6	≈ 62	10-20 volt
9 BE: Al 1%Si+ Ti+TiN TE: Cr / Au	PECVD Oxide LF Top	78	C2	≈ 47	?
			C3	< 40	?
			C5	< 40	?
10 BE: Al 1%Si+ Ti+TiN TE: Cr / Au	PECVD Nitride HF	100	C1	?	> 10 volt
			C6	45	> 10 volt
11 BE: Al 1%Si+ Ti+TiN TE: Cr / Au	PECVD Nitride LF	87	C6	35	> 10 volt

Table 1. Full list of the measured devices. The wafer #, with the bottom electrode (BE) and the top electrode (TE) are given, with the dielectric and deposition technique. P is for Poly-silicon. The thickness is in nm ± 2%. The breakdown voltage V_B is also shown, when it was possible to measure it. Charge injection is almost immediately recorded in many cases. HF and LF stand for high frequency and low frequency of deposition respectively.

Several wafers have been characterized, with repeated structures like those shown in Fig. 14. Actually, TEOS, LTO and Nitride (Si_3N_4) deposited following different methods have been obtained. Top and bottom electrodes have been changed too. The material and structural parameters are summarized in the following Table 1. Ramps of 0.05 and 0.1 V/s have been imposed. In particular, Wafer from #1 to #5 emulate the structure of the actuation pads, while from wafer #7 to #11 the situation of the underpass in the area of the bridge is proposed.
A selection of the measurements performed on the samples is given in the following figures. The findings in Fig. 16 have been interpreted as the contribution of the electric field generated by: (i) trapped charges, and (ii) interface states. Both effects contribute in the opposite way

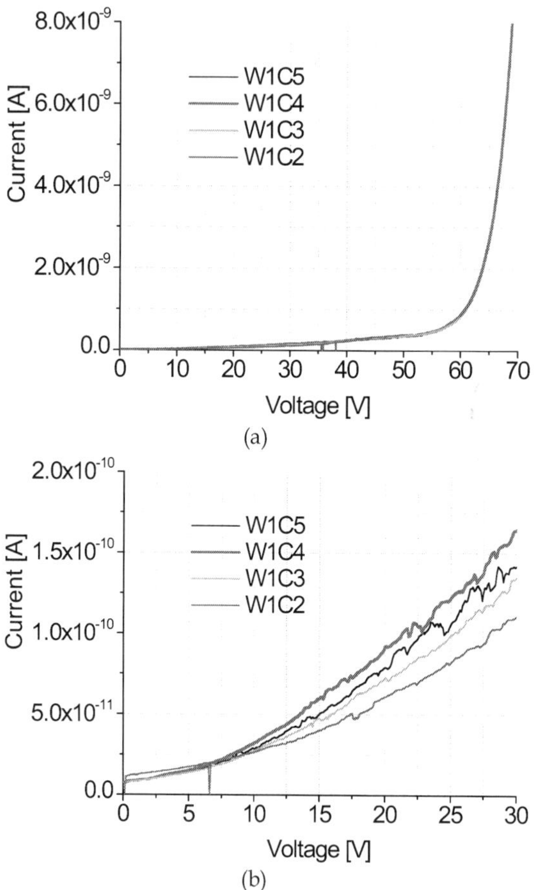

Fig. 15. I vs V for wafer #1 (W1). The same response (a) is obtained for different samples (C2, C3, C4, C5) having the same geometry and dielectric (nitride, Si_3N_4). Small differences can be seen only at low voltage and current values in (b) and can be attributed to the technological reliability. Actually, a charge injection is anyhow measurable, as the current response is not flat as a function of the applied voltage.

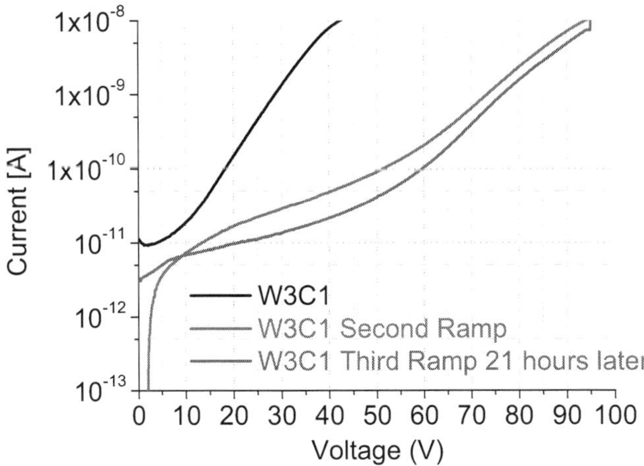

Fig. 16. I vs V for sample C1 in wafer #3 (W3C1, TEOS). A second ramp has been imposed after one minute, with clear evidence for the sample charging. The same behavior is exhibited by the other TEOS devices. The measurement has been repeated almost one day after the first one (21 hours later). In this case, the initial conditions are not yet restored, as it has been measured in other samples too.

with respect to the external DC field due to the actuation voltage. As a result, the sample experiences a decrease in the current flowing through the device. More in detail, the response of the dielectric is characterized, when the second ramp is applied, by a negative current for a relatively long time (40 sec ca. for a bias sweep rate of 50 mV/sec). This finding can be explained in terms of the additional contribution of the interface states, providing an increase in the number of charges. Actually, the trapping mechanism does not allow the injection of further charges, whereas the interface states can provide such an additional current, always opposite with respect to that induced by the external DC bias. This behavior has not been experienced 21 hours after because this long time allows the natural discharging process of the interface states. On the other hand, in almost one day, the trapped bulk charges had not the time for a full restoring of the initial conditions. In fact, during the third ramp a further positive shift of the voltage necessary for the onset of the charging process has been measured, in spite of the long time passed between the second and the third ramp.

The contribution of the electric field generated by the trapped charges and by the interface states is also evidenced in the plot of Fig. 17 for wafer #1, but a bi-polar actuation scheme has been adopted, instead of the uni-polar one used for the previous measurement. The results in Fig. 17 have been interpreted as it follows: from 0 to 80 V, during the first ramp, the dielectric is charged. From 80 V to 0 it is like to impose a second ramp (negative or positive slope it does not matter) and the current is down-shifted. In the third ramp it looks like to have the dielectric fully de-charged because of the second ramp, as the current response is symmetric with respect to the first ramp. During the fourth ramp, the current is increased in absolute value, with a peak probably due to a "frozen" charge. After that, the fifth ramp gives a response qualitatively similar to the previous plot, but higher values are recorded because the residual current is in the same sense with respect to the imposed one. It is worth noting that the measurements have been re-normalized to the first quadrant, as negative currents correspond to negative voltages. The findings in Fig. 17 have been

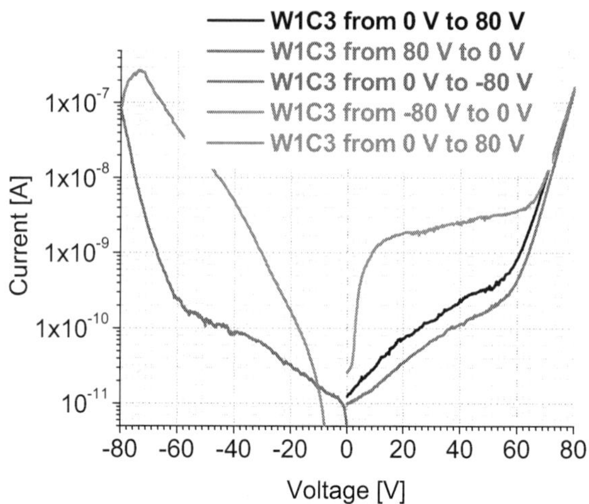

Fig. 17. The sample C3 from wafer #1 (W1C3, Si_3N_4) has been subjected to voltage ramps going forth and back up to a maximum value of ±80 V. As a result, a down shift of the current response is obtained by applying a voltage from 0 to 80 V and back from 80 V to 0. Then, an almost symmetric response is obtained when a negative bias is imposed.
A completely different trend is measured by decreasing the applied voltage to 0, and finally a current increase is experienced going again to 80 V. Similar responses have been obtained for other samples.

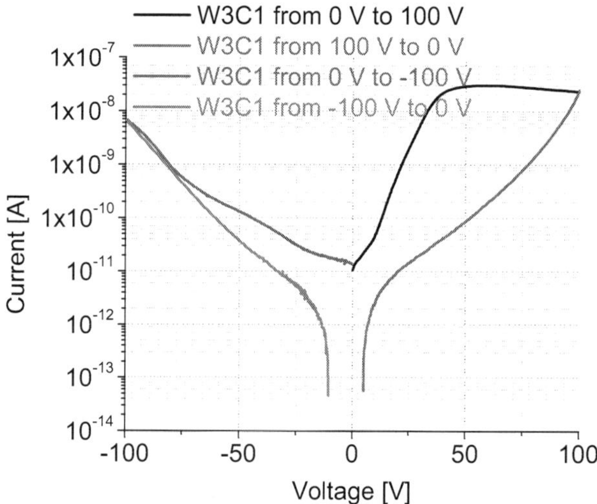

Fig. 18. The sample C1 of wafer #3 (TEOS) was measured by imposing a full cycle from positive to negative values and back to zero as it was in the data of Fig. 17.

interpreted again as the contribution of the electric field generated by the trapped charges and by the interface states, but in this case the field is in the same way with respect to the external one. Moreover, LPCVD Si_3N_4 shows a better response in terms of charge injection, because it happens at higher voltage values with respect to SiO_2. When the second ramp is imposed (Fig. 17) the absolute value of the current is higher with respect to the first one. At this stage, a de-charging effect is experienced, and the new charging process is evidenced at about 60 V, like in the first ramp, when it occurred at -60 V. It means that when a bi-polar scheme for the actuation is imposed, a fast de-charging is experienced, similarly to what occurs in the case of RF MEMS switches. In Fig. 18, the results for a TEOS based MIM in wafer #3 are shown.
Looking at Fig. 18, a peak similar to the application of a negative bias experienced by the Si_3N_4 in the previous measurements (but less pronounced) is recorded also for TEOS during

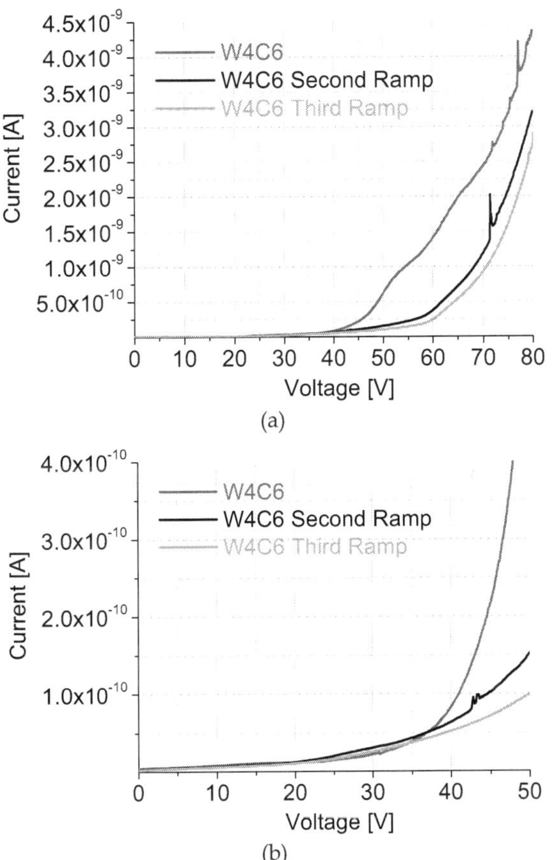

Fig. 19. The sample C6 of wafer #4 (W4C6, LTO) was measured by repeating the ramp three times (a). The charging process is enhanced, but the effect is less important the third time, thus suggesting the possibility for a saturation of the charge injected in the sample, which is not visible in (b), being the voltage below the threshold for the onset of the charging effect.

the first ramp (0 - 100 V), probably due to the same proposed effect of "charge freezing" for the previous material. A negative current is obtained by means of the second ramp (100 - 0 V), increasing the absolute value. When V_g is low and the capacitor is almost de-charged, current is injected in the opposite way, changing the slope with respect to the first ramp. During the fourth ramp (-100 - 0 V) the sample is de-charged again and the charge is newly injected at low voltage values. Ramps have been imposed again on an LTO based MIM from wafer #4, and plots in Fig. 19 and 20 give evidence for charging mechanisms when always the positive voltage is applied in successive ramps. As expected, the charging process is enhanced, but the effect is less important the third time, thus suggesting the possibility for a saturation of the charge injected in the sample.
For the wafer #5 and #7 the measured current increases with respect to the first ramp, as it is shown from Fig. 20 to Fig. 22.
In the case of PECVD Oxide LF Top in wafer #9 the response is the same recorded for wafers from #1 to #5. As a further characterization, one sample was subjected to DC cycling in a way analogous to that used for real RF MEMS switches. Specifically, the sample C2 belonging to Wafer #1, was measured after imposing a uni-polar train of 10^4 pulses with amplitude V_g=50 V, having a pulse-width τ=250 ms and a period T=500 ms. Since the data obtained on this wafer are superimposed for all the measured samples, we used one C3 device, exactly equal to C2, as a reference structure. The result was an almost ideal dielectric response for low voltage values, i.e. a constant value for the current as a function of the applied voltage. The C3 structure, which was not "stressed" in the same way, behaves exactly as it was in the previous measurements, with a linear response of the current as a function of the voltage, starting from the very beginning. We believe that the above treatment was useful for helping the re-combination of charges left free from the technological processes at the interface metal-dielectric, which were sensitive to the voltage gradient experienced during the application of the train of pulses, and especially to the sudden gradient imposed in correspondence of the trailing and leading edge of the pulses. The result is presented in Fig. 23.

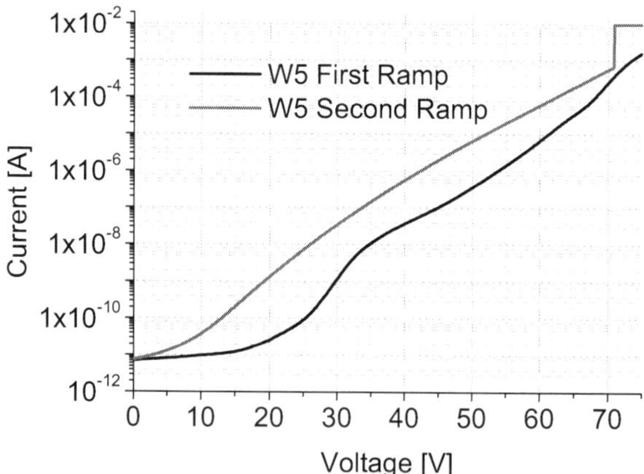

Fig. 20. Wafer #5 (PECVD Nitride HF). Shift of the current by using the same ramp and maximum value of the applied voltage, Vmax, in two successive measurements separated by one minute ca. After the first ramp the nature of the dielectric was dramatically changed, thus exhibiting an up-shift of the measured current. The second time we are almost at the breakdown voltage, around 70 V.

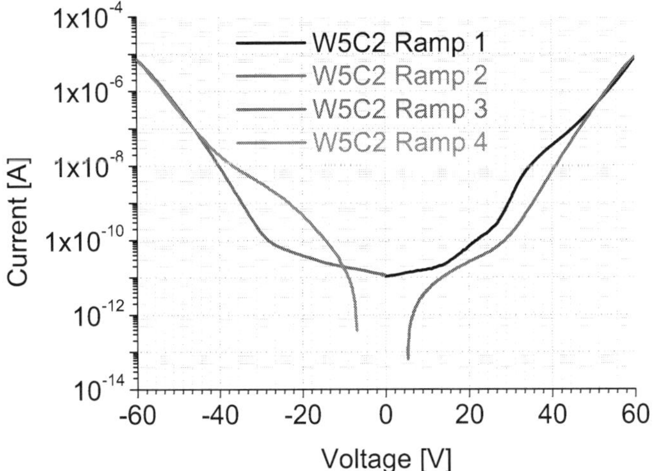

Fig. 21. I vs V for wafer #5, sample 2 (W5C2, PECVD Nitride HF). It is worth noting that there is not serious current reversal as it happened to TEOS. Actually, the difference with respect to the results for TEOS could be due to a higher densification temperature during the film preparation, reducing the contribution of free charges.

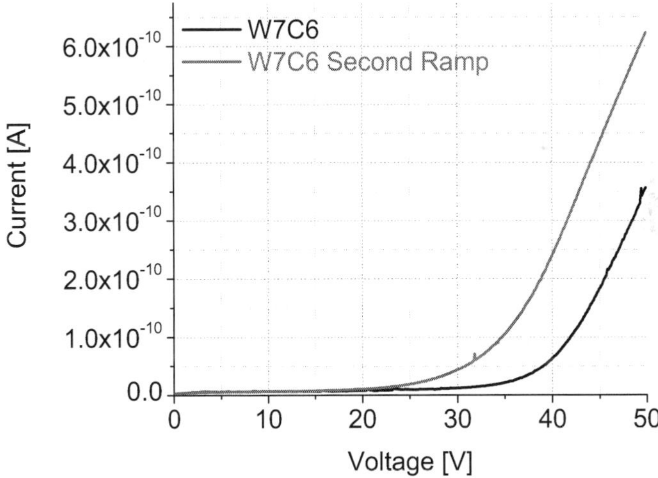

Fig. 22. I vs V for sample C6 in wafer #7 (W7C6, LTO). In this case the same structure present in the centre of the bridge is realized, with a multilayer as a bottom electrode and gold as the top one. Actually, an up-shift of the current is measured. As in the case of wafer #5 in Fig. 20, we were very close to the breakdown, around 62 volt, and this could change the general characteristics of the sample when the second ramp is used.

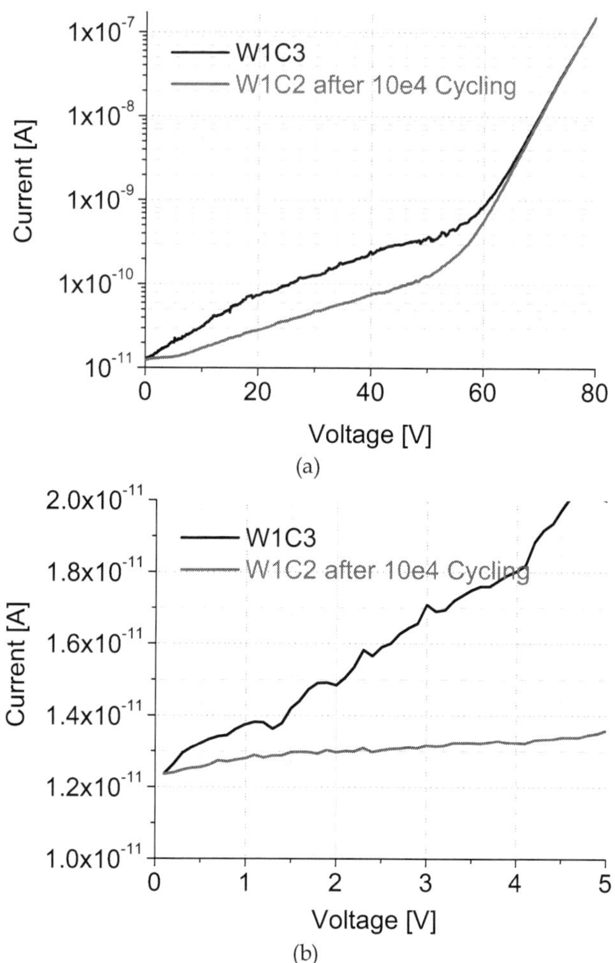

Fig. 23. Comparison between the I vs V curves of a sample in wafer #1 (Nitride) before (reference sample C3) and after (sample C2) imposing 10^4 cycles of a DC train at 50 V.

From the analysis of data plotted in Fig. 22 and in Fig. 23, it turns out that the response with poly-silicon is still affected by charge injection also after the described processing for Vg>5 V, thus giving evidence for a residual contribution coming directly from the interface between doped poly-silicon and dielectric. In fact, the recorded curves are different with respect to the behaviour of MIMs manufactured by using top and bottom metal electrodes, because the poly-silicon electrodes are always characterized by a ramp behaviour in the first region. It is also worth noting that by using a DC train with a voltage value less than that needed for the charge injection onset (60 V), no shift is recorded (see Fig. 22).
From the I vs V plots and from data recorded in Table 1, we can draw the following general conclusions:

The breakdown is not critical for structures with Poly-silicon electrodes. Usually VB ≥ 100 volt is measured. On the other hand the dielectric looks like not ideal, because a linear response of the current vs the applied voltage is recorded already at low voltage levels, thus demonstrating a not negligible resistive contribution of the bulk of the capacitor. Another possible mechanism for conduction could be due to the presence of Poly-silicon: the dielectric interface can probably be considered as a sort of MOS with a poly-silicon p-doped and a thin non-ideal dielectric layer.
Charging of the samples is obtained when successive ramps are applied, as evidenced from the shift of the I vs V characteristics when the measurement is repeated, in times shorter or in the order of one minute, in the same direction (positive or negative voltages). Moreover, the de-charging is very slow, and also after one day there is not a complete spontaneous restoring of the initial conditions.
Partial de-charging occurs when ramping the sample with positive and negative voltages, and re-combination of the charges is obtained, but the initial conditions are never re-obtained also by using this treatment.
The measured trend of the current is never ideal for the exploited samples, and a linear response is always obtained as a function of the applied voltage, while a constant value is expected for an almost ideal dielectric material. So far, the second term in Eq. (2) is always present. By cycling one sample with pulses as high as 50 V such a response is flattened, maybe due to the re-combination of residual charges belonging to defects of the material coming out from the technological process.
In the structures measured on wafer #7 to #11 some criticality in the measurements is evidenced, because of the small thickness of the metal contact, due to the pressure to be exerted by the probes.
In the case of the sample C3 belonging to wafer #1, with Si_3N_4, a linear fit has been superimposed to the I vs V curve to evaluate the resistance of the sample. The result is presented in Fig. 24, from which it turns out a slope of $0.4\times10^{-11}\ \Omega^{-1}$, i.e. $Ra=2.5\times10^{11}\ \Omega$.

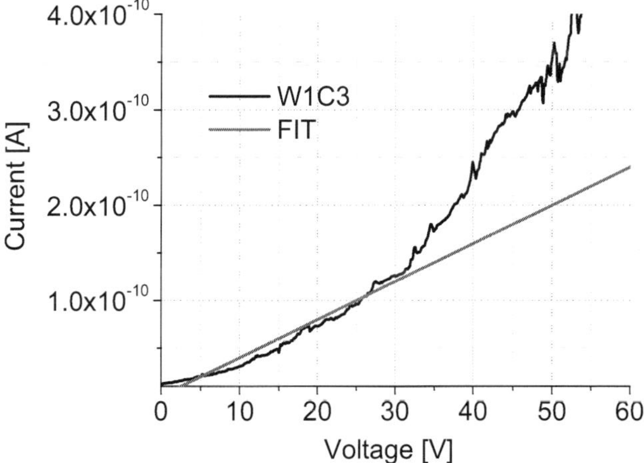

Fig. 24. Linear Fit to evaluate the resistance offered by the MIM material, namely Si_3N_4, before the onset of the Poole-Frenkel effect. A slope of $0.4\times10^{-11}\ \Omega^{-1}$ is obtained.

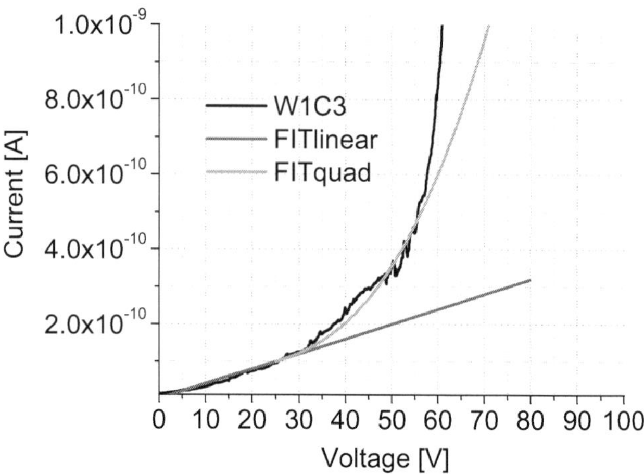

Fig. 25. Linear and quadratic fit for the measured current vs the applied voltage when the sample C3 belonging to wafer #1 (W1C3, Si_3N_4) is biased. Up to 25-30 V a linear dependence is obtained, while an almost quadratic law is found for V_g>25 V.

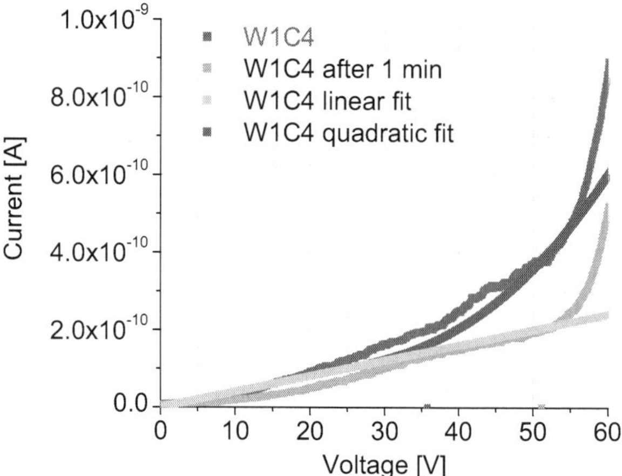

Fig. 26. Measurement of the C4 sample belonging to wafer #1. The red curve is for the first ramp imposed to the sample, and the green one is the response after the second ramp, one minute after the first one. The azure and blue curves refer to the linear and quadratic response respectively (low and high voltage values). The green curve is shifted due to the onset of the Poole-Frenkel effect, which lowers the current response at the same voltage.

Considering the area of the MIM $A=(440\times10^{-6})^2$ m^2 and the thickness $d=0.1\times10^{-6}$ m, the resistivity of the material will be $\rho=A\times Ra/d=4.84\times10^{11}$ Ωm, or $\sigma=2.07\times10^{-12}$ Ω^{-1}m^{-1}, thus confirming the high resistivity of the material, but a non-ideal response in terms of dielectric behavior [67].

Actually, in Figure 25, the linear fit is compared with the quadratic one, obtained by means of the formula f(Vg) = 0.4×10^{-11}Vg+0.5×10^{-13}(Vg-25)$^{2.5}$. This result is a correction with respect to the simple quadratic law in [61].
As a final comparison, the sample C4 in wafer #1 has been subjected to a measurement I vs V and fitted following the law $(V_g\text{-}25)^{2.5}$. The result is shown in Fig. 26, where the displacement due to charging is evidenced, and it has to be attributed to the Poole-Frenkel Effect.

5. Dielectric polarization and Poole-Frenkel effect in RF MEMS and MIM

On the time scale of interest to the RF-MEMS capacitive switches response (i.e. greater than 1 μsec) an electric field can interact with the dielectric film in two primary ways. These are: (i) the re-orientation of defects having an electric dipole moment, such as complex defects, and (ii) the translational motion of charge carriers, which usually involve simple defects such as vacancies, ionic interstitials and defect electronic species. These processes give rise to the dipolar (P_D) and the intrinsic space charge ($P_{SC\text{-}i}$) polarization mechanisms, respectively. Moreover, when the dielectric is in contact with conducting electrodes charges are injected through the trap assisted tunneling and/or the Poole-Frenkel effect [69] giving rise to extrinsic space charge polarization ($P_{SC\text{-}e}$) whose polarity is opposite with respect to the other two cases. In RF-MEMS capacitive switches during the actuation all the polarization mechanisms occur simultaneously and the macroscopic polarization is given by

$$P_{tot} = P_D + P_{SC-i} - P_{SC-e} \tag{4}$$

Now, from elementary physics it is known that the electric displacement *D*, defined as the total charge density on the electrodes, will be given by $D = \varepsilon_0 E + P$, where *E* is the applied field and *P* the dielectric material polarization. The resulting polarization *P* may be further divided into two parts according to the time constant response [70]:
a) An almost instantaneous polarization due to the displacement of the electrons with respect to the nuclei. The time constant of the process is about 10^{-16} sec and defines the high frequency dielectric constant that is related to the refractive index.
b) A delayed time dependent polarization *P*(*t*), which determines the dielectric charging in MEMS, starting from zero at *t*=0, due to the orientation of dipoles and the distribution of free charges in the dielectric, the dipolar and space charge polarization respectively. Moreover the growth of these polarization components may be described in the form of $P_j(t) = P_{j0}\cdot\left[1 - f_j(t)\right]$. The index *j* refers to each polarization mechanisms, and $f_j(t)$ are exponential decay functions of the form $\exp\left[-\left(\frac{t}{\tau}\right)^{\beta}\right]$. Here τ is the process time and β the stretch factor. If β=1 the charging/discharging process is governed by the Debye law. In disordered systems like the amorphous oxides, which possess a degree of disorder, β<1 and the charging/discharging process is described by the stretched exponential law.
In the case of a MEMS switch that operates under the waveforms in Fig. 4, the dielectric is subjected to charging when the bridge is in the DOWN position and discharging in the UP position, independently of the ON or OFF functionality of the device. More specifically, when a uni-polar pulse train is applied (Fig. 4 (a)) then the device is subjected to contact-less

charging below pull-in and pull-out. Above pull-in and pull-out the device is subjected to contact charging.

If we assume that at room temperature the density of free charges in the LTO, i.e. SiO_2 deposited at low temperature, is very low we can re-write Eq. (4) as:

$$P = P_D - P_{SC} \tag{5}$$

where P_{SC} is the space charge polarization of extrinsic origin. When we apply a pulse train the following will occur:

- during the contact-less charging the electric field increases the dipolar polarization and assists to re-distribution and dissipation of injected charges
- during the contact charging the high electric field causes a further increase of the dipolar polarization, and through the charge injection contributes to the build-up of space charge polarization

Due to the dielectric film polarization the pull-in and pull-out voltages will be determined by:

$$V_{pi} = \sqrt{\frac{8kz^3}{27\varepsilon_0 A}} - \frac{z_1 P_{pi}}{\varepsilon_0} \ ; \ V_{po} = \sqrt{\frac{2kz_1^2(z - z_1)}{\varepsilon_0 A}} - \frac{z_1 P_{po}}{\varepsilon_0} \tag{6}$$

In the Si_3N_4 dielectric it has been shown that, at room temperature, the space charge polarization induced by the charge injection is the dominant mechanism [71][72]. If we assume that the same effect holds for SiO_2 we are led to the conclusion that the pull-out voltage will increase with time when a uni-polar pulse train is applied.

The dependence of the actuation and de-actuation voltages on the number of cycles was fitted for the exploited RF MEMS devices S1 and CL studied in the previous sections, by assuming that the charging process follows the stretched exponential law.

The fitting of data has been performed as a function of the number of cycles (N), since each cycle maintains a constant shape and represents a certain effective ON and OFF time. This is particularly useful in actual devices, when the reliability can be determined by the number of total actuations as well as the total time during which the RF MEMS switch remains actuated. The differences in the effective ON and OFF times will reflect in the number of cycles (N^*) that corresponds to the process time τ. According to Eq. (6), and in agreement with the above discussed growth for the polarization, we can apply the following equation to describe the evolution of the pull-in and pull-out voltages as a function of time/number of cycles.

$$V_j(N) = V_{0,j} - \frac{z_1 P_j}{\varepsilon_0} \cdot \left\{ 1 - \exp\left[-\left(\frac{N}{N_j^*} \right)^{\beta} \right] \right\} \tag{7}$$

where z_1 is the dielectric thickness, j an index that stands for actuation (pull-in) and de-actuation (pull-out) while $V_{0,j}$ represents the pull-in and pull-out voltages that are determined by the electro-mechanical model.

The fitting results show excellent agreement with the experimental data, and the fitting parameters are listed in Table 2, with reference to Fig. 5 and Fig 6 of the current contribution.

Here it must be pointed out that:

$$\Delta V = -\frac{z_1 P_j}{\varepsilon_0} = -\frac{z_1 \cdot \left(P_{D,j} - P_{SC,j}\right)}{\varepsilon_0} \tag{8}$$

		V_0	ΔV	β	N*
Fig. 5	Act	13.5	54.4	0.69	1.67
	Deact	29.5	33.2	0.96	1.96
Fig. 6	Act	27.0	54.5	1	1.67
	Deact	36.8	22.4	0.83	2.5

Table 2. Fitted values for the exponential trend of the actuation (Act) and de-actuation (Deact) of both S1 and CL devices.

The fitting results reveal that the dominant mechanism is the space charge polarization (P_j<0). Moreover, it is worth noting that the actuation voltage increases faster than the de-actuation one. Such a behavior can result from a faster increase of space charge polarization or decrease of dipolar polarization when the bridge is non-actuated. Taking into account that the dipolar polarization in SiO_2 is characterized by long time constants (fig. 4 of [72]), we are led to the conclusion that the differences in the increase of actuation and de-actuation voltages arise from the mechanisms involved in charge injection and collection, respectively. This can be easily understood if we bear in mind that the charge injection occurs under high electric field and that the trap assisted tunneling charging process gives rise to a spatial distribution of charges in the vicinity of the injecting contacts. During the OFF state, where no bias is applied, the trapped charges (located at the dielectric free surface) are emitted and finally collected by the bottom electrode. The charge collection, which takes place through diffusion and drift in the presence of local electric fields, is complex owing to multi trapping processes. Regarding the charges that are injected from the bottom electrode, they are collected much sooner than the top ones. Moreover, these charges have a small influence on the dielectric polarization.

Another possible situation could be the charging between the actuation pads and the ground plane of CPW across the substrate dielectric [73]. This charging process gives rise to a longitudinal polarization across the substrate oxide that behaves like the dipolar polarization. The values of *N** for actuation and de-actuation agree with the presence of both mechanisms, which is a slower build-up of space charge polarization and competition from a longitudinal polarization across substrate.

Applying a bipolar bias scheme we observe that both actuation and de-actuation voltages do not vary significantly with time, as it is also the case for other recently considered RF MEMS switch configurations, like the miniature one [74]. This can be easily attributed to the field induced charging/discharging processes. A significant difference that arises from the bipolar actuation is the reversal of magnitude of actuation and de-actuation voltages. This behavior has been occasionally observed but not investigated in depth. If we take into account that in MEMS switches the charging is asymmetric, a reason that leads to stiction even under bipolar actuation, we may assume that this effect is probably responsible for the observed reversal. In such a case the value of P_{po} will change polarity and magnitude after each change of the actuation voltage polarity. In any case the observed behavior is under investigation.

Concerning the charging effects in a MIM capacitor, let' analyse as an example experimental results based on silicon dioxide deposited by means of a low temperature process (LTO=low temperature oxide), as it is shown in Fig. 19 (a).

The MIM was made by a poly-silicon layer as the bottom electrode, with metal on the top side (top electrode) and LTO as a dielectric layer. The structure emulates the situation of a fully collapsed bridge by means of a lateral actuation, where poly-silicon is used as the material for the feeding lines and for the pad under bridge, while LTO is deposited on the top of it to provide an electrical isolation; the metal on the top is equivalent to the bridge touching the actuation electrode when the voltage is applied. Such an arrangement, i.e. a multilayer polysilicon/dielectric/metal, is also a source of further injection of charges, because polysilicon is not just a bad conductor and it can also contribute at the interface polysilicon/dielectric. The sample was measured by repeating a slow voltage ramp three times and measuring the corresponding current. In particular, a ramp rate dV/dt=0.05 V/sec and a maximum voltage of 80 volt were imposed. As expected, the charging process is enhanced, and this is evidenced by the current drop after each ramp, but the effect is less important the third time, thus demonstrating the saturation of the charge injected in the sample, as also experienced in the real MEMS switches already discussed in this paper.
As already outlined, the measured trend of the current is not ideal for the exploited sample and for similar ones based on silicon nitride, and a linear response is always obtained as a function of the applied voltage, while a constant value is expected for an almost ideal dielectric material at low voltage values, i.e. in a range up to, at least, 20-25 volt for typical dielectric materials used in microelectronics. The same data of Fig. 19 (a) have been plotted in Fig. 27, by using I/V vs $V^{1/2}$, to check the Poole-Frenkel effect. Actually, the current dependence on bias seems to be determined by the Poole-Frenkel effect when the applied bias exceeds the value of 50 volt:

$$I(V) = I_0 E \exp\left(b_{PF}^{*} \sqrt{E} - \frac{\Phi_n}{kT} \right) \text{ where } b_{PF}^{*} = \frac{1}{kT}\sqrt{\frac{q^3}{\pi \varepsilon \varepsilon_0}} \tag{9}$$

The change of b_{PF} in SiN has been investigated by S.P. Lau et al. and attributed to large concentration of defects in SiN and the formation of defect band. Taking into account the increase of b_{PF} with the applied electrical stress we are led to the conclusion that the latter decreases the density of traps in the SiN film [75].

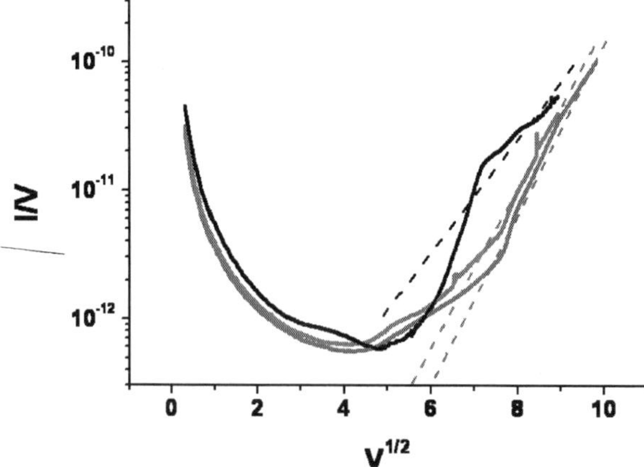

Fig. 27. I/V curve as a function of the $V^{1/2}$ by using data from Fig. 19 (a).

To better investigate this aspect, an additional characterization was performed on a sample with the same structure for the bottom and for the top electrodes and with Si_3N_4 as dielectric. Two samples have been measured: (i) the first one in the usual way, by means of a slow voltage ramp, and (ii) another one by imposing a typical stress used for the switches, subjecting it to a high number of DC pulses and measuring the characteristic current vs voltage after that. Actually, 10^4 pulses with a pulse-width τ = 250 ms and with a period T = 500 ms (duty cycle = τ/T = 50%), with a voltage V=50 volt, have been used. As a result, the low voltage response has been "rectified" as it is shown in Fig. 28, where the initial behavior is almost constant, as expected by a dielectric material without free charges incorporated. We believe that such a trend can be justified by the neutralization of surface free charges at the interface between the dielectric layer and the top metal, where, due to the roughness, charges are trapped but free to contribute when a DC field is imposed. The energy released by the DC input pulses, provided quickly with respect to the time constants for the material de-charging, was high enough to favor the re-combination of the charges, thus locally improving the quality of the dielectric material.

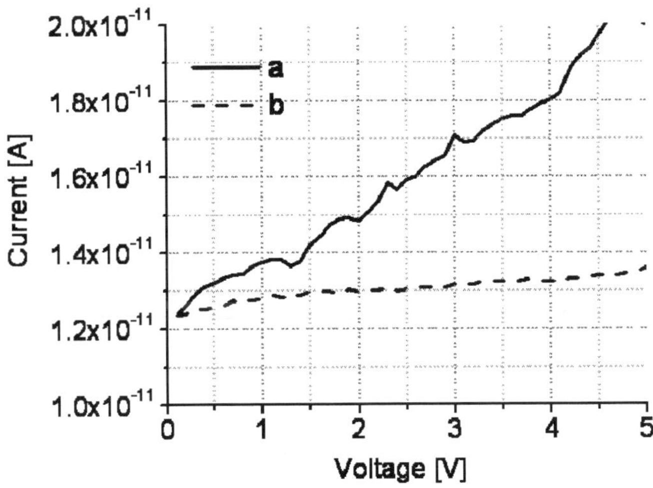

Fig. 28. Measured trend of the current as a function of the applied voltage for a MIM made by Si_3N_4 before (curve a) and after (curve b) cycling the sample with pulses as high as 50 volt. Generally, a linear response is always obtained as a function of the applied voltage, while a constant value is expected for an almost ideal dielectric material. By cycling the sample such a response is flattened, maybe due to the re-combination of residual charges belonging to defects of the material surface coming out from the technological process. Actually, a comparison has been done between the charging response of MIM capacitors and RF MEMS switches, and the differences coming from such an analysis have been discussed with emphasis on the different times needed for re-storing the initial conditions or for preventing the charging itself.

6. Conclusion

In conclusion, this chapter has been organized describing the technological aspects for manufacturing both MIMs and RF MEMS switches, and discussing, on the base of several experimental findings, the theoretical framework for the interpretation of the measured charging effects. In particular, the theoretical approach for charging occuring in the exploited devices has been based on the Poole-Frenkel effect and it has been related to the involved polarization mechanisms. Many structures have been studied, looking for the most promising ones to be used for the actuation of RF MEMS switches, minimizing the charging effects.

Two configurations of RF MEMS switches using electrostatic actuation, and several MIMs devices simulating the RF MEMS actuation pads, with various dielectric materials and electrodes, have been measured.

As experienced in the RF MEMS measurements, and well established in literature, the charge stored in the dielectric material used for the actuation pads creates an electric field that is always opposite with respect to the electric field generated by the actuation voltage. This is evident in the case of an uni-polar actuation signal, with an increase in the actuation voltage for the switch, and it was confirmed in our measurements.

Mainly, charging is responsible for sticking, and it is also related to the increase of the actuation voltage, especially under uni-polar DC biasing. By using lower actuation voltages or a bi-polar scheme this effect is more under control and compliant with ground and space applications, which should not overcome 50 volt of bias to be really appealing in several sub-systems.

The process necessary to trap and de-trap the carriers in the uni-polar scheme can be described mainly by the Poole-Frenkel effect; it is very slow, and the initial conditions for the device should need long times to re-obtain the same actuation voltage. To accelerate the restoring mechanism, a bi-polar actuation scheme was applied to the same devices, and from the experiment it turns out that the gradient experienced by the switch under test helps a faster de-trapping mechanism, giving back the initial value of the actuation conditions. Actually, the voltage difference necessary for the successive actuations in the bi-polar scheme is always constant and the absolute value of the actuation and de-actuation voltages too, at least for a limited number of actuations.

For the MIM structures, a comparison has been performed between different materials and electrodes to simulate the RF MEMS actuation pads. From the measurements, it turns out that the change of interface and of the dielectric material, as well as the deposition technique used for obtaining the dielectric layer, are critical choices to activate charging mechanisms.

The breakdown is not critical for structures with Poly-silicon electrodes. Usually $V_B \geq 100$ V is measured, while metal bottom electrodes have $V_B \leq$ 50-60 V. On the other hand all the exploited dielectric materials look like not ideal, as a linear response of the current Vs the applied voltage is recorded already at low voltage levels, thus demonstrating a not negligible resistive contribution of the bulk of the capacitor. Another possible mechanism for conduction could be the presence of Poly-silicon: the dielectric interface can probably be considered as a sort of MOS with a poly-silicon p-doped and a thin non-ideal dielectric layer.

Charging of the samples is obtained when successive ramps are applied, as evidenced from the shift of the I Vs V characteristics by means of the application of positive and negative voltages. Moreover, the de-charging of the MIM is very slow, and also after one day there is

not a complete spontaneous restoring of the initial conditions, against the previous finding for RF MEMS switches. *This could be an evidence that the charging effects occurring in the actual MEMS device cannot be completely emulated by a MIM structure, as the times for restoring the initial conditions are quite different between them. Anyway, in spite of a possible indication for different processes, due to the actuation itself, the charging properties of the material used for the actuation pads will be always present.* In the case of the measured switches, TEOS was used for the actuation pads, which exhibits quite pronounced charging effects as evidenced also in MIM structures (see Fig. 16).
Moreover, better performances in the I Vs V response can be obtained when the MIM is subjected to several pulses, analogously to those used in operating conditions for RF MEMS, maybe due to recombination of charges (left free from the technological process) when subjected to such an electrical stress.
Concerning the materials and the deposition techniques, from the results shown in Table 1 and from the plots is difficult to draw a final conclusion, but one can see that generally Si_3N_4 exhibits an almost linear response for the current as a function of the applied voltage in a voltage range wider with respect to SiO_2 (LTO, TEOS). Moreover, the PECVD HF Nitride deposited at 300 °C looks like better also in terms of current reversal with respect to TEOS, and it is attributed to a higher densification temperature (Fig. 21). Actually, charge injection is present in both materials owing to the non-ideal response of the I Vs V curve, which should be flat at low voltages, but a strong non-linear behaviour due to the Poole-Frenkel effect is obtained only for $V > 50$-60 V for Si_3N_4 and for $V > 20$-30 V for SiO_2.

7. Acknowledgment

Work partially funded by the European Space Agency (ESA) Contract 20847/07/NL/GLC "High Reliability MEMS Redundancy Switch".
Adriano Cola from CNR-IMM Lecce and Luigi Mariucci from CNR-IMM Roma are kindly acknowledged for helpful discussions on charge effects in MIM structures.

8. References

[1] Hopkinson, J.; Wilson, E. On the capacity and residual charge of dielectrics as affected by temperature and time. *Phil. Trans. Roy. Soc. London. A* 1897, 189, 109-135.
[2] Binet, G.; Freire, M.; Van Eesbeek, M.; Daly, E.; Drolshagen, G.; Henriksen, T.; Thirkettle, A.; Poinas, P.; Eiden, M.; Guglielmi, M. *Space specifications check list*; ESA-ESTEC: Noordwijk, Netherlands, 2006, https://iti.esa.int/iti/resource/Space_Specifications_Checklist.doc.
[3] Asokan, T. Ceramic dielectrics for space applications. *Curr. Sci.* 2000, 79, 348-351.
[4] Nguyen, C.T.-C.; Katehi, L.P.B.; Rebeiz, G.M. Micromachined devices for wireless communications. *Proc. IEEE* 1998, 86, 1756-1768.
[5] De Los Santos, H.J. *Introduction to Microelectromechanical (MEM) Microwave Systems*, Artech House, Boston, 1999.
[6] Senturia, S. *Microsystem Design*, Springer, New York, 2001.
[7] De Los Santos, H.J. *RF MEMS Circuit Design for Wireless Communications*, Artech House, Boston, 2002.
[8] Rebeiz, G. M. *RF MEMS Theory, Design, and Technology*, 1st Ed.; John Wiley & Sons: Hoboken, New Jersey, USA, 2003.

[9] Maluf, N.; Williams, K. *An Introduction to Microelectromechanical Systems Engineering*, 2nd Ed.; Artech House, Boston, 2004.

[10] Joung, J.; Shen, J.; Grodzinski, P. Micropumps based on alternating high-gradient magnetic fields. *IEEE Trans. Magn.* 2000, 36, 2012–2014.

[11] Yan D.; *Mechanical Design and Modeling of MEMS Thermal Actuators for RF Applications*, thesis on Master of Applied Science in Mechanical Engineering, http://resonance.uwaterloo.ca/students/dyan/thesis_winter_master.pdf, Waterloo, Ontario, 2002

[12] Lee, H.-C.; Parkand, J.-Y.; Bu, J.-Uk. Piezoelectrically Actuated RF MEMS DC Contact Switches With Low Voltage Operation *IEEE Microwave and Wireless Components Lett*, 2005, 15, 202-204.

[13] De Los Santos, H.; Fischer, G.; Tilmans, H.A.C.; van Beek, J.T.M. RF MEMS for Ubiquitous Wireless Connectivity Part 1-Fabrication and Part 2-Application. *IEEE Microwave Magazine*, 2004, 5, 36-65

[14] ESA/ESTEC Project No. 14628/NL/CK-MEM Switch on: MICROWAVE ELECTROSTATIC MICRO-MACHINED DEVICES FOR ON-BOARD APPLICATIONS

[15] ESA-ESTEC Project MEDINA No. 14627/00/NL/WK

[16] Fernández-Bolaños, M.; Lisec, T.; Dainesi, P.; Ionescu, A. M. Thermally Stable Distributed MEMS Phase Shifter for Airborne and Space Applications. *Proceedings of the 38th European Microwave Conference*, 2008, October 2008, Amsterdam, The Netherlands, 100-103.

[17] Dussopt, L.; Rebeiz, G. M. Intermodulation distortion and power handling in RF MEMS switches, varactors, and tunable filters. *IEEE Trans. Microw Theory Tech.*, 2003, 51, 1247–1256.

[18] Girbau, D.; Otegi, N.; Pradell, L. Study of Intermodulation in RF MEMS Variable Capacitors. *IEEE Trans. Microw Theory Tech.*, 2006, 54, 3, 1120-1130.

[19] Mercier, D.; Blondy, P.; Barataud, D.; Cros, D.; Guillon, P.; Champeaux, C.; Tristant, P.; Catherinot, A. Model for MEMS Switches Power Handling and Phase Noise. *Proc. of the European Microwave Week* 2002, Milano, Italy, 1-4.

[20] Peroulis, D.; Pacheco, S. P.; Katehi, L. P. B. RF MEMS Switches With Enhanced Power-Handling Capabilities. *IEEE Trans. Microw Theory Tech.*, 2004, 52, 50-68.

[21] Choi, Joo-Young; Ruan, Jinyu; Coccetti, Fabio; Lucyszyn, Stepan, Three-Dimensional RF MEMS Switch for Power Applications, *IEEE Trans. on Ind. Electronics*, Vol. 56, No. 4, April 2009, 1031-1039.

[22] Mardivirin, D.; Pothier, A.; Orlianges, J.C.; Crunteanu, A.; Blondy, P. Charging Acceleration in Dielectric Less RF MEMS Switched Varactors under CW Microwave Power, *Proc. of Int. Microwave Symposium*, 2009.

[23] TERAVICTA DATA Sheet on *"SP4T 7GHz RF MEMS Switch"*, http://www.teravicta.com/site/images/pdf/TT1414/DS-TT1414_1.3.pdf (2007).

[24] Di Nardo, S.; Farinelli, P.; Giacomozzi, F.; Mannocchi, G.; Marcelli, R.; Margesin, B.; Mezzanotte, P.; Mulloni, V.; Russer, P.; Sorrentino, R.; Vitulli, F.; Vietzorreck, L. Broadband RF-MEMS Based SPDT; In *Proceedings of the 36th Microwave Conference*, Manchester, UK, 10-15, September 2006; pp. 1727 – 1730.

[25] McErlean, E.P.; Hong, J.-S.; Tan, S. G.; Wang, L.; Cui, Z.; Greed, R. B.; Voyce, D.C. 2x2 RF MEMS switch matrix. *Microwaves, Antennas and Propagation, IEE Proceedings* 2005, 449 - 454.

[26] Catoni, S.; Di Nardo, S.; Farinelli, P.; Giacomozzi, F.; Mannocchi, G.; Marcelli, R.; Margesin, B.; Mezzanotte, P.; Mulloni, V.; Pochesci, D.; Sorrentino, R.; Vitulli, F.; Vietzorreck, L.: RF MEMS Matrices for Space Applications, In Proceedings of the 2007 MEMSWAVE Workshop, 8th International Symposium on RF MEMS and RF Microsystems, Barcelona, Spain, 26-29 June 2007.

[27] Barker, S.; Rebeiz, G. M. Distributed MEMS true-time delay phase shifters and wide-band switches, *IEEE Trans. Microw Theory Tech.*, 1998, 46, 1881-1890

[28] Buttiglione, R.; Dispenza, M.; Fiorello, A. M.; Tuominen, J.; Kautio, K.; Ollila, J.; Jaakola, T.; Rönkä, K.; Catoni, S.; Pochesci, D.; Marcelli, R. Fabrication of high performance RF-MEMS structures on surface planarised LTCC substrates. In Proceedings of EMPC2007, European Microelectronics and Packaging Conference and Exhibition , Oulu, Finland, 17-20 June, 2007.

[29] Rebeiz, G. M.; Tan, G.-L.; Hayden, J. S. RF MEMS Phase Shifters: Design and Applications. *IEEE Microwave Magazine*, 2002, 3,72-81.

[30] Bartolucci, G.; Catoni, S.; Giacomozzi, F.; Marcelli, R.; Margesin, B.; Pochesci, D. Realization of a distributed RF MEMS Phase Shifter with a very low number of switches. *Electron. Lett.*, 2007, 43, 1290 - 1291.

[31] Zhou, L.; RF MEMS DC Contact Switches for Reconfigurable Antennas. Thesis on Master of Science in Electrical Eng., San Diego State University, http://digitaladdis.com/sk/Lei_Zhou_Thesis_RF_MEMS.pdf (2006)

[32] Kornrumpf, W. P.; Karabudak, N. N.; Taft, W. J. RF MEMS PACKAGING FOR SPACE APPLICATIONS. In Proc. of 22nd AIAA International Communications Satellite Systems Conference & Exhibit, Monterey, California, 9 - 12 May 2004.

[33] Goldsmith, C.; Ehmke, J.; Malczewski, A.; Pillans, B.; Eshelman, S.; Yao, Z.; Brank, J.; Eberly, M.; Lifetime characterization of capacitive RF MEMS switches. *Proc. of IEEE MTTS Int Microw Symp*, 2001, 227-230.

[34] Yuan, X.; S. Cherepko, V. J.; Hwang, C. M.; Goldsmith, C. L.; Nordquist, C.; Dyck C. Initial observation and analysis of dielectric-charging effects on RF MEMS capacitive switches, *Proc. of IEEE MTTS Int Microw Symp*, 2004, 1943-1946.

[35] Van Spengen, W.M.; Puers, R.; Mertens, R.; De Wolf, I. A comprehensive model to predict the charging and reliability of capacitive RF MEMS switches *J. Micromech. Microeng.* 2004, 14, 514–521.

[36] Patton, S. T.; Zabinski, Jeffrey, S. Effects of dielectric charging on fundamental forces and reliability in capacitive microelectromechanical systems radio frequency switch contacts. *J. Appl. Phys.*, 2006, 99, 94910-94910-11

[37] Peng, Z.; Yuan,X.; Hwang, J. C. M.; Forehand, D. I.; Goldsmith, C L. Superposition Model for Dielectric Charging of RF MEMS Capacitive Switches Under Bipolar Control-Voltage Waveforms *IEEE Trans. Microw Theory Tech.*, 2007, 55, 2911-2918.

[38] Peng, Z.; Palego, C.; Hwang, J. C. M.; Moody, C.; Malczewski, A.; Pillans, B. W.; Forehand, D. I.; Goldsmith, C.L. Effect of Packaging on Dielectric Charging in RF MEMS Capacitive Switches *Proc. of IEEE MTTS Int Microw Sym.*, 2009, 1637-1640.

[39] Marcelli, R.; Papaioannu, G.; Catoni, S; De Angelis, G.; Lucibello, A.; Proietti, E.; Margesin, B.; Giacomozzi, F.; Deborgies, F.; Dielectric Charging in Microwave

Micro-electro-mechanical Ohmic Series and Capacitive Shunt Switches. *J Appl Phys* 2009, 105, 114514-1 - 114514-10.

[40] Wang, G.; *RF MEMS Switches with Novel Materials and Micromachining Techniques for SOC/SOP RF Front Ends* thesis on School of Electrical and Computer Engineering of the Georgia Institute of Technology, http://smartech.gatech.edu/handle/1853/14112 2006.

[41] Theonas, V. G.; Exarchos, M.; Konstantinidis, G.; Papaioannou, G.J. RF MEMS sensitivity to electromagnetic radiation. *J Phys* 2005, *Conference Series* 10, 313–316

[42] Tazzoli, A.; Peretti, V.; Autizi, E.; Meneghesso, G. EOS/ESD Sensitivity of Functional RF-MEMS Switches, *Proc. of EOS/ESD Symposium* 2008, 272-280

[43] Ruan, J.; Papaioannou, G.J.; Nolhier, N.; Bafleur, M.; Coccetti, F.; Plana, R. ESD Stress in RF-MEMS Capacitive Switches: The Influence of Dielectric Material Deposition Method *IEEE CFP09RPS-CDR 47th Annual International Reliability Physics Symposium*, Montreal, 2009, 568-572

[44] Papandreou, E.; Lamhamdi, M.; Skoulikidou, C.M.; Pons, P.; Papaioannou, G.; Plana, R.; Structure dependent charging process in RF MEMS capacitive switches. *Microelectron Reliab* 2007, 47, 1812–1817.

[45] Mardivirin, D.; Pothier, A.; Crunteanu, A.; Vialle, B.; Blondy, P. Charging in Dielectricless Capacitive RF-MEMS Switches, *IEEE Trans. on Microwave Theory and Tech.*, Vol. 57, No. 1, January 2009, 231-236.

[46] Peng, Z.; Palego, C.; Halder, S.; Hwang, J. C. M.; Jahnes, C. V.; Etzold, K. F.; Cotte, J. M.; Magerlein, J. H. Dielectric Charging in Electrostatically Actuated MEMS Ohmic Switches, *IEEE Trans. on Device and Materials Reliability*, Vol. 8, No. 4, December 2008, 642-646.

[47] Yuan, X.; Peng, Z.; Hwang, J. C. M.; Forehand, D.; Goldsmith, C. L. Acceleration of Dielectric Charging in RF MEMS Capacitive Switches, IEEE Trans. on Device and Materials Reliability, 2006, 6, 556-563.

[48] Zaghloul, U. ; Belarni, A. ; Coccetti, F.; Papaioannou, G.J.; Bouscayrol, L.; Pons, P.; Plana, R. A Comprehensive Study for Dielectric Charging Process in Silicon Nitride Films for RF MEMS Switches using Kelvin Probe Microscopy, *Proc. of Transducers 2009*, Denver, CO, USA, June 21-25, 2009, 789-793.

[49] Broué, A.; Dhennin, J.; Seguineau, C.; Lafontan, X.; Dieppedale, C.; Desmarres, J.-M.; Pons, P.; Plana, R. Methodology to Analyze Failure Mechanisms of Ohmic Contacts on MEMS Switches *Proc. of IEEE CFP09RPS-CDR 47th Annual International Reliability Physics Symposium*, Montreal, 2009, 869-873.

[50] Czarnecki, P.; Rottenberg, X.; Soussan, P.; Nolmans, P.; Ekkels, P.; Muller, P.; Tilmans, H.A.C.; De Raedt, W.; Puers, R.; Marchand, L.; De Wolf, I. New Insights into Charging in Capacitive RF MEMS Switches *Proc. of IEEE CFP08RPS-CDR 46th Annual International Reliability Physics Symposium*, Phoenix, 2008, 496-505.

[51] Richard Daigler, Eleni Papandreou, Matroni Koutsoureli, George Papaioannou , John Papapolymerou, Effect of deposition conditions on charging processes in SiNx: Application to RF-MEMS capacitive switches, *Microelectronic Engineering* 86 (2009) 404–407.

[52] Romolo Marcelli, Giancarlo Bartolucci, George Papaioannu, Giorgio De Angelis, Andrea Lucibello, Emanuela Proietti, Benno Margesin, Flavio Giacomozzi, François

Deborgies, Reliability of RF MEMS Switches due to Charging Effects and their Circuital Modelling, *Microsystem Technologies*, Vol. 16, pp. 1111-1118 (2010).

[53] Catoni, S.; Di Nardo, S.; Farinelli, P.; Giacomozzi, F.; Mannocchi, G.; Marcelli, R.; Margesin, B.; Mezzanotte, P.; Mulloni, V.; Sorrentino, R.; Vitulli, F.; Vietzorreck, L. Reliability and Power Handling Issues in Ohmic Series and Shunt Capacitive RF MEMS Switches *Proceedings of the 2006 MEMSWAVE Workshop*, 7th International Symposium on RF MEMS and RF Microsystems, Orvieto, Italy, 26-29 June, 2006.

[54] Melle, S.; De Conto, D.; Mazenq, L.; Dubuc, D.; Poussard, B.; Bordas, C.; Grenier, K.; Bary, L.; Vendier, O.; Muraro, J.L.; Cazaux, J.L.; Plana, R. Failure Predictive Model of Capacitive RF-MEMS. *Microelectron Reliab* 2005, 45, 1770–1775.

[55] Vandershueren, J. and J. Casiot in *Thermally stimulated relaxation in solids;* Braunlich, P. (Ed.); Springer-Verlag, Berlin, Germany, 1979, volume 37

[56] Papaioannou, G.; Papapolymerou, J.; Pons, P.; Plana, R.; *Appl Phys Lett* 2007, 90, 233507

[57] Papaioannou, G.; Giacomozzi, F.; Papandreou, E.; Margesin, B.; Charging Processes in RF-MEMS Capacitive Switches with SiO_2 Dielectric *Proceedings of the 2007 MEMSWAVE Workshop,* 8th International Symposium on RF MEMS and RF Microsystems, Barcelona, Spain, 26-29 June 2007.

[58] Czarnecki, P.; Rottenberg, X.; Soussan, P.; Ekkels, P.; Muller, P.; Nolmans, P.; De Raedt, W.; Tilmans, H.A.C.; Puers, R.; Marchand, L.; De Wolf, I.; Influence of the substrate on the lifetime of capacitive RF MEMS switches. *Proc. MEMS-2008.*

[59] Xiaobin,Y.; Zhen, P.; Hwang, J.C.M.; Forehand, D.; Goldsmith, C.L.; A transient SPICE model for dielectric-charging effects in RF MEMS capacitive switches *IEEE Transactions on Electron Devices*, 2006, 53, 2640 - 2648.

[60] Melle, S.; De Conto, D.; Mazenq, L.; Dubuc, D.; Poussard, B.; Bordas, C.; Grenier, K.; Bary, L.; Vendier, O.; Muraro, J.L.; Cazaux, J.L.; Plana, R. Failure Predictive Model of Capacitive RF-MEMS *Microelectron Reliab* 2005, 45, 1770–1775.

[61] Franclov´a, J.; Ku˘cerov´a Z.; Bur˘s´ıkov´, V.; Electrical Properties of Plasma Deposited Thin Films *WDS'05 Proceedings of Contributed Papers* ,2005, Part II, 353–356.

[62] Harrell, W.R.; Frey, J.; Observation of Poole-Frenkel effect saturation in SiO2 and other insulating films, *Thin Solid Films*, 1999, 352, 195-204.

[63] Lamhamdi, M.; Guastavino, J.; Bpudou, L.; Segui, Y.; Pons, P.; Bouscayrol L.; Plana, R. Charging-Effects in RF Capacitive Switches Influence of insulating layers composition, *Microelectron Reliab* 2006, 46, 1700-1704.

[64] Gupta, D. K.; Doughty, K.; Brockley, R.S. Charging and discharging currents in polyvinylidenefluoride, *J Phys D Appl Phys* 1980, 13, 2101-2114.

[65] Wigner, E. On the constant A in the Richardson's Equation. *Phys. Review*, 1936, 49, 696-700

[66] Schug, J. C.; Lilly A. C.; Lowitz, D. A. Schottky Currents in Dielectric Films, *Phys. Rev. B*, 1970, 1, 4811-4818.

[67] http://www.memsnet.org/material/silicondioxidesio2bulk/

[68] http://www.siliconfareast.com/sio2si3n4.htm

[69] Melle, S.; De Conto, D.; Mazenq, L.; Dubuc, D.; Poussard, B.; Bordas, C.; Grenier, K.; Bary, L.; Vendier, O.; Muraro, J.L.; Cazaux, J.L.; Plana, R.; Failure predictive model of capacitive RF-MEMS, *Microelectronics Reliability*, 45, 1770 (2005)

[70] J. Vandershueren and J. Casiot in: Braunlich P (Ed.) Topics in Applied Physics: *Thermally stimulated relaxation in solids*, vol. 37, ch.4, pp 135-223, Springer-Verlag, Berlin, (1979)

[71] G. Papaioannou, J. Papapolymerou, P. Pons and R. Plana, *Appl. Phys. Letters* 90, 233507, (2007)

[72] G. Papaioannou, F. Giacomozzi, E. Papandreou and B. Margesin, Proceedings of the 2007 MEMSWAVE Workshop, *8th International Symposium on RF MEMS and RF Microsystems*, Barcelona, Spain, (2007).

[73] P.Czarnecki, X. Rottenberg, P. Soussan, P. Ekkels, P. Muller, P. Nolmans, W. De Raedt, H.A.C. Tilmans, R. Puers, L. Marchan3 and I. De Wolf, *Proceed. of MEMS2008 Conference* (2008).

[74] Balaji Lakshminarayanan, Denis Mercier, and Gabriel M. Rebeiz, *IEEE Trans. on Microwave Theory and Tech.*, 56, 971 (2008).

[75] S.P. Lau, J.M. Shannon and B.J. Sealy, *Journal of Non-Crystalline Solids* 277, 533, (1998)

11

Plasma Based Dry Release of MEMS Devices

Hamood Ur Rahman
College of Electrical and Mechanical Engineering
National University of Sciences and Technology (NUST)
Islamabad,
Pakistan

1. Introduction

Microelectromechanical Systems (MEMS) are constructed to achieve a certain engineering function or functions by electromechanical or electrochemical means. Very intricate moveable structures can be fabricated using the sacrificial layer (Madou, 2002). These moveable suspended structures are micro-bridges fabricated using the sacrificial layer. These structures can be cantilever beam bridges fixed at one end or membrane bridges fixed at both ends. In the field of RF MEMS these two types of bridges are used to fabricate the series or shunt switches.

In a fabrication process, the final release of a MEMS device is the most crucial step. Surface micromachining process relies on both wet and dry etching techniques. Wet etching has been widely used for pattern delineation. In wet etching, liquid etchant dissolves away the exposed film and attacks isotropically, resulting in loss of pattern definition due to undercutting and rounding of film features. This chapter suggests a solution to the problem of stiction by avoiding the wet release and in the absence of Critical Point Dryer (CPD). A dry release technique is presented for the RF MEMS structures, that combines the removal of sacrificial layer through wet etching and its substitution with standard photoresist. After coating, this photoresist acts as a supporting layer under the structure and rejects the structure to attack the substrate (Orpana & Korhonen, 1991). During this complete process wafers are not allowed to dry at any moment of time, otherwise structures may be permanently bonded with the substrate. The supporting layer is removed by oxygen plasma using the Reactive Ion Etching (RIE). In the dry etching, residual waste is off concern which effect the reliability of the MEMS structures. Motivation for this unique process was that some left over residues were observed after the single step or traditional RIE process. Secondly, this process was more cost effective as compared to a wet release CPD technique using CO_2 dryer. The process not only produced less residual waste but achieved a clean dry release.

2. Need for dry etching

A serious limitation of suspended MEMS structures is that they tend to deflect through stress gradient or surface tension induced by trapped liquids during the final rinsing and drying step. Problems like stiction and bridge collapse are associated with producing a free standing structure. The stiction is described as a process of bonding the top and bottom electrodes together by a microscopic surface due to the planner nature of the electrodes. Stiction of MEMS is a common concern. When a sacrificial layer is removed and rinsed in deionized water, the surface tension of rinse water pulls the delicate micro structure to the substrate as the wafer dries. Risk of stiction is caused by the capillary forces originating from the dehydration of meniscuses, van der Waals force or the electrostatic force formed between the suspended beam structures and the substrate following the wet etching (Madou, 2002). These forces keep the structure firmly attached with the substrate. Stiction remains a reliability issue due to contact with adjacent surfaces after release.

Stiction is an inevitable problem we deal with for achieving the working RF MEMS devices. With increase in cantilever length, its flexibility perpendicular to the substrate increases which also increases the susceptibility to stiction. When the structure gets attached with the substrate due to stiction, the mechanical force required to dislodge it from the surface is large enough resulting in damage to MEMS structure (Modou, 2002). The surface morphology has a strong influence on stiction and is a serious problem particularly in metal to metal contact switches (Varadan et al., 2003).

In order to achieve a released structure, contact between the structural elements and the substrate should be avoided during processing. Etching can be done by physical damage, chemical damage or combination of both. Release of these suspended beam structures can be done either through wet etching or dry etching. Etching in a plasma environment has several advantages as compared to wet etching. In the wet etching, this may become impossible or very difficult due to large surface tension forces. Moreover, if a MEMS structure is left too long in the etchant, the structure can be over etched and damaged (Harsh et al., 1999). Plasmas are easier to start and stop than simple immersion wet etching (Campbell, 1996). Also sensitivity of plasma etch is less prone to small changes in the temperature of the wafer. Above mentioned factors make plasma etching more repeatable than wet etching.

Different techniques over a period of time have been used to avoid stiction. Method of creating stand-off bumps on the underside of a polysilicon plate was introduced (Abe et al., 1995) which added meniscus shaping microstructures to the perimeter of the microstructure for reducing the chance of stiction. To avoid stiction critical point drying technique using CO_2 dryer is used (Chan et al., 2007) to release the structures.

3. Mitigation of stiction

3.1 Causes of stiction

The gap between two metal surfaces or device to substrate is so small that strong capillary forces can be developed during the dehydration which may lead to the adhesion of two surfaces. The same adhesion can occur when device is exposed to high humid conditions which lead to capillary condensation. Microstructures which contaminate the contact surface if stiction occurs, are in fact the synthetic particles of the metals (Alley et al., 1992).

The adhesion may occur due to solid bridging or liquid bridging. In solid bridging, the non volatile impurities present in the drying liquid are deposited on solid surfaces if drying by evaporation is conducted. These impurities may be introduced due to dissolution of the particles or substrate materials by liquid or through dissolution of residues distributed uniformly on the surface of the substrate. The deposition of impurities is pronounced in narrow spaces and between the two metal contacting surfaces upper and lower. This results in adhesion between the metal surfaces. The adhesion strength through solid bridging is

difficult to estimate because of the variation in deposition process or the density of deposited material. In any case, the adhesion strength tends to be significant.
The liquid bridging occurs due to the surface tension of the trapped capillary liquids. The drying of this trapped liquid is difficult due to the presence of concentrated soluble impurities. These trapped impurities increase surface tension while decreasing the vapor pressure. A third possible adhesion cause can occur if suspended membrane is placed in contact with the lower contact surface due to some external force. This adhesion can occur due to deliberate placement of collapsing forces or can be due to shock effect (Mastrangelo, 2000).

3.2 Stiction due to capillary forces

The removal of sacrificial layer to achieve a suspended microstructure is the final step in the surface micromachining process. This process mostly requires a wet etching for removal of sacrificial layer. In some cases the removal is also done using plasma etch when sacrificial layer is other than a metal layer like polyimide or photo-resist. After the wet etching the microstructure is rinsed using DI water to remove the residues left during the etching. When the microstructure is pulled out of DI water a strong capillary force develops. A meniscus forms at the interface under the microstructure when the microstructure is pulled out of water. The curved interface creates a pressure called Laplace pressure which is given by (Israelachvili, 1991)

$$P_L = \gamma_l \left(\frac{1}{r_a} + \frac{1}{r_b} \right) \tag{1}$$

The liquid surface tension is denoted by γ_l and two radii of curvature of liquid surfaces are given as r_a (parallel to surface normal of the substrate) and r_b (in the plane of the substrate). In most cases, the liquid droplet on the surface of the substrate will not wet it. It will present a definite angle of contact between the liquid and the substrate as shown in figure 1.

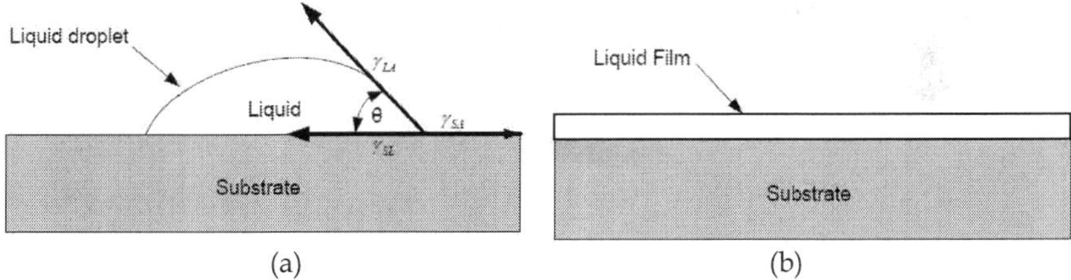

Fig. 1. Contact angle at solid liquid interface of (a) non-spreading (b) spreading liquid

In equilibrium condition, the contact angle between liquid and solid is determined by the balance between the surface tension of the three interfaces. The contact angle θ at the junction of three interfaces is defined as the angle formed between solid-air, liquid-air and liquid-solid interfacial tensions in equilibrium. The contact is given by the Young's equation (Israelachvili, 1991) as

$$\gamma_{SA} = \gamma_{SL} + \gamma_{LA}\cos\theta \quad 0 < \theta < \pi \tag{2}$$

If the γ_{SA} surface tension is smaller than the sum of γ_{SL} and γ_{LA} surface tensions, then the contact angle is larger than 0^0 and liquid will be non spreading as shown in figure 2(a). If the γ_{SA} surface tension is larger, than the sum of γ_{SL} and γ_{LA} surface tensions, it will spread the liquid energetically. Then the contact angle is equal to 0^0 and liquid will spread thus forming a drop bridging between the two surfaces as shown in figure 2(b). The total surface energy of the area between the metal contacting parts can be calculated by adding the surface tensions of all the three interfaces (Mastrangelo & Hsu, 1993).

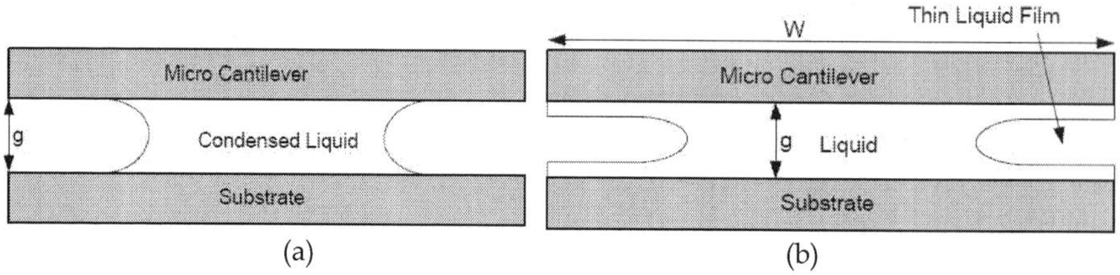

Fig. 2. The capillary condensation phenomenon of (a) non-spreading (b) spreading liquid showing underneath of a cantilever beam (front view)

Because the lateral dimesions of microstructures like cantilever beams are much larger than the vertical gap spacing (g) due to liquid layer thickness, i.e. $r_b \gg r_a$, therefore we may write (1) as,

$$P_L = \frac{2\gamma_l cos\theta}{g} \tag{3}$$

where θ is the contact angle of the liquid at the surface of the substrate and g is the gap height between the cantilever and the substrate which is equal to $2r_a \, cos\theta$.
The shape of the meniscus will be concave ($r_b < 0$) under a cantilever structure on a hydrophilic surface (silicon or any metal) which forms a quite shallow contact angle i.e., $cos\theta \approx 1$ as shown in figure 2(a), so the resulting Laplas pressure is negative. This will create sufficient attractive capillary force that will pull the cantilever beam structure down into contact with lower metal surface or substrate as shown in figure 3. Hence the cantilever beam falls into adhesion or stiction with substrate or metal surface following the drying process.

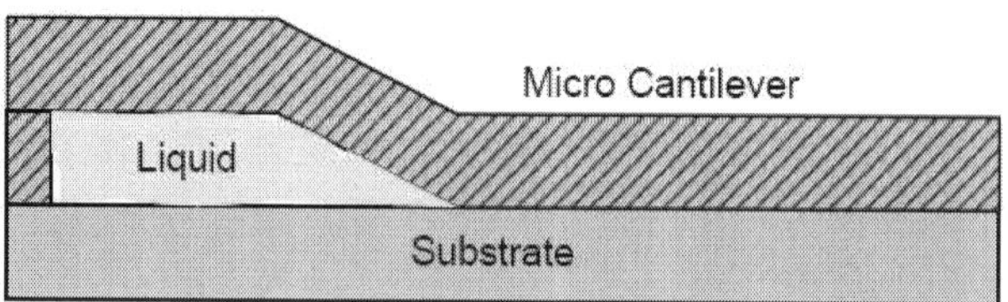

Fig. 3. The process of the microstructure drying that leads to the adhesion of micro cantilever to adjacent surfaces

The stiction between the two metal surfaces due to capillary forces looks quite similar to solid bridging. In the solid bridging, nonvolatile impurities are deposited on the solid surface causing the adhesion during the drying whereas in the capillary forces adhesion of a thin liquid layer works as an adhesion force between the two solid surfaces. If the contact angle θ between the solid and liquid is less than 90^0 (figure 2(a)) then the pressure inside the liquid drop will be less than outside. This results in a net attractive force between the two contacting plates. Figure 4 shows the SEM image of the cantilever beam structure stuck with lower metal surface due to capillary force. The adhesion of the beam was strong. An attempt was made to release the front part of the cantilever using micro probe. This resulted in breaking of the cantilever beam.

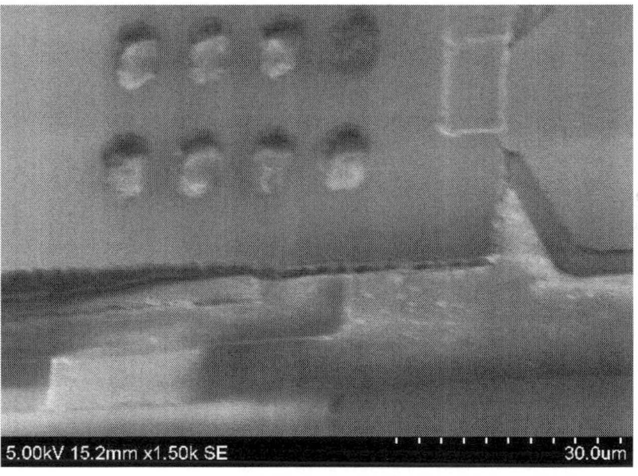

Fig. 4. SEM of cantilever beam held with stiction due to capillary forces

3.3 Drying method of cantilever beams

In this study, fabrication of RF MEMS is done using the CPW structure. When liquid attaches to the long cantilever beam, separation between the beam and CPW is a function of position whereas gap is smaller near the tip than near the anchor. As $r_a \rightarrow \infty$, the radius r_b remains constant as the droplet pressure is constant. As the liquid dries, the decrease in surface area drives the radius r_b to smaller values. The droplet will form an inside meniscus near the base which forces the droplet to neck down in this region resulting in a negative value of r_a. The necking continues with lower negative value of r_a until the two meniscuses on either side of the beam meet and pinch off a separated droplet. This technique for drying of water activates the capillary forces which can lead to adhesion. However, for applications discussed here the drying method has been used while avoiding the meniscus so that capillary force does not come into play at the time of drying.

In this drying method, the sacrificial layer was removed by wet etching. During rinsing with DI water when capillary forces could act for stiction during drying, we dipped the device in acetone till the time whole water under the cantilever beam was removed. The acetone dip also removed the diluted impurities which were present in the DI water while removing the sacrificial layer. Drying does not takes place with acetone, but device was dipped into a thin photo-resist (AZ5214E) which took a place underneath the cantilever beam as a supporting layer. The acetone evaporates quickly, therefore at no stage sample was exposed to the air. The photo-resist became a concentrated resist as the acetone evaporates leaving only the supporting layer of photo-resist. Now drying of photo-resist with hard-bake will remove the solvents from the photo-resist and then etch the photo-resist supporting layer with O_2 plasma using RIE. This drying method resulted in clean surface without residues and no stiction was observed during the process.

3.4 Stiction by contact adhesion

Another phenomenon which can produce adhesion between the two surfaces is an inter-solid adhesion which can overcome the restoring force of the elastic beam. Figure 5 shows the cross section of a cantilever beam with length *L*, width *W*, height *g*, and thickness *t*. The Young's modulus of the beam is represented by *E* which is 78GPa in terms of Au metal used for the fabrication of the switch. The figure shows that beam is adhering to the substrate at a distance $d = (L-x)$ from its tip.

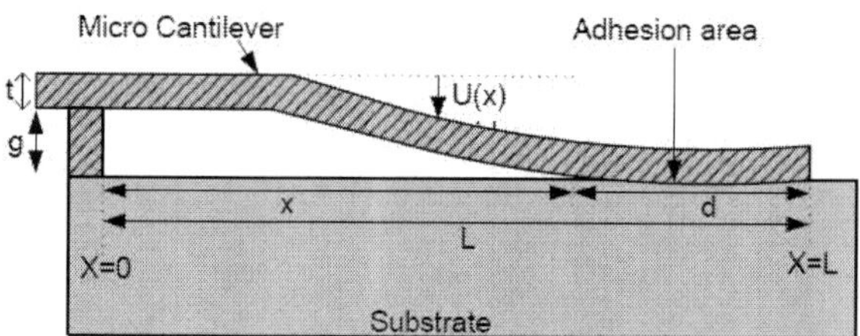

Fig. 5. Schematic of a cantilever beam adhering to the substrate

We can calculate the total energy of the system which is sum of the elastic and surface energies and is given (Mastrangelo, 2000) by

$$U_T = U_E + U_S = \frac{6EIg^2}{x^3} - \gamma_s Wd \tag{4}$$

where U_E is the bending energy stored in the beam and U_S is the interfacial energy of the contact area. Shear deformations are particularly important for $x \to L$, as $d = (L-x)$ is very small and tip of cantilever beam changes its elastic energy substantially just before detachment. This causes the beam to detach from substrate (Mastrangelo &Hsu, 1993) when $L = x = (3Et^3h^2/8\gamma_x)^{1/4}$ where γ_x is the surface energy which is determined from the detachment length and beam dimensions.

3.4.1 Inter solid adhesion reduction method

To eliminate the chance of permanent adhesion failure between the two solid surfaces, an inter-solid surface adhesion reduction is required. This can be done using techniques such as use of textured surfaces and posts, low energy molecular coatings and fluorinated coatings. The textured surfaces and posts approach has been used for the method presented here. Contact area between the elastic cantilever beam and the lower metal contact area on substrate was reduced which in turn reduced the adhesion forces.

The surface roughness of the upper contact area and lower contact area is rough enough, thus generates a rough interface between the two contact areas. The measured surface roughness of the upper contact area is 18nm and lower contact area is 22nm, respectively. This accomplishes the texturing of contact surfaces.

Texturing of the two solid surfaces was enhanced deliberately by introducing the construction of a small supporting post. In this approach, a dimple was introduced under the front tip of the cantilever. The dimple was constructed by making an extra mask layer in the fabrication process before patterning the cantilever beam. The dimensions of the dimple are 10×20μm with the height 0.9μm. Figure 6 shows the SEM of the dimple. It was taken by manually turning the cantilever beam using a lift out microscope.

Fig. 6. SEM of the dimple under the front end of the cantilever beam

The contact adhesion was also investigated by using a sharp Tungsten probe tip. The radius of curvature of the probe tip was 1μm. The sharp tip of the probe was used to pull the beams down under a high magnification microscope to ensure that cantilever tip has made contact with lower surface. When the tip of the cantilever touched the lower CPW surface, the probe tip was removed and cantilever beam started peeling off the surface. During this experimentation, there were only two options available to verify that either the beam would stick to lower surface or the beam would come off without stiction. A number of samples were tested after the release and no inter-solid adhesion was observed in these samples.

4. Dry etch process optimization using RIE

The first batch of wafers was used to optimize the release of the final device. For this purpose whole fabrication process was skipped and only the mask layers which were required for the fabrication of cantilever beams were used. For optimizing the release process Aluminum (Al) metal was used as it was readily available rather than expensive Au metal. Once the release process was carefully optimized, Au layer was used for the final fabrication of RF MEMS switches.

A 2.5μm thick layer of photo-resist (AZ6612) was deposited and then patterned for anchor. A 1.5μm thick layer of evaporated Al was deposited using e-beam evaporator. A layer of 1.0μm of photo-resist was deposited to pattern the cantilever beam. Two approaches were used to dry release the cantilever beams. First, one wafer was used to directly dry etch the sacrificial layer of photo-resist without using any other process on it. In this approach, it was observed that RIE tool was not able to remove the sacrificial layer material sufficiently from the devices. Some leftover metal residues were also observed which could not be cleaned even after extensive DI water rinsing of the wafer. A prolonged exposure of anisotropic RIE also damaged the cantilever beam structures.

In the second approach, a combination of wet and dry release was used to remove the leftover metal residues after etching while replacing a new layer of photo-resist as a supporting layer. The sample was inspected under microscope followed by SEM and in this case metal residues were not observed on the sample or supporting layer. Wafer was exposed under O_2 plasma in RIE chamber for dry etch. The wafer was exposed to high RF power (70W) and high pressure (30Pa). The high RF power generated an intense bombardment of plasma atoms with high pressure.

Devices were inspected under the microscope after the first etching exposure. It was observed that although plasma etched the photo-resist from top and sides of the device a significant amount of resist was observed under the cantilever beam. A second exposure of plasma was given again to the samples. After second exposure of plasma it was observed that resist was still visible under the beam. However, the beam structures were discoloured. It was assumed that some resist was still on the beam which created this discolourization. However, when the samples were observed under SEM, it showed that this discolourization was not due to resist but the beam structures were damaged due to high power plasma particles.

Figure 7(a) shows the SEM image of bridge over bias line with damaged surface. A metal peeling from some parts of the bridge is also visible. One can observe that high power bombardment of plasma atoms has damaged the metal layers on the device. Figure 7(b) shows the cantilever beam structure damaged due to RIE plasma while optimizing the release process.

After a number of iterations, it was revealed that power and pressure were the main factors for the optimization of dry release process. Variation of power and pressure from high to low and vise versa can change the plasma behavior inside the chamber. The voltage bias was also controlled once these parameters were changed. With high RF power and low pressure we achieved a bias of 232±6V which indicated that plasma particles generated inside the chamber strike the surface of the substrate with more power giving anisotropic etching behavior. With low power and high pressure the bias changed to as low as 90±6V which changed the plasma atoms behavior from anisotropic to isotropic giving etch profile below the surface of the cantilever beam also.

Three wafer samples were used to optimize the RIE process using the supporting layer technique. In this case, a careful shifting of power and pressure parameters was done. Once sample was ready with supporting layer of photo-resist underneath the cantilever beam, the sample was exposed to high power and low pressure for one hour. In this case, the plasma particles struck the wafer surface with high power but under low pressure. It did not damage the device surface. The sample was observed under the microscope and a significant amount of photo-resist was observed on the sample as well as under the beam. The sample was exposed to plasma for 30min and then inspected again under microscope. It was observed that much of the resist from top of the beam and sides was removed but a small amount of resist was still visible on the beam and significant amount of resist was observed under the beam. Sample was exposed to another 30min exposure which cleaned the resist from top and sides of the beam but under the beam resist was observed. Now, the power and pressure parameters of the RIE tool were changed from high power to low power and from low pressure to high pressure. This created an isotropic behavior of the plasma instead of anisotropic behavior which was observed in the first setting. Sample was exposed to plasma for one hour which resulted in a clean release of the structures.

(a) (b)

Fig. 7. SEM of discolored/damaged areas after RIE (a) bias line bridge (b) cantilever beam

5. Fabrication process

The fabrication of the RF MEMS switches is a six mask all metal fabrication process, as shown in figure 8. All processing steps are developed on the basis of standard CMOS processing. A standard one-layer photo-resist was used as a mask during the fabrication process to provide precise pattern definition. However, during release step the photo-resist was also used as a sacrificial layer. The photo-resist (AZ6612) was a positive photo-resist sensitive to ultraviolet (UV) radiation and can be developed with AZ-300 MIF solution. Throughout the fabrication process, alignment was performed with Quintal Q-6000 mask aligner with UV light exposure. In order to achieve good RF performance device, the switch was fabricated on a low loss alumina substrate with dielectric constant 9.9.

The fabrication is a six mask all metal process. The process started with the standard wafer cleaning process. DC bias lines and actuation pads were defined by evaporating 0.04μm layer of Cromium (Cr). This layer was then patterned with mask one. The Cr metal evaporation was done using Lesker evaporator operating in 10^{-6} Torr range. An insulator layer in a series switch served as a mechanical connection as well as electrical isolation between the actuator and the contact. Since the switch was made of metal, the insulator layer also acted as a dielectric layer which was needed to prevent direct contact between the metal cantilever bridge and the actuation pad.

A 0.75μm thick layer of silicon nitride was deposited as dielectric layer using PECVD and patterned with mask two. The deposition of Si_3N_4 was performed using VACUTEC-1500 series PECVD equipment. The CPW lines were defined by evaporating/RF sputtering of 0.04/1.0μm thick layer of Cr/Au and patterned with mask three. Cr was used as an adhesion layer between the Au and substrate. The sputtering was performed using Edwards E-306 series sputtering tool which was used for RF sputtering of the Au film.

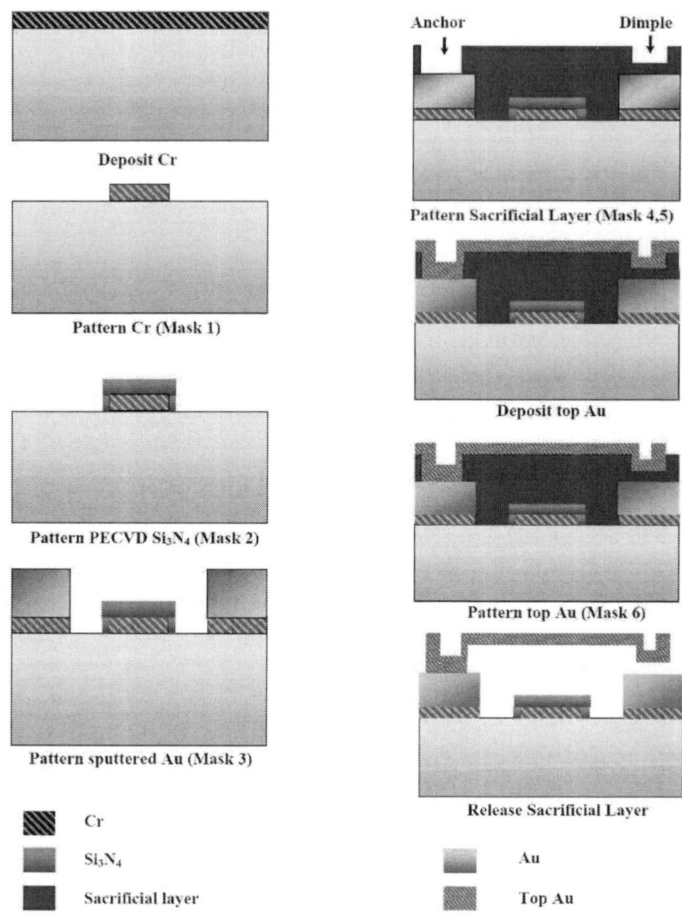

Fig. 8. Six mask fabrication process for RF MEMS switches

Then a 2.5μm thick layer of photo-resist (AZ-6632) was deposited as sacrificial layer and patterned for anchor and dimple with mask four and five respectively. While defining the anchor and dimple full dark masks were used to expose only the anchor and dimple areas. This was followed by a 1.5μm thick layer of RF sputtered Au which was patterned with mask six to form the cantilever beam. Finally, the bridge structure was released using a unique dry release process.

6. Dry release process

During fabrication of RF MEMS switches both dry and wet release methods were applied. The yield of wet release was very low and no working prototype was achieved. Problems related to wet release and stiction have already been discussed in the previous sections. As explained in last section, in single dry release process, the problem of left over residues of metal after etching the metal layer was experienced. So a unique dry release process was developed with a combination of wet and dry release to achieve better results.

6.1 Dry release model

Motivation for this unique process was that some left over residues were observed after the single step or traditional RIE process. Secondly, this process was more cost effective as compared to a wet release CPD technique using CO_2 dryer. The process not only produced less residual waste but achieved a clean dry release. The steps for dry release process are described in figure 9.

First, the sacrificial layer was removed using acetone. This also included the removal of some Au leftover residues on photo-resist from the previous wet etching with mask 6 [figure 9(a)]. After this, sample was dipped again into clean acetone for 30 min for final cleaning. Then the structure was immediately dipped into another resist (AZ5214E), until all the liquid covering the sample was concentrated resist [figure 9(b)] (Forsen et al., 2004 & Orpana & Korhonen, 1991). The resist covered sample was spun at 2500 rpm to achieve uniform layer of resist and then soft backed at 90°C resulting in a thick layer of photo-resist fully encapsulating the suspended beam as a supporting layer.

It must be noted that the wafer was never allowed to dry during the the process or else structure would be permanently bonded to the substrate. The structure was then dry released by Oxygen plasma using the single process RIE in two steps. In step one the etching was done using high power and low pressure (15sccm O_2, 180 W, 8 Pa) giving an anisotropic etch of the photo-resist [figure 9(c)]. In step two low power and high pressure (15sccm O_2, 50 W, 40 Pa) was used. This resulted in isotropic etching of the photo-resist thus giving a free standing structure at the end [figure 9(d)].

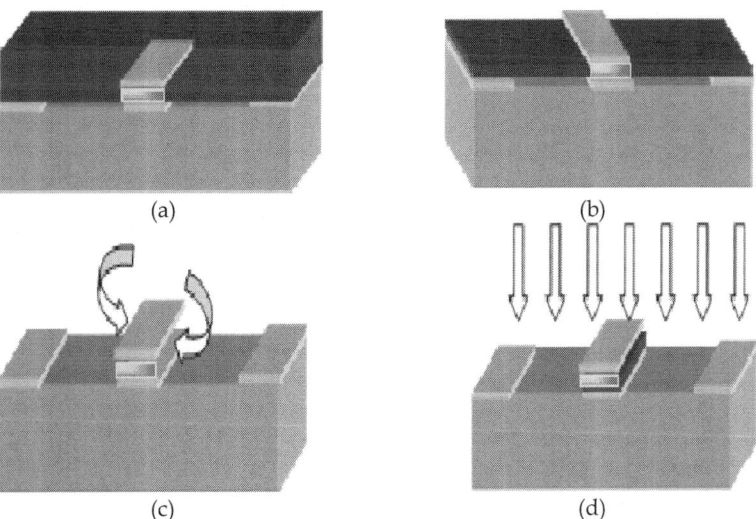

Fig. 9. Schematic representation of process steps involved in dry release process of MEMS structures (a) patterned cantilever beam over sacrificial layer of AZ-6612 (b) cantilever beam dipped in structural layer of AZ-5214E (c) anisotropic etching (d) isotropic etching

6.2 Dry release using RIE

Figure 10 displays a SEM image of the fabricated switch. The sacrificial layer (AZ-6612) has been removed after two dips in acetone; supporting layer below the structure has been made with another photoresist (AZ5214E). It can be observed that structure has got a clear standing on the supporting layer. There is no indication of left over residues of the Au after acetone cleaning.

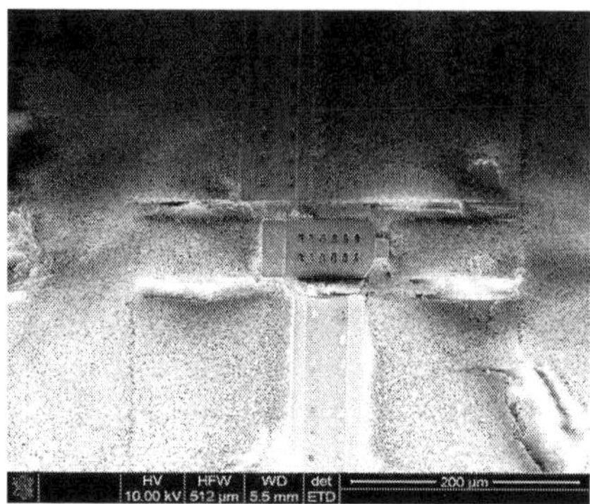

Fig. 10. SEM of the RF MEMS switch with supporting layer

Figure 11 shows a SEM image of the structure after anisotropic etching during the first step of single process RIE. The structure rests on supporting layer. Some leftover parts of the chemical waste are also visible. The chemical waste observed during the dry etching was comparatively less than as seen in the wet etching.

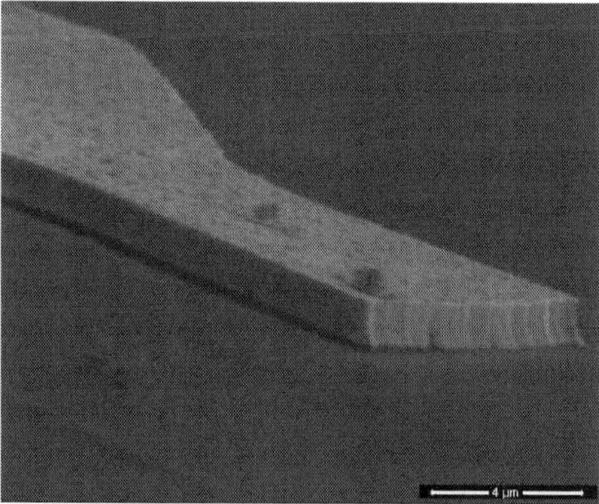

Fig. 11. SEM of the cantilever beam structure resting on supporting layer after anisotropic etch

During the isotropic etching step of RIE, plasma moves in all directions and etches the photo-resist layer located below the cantilever beam structure. Figure 12 shows the released RF MEMS cantilever beam structures. The clean standing structure of the MEMS bridge can be observed. The release of structure was clean and results achieved by this process technique were satisfactory.

Fig. 12. Released RF MEMS switch cantilever beam (front tip view)

From the SEM images and optical microscopy it was observed that the released beam structure showed higher curling up trend. This was due to residual gradient stress in the film and lead to the increase in the actuation voltage. The stress gradient lead to the lift of beam around 1µm after the release of structure. The measured lift of cantilever front end is 4.3µm after release. Figure 13 shows a DEKTAK profile of the unreleased and released beam tip of RF MEMS switch. In figure 13(a) DEKTAK profile indicates the beam height after patterning mask six which also confirms the gap height distance of 2.5µm. When beam structure was released using RIE plasma technique, the lift of the front tip of the cantilever beams was measured again which confirms the curling up trends of the beam stated above.

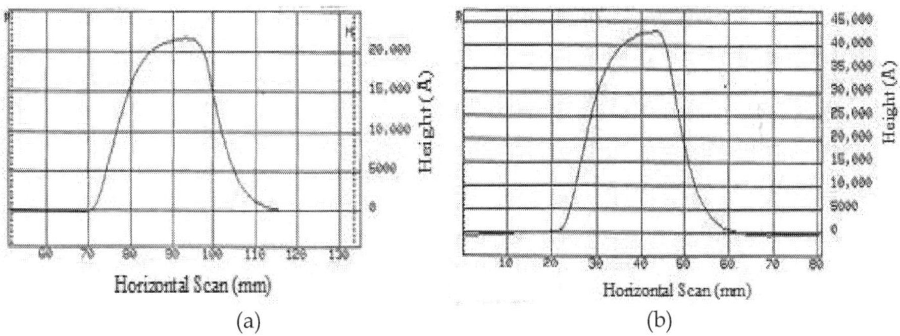

Fig. 13. DEKTKK image of beam tip (a) before release (b) after release

6.2.1 Yield

The yield of the released structures on the wafer was measured using visual examination and SEM. No stiction was observed with new release process. However presence of some residues was observed on the outer samples of the substrate. This was due to non uniform plasma distribution during the RIE. The yield of the release process was worked out on full cleaned samples. A yield of more than 70% was achieved with contact resistance of less than 2.7Ω.

7. Fabricated RF MEMS switches

Due to better flexibility for large systems and wide band applications, metal to metal contact switch was chosen over the capacitive switch. The CPW centre conductor was 60µm wide, 20µm gap and 210µm ground widths which resulted in characteristic impedance of 50Ω. The beam was suspended 2.5µm from the substrate. The ground planes around the beam were suspended as it provided easy access to beam and electrode when being used in biasing systems. The switches were fabricated using the developed six mask all metal process. A dimple was used at the bottom of front centre tip of the cantilever beam to reduce the stress sensitivity of the beam. Front tip contact area was small as compared to conventional cantilever beams because of following reasons. First, small contact points would reduce the metal-to-metal stiction and would increase the contact pressure. Secondly, it gave better isolation.

Figure 14 displays the proposed arrow beam design. The length of the beam is 120µm and the width of the beam is 60µm. The beam has been curved inside from the front with a front tip 20µm in width.

Fig. 14. SEM of fabricated arrow beam cantilever based RF MEMS switch

Figures 15 and 16 show the two proposed cantilever beam designs. In figure 15, the beam labeled Design-1 has three supported cantilever bars which behave like three springs while moving the beam during the actuation. The length of each cantilever bar is 20µm and the width is 10µm. The gap between each cantilever bar is 15µm and provides symmetry to the beam structure. All three cantilever bars are connected with an anchor which is 20µm in length and 60µm wide. The supported cantilever bars are then connected with a beam of length 100µm and width of 60µm.

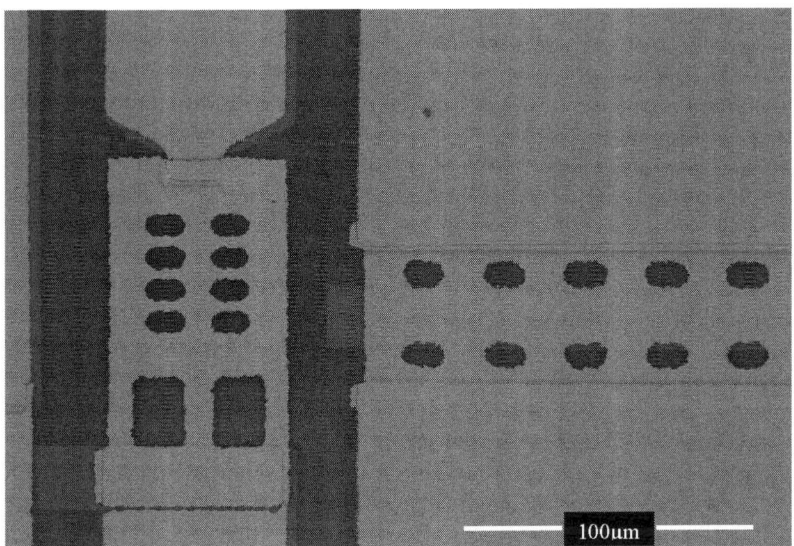

Fig. 15. SEM of fabricated RF MEMS switch

Figure 16 shows the beam labeled Design-2, with three supported cantilever bars and an extended cantilever at the front. The dimensions of the three supported cantilever beams are the same as that for Design-1, with the centre 60μm×60μm and the extended cantilever at the front 40μm×20μm.

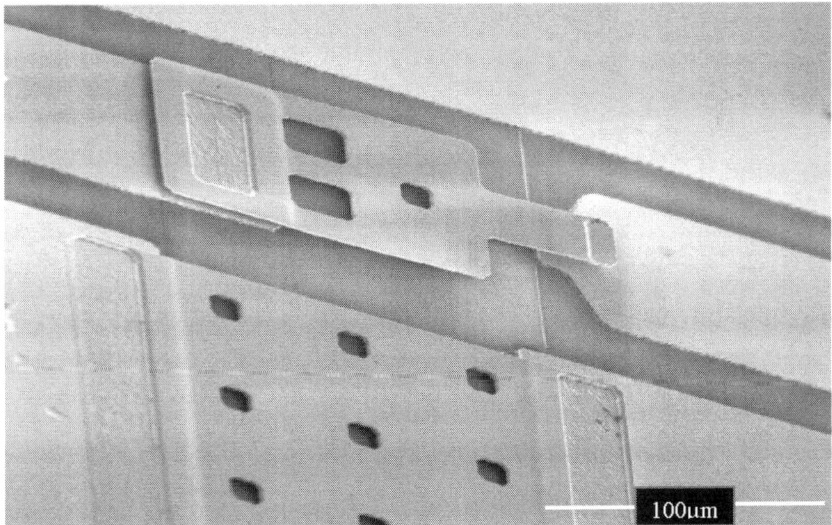

Fig. 16. SEM of fabricated RF MEMS switch

Fig. 17. SEM of fabricated RF MEMS switch

Figure 17 shows the RF MEMS switch which has the same dimensions of beam as explained in figure 16, instead of three supporting bars, has two supporting bars with a single cross bar link intended to increase the strength of the two low spring constant supporting bars.

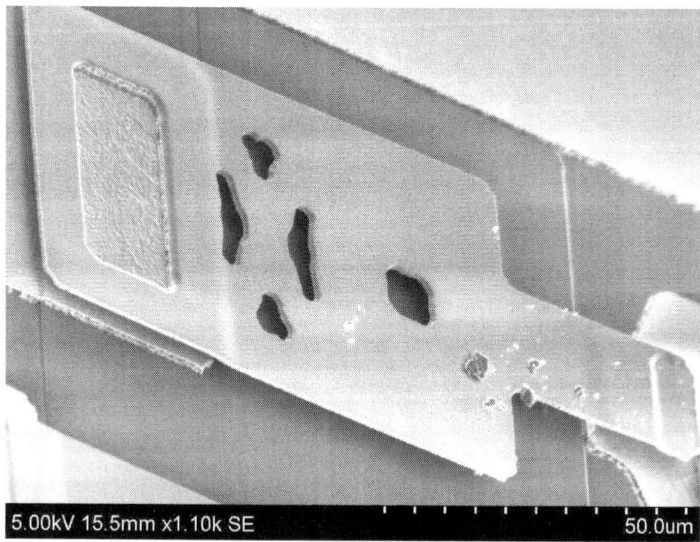

Fig. 18. SEM of fabricated novel RF MEMS switch

Figure 18 shows another design of switch which has a standard cantilever with dimensions of 80μm×60μm at the rear and an extended cantilever at the front with 40μm×20μm. Some metal particles can be seen on front portion of the cantilever which came after testing of the device while dragging the probe for contact.

8. Experimental results

The measurement setup for actuation voltage and RF performance was employed using two test configurations, i.e, preliminary screening and RF characterization. Preliminary screening was made using Cascade Microtech 10000 probe station with Tungsten needle connected to a Sony Tektronix 370 Programmable Curve Tracer. No RF performance was analyzed at this stage. The curve tracer was programmed to 0-100V DC signal with a step of 2V increment. The switches actuated at 19V and 23V. At this stage, the contact confirmation was made between two surfaces while measuring the contact resistance.

For RF performance, a two port on wafer measurement of the RF MEMS switches was performed using HP-8510 vector network analyzer (VNA) from 0-40 GHz. A Cascade Microtech 10000 probe station was used. RF probing was done using Cascade Microtech GSG RF probes with a pitch of 100μm. The SOLT (short-open-line-through) method was used for the calibration of the system before each test sequence.

An HP 4140 DC voltage source was used to actuate the switch during RF characterization. The actuation voltage for the RF MEMS switches was applied between the cantilever beam and the lower actuation pad. Two Picosecond Pulse Labs 5590 DC blocks were connected between VNA and RF cables connected with RF probes.

8.1 Electrical performance

A two port on wafer measurement of the RF MEMS switches have been performed from 0-40 GHz. When our switches were unactuated and beam was in up position, switches were in OFF state.When switches were actuated and the beam was pulled down, switches were in ON state.

8.1.1 Isolation

In order to determine the RF performance of the switch the insertion loss, return loss and isolation of the switches were measured. Isolation of the switch was measured when signal was in OFF state. Figure 19 and 20 illustrate the measured S-parameters for Design-1 and Design-2 respectively. As shown, Design-1 had an isolation of 28dB at 20GHz and better than 23dB at 40GHz. For Design-2, the isolation of the switch was 30dB at 20GHz and better than 28dB at 40GHz.

8.1.2 Insertion loss and return loss

The return loss and insertion loss of the switches were measured when signal passed through the ON state. Design-1 has a return loss better than 22dB at 20GHz and 19dB at 40GHz, for Design-2 it was better than 20dB at 20GHz and 18dB at 40GHz. This reveals good impedance matching to 50Ω of our RF MEMS designs.

Insertion loss for Design-1 was 0.75dB at 20GHz and 1.15dB at 40GHz where as insertion loss for the Design-2 was 0.8dB at 20GHz and 1.3dB at 40GHz. Higher insertion loss was

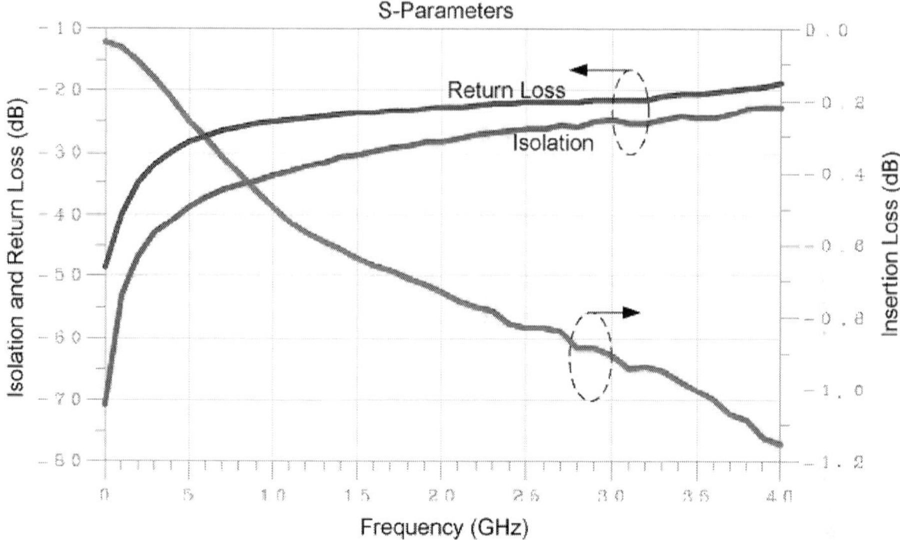

Fig. 19. Measured S-parameters of the switch using Design-1

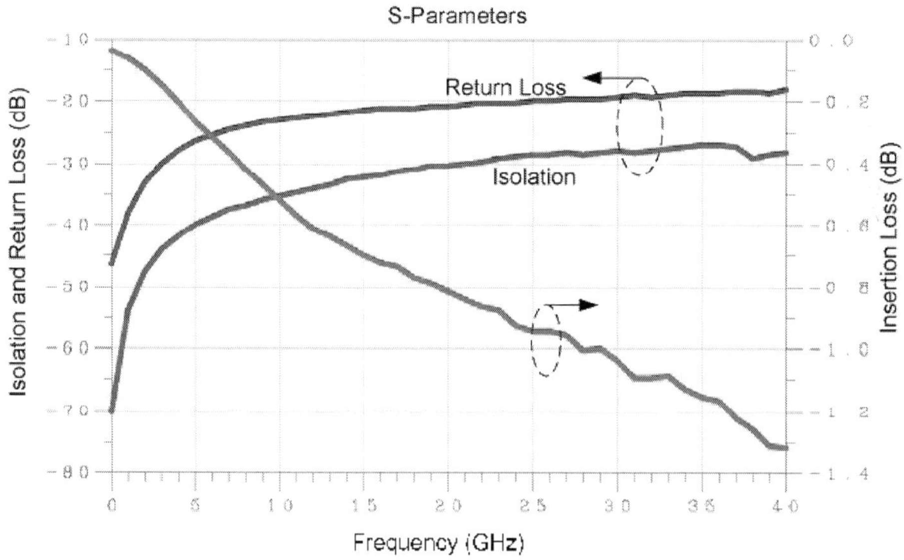

Fig. 20. Measured S-parameters of the switch using Design-2

attributed to following reasons. First, the higher contact resistance was achieved which was due to high surface roughness of the metal surface and smaller contact area. The surface roughness value is 18nm for dimple and 22nm for signal line contact area which showed that surfaces of both contact points are quite rough.

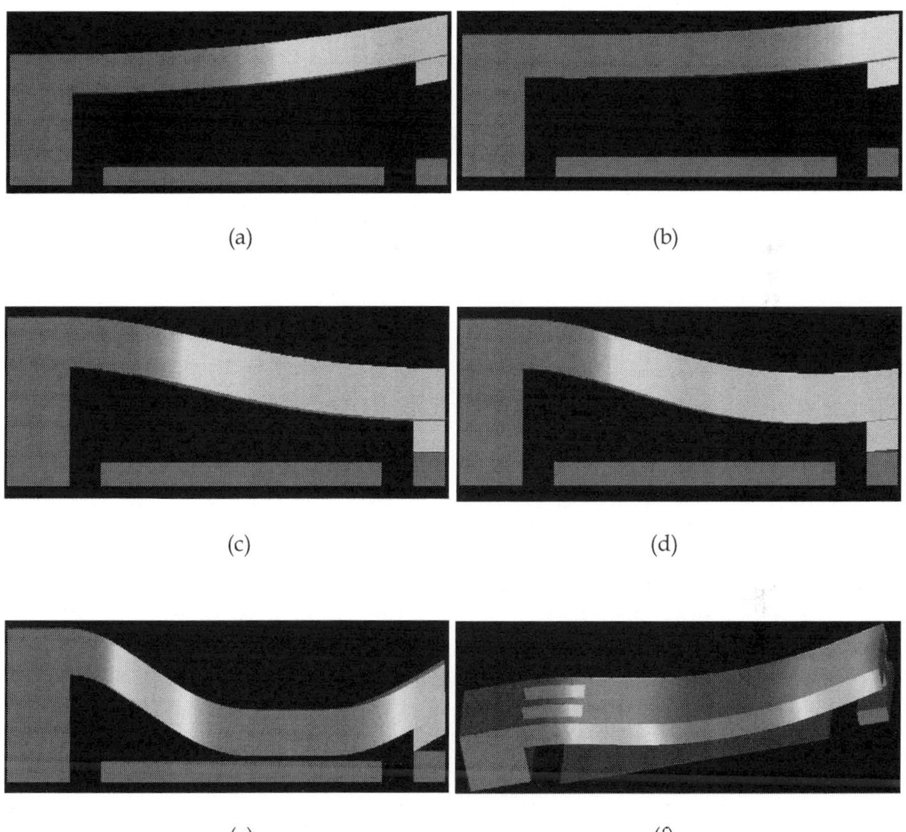

(a) (b)

(c) (d)

(e) (f)

Fig. 21. Simulated view of dimple contact during different stages of actuation

Secondly, dimple surface might have an uneven surface contact with signal line contact area. To validate this observation, a simulation test was conducted to see the dimple movement in different stages of the actuation. Figure 21 showed the movement of the cantilever beam with dimple in different stages when actuation bias was applied. The dimple made a perfect smooth contact with lower contact surface as shown in figure 21(c) but at this point full boundary conditions were not enforced. When boundary condition were fully enforced and beam was placed in the hold down position the complete surface of dimple was not in contact and front surface of dimple has lifted up as shown in figure 21(e). This phenominon lead to higher insertion loss.

8.1.3 Actuation voltage

The measured actuation voltage of the Design-1 is 19V and Design-2 is 23V. A number of release holes can be observed in the fabricated switches. The effect on electrostatic force due to release holes had already been rationalized with inclusion of 40% of fringing field effect during simulation of spring constant of the beam designs (Rebeiz, 2003).

9. Conclusion

Bulk and surface micromachining are the two most widely used techniques for fabrication of MEMS devices. Wet and dry etching is used to achived the final release of the MEMS devices. However, in surface micromachining a combination of wet and dry etching techniques is used. Pros and cons of both the techniques have been disscused in the chapter. In wet etching capilary forces developed during the process leads to stiction and permanent adhesion of the MEMS devices. To eliminate the chance of stiction and permanent adhesion between the two solid surfaces, an inter-solid surface adhesion reduction is required. The textured surfaces and posts approach has been used for to eliminate the inter-solid surface adhesion. This was done firstly, by introducing a dimple as a supporting post between two solid surfaces and secondly, by reducing the contact area between the two metal surfaces. Both the methods were used to design improved switches which include the stiction mitigation structures. In dry etching, a dry release method with better solution for release of RF MEMS structures has been used. The technique was developed using RIE instead of CO_2 dryer or critical point drying technique. This process may lead to long term storage of the MEMS devices. Finally, fabricated novel switches have been presented validating the developed fabrication process.

A six mask all metal fabrication process was used for fabrication of RF MEMS switches. The experimental RF performance of the two fabricated swithes achieved a measured actuation voltages of 19V and 23V, respectively. Both switches showed good RF performance. Design-1 exhibited an isolation of 28dB at 20GHz and better than 23dB at 40GHz whereas, Design-2 exhibited an isolation of 30dB at 20GHz and better than 28dB at 40GHz. Both RF MEMS designs showed good impedance matching to 50Ω as deducted from the ON state S-parameter measurements. A return loss better than 22dB at 20GHz and 19dB at 40GHz was measured for Design-1 whereas, Design-2 exhibited a return loss of better than 20dB at 20GHz and 18dB at 40GHz. The insertion loss was 1.15dB and 1.3dB respectively, for all frequency band of interest.

10. Acknowledgment

This research was carried out at Centre for Quantum Computer Technology (CQCT) Micro Fabrication Laboratory, University of New South Wales (UNSW), Australia. The author wishes to thank Professor Rodica Ramer for continuous support and supervision during the whole research The helpful advice by Dr Eric Gauja during the fabrication is greatly appreciated.

11. References

Abe, T ; Messner, W. C. & Reed, M. L. (1995). Effective Methods to Prevent Stiction during Post Release Etch Processing, *Proceedings of International Conference on Micro Electro Mechanical Systems (IEEE MEMS 95)*, pp. 94-99, ISBN 0-7803-2503-6, 29th Jan - 2nd Feb 1995. Netherland

Alley, R. L.; Cuan, G. J.; Howe, R. T. & Momvopoulos, K. (1992). The Effect of Release Etch Processing on Surface Microstructure Stiction, *Proceedings of IEEE 5th Technical Digest on Solid State Sensors and Actuators Workshop*, pp. 202-207, ISBN 0-7803-0456-X, 22-25 June 1992, SC, USA

Campbell, S. A. (1996). *The Science and Engineering of Microelectronic Fabrication*, Oxford University Press, ISBN 0-19-510508-7, New York USA

Chan, K. Y.; Daneshmand, M.; Mansour, R. R. & Ramer, R. (2007). Novel Beam Design for Compact RF MEMS Series Switches, *Proceedings of Asia Pacific Microwave Conference (APMC 2007)*, pp. 229-232, ISBN 978-1-4244-0749-4, 11-14 Dec 2007, Bangkok Thailand

Forsen, E.; Davis, Z. J.; Dong, M.; Nilsson, S. G.; Montelius, L. & Boisen, A. (2004). Dry Release of Suspended Nanostructures, *Journal of Microelectronic Engineering*, No. 73-74, pp. 487-490, doi: 10.1016/j.mee.2004.03.022, ISSN 0167-9317

Harsh, K. F.; Zhang, W.; Bright, V. M. & Lee, Y. C., (1999). Flip Chip Assembly for Si Based RF MEMS, *Proceedings of 12th International Conference on Micro Electro Mechanical Systems (IEEE MEMS 99)*, pp. 273-278, ISBN 0-7803-5194-0, 17-21 Jan 1999, Orlando, FL USA

Israelachvili, J. N. (1991). *Intermolecular and Surface Forces*, Academic Press Ltd, ISBN 0-12-375181-0, London UK

Madou, M. J. (2002). *Fundamentals of Microfabrication*, pp. 276-277, CRC Press, Florida, ISBN 0-8493-0826-7, USA

Mastrangelo, C. H. (2000). Suppression of Stiction in MEMS, *Proceedings of Material Research Society Symposium*, Vol. 605, pp. 105-116

Mastrangelo, C. H. & Hsu, C. H. (1993). Mechanical Stability and Adhesion of Microstructures under Capillary Forces - Part I: Basic Theory, *Journal of Microelectromechanical Systems*, Vol. 2, No. 1, pp. 33-43, ISSN 1057-7157

Mastrangelo, C. H. & Hsu, C. H. (1992). A Simple Experimental Technique for Measurement of the Work of Adhesion of Microstructures, *Proceedings of IEEE 5th Technical Digest on Solid State Sensors and Actuators Workshop*, pp. 2008-2012, ISBN 0-7803-0456-X, SC, USA

Orpana, M. & Korhonen, A. O. (1991). Control of Residual Stress of Polysilicon Thin Films by Heavy Doping in Surface Micromachining, *Proceedings of International Conference on Solid State Sensors and Actuators Transducers 91*, Vol. 23, pp. 957-960, ISBN 0-87942-585-7, 24-27 June 1991, San Francisco, CA USA

Rebeiz, G. M. (2003). *RF MEMS : Theory Design and Technology,* John Wiley and Sons Ltd, ISBN 0-471-20169-3, New York USA

Varadan, V. K.; Vinoy, K. J. & Jose, K. A. (2003). *RF MEMS and Their Applications,* John Wiley and Sons Ltd, ISBN 0-470-84308-X, New York USA

12

Optical MEMS

Wibool Piyawattanametha[1,2] and Zhen Qiu[3]
[1]*Advanced Imaging Research (AIR) Center, Faculty of Medicine, Chulalongkorn University, Pathumwan,*
[2]*National Electronics and Computer Technology Center, Pathumthani,*
[3]*University of Michigan, Biomedical Engineering, Ann Arbor, Michigan*
[1,2]*Thailand*
[3]*USA*

1. Introduction

In 1989, a group of scientists and engineers in Salt Lake City started a workshop called Micro-Tele-Operated Robotics Workshop. There, the acronym for Microelectromechanical systems (MEMS) was officially adopted. However, MEMS technology has already had a head start since at least 7 years ago from the classic work published by Petersen in 1982 [1]. Twenty years later, MEMS technology has started major novel innovations in several scientific fields and created highly promising market potential. In 2003, the most conservative market studies predict a world MEMS market in excess of $8 billion.

Optics and photonics are among these research fields impacted by MEMS techniques. Optical MEMS has created a new fabrication paradigm for optical devices and systems. These micro optical devices and systems are inherently suited for cost effective wafer scale manufacturing as the processes are derived from the semiconductor industry. The ability to steer or direct light is one of the key requirements in optical MEMS. In the past two decades since Petersen published his silicon scanner [1], the field of optical MEMS has experienced explosive growth [2,3]. In the 80's and early 90's, displays were the main driving force for the development of micromirror arrays. Portable digital displays are commonplaces and head mount displays are now commercially available. In the past decade, telecommunications have become the market driver for Optical MEMS. The demand for routing internet traffic through fiber optic networks pushes the development of both digital and scanning micromirror systems for large port-count all-optical switches with the ability to directly manipulate an optical signal, Optical MEMS systems eliminate unnecessary optical-electrical-optical (O-E-O) conversions. In the biomedical arena, micro-optical scanners promise low-cost endoscopic three-dimensional imaging systems for *in vivo* diagnostics.

This chapter summarizes the state of the art of Optical MEMS technology by describing basic fabrication processes to derive with actuation mechanisms and select examples of devices that are either commercially available, or show great promise of becoming products in the near term. The chapter is organized into the following sections: Section 2 describes the generic actuation mechanisms commonly used for MEMS devices. Section 3 discusses the applications.

2. Mechanisms of actuations

Since the actuators are important engines for Optical MEMS, we will introduce different kinds of actuation mechanisms first and explain the working principles in detail. Furthermore, for more specific applications, there are many variants or requirements based on these basic actuation mechanisms.

2.1 Electrostatic actuation

Electrostatic MEMS devices work at different motion modes. Here we mainly introduce the MEMS devices with torsional rotation. Electrostatic MEMS devices with torsional rotation can be described as follows: when voltage is applied between the movable and the fixed electrodes, the moving part rotates about the torsion axis until the restoring torque and the electrostatic torque are equal. The torques can be expressed as:

$$T_e(\theta) = \frac{V^2}{2}\frac{\partial C}{\partial \theta} \tag{1}$$

$$T_r(\theta) = k\theta \tag{2}$$

where V is the applied voltage across the fixed and movable electrodes, C is the capacitance of the actuator, θ is the rotation angle, and k is the spring constant. The capacitance is determined by the area of the electrode overlap and the gap between the electrodes. For simple parallel plate geometry, the capacitance can be expressed by

$$C = \frac{\varepsilon_0 A}{g} \tag{3}$$

where ε_0 is the permittivity of free space, A is the area of electrode overlap, and g is the gap between fixed and moving electrodes.

There are two major types of electrostatic actuators. The first is based on parallel-plate capacitance, and the other is based on comb-drive capacitance. For the parallel-plate type devices (Figure 2-1), the area of the electrode overlap is essentially the area of the fixed electrode. The gap for the parallel-plate actuator is a function of the rotation angle and there is a tradeoff as the initial gap spacing needs to be large enough to accommodate the scan angle, but small enough for reasonable actuation voltage. The stable scan range is further limited by a pull-in phenomenon to 34-40% of the maximum mechanical scan angle [4,5].

Another type of actuator is based on the vertical combdrive. The vertical combdrive offers several advantages: (1) the structure and the actuator are decoupled; (2) the gap between the interdigitated fingers of the combdrive is typically quite small, on the order of a couple of microns [6,7]. Large rotation angle and low actuation voltage can be achieved simultaneously. In the combdrive, the gap is constant and the area of the electrode overlap is a function of the rotation angle. Typically, there are three approaches to realize the vertical combdrive actuators. The first and popular one is the staggered vertical combdrive actuators [46]. Figure 2-2 shows the schematic of a vertical combdrive actuator. The second type is the angular vertical combdrive actuators [8,58,59] (Figure 2-3). Both staggered and angular vertical combdrive actuators can work at either resonant tilting mode or quasi-static (or called DC tilting) mode. At quasi-static mode, the maximum rotation angle usually is limited because it is typically the point where the overlap area of electrodes reaches the

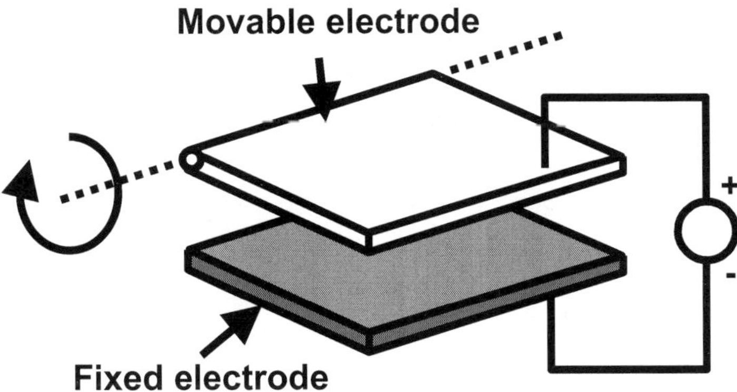

Fig. 2.1. Schematic of a parallel-plate actuator

maximum. In addition to the staggered and angular vertical combdrive actuators, in-plane configuration vertical combdrive actuator is the third approach firstly demonstrated by Harald [8] at Fraunhofer IPMS Dresden, Germany. The in-plane configuration vertical combdrive (Figure 2-4) means that the moving combdrive is on the same plane of the static combdrive, which is based on the parametric resonant working principle. The resonantly excited actuator has several advantages as below: (1) it can achieve very large tilting angle with low driving voltage; (2) only one-time DRIE is needed for combdrive and gimbal frame structure during micro-machining which eliminates wafer bonding or accurate alignment. However, the vertical combdrive actuators with in-plane configuration only work at resonant mode. Recently, researchers at IPMS also have made effort to realize quasi-static rotation mode with vertical out-of-plane combdrive using leverage structures [9].

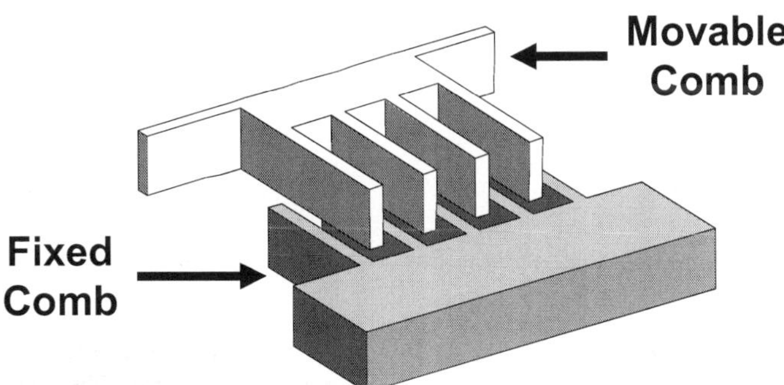

Fig. 2.2. Schematic of a vertical combdrive actuator

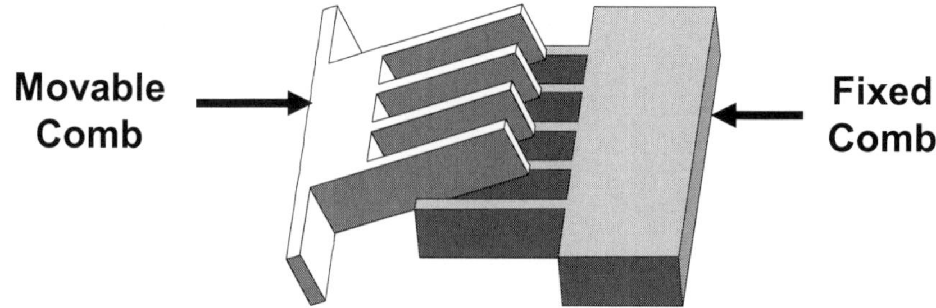

Fig. 2.3. Schematic of an angular combdrive actuator

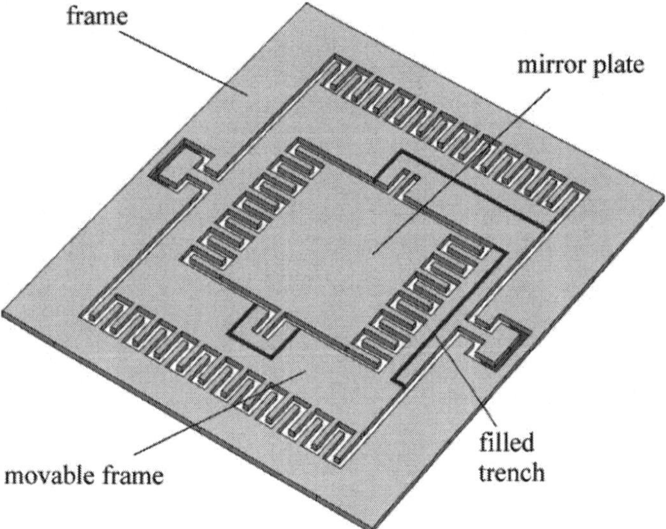

Fig. 2.4. Schematic of an in-plane configuration combdrive actuator

2.2 Magnetic actuation

Magnetic actuation is practical when the structural dimensions are on the millimeter scale since the magnetic torque (generated by the magnetic device interacting with an external magnetic field) scales with volume for permanent magnetic materials and with total coil area for electromagnets. For an analysis of magnetic torque see Judy and Muller [10]. The overall system size must accommodate the magnets (permanent or electric coils) used to generate the external magnetic field. Therefore, the motivations for this type of scanner are usually cost reduction through batch fabrication and lower power consumption rather than miniaturization. In addition, magnetic actuation also has the advantage of operating in liquid environment.

Magnetic field can be induced by electrical current. This current-induced magnetic field can generate the force exerted on the moving magnetic material [10]. While the moving structure is not made of magnetic material, electromagnetic coils can be integrated on the movable part, making it quasi-magnetic by current injection [11]. Figure 2-5 shows an example of the electromagnetic scanner that is being used in table-top confocal microscopes. Researchers at the University of Michigan have demonstrated a miniature magnetic 3D scanner for optical alignment [12] shown in Figure 2-6.

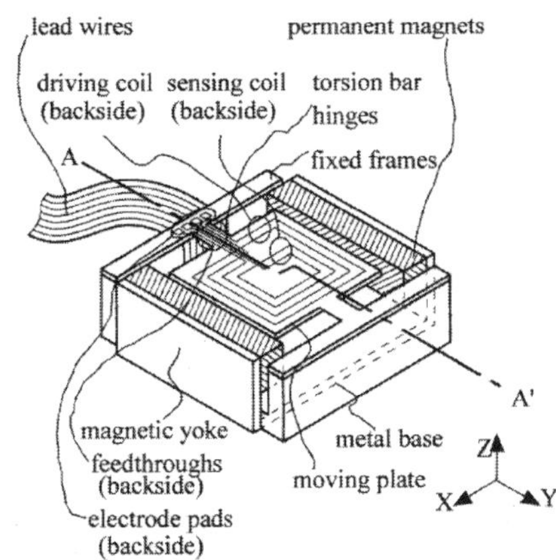

Fig. 2.5. (a) Schematic and (b) photograph of packaged electromagnetic 1D scanner in [11] (Picture courtesy of Hiroshi Miyajima).

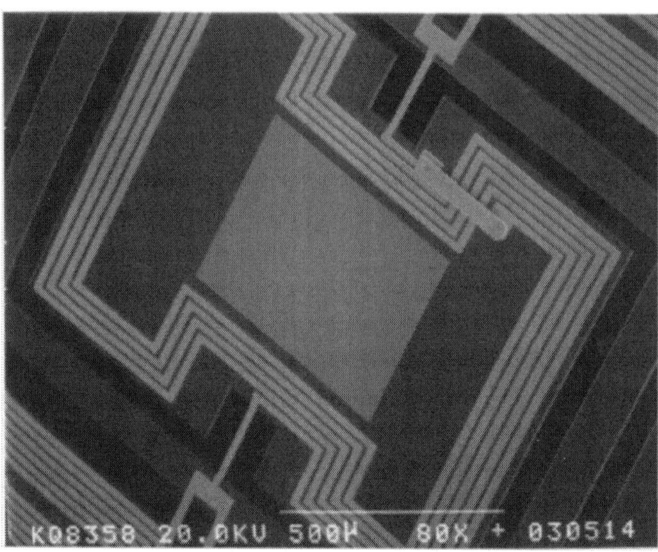

Fig. 2.6. SEM of the miniature electromagnetic 3D scanner in [12] (Picture courtesy of Il-Joo Cho).

2.3 Thermal actuation

Thermal actuation utilizes the mismatch between thermal expansion coefficients of materials, which yields structural stress after temperature change. The structure deforms due to this built-in stress. The major advantage of thermal actuation is its ability to generate large deflection. Electrical current injection is one of the common mechanisms used for heating up the structure. However, temperature control and power consumption are issues for this type of actuators. Several new electro-thermal micromirror and actuators have been reported [13,14,15,16,17,18] by researchers from University of Florida. For example, shown in Figure 2-7, the electrothermal bimorph structure in [17] is based on two thin-film materials with different coefficients of thermal expansion(CTE), like Aluminum (high CTE) and SiO2 (low CTE). The heater material can be polysilicon or Pt. The induced thermal bending arc angle (θ_T) of the bimorph beam can be expressed as below

$$\theta_T = \theta_0 - \theta_1 = \beta \cdot \frac{l_b}{t_b} \cdot \Delta\alpha \cdot \overline{\Delta T}$$

Where θ_0 and θ_1 are the arc angles before and after the temperature increase, t_b and l_b are the bimorph thickness and length, $\Delta\alpha$ is the CTE difference of the two bimorph materials, β is the actuation coefficient related to the thickness and biaxial Young's Modulus ratios of the two materials, and $\overline{\Delta T}$ is the average temperature change on the bimorph.

The electrothermal microactuator in [16], shown in Figure 2-8, using three-bimorph actuation mechanism can achieve large lateral-shift-free (LSF) piston motion at low driving voltage (5.3V) with thermal response time around 25ms. The design is potentially suitable for both pure z-axis displacement actuator and large angle tilting mirror at low speed application.

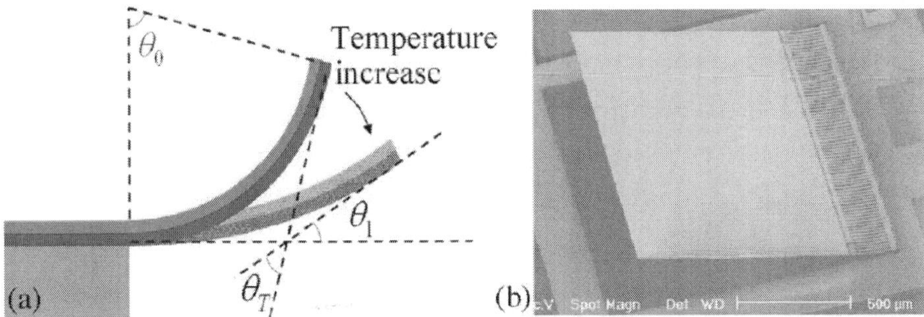

Fig. 2.7. Schematic show actuation principle of (a) a single bimorph and (b) SEM of a bimorph-based micromirror [17] (Picture courtesy of Huikai Xie).

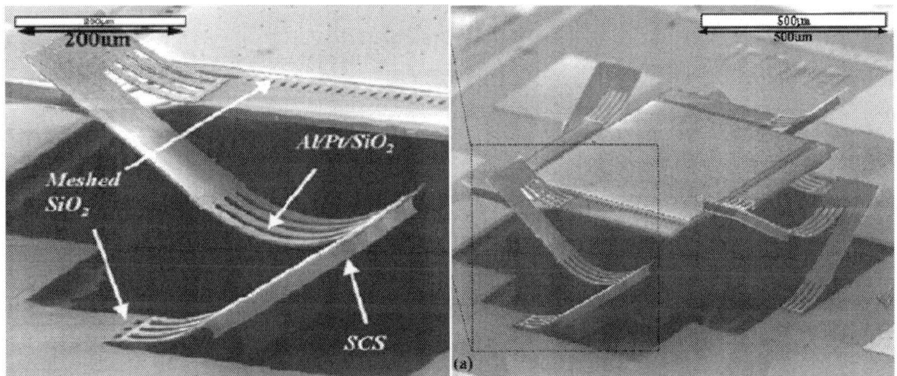

Fig. 2.8. Electrothermal bimorph microactuator with initial elevation in [16] (Picture courtesy of Huikai Xie)

2.4 Piezoelectric actuation mechanisms

Piezoelectric material deforms when electric field is applied across the structure. This property can be used as the driving mechanism in MEMS and NEMS. Different kinds of piezoelectric scanners and actuators have been demonstrated [19,20,21,22]. Using piezoelectric thin films, researchers at the University of Michigan have developed a novel thin-film lead zirconate titanate (PZT) based large displacement (around 120um) vertical translational microactuator [19]. The microactuator consists of four compound bend-up/bend-down unimorphs to generate z-axis motion of a moving stage. Figure 2-9 and Figure 2-10 show an example of the thin-film PZT based microactuator with large displacement. The large displacement within small footprint and high bandwidth (fast response) of the actuators at low-voltage and low-power levels should make them useful to a variety of optical applications, like endoscopic microscopy.

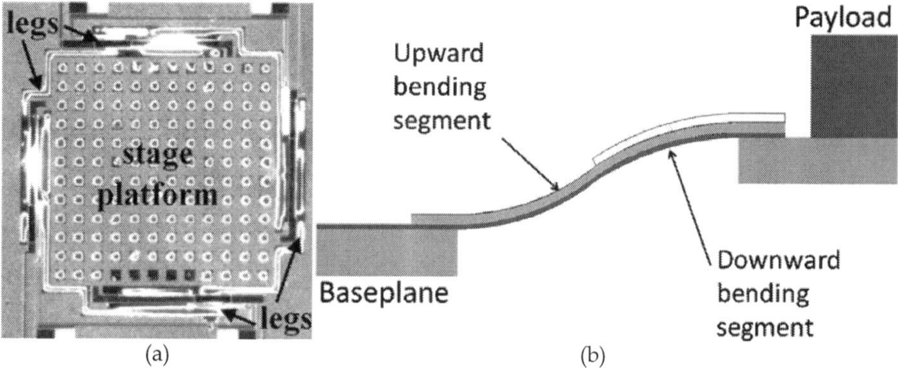

Fig. 2.9. (a) Top-view, piezoelectrically actuated vertical stage, (b) Schematic cross-section of a vertical translational stage in [19] (Picture courtesy of Kenn Oldham).

Fig. 2.10. (a) Top view and oblique view of the corner of a prototype vertical actuator under 0 and 20 V. (b) Scanning-electron microscope (SEM) image of end of an actuation leg (full actuator inset), showing the XeF2 etching profile below the device layer

Aluminium nitride (AlN) is another promising potential piezoelectric material for future Optical MEMS applications [21,22]. One of the important advantages of AlN is that it is very suitable for full CMOS compatible MEMS processes. Currently, researchers from IPMS Germany are developing quasistatic deformable mirrors by actively coupling lateral strain in micro machined AlN based membranes.

2.5 Other actuation mechanisms

Magnetostrictive materials transduce or convert magnetic energy to mechanical energy and vice versa. As a magnetostrictive material is magnetized, it strains; i.e., it exhibits a change in length per unit length. Conversely, if an external force produces a strain in a magnetostrictive material, the material's magnetic state will change. This bidirectional coupling between the magnetic and mechanical states of a magnetostrictive material provides a transduction capability that is used for both actuation and sensing devices. It has the advantage of remote actuation by magnetic fields. 2D optical scanners using magnetostrictive actuators have been reported [23].

3. Applications

3.1 Display, imaging, and microscopy

3.1.1 Texas Instruments' Digital Micromirror Device (DMD)

The Digital Micromirror Device (DMD) started in 1977 by Texas Instruments. The DMD technology has helped revolutionize projector systems by dramatically decreasing costs and increasing performance from traditional projector systems which are based on LCD technology. The research initially focused on deformable mirror device. Eventually DMD becomes the preferred device. TI uses Digital Light Processing™ (DLP) to denote optical projection displays enabled by the DMD technologies [24].

The DMD is a reflective spatial light modulator (SLM) which consists of millions of digitally actuated micromirrors. Each micromirror is controlled by underlying complementary metal-oxide-semiconductor (CMOS) electronics, as shown in Figure 3-1. A DMD panel's micromirrors are mounted on tiny hinges that enable them to tilt either toward the light source (ON) or away from it (OFF) depending on the state of the static random access memory (SRAM) cell below each micromirror. The SRAM voltage is applied to the address electrodes, creating an electrostatic attraction to rotate the mirror to one side or the other. The details of operating principle, design, fabrication, and testing of DMD have been discussed in [25] and will not be repeated here.

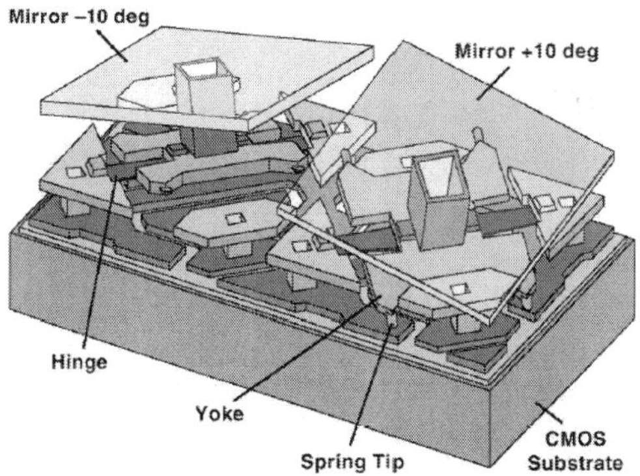

Fig. 3.1. Schematic drawing of two DMD mirror with underlying structures (Picture courtesy of TI. Reprinted from [24] with permission).

In projection systems, brightness and contrast are the two primary attributes that impact the quality of the projected image. The DMD has a light modulator efficiency in the range of 65%, and enables the contrast ratio ranging from 1000:1 to 2000:1. Because of the fast switching speed of the mirror, it enables the DLP to have a wide range of applications in video and data projectors, HDTVs, and digital cinema. Though, DMD was developed primarily for projection display applications, there are some interesting non-display applications. An emerging new DMD application is volumetric display, in which DMDs are used to render three-dimensional images that appear to float in space without the use of encumbering stereo glasses or headsets. It is realized by using 3 DMD's to create 3D images viewed without glasses or headsets [26,27]. DMD also has applications in maskless lithography and telecom. Traditionally, the patterns in lithographic applications, such as print settings, printed circuit board (PCB) and semiconductor manufacturing, have been provided via film or photomasks. However, it is desirable to directly write on the UV-sensitive photoresist directly from digital files. DMD can be used as the spatial light modulator to generate the designed patterns [28]. For maskless lithography in sub-100nm semiconductor manufacturing, analog micromirror arrays with either tilting or piston motions are needed. Smaller mirror size is also desired.

DMD also has interesting applications in microscopy and spectroscopy. In microscopy application, DMD is used as a spatial modulator of the incident or collected light rays. It replaces the aperture in conventional optical system. The DMD can shape or scan either the illumination or collection aperture of an optical microscope thus to provide a dynamic optical system that can switch between bright field, dark field and confocal microscopy [29,30,31,32]. In spectroscopy application, the DMD is used as an adaptive slit selectively routing the wavelength of interest to a detector. It can also chop the light reaching the detector to improve detection sensitivity [33].

3.1.2 GLV display

The schematic of the Grating Light Valve™ (GLV™) shown in Figure 3-2 is a diffractive spatial light modulator [34]. The GLV device switches and modulates light intensities via diffraction rather than by reflection. Distinct advantages of GLV include high speed modulation, fine gray-scale attenuation, and scalability to small pixel dimensions.

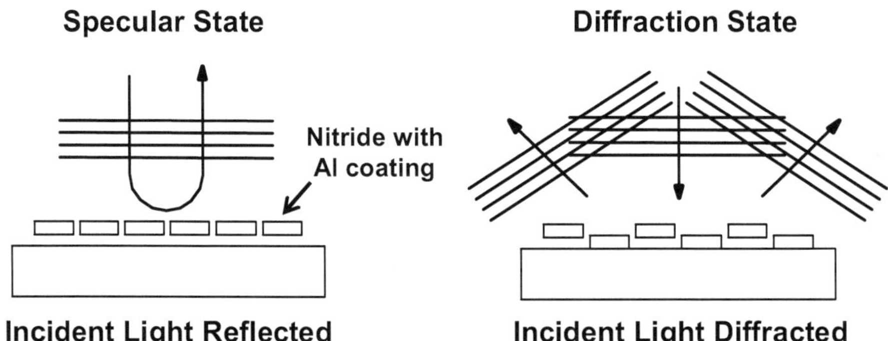

Fig. 3.2. Cross-section of the GLV device showing the specular and diffraction states (after [34]).

The GLV device is built on a silicon wafer and is comprised of many parallel micro-ribbons that are suspended over an air gap above the substrate. Alternative rows of ribbons can be pulled down approximately ¼ wavelength to create diffraction effects on incident light by applying an electrical bias between the ribbons and the substrate. When all the ribbons are in the same plane, the GLV device acts like a mirror and incident light is reflected from their surfaces. When alternate ribbons are deflected, the angular direction in which incident light is steered from the GLV device is dictated by the spatial frequency of the diffraction grating formed by the ribbons. As this spatial frequency is determined by the photolithographic mask used to form the GLV device in the fabrication process, the departure angles can be very accurately controlled, which is useful for optical switching applications. The linear deflection of the GLV is quite small, with no physical contact between moving elements, thus avoiding wear and tear as well as stiction problems. There are also no physical boundaries between the pixel elements in the GLV array. When using as a spatial light modulator in imaging applications, this seamless characteristic provides a virtual 100% fill-factor in the image.

The ribbons are made of suspended silicon nitride films with aluminum coating to increase its reflectivity. The silicon nitride film is under tensile stress to make them optically flat. The tension also reduces the risk of stiction and increases their frequency response. The GLV materials are compatible with standard CMOS foundry processes. GLV can be made into one-dimensional or two-dimensional arrays for projection display applications. Today, the GLV technology is used in high resolution display, digital imaging systems and WDM telecommunications [34].

3.1.3 Microvision retinal display and pico-projector

Retinal scanning display (RSD) uses a different approach from other microdisplays. Rather than a matrix array of individual modulator or source for each pixel as seen in liquid crystal display (LCD), organic light-emitting diodes (OLED), and DMD microdisplays, a RSD optimizes a low power light source to create a single pixel and scans this pixel with a single mirror to paint the displayed image directly onto the viewers' retina. With this technique, it offers high spatial and color resolution and very high luminance. There are several papers that provide an overview of the RSD and its applications [35,36]. This technology is developed by Microvision. The RSD systems typically employ two uniaxial scanners or one biaxial scanner. The combinations of two actuation mechanisms, electrostatic (for faster response) and electromagnetic (for larger force) actuations, were selected for a MEMS scanner [37]. Figure 3-3 shows a schematic drawing of the MEMS scanner.

The horizontal scanner (the inner mirror axis) is operated at resonance by using electrostatic actuation. The drive plates are located on the substrate below the MEMS mirror. The inner mirror axis has the resonant frequency of 19.5 kHz with the maximum mechanical scan excursion of 13.4 degrees. The vertical scanner (the outer mirror axis) is magnetically driven by means of permanent magnets within the package and cols with a 60 Hz linear ramp waveform. The magnets need to be positioned carefully and provide sufficient magnetic field to move the mirror to the desired angular deflection. The maximum mechanical scan excursion of 9.6 degrees was achieved on the outer mirror axis. The devices were bulk micromachined utilizing both wet and dry anisotropic etching and electroplating was used to form electromagnetic coils on the outer frame. These scanners must be stiff to remain flat and withstand the forces developed in resonant scanning mode. The dynamic mirror flatness of 0.05 microns rms was measured. The scanners also incorporate piezoresistive strain sensors on the torsion flexures for closed loop control. The scanners are designed to meet SVGA video standards that require 800 x 600 resolution. The design, fabrication, and control details of this bi-axial scanner can be found in [37] and [38].

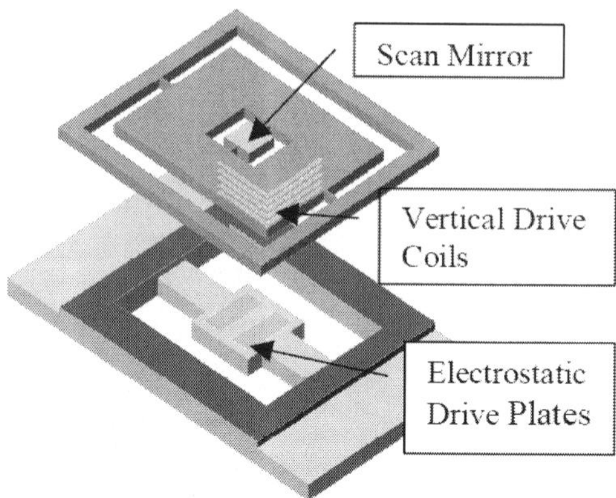

Fig. 3.3. Schematic drawing of the electrostatic/electromagnetic scanner (Picture courtesy of H. Urey. Reprinted from [37] with permission).

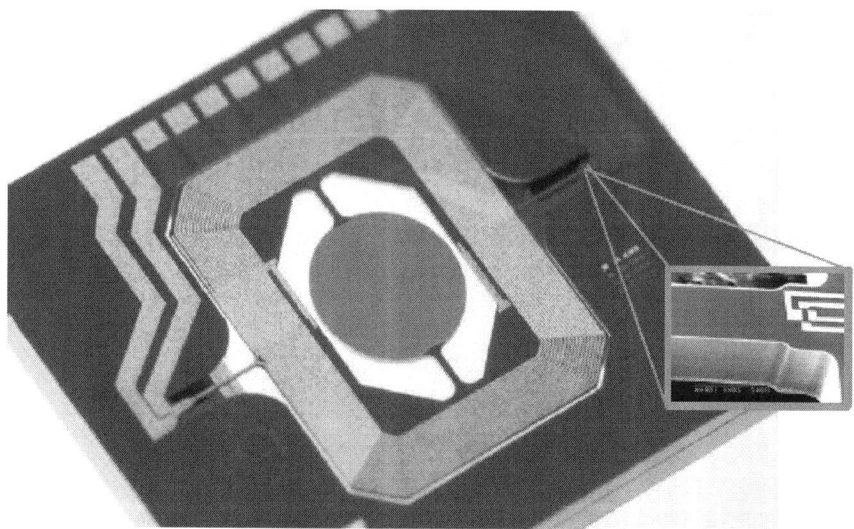

Fig. 3.4. MEMS scanning mirror die and close up of PZR strain sensor used for vertical scan position feedback. (Picture courtesy of Davis, W.O.. Reprinted from [39] with permission).

A MEMS scanner based pico-projector system has been commercialized by Microvision [39]. Figure 3-4 shows a real magnetic dual-axis scanner die with piezo-resistor(PZR) strain sensor for 2D raster scanning engine system.

3.1.4 Confocal microscopy and OCT (Optical Coherence Tomography)

Confocal microscopy offers several advantages over conventional optical microscopy, including controllable depth of field, the elimination of image degrading out-of-focus information, and the ability to collect serial optical sections from thick specimens. The concept was introduced by Marvin Minsky in the 1950's when he was a postdoctoral fellow at Harvard University. In 1957, he patented his "double-focusing stage-scanning microscope" in 1957 [40] which is the basis for the confocal microscope.

In a conventional widefield microscope, the entire specimen is bathed in light from a mercury or xenon source, and the image can be viewed directly by eye or projected onto an image capture device or photographic film. In contrast, the method of image formation in a confocal microscope is fundamentally different. Figure 3-5 shows the schematic drawing of the confocal imaging system. Illumination is achieved by scanning one or more focused beams of light, usually from a laser or arc-discharge source, across the specimen. This point of illumination is brought to focus in the specimen by the objective lens, and laterally scanned using some form of scanning device under computer control. The sequences of points of light from the specimen are detected by a photomultiplier tube (PMT) through a pinhole (or in some cases, a slit), and the output from the PMT is built into an image and displayed by the computer.

The scanning confocal optical microscope has been recognized for its unique ability to create clear images within thick, light scattering objects. This capability allows the confocal microscope to make high resolution images of living, intact tissues and has led to the expectation that confocal microscopy has become a useful tool for in vivo imaging.

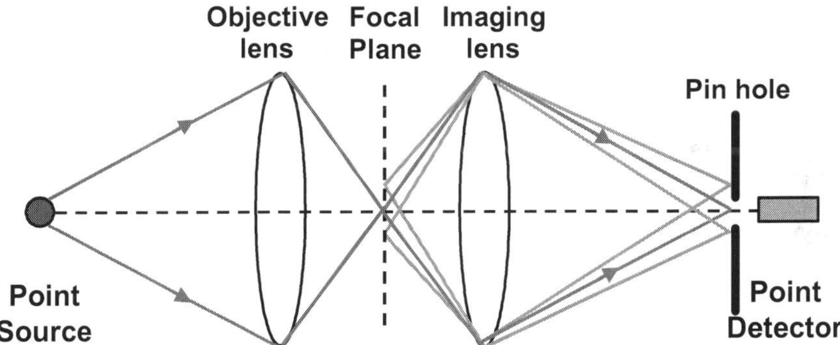

Fig. 3.5. Schematic drawing shows the concept of the confocal imaging system.

The first compact rectangle shape endoscope (2.5 mm (w) x 6.5 mm (l) x 1.2 mm (t)) based on MEMS scanning mirrors was developed by D.L. Dickensheets, et al. [41]. The architecture of the micromachined confocal optical scanning microscope, illustrated in Figure 3-6, consists of a single-mode optical fiber for illumination and detection, two cascaded one-dimensional bulk micromachined electrostatic scanners with orthogonal axes of rotation to accomplish *x-y* scanning, and a binary transmission grating as the objective lens. The maximum mechanical scanned angle is ±2 degrees. The resonant frequencies of both axes are over 1 kHz.

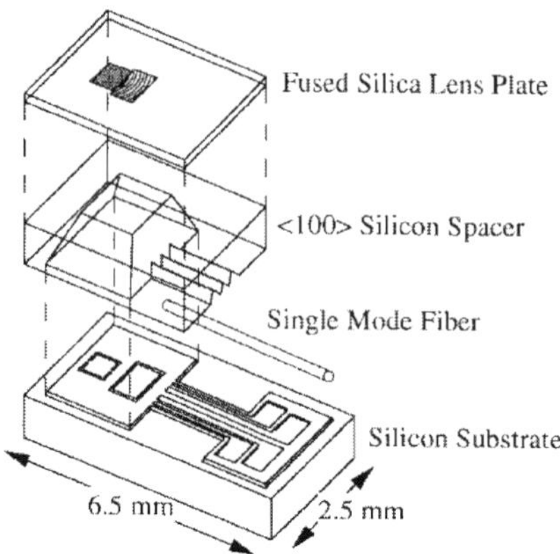

Fig. 3.6. Schematic drawing of the endoscope head showing various components of the assembly (Picture courtesy of D.L. Dickensheets. Reprinted from [41] with permission).

Later, Olympus Optical Company, Ltd. developed the first commercialized cylindrical shape confocal endoscope with an outside diameter of 3.3 mm and a length of 8 mm [42]. Figure 3-7 shows a cross sectional drawing of the endoscope head. The scanner is a gimbal based two-dimensional bulk micromachined electrostatic scanner [43] with the size of 1.3 mm x 1.3 mm. The mirror has the diameter of 500 um and resonant frequency of 3 kHz. The maximum mechanical scanned angle is ±3 degrees.

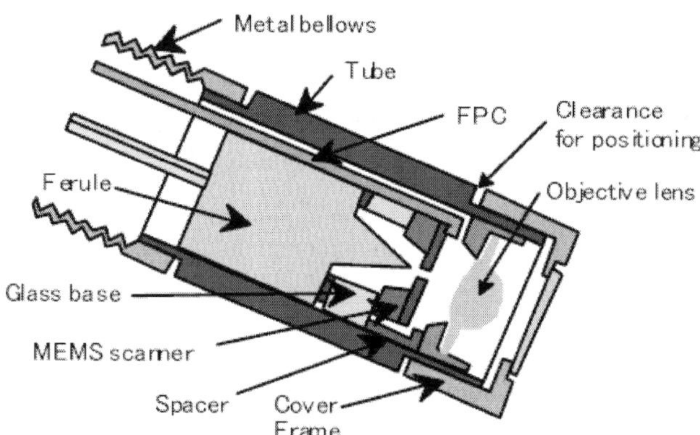

Fig. 3.7. A cross sectional drawing of the endoscope head (Picture courtesy of Olympus Optical Company, Ltd. Reprinted from [42] with permission).

Recently, researchers developed the dual-axis confocal (DAC) configuration,in Figure 3-8, to overcome conventional confocal microscope's limitations for endoscope compatibility and in vivo imaging by utilizing two optical fibers oriented along the intersecting optical axes of

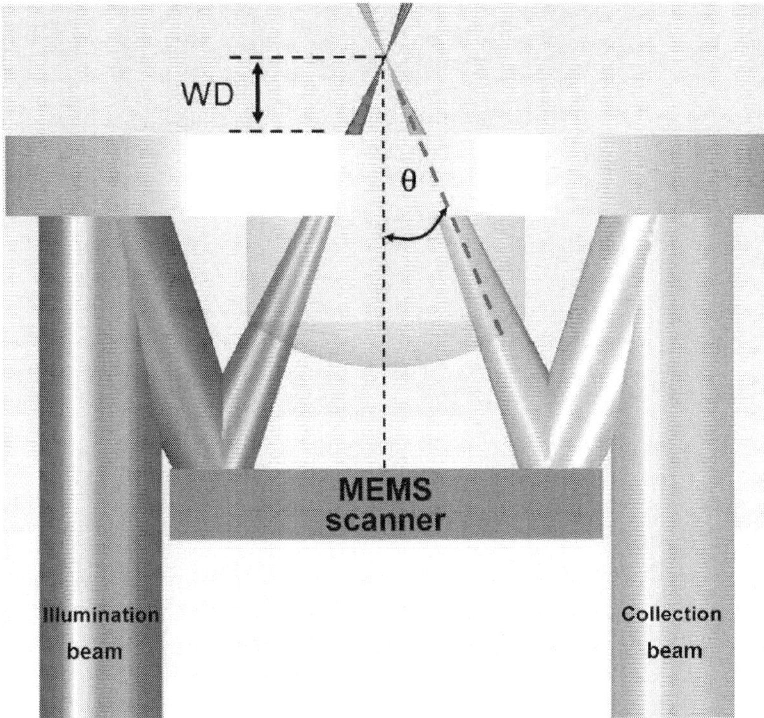

Fig. 3.8. DAC microscope architecture (Reprinted from [49] with permission).

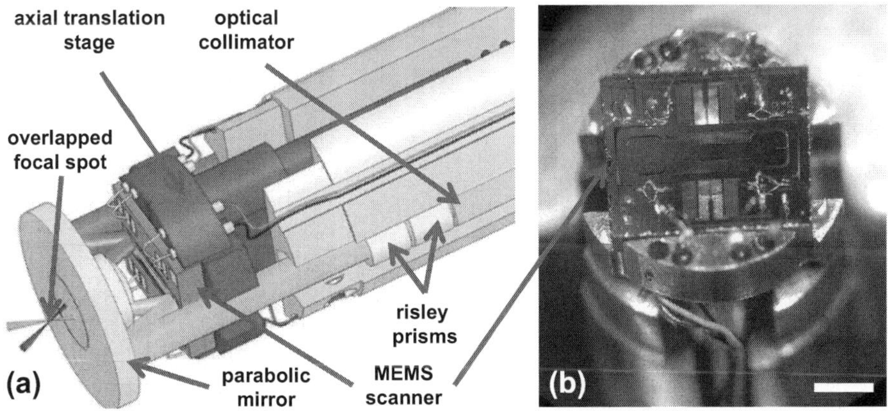

Fig. 3.9. A 5.5-mm diameter DAC microendoscope scanhead. (a) Two collimated beams are focused by a parabolic mirror. Real-time en face scanning is performed by a 2D MEMS scanner. (b) A photograph of the microendoscope without its cap shows a 2D MEMS scanner mounted on the axial translation stage, scale bar 3mm (Reprinted from [49] with permission).

two low-NA objectives to spatially separate the light paths for illumination and collection [44,45,46,47,48]. A state-of-the-art miniature (OD 5mm) 2D MEMS scanner based near-infrared dual-axis confocal microscopy system with z-axis focusing has been demonstrated, using 2D MEMS scanner. Figure 3-9 shows the schematic drawing of the whole system. It is the first time that MEMS device based endoscopy system, shown in Figure 3-10, has been used for in vivo imaging in human patients [49].

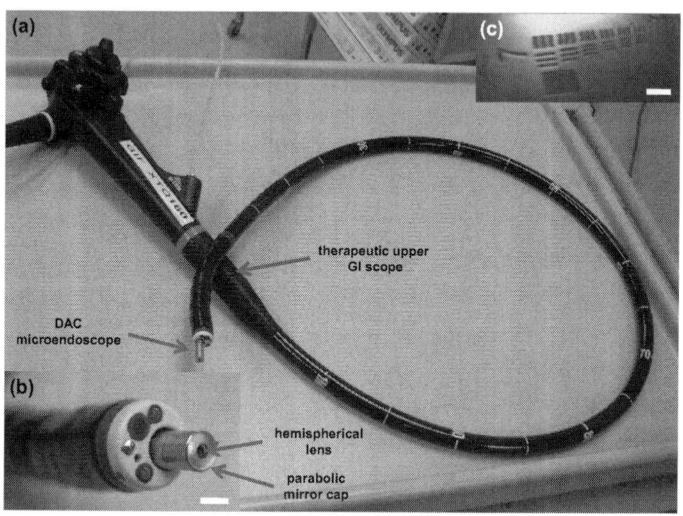

Fig. 3.10. (a) A DAC microendoscope is passed through the instrument channel of an Olympus XT-160 therapeutic endoscope that has a 6-mm diameter instrument channel. (b) Distal end of an endoscope shows the protruding DAC microendoscope, scale bar 5 mm. (c) Cropped reflectance image of a 1951 United States Air Force resolution test chart collected with the DAC microendoscope shows a transverse resolution of 5 μm, scale bar 20 μm (Reprinted from [49] with permission).

Optical coherence tomography (OCT) is an optical imaging technique that is analogous to B-mode medical ultrasound except that it uses low coherent light (low coherence interferometry) instead of sound. Generally, OCT imaging is performed using a fiber-optic Michelson interferometer with a low-coherence-length light source. Figure 3-11 shows the schematic drawing of the Michelson-type interferometer. One interferometer arm contains a modular probe that focuses and scans the light onto the sample, also collecting the backscattered light. The second interferometer arm is a reference path with a translating mirror or scanning delay line. Optical interference between the light from the sample and reference paths occurs only when the distance traveled by the light in both paths matches to within the coherence length of the light [50]. The interference fringes are detected and demodulated to produce a measurement of the magnitude and echo delay time of light backscattered from structures inside the tissue. The obtained data constitute a two-dimensional map of the backscattering or back reflection from internal architectural morphology and cellular structures in the tissue. Image formation is obtained by perform repeated axial measurement at different transverse positions as the optical beam is scanned across the tissue.

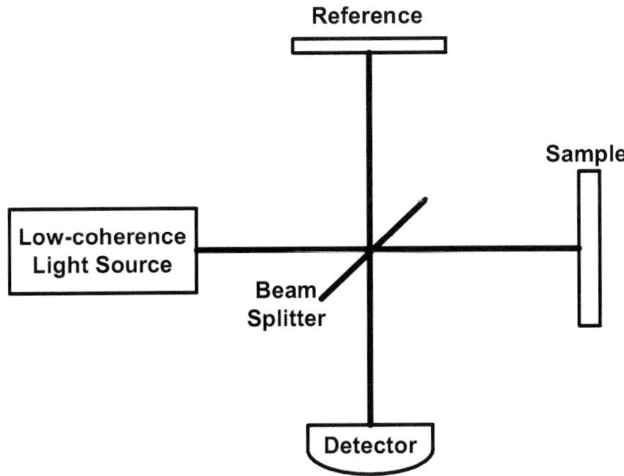

Fig. 3.11. Schematic drawing of a Michelson-type interferometer.

Since its initial use for imaging the transparent and low-scattering tissue of eyes [51], OCT has become attractive for noninvasive medical imaging. Real time in vivo endoscope based OCT imaging systems [52] of the gastrointestinal and respiratory tracts of a rabbit were demonstrated with an axial resolution of 10 um and sensitivity of more than 100 dB. The catheter-endoscope consisted of an encased, rotating hollow cable carrying a single-mode optical fiber.

Previously, the scanning element inside the OCT probe head used in clinical trials uses a spinning reflective element to scan the light beam across the tissue in circumferential scan geometry [52, 53]. This scanning arrangement allows the imaging probe to view only targets that are directly adjacent to the probe. The scan control of the probe is located outside the probe (proximal actuation). This type of actuation has some drawbacks such as a non-uniform and slow speed scanning. In addition, by applying a rotating torque on the optical fiber, it can cause unwanted polarization effects that can degrade image quality.

By using MEMS scanning mirrors, the scan control is located inside the probe head (distal actuation) which can reduce the complexity of scan control and potentially have a lower cost. Because of the scanner's miniature size, the overall diameter of the endoscope can be very small (< 5 mm). High speed and large transverse scan can also be achieved which enables real time in vivo imaging and large field of view, respectively.

Therefore, a need for compact, robust, and low cost scanning devices for endoscopic applications has fueled the development of MEMS scanning mirrors for OCT applications. Y. Pan, et al. developed a one-axis electro-thermal CMOS MEMS scanner for endoscopic OCT [54]. The mirror size is 1 mm by 1 mm. The SEM is shown in Figure 3-12. The bimorph beams are composed of a 0.7-um-thick Al layer coated on top of a 1.2-um-thick SiO_2 layer embedded with a 0.2-um-thick poly-Si layer. The mirror is coated with a 0.7-um-thick Al layer, and the underlying 40-um-thick single-crystal Si makes the mirror flat. The maximum optical scanned angle is 37 degrees (only in one direction).

Later, J.M. Zara, et. al fabricated one dimensional bulk micromachined MEMS scanner [55]. The scanner (1.5 mm long) is a gold-plated silicon mirror bonded on a 30-um-thick flat polyimide surface (2 mm long and 2.5 mm wide) that pivots on 3-um-thick polyimide torsion hinges. Figure 3-13 shows an optical image of the endoscope head. The actuator used to tilt the mirror, the integrated force array (IFA), is a network of hundreds of thousands of micrometer-scale deformable capacitors. The capacitive cells contract because of the presence of electrostatic forces produced by a differential voltage applied across the capacitor electrodes.

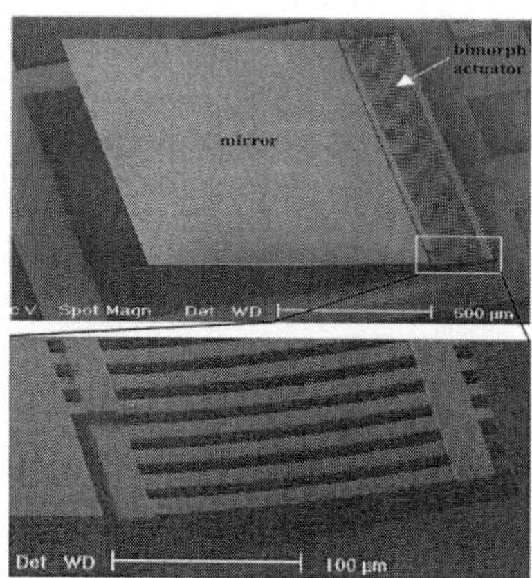

Fig. 3.12. SEM of electro-thermal CMOS MEMS scanner with an inset shows a close-up view of the bending springs (Picture courtesy of T. Xie. Reprinted from [54] with permission).

Fig. 3.13. An optical image of the endoscope head (Picture courtesy of J.M. Zara. Reprinted from [55] with permission).

Researchers at MIT and UCLA [56] have developed the first two-dimensional endoscopic MEMS scanner for high resolution optical coherence tomography. The two dimensional scanner with angular vertical comb actuators (AVC) is fabricated by using surface and bulk micromachining techniques [57]. The angular vertical comb (AVC) bank actuators provide high-angle scanning at low applied voltage [58]. The combination of both fabrication techniques enables high actuation force, large flat micromirrors, flexible electrical interconnect, and tightly-controlled spring constants [58,59]. The schematic drawing of the 2D scanner is illustrated in Figure 3-14. An single-crystalline silicon (SCS) micromirror is suspended inside a gimbal frame by a pair of polysilicon torsion springs. The gimbal frame is supported by two pairs of polysilicon torsion springs. The four electrically isolated torsion beams also provide three independent voltages (V_1 to V_3) to inner gimbals and mirrors. The torsion spring is 345 μm long, 10 or 12 μm wide, and 3.5 μm thick. The scanner has 8 comb banks with 10 movable fingers each. The finger is 4.6 μm wide, 242 μm long, and 35 μm thick. The gap spacing between comb fingers is 4.4 μm. The mirror is 1000 μm in diameter and 35 μm thick. The AVC banks are fabricated on SCS. The movable and fixed comb banks are completely self-aligned [58].

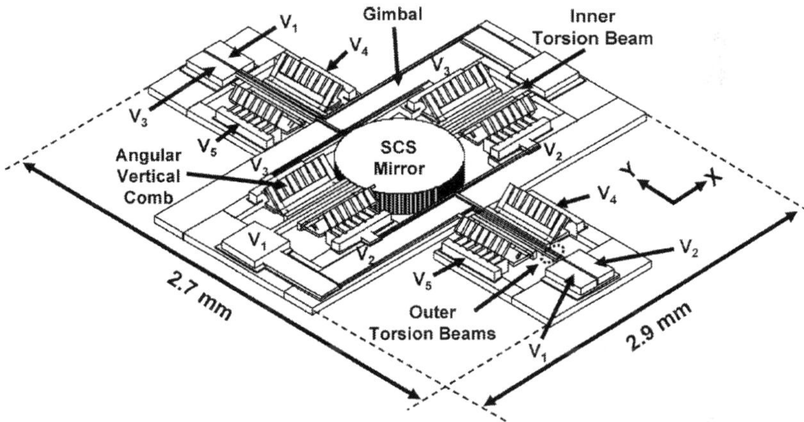

Fig. 3.14. Schematic of 2D AVC gimbal scanner (Reprinted from [58] with permission).

The endoscope head is 5-mm in diameter and 2.5-cm long, which is compatible with requirements for minimally invasive endoscopic procedures. Figure 3-15 shows a schematic of the fiber coupled MEMS scanning endoscope. The compact aluminum housing can be machined for low cost and allows precise adjustment of optical alignment using tiny set screws. The optics consists of a graded-index fiber collimator followed by an anti-reflection coated achromatic focusing lens producing a beam diameter of ~ 12 um [56,57].

The 2D MEMS scanner is mounted at 45 degrees and directs the beam orthogonal to the endoscope axis in a side-scanning configuration similar to those typically used for endoscopic OCT procedures. Post-objective scanning eliminates off-axis optical aberration encountered with pre-objective scanning. Figure 3-16 shows a scanning electron micrograph of the 2D AVC scanner located inside the endoscope package. The large 1-mm diameter mirror allows high-numerical-aperture focusing.

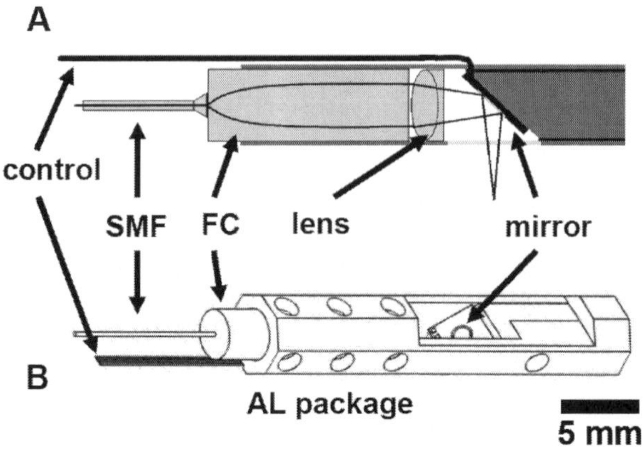

Fig. 3.15. Schematic drawing of the endoscope head (Reprinted from [56] with permission).

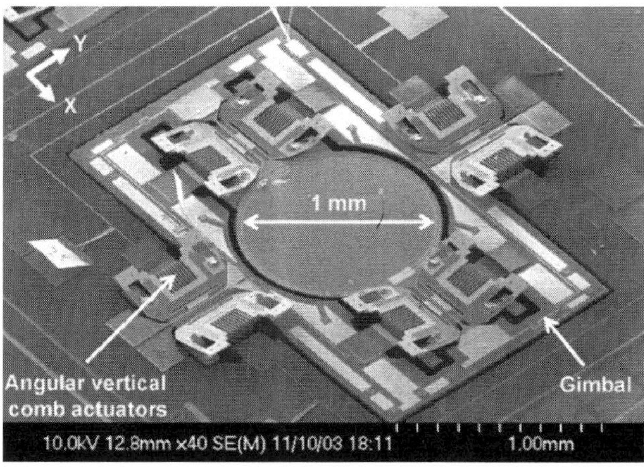

Fig. 3.16. SEM of 2D AVC gimbal scanner (Reprinted from [58] with permission).

Recently, Huikai Xie et. al developed a miniature endoscopic optical coherence tomography probe system employing a two-axis microelectromechanical scanning mirror with through-silicon vias (TSV). The new scanner's design (Figure 3-17) improves the assembly and package of the probe.

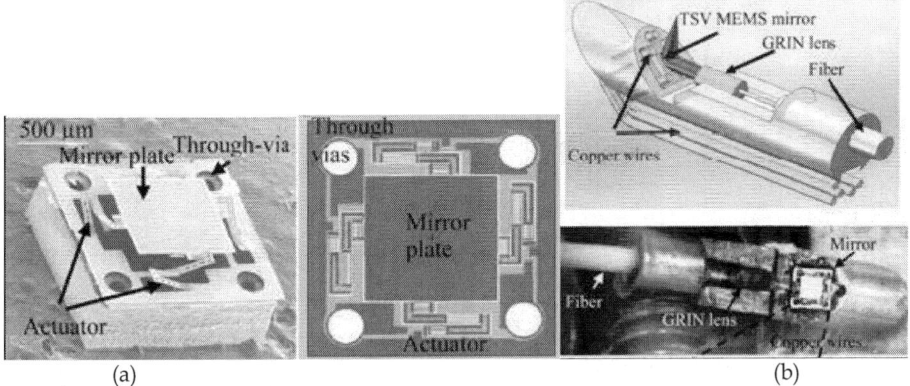

(a) (b)

Fig. 3.17. Electrothermal MEMS OCT reported by Liu et al. (a) SEM and schematic of through vias electrothermal MEMS mirror, (b) 2.6mm probe 3D model and assembled probe, (Picture courtesy of Huikai Xie. Reprinted from [18] with permission).

3.1.5 Adaptive optics

Optical MEMS devices offer a promising alternative to piezoelectric and other deformable mirror types used in adaptive optics applications. Adaptive optics refers to optical systems which adapt to compensate for undesirable optical effects introduced by the medium between the object and its image. It provides a means of compensating for these effects, leading to appreciably sharper images approaching the theoretical diffraction limit. These efforts include wave front correction, aberration cancellation etc. While sharper images come with an additional gain in contrast for astronomy, where light levels are often very low, this means fainter objects can be detected and studied [60, 61]. Other interesting applications used adaptive optics are confocal microscopy [62], adaptive laser wavefront correction [63], cryogenic adaptive optics [64], and human vision [65]. Several groups have developed MEMS deformable mirrors.

A two-level silicon surface micromachining approach was employed by researchers at Boston University to produce MEMS deformable mirrors using an original architecture described in Figure 3-18 [66,67]. The kilo-pixel spatial light modulator is made up of 1024 individually addressable surface-normal electrostatic actuators with center posts that support individual optical mirror segments. Each electrostatic actuator consists of a silicon membrane anchored to the substrate on two sides above a silicon electrode. These devices were manufactured at a commercial MEMS foundry [68]. A post centered on each actuator supports a 338μm x 338μm x 3μm optically coated mirror segment. The spatial light modulator has an aperture of 10 mm, an actuator stroke of 2 μm, and a position repeatability of 3 nm. The resonant frequency of the mirror is around 60 kHz.

At Jet Propulsion Laboratory (JPL), researchers developed a single crystal silicon continuous membrane deformable mirror with underlying piezoelectric unimorph actuators as shown in Figure 3-19 [69]. A PZT unimorph actuator of 2.5 mm in diameter with optimized PZT/Si thickness and design showed a deflection of 5 μm at 50V. Deformable mirrors consisting of 10μm thick single-crystal silicon membranes supported by 4×4 actuator arrays were fabricated and optically characterized. An assembled deformable mirror showed a stroke of 2.5μm at 50V with a resonant frequency of 42 kHz.

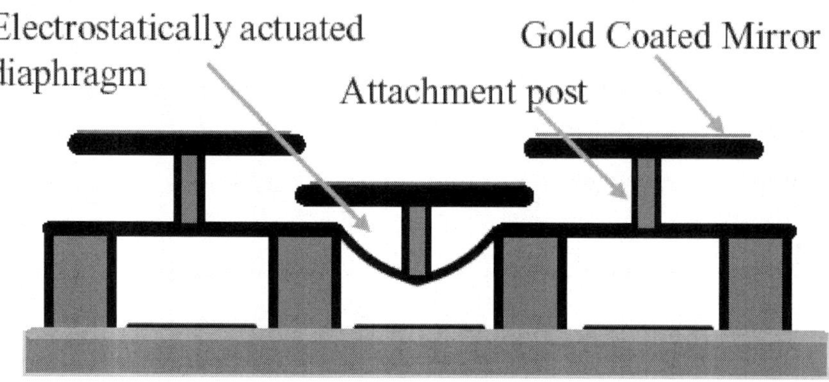

Fig. 3.18. Schematic cross section of a gold coated spatial light modulator with a central deflected actuator (Picture courtesy of Thomas G. Bifano. Reprinted from [66] with permission).

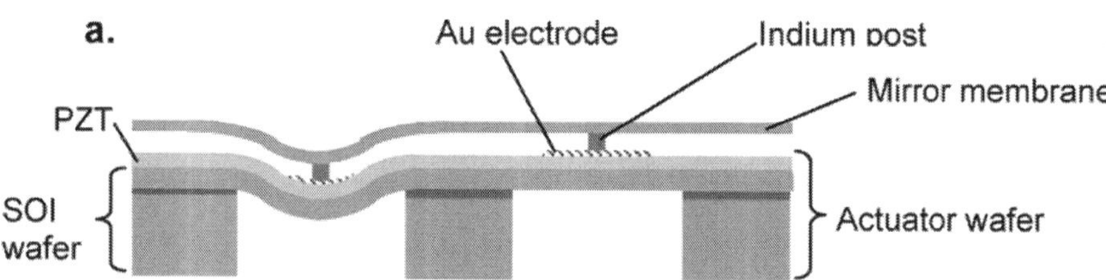

Fig. 3.19. Cross-sectional drawing of the deformable mirror (Reprinted from [69] with permission).

Another MEMS adaptic optical system has been developed by IRIS AO Inc. [70] using Au-coated dielectric-coated segmented MEMS deformable mirrors (DM) for Optical MEMS application. The AO system (Figure 3-20) consists of piston-tip-tilt (PTT) segmented MEMS deformable mirrors (DM) and adaptive optics controllers for these DMs, shown in Figure 3-20. Researchers from IRIS AO Inc. have made many improvements for mirrors to realize open-loop flatten up to 4 nm rms.

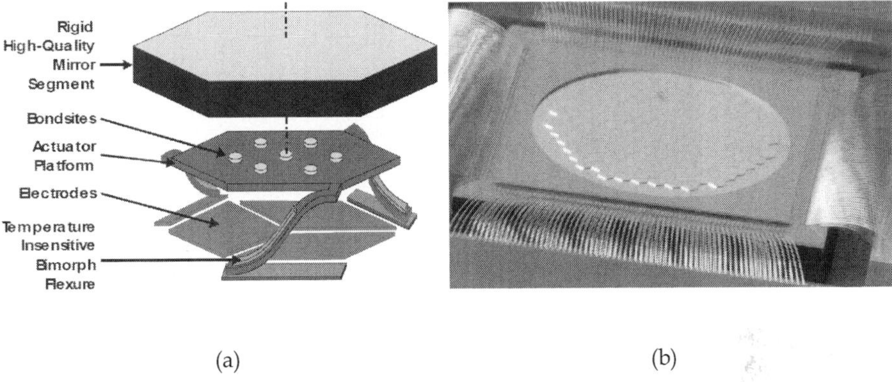

(a) (b)

Fig. 3.20. a) Schematic diagram of a 700 µm diameter (vertex-to-vertex) mirror segment. Scaling is highly exaggerated in the vertical direction. b) Die photograph of a 489-actuator 163-piston/tip/tilt-segment DM with 7.7 mm inscribed aperture (Product name: PTT489-5). (Reprinted from [70] with permission).

3.1.6 Other examples

Researchers at UCLA fabricated two cascaded two dimensional scanners for optical surveying instruments [71]. Currently, cascaded acousto-optic deflectors are used the optical surveying instruments. MEMS scanners are very attractive candidates for replacing those scanners. They offer many advantages, including lower power consumption, smaller size, and potentially lower cost. Optical surveying instruments require mirrors with reasonably large scan range (~ ±6° mechanical), high resonant frequencies (5-10 kHz for fast axis), large radius of curvature, and low supply voltage (< 50V).

Typically, the target, a highly reflective surface consisting of corner cubes, is located several meters to several kilometers away from the instruments. The required scanning angular range is relatively small, on the order of a few degrees. The angular divergence of the measurement laser beam is typically a few milliradians or narrower. Hence the target search system needs to resolve several tens of spots in the entire scan range. Raster scanning has been used because the laser beam needs to search the entire area within the field of view to find the target. A combination of fast and slow scanning scanners has been employed. Both are fabricated by using bulk micromachining technique with 25-µm-thick SOI wafer. The fast scanner has a circular shape and achieved a resonant frequency of 7.5 kHz with the maximum mechanical rotation in DC mode of ±3.2 degrees. The slow scanner has an elongated circular shape and achieved a resonant frequency of 1.2 kHz with the maximum mechanical rotation in DC mode of ±0.74 degree. Figure 3-21 shows the SEM of the fast mirror.

Fig. 3.21. SEM of the fast mirror (Reprinted from [71] with permission)

3.2 Optical communication

3.2.1 2D MEMS switch

The two-dimensional (2D) MEMS optical switch is basically an optical crossbar switch with N^2 micromirrors that can selectively reflect the optical beams to orthogonal output ports or pass them to the following mirrors. They are often referred to as "2D switches" because the optical beams are switched in a two-dimensional plane. This is in contrast to the 3D switch (discussed in the following section) in which the optical beams are steered in three-dimensional space. A generic configuration of the 2D switch is shown in Figure 3-22. The core of the switch is an NxN array of micromirrors for a switch with N input and N output fibers. The optical beams are collimated to reduce diffraction loss. The micromirrors intersect the optical beams at 45°, and can be switched in and out of the optical beam path. The micromirrors are "digital", that is, they are either in the optical beam path (ON) or completely out of the beam path. When the mirror in the i-th row and j-th column (M_{ij}) is ON, the i-th input beam is switched to the j-th output fiber. Generally, only one micromirror in a column or a row is ON. Thus during operation of an NxN switch, only N micromirrors are in the ON position while the rest of the micromirrors are in the OFF position. MEMS 2D switches were first reported by Toshiyoshi and Fujita [72]. Several different ways of switching micromirrors have been reported, including rotating, sliding, chopping, and flipping motions. The switches are usually actuated by electrostatic, electromagnetic, or piezoelectric mechanisms.

2D switches using various types of flip-up (or pop-up) mirrors have been reported [72,73,74,75]. They are realized by either the bulk- [72] or the surface-micromachining [73] technology, or the combination of both [74]. The mirror lies in the plane of the substrate during OFF state and pops up in the ON state. Since the mirror angle is changing continuously during switching, a common challenge for this type of switch is the reproducibility of mirror angle. This is a critical issue as the mirror angles and their uniformity play a critical role in the performance and the scalability of 2D switches. The reproducibility of mirror angles over switching cycles determines the repeatability of insertion loss.

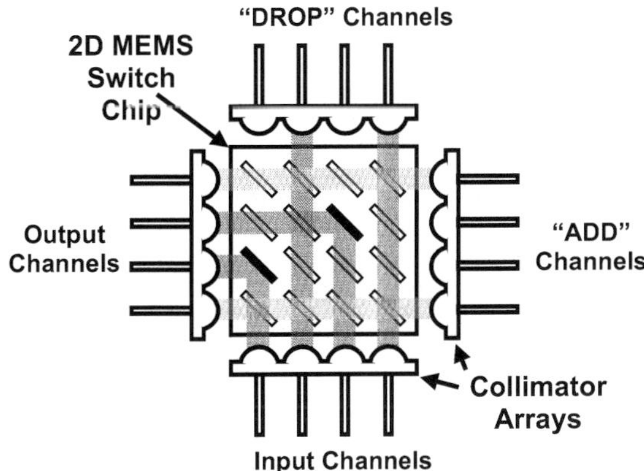

Fig. 3.22. Schematic of 2D MEMS optical switches.

The first 2D matrix optical switch reported by Toshiyoshi and Fujita employed pop-up-type switching elements, as shown in Figure 3-23 [72]. It consists of two bonded wafers. The mirrors are suspended by a pair of torsion beams in the plane of the top wafer. The biasing electrodes are fabricated on the bottom wafer. When a voltage is applied between the mirror and the bottom electrode, the mirror rotates downward by 90° by electrostatic actuation. The mirror angle in the ON (down) state is controlled by a stopper on the bottom wafer. Since the mirror angle is defined by the relative positions of two wafers, precise alignment is necessary to achieve accurate and uniform mirror angles. A single-chip electrostatic pop-up mirror has recently been reported [76]. The actuation and mechanical stopper are realized between a back-flap and a vertical trench etched in the silicon substrate. The angular accuracy and uniformity of their mirrors depend on the etched sidewall profile and the lithographic alignment accuracy.

AT&T Lab has reported surface-micromachined 2D switches with free-rotating hinged mirrors [73,77,78]. The schematic drawing and the SEM of the switch are shown in Figure 3-24. It is fabricated using the MUMPs process. The mirror is pivoted on the substrate by microhinges. A pair of pushrods is used to convert in-plane translations into out-of-plane rotations of the mirrors. The switch is powered by scratch drive actuators [79]. Though scratch drive actuators do not move at high speed, fast switching time is achieved (700 μsec) because only a short traveling distance (22 μm) is needed for the mirror to reach 90°. The free-rotating microhinges have an inherent 0.75-μm clearance between the hinge pins and the staples, which could result in large variations in mirror angles. Using improved design and mechanical stoppers that are insensitive to lithographic misalignment during fabrication, mirror angular repeatability of better than 0.1° was experimentally demonstrated [80]. The mirror flatness was improved by using a multi-layer structure with phosphosilicate glass (PSG) sandwiched between two polysilicon layers. The largest switch size demonstrated is 8x8 due to the foundry-imposed chip-size limits of 1 cm x 1 cm. One of the potential issues is the constant tear and wear of the free-rotating hinges and actuators.

This might affect the reliability of the switch and the accuracy and uniformity of mirror angles over many switching cycles.

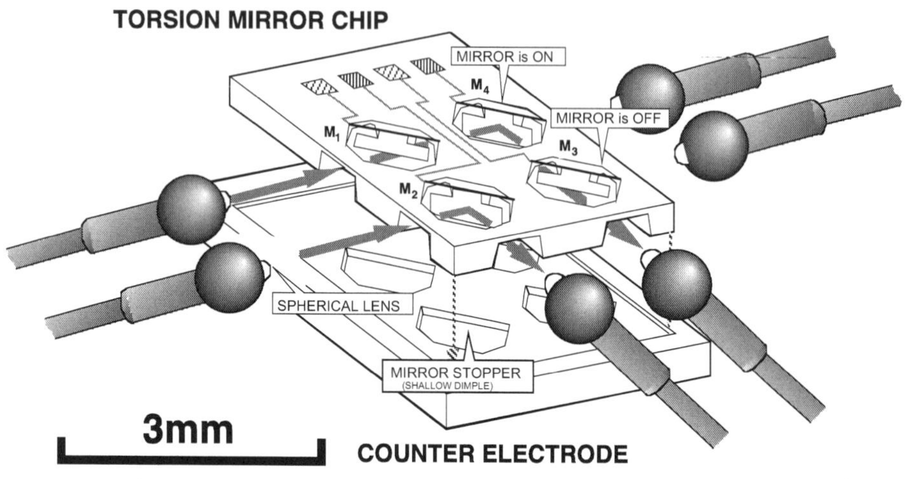

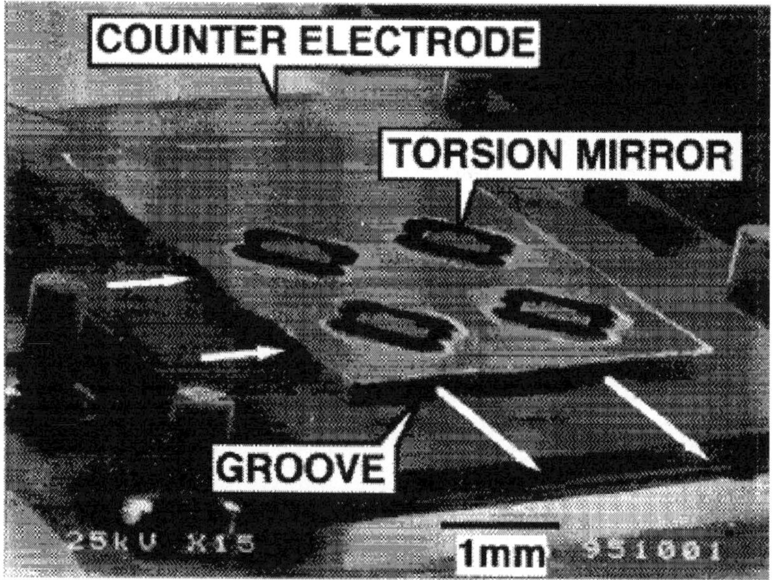

Fig. 3.23. Schematic and SEM of bulk-micromachined 2D switch with free-rotating torsion mirrors (Pictures courtesy of Hiroshi Toshiyoshi. Reprinted from [72] with permission).

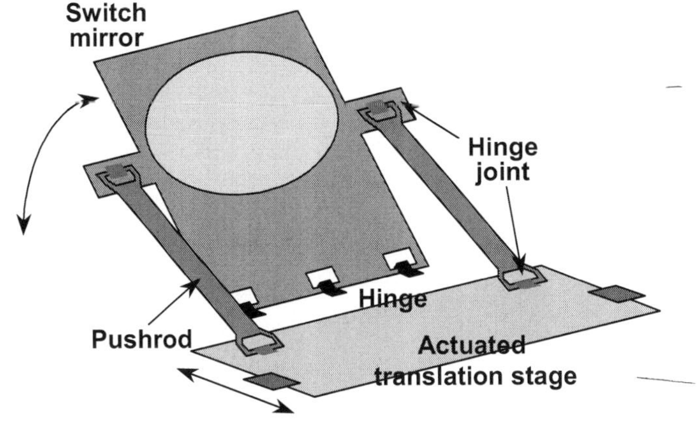

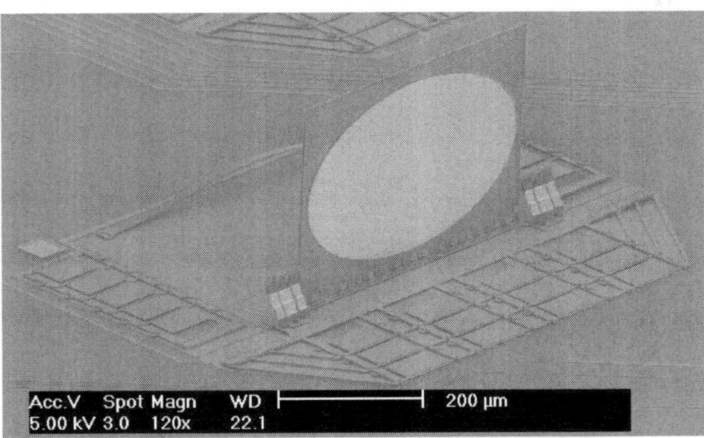

Fig. 3.24. Schematic and SEM of the surface-micromachined free-rotating hinged mirrors reported by AT&T (Picture courtesy of Lih Y. Lin. Reprinted from [73,77] with permission).

On the other hand, the "chopper type" 2D switch employs a vertical mirror whose height can be changed by MEMS actuators [81,82]. The mirror angle is fixed during switching, and excellent repeatability of insertion loss has been reported. Figure 3-25 shows the schematic diagram of OMM's 2D switch [81]. The mirror is assembled vertically at the tip of a long actuator plate. The plate is tilted upward and fixed by micro-latches to raise the mirror height. Large traveling distance is achieved by extending the actuator arm. Several hundred microns displacement can be achieved with this configuration. The switch is actuated electrostatically by applying a voltage between the actuator plate and a bottom electrode on the substrate. The mirror moves in the vertical direction and the mirror angle is maintained at 90° during the entire switching cycle. The actuator is basically a gap-closing actuator. A mechanical stopper defines the lower position of the mirror. OMM employs a curved landing bar with a single point contact to minimize stiction and increase reliability (see Figure 3-25). More than 100 million cycles have been demonstrated with repeatable mirror angle and performance. The landing bar also provides a cushion that helps reduce mirror ringing and improve switching time. They have demonstrated a switching time of 12 ms using a square-wave driving voltage without pre-shaping the waveform. OMM's switch is fabricated with polysilicon surface-micromachining technology. The mirrors and the actuators are batch-assembled into the 3D structures. Figure 3-26(a) shows the SEM of a 16x16 switch. The distribution of mirror angles is shown in Figure 3-26(b). The uniformity is better than ±0.1 degrees for 256 mirrors. The switch is hermetically packaged with optical collimator arrays. Extensive testing has been performed for the packaged switches. The maximum insertion loss is less than 3.1 dB and the crosstalk is less than - 50 dB. Loss variation over the wavelength range of 1280 - 1650 nm range is less than1 dB. Return loss is greater than 50dB, and maximum temperature variation is < 1 dB over a temperature range of 0 - 60°C. Polarization dependent loss (PDL) is < 0.4 dB and polarization mode dispersion (PMD) is < 0.08 ps. Vibration tests show < 0.2 dB change under operation, and 3 axis shock tests confirm no change of operational characteristics under 200 G.

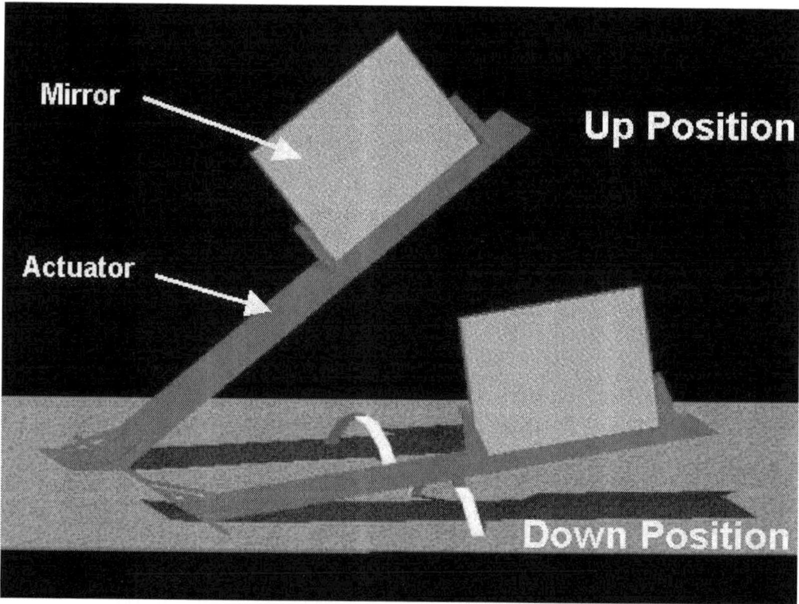

Fig. 3.25. Schematic of a switching element in OMM's 2D switch (Picture courtesy of Li Fan. Reprinted from [81] with permission).

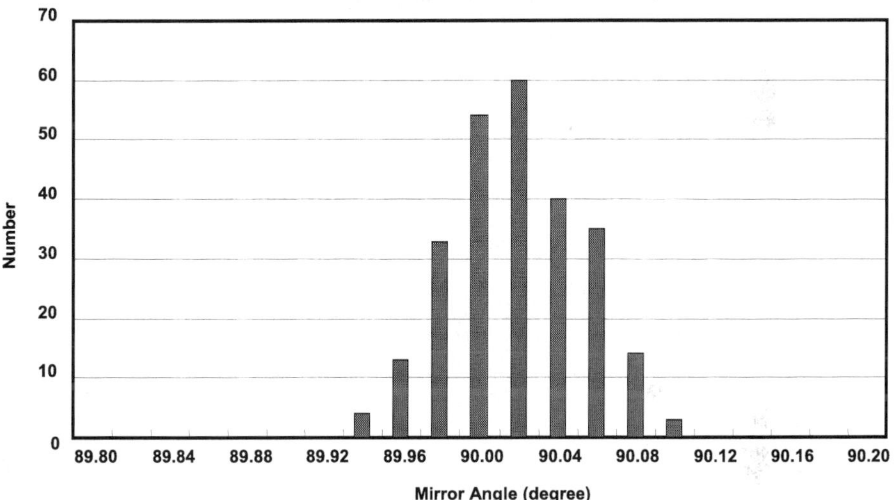

Fig. 3.26. (a) SEM of OMM's 16x16 switch. (b) Measured distribution of mirror angles for the 16x16 switch (Picture courtesy of Li Fan. Reprinted from [81] with permission).

The structure of a 2x2 2D MEMS switch can eventually be simplified so that it requires only one single micromirror between two pairs of orthogonal fibers. A simple, elegant solution for 2x2 switches is using SOI-based Optical MEMS [83,84]. The schematic and SEM picture (only fiber grooves shown) of the 2x2 switch are shown in Figure 3-27. Electrostatic comb drive actuators and vertical micromirrors are fabricated on 75-μm SOI wafers. The mirrors can be coated with metal by angle evaporation to increase their reflectivity. The most critical part of the process is the etching of thin (< 2 μm) vertical mirrors with smooth sidewalls. Thin vertical mirror is required for such a 2x2 switch because the offset of the reflected optical beams from the opposite sides of the mirror caused by the finite thickness of the mirror will introduce additional optical loss. The Institute of Microtechnology (IMT) at

Neuchatel has perfected the mirror etching technology by DRIE [85]. A surface roughness of 36 nm has been achieved. The switch has excellent optical performance: 0.3-0.5 dB optical insertion loss and 500 μsec switching time, and very low polarization dependence [84].

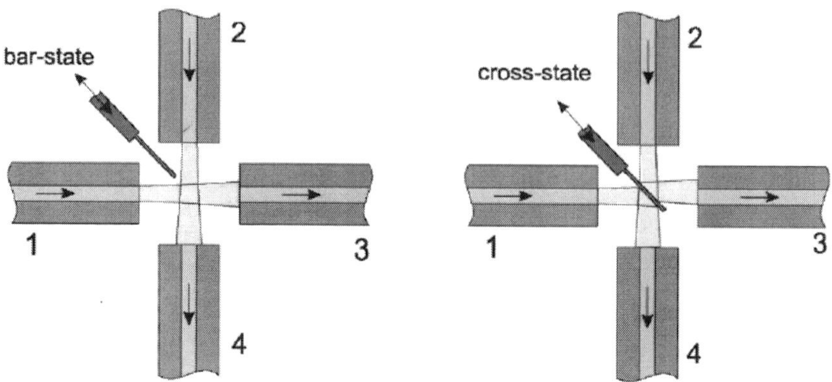

Fig. 3.27. Schematic and SEM of 2x2 switch fabricated by SOI MEMS (Picture courtesy of Nico de Rooij. Reprinted from [83] with permission).

3.2.2 3D MEMS switch

Telecommunication switches with large port count have been the main driver for the two-axis scanner in the past few years. With increasing number of wavelengths and bandwidth in dense wavelength-division-multiplexed (DWDM) networks, there is a need for optical crossconnect (OXC) with large port count [86,87,88]. The dual axis analog scanning capability is the key for these applications since each mirror associated with the input fiber array can point to any mirror associated with the output fiber array. Implementation of NxN OXC using two arrays of N analog scanners is illustrated in Figure 3-28. Even though the implementations may vary, we can always conceptually refer to this illustration. This switch is often called a 3D MEMS switch because the optical beams propagate in three-dimensional space. 3D switch is a better choice for larger port-count NxN OXC compared to 2D switch as the number of mirrors for a 2D switch is N^2. Since the optical path length is independent of the switch configuration, uniform optical insertion loss (2 to 3 dB) can be achieved. The port count of the 3D MEMS switch is limited by the size and flatness of the micromirrors, as well as their scan angle and fill factor. For a complete discussion of scaling laws for MEMS free space optical switches, see Syms [88]. For large port count (approaching 1000 x 1000), single crystal micromirror scanners are necessary to achieve large mirror size with required flatness.

During the telecom boom, several companies have invested heavily to develop 3D MEMS OXC's. These companies include (but not limited to) Lucent Technologies [86,88,89,90,91,92,93,94], Corning [95,96,97], NTT Corp. [98,99], Fujitsu Laboratories, Ltd. [100,101], Tellium, Inc. [102,103,104], and Calient Networks [105]. They have demonstrated various designs of two-axis scanners, which are the key components of these 3D MEMS switches.

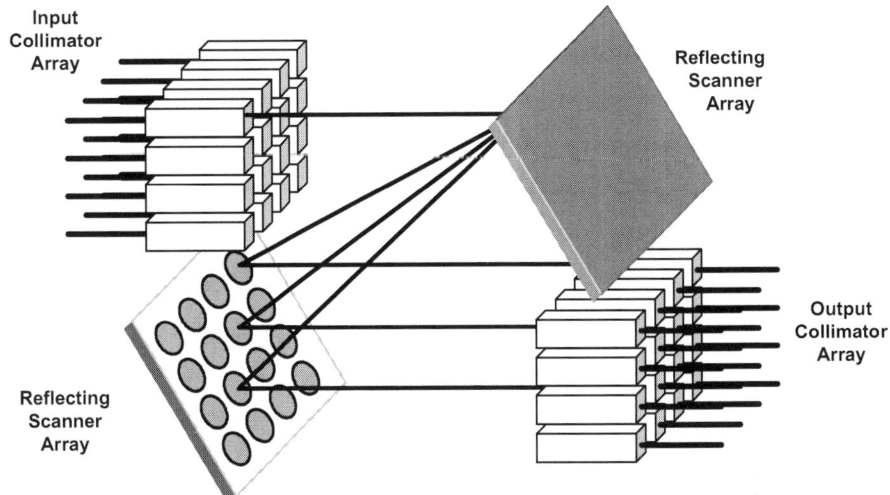

Fig. 3.28. Configuration for 3-D optical switch (NxN) with 2N analog scanning mirrors.

Lucent technologies employed a self-assembly technique which was driven by the residual stress in deposited thin films (Cr/Au on polysilicon) to raise two-axis polysilicon scanners (500 μm mirror diameter) to a fixed position 50 μm above the substrate [106,107]. The scanning electron micrograph (SEM) is shown in Figure 3-29. Two-axis scanning is achieved

Fig. 3.29. SEM of surface-micromachined 2-axis scanners Lucent Technologies (Courtesy of [109] Lucent Technologies Inc. © 2003 Lucent Technologies Inc. All rights reserved.)

by electrostatic force between the mirror and the quadrant electrodes on the substrate. SCS two-axis scanner with long-term stability and high shock resistance has also been developed by Lucent Technologies for 3D MEMS switches [108, 109]. SCS is used to improve the mirror flatness. The long-term stability is achieved by the removal of exposed dielectric to avoid electrostatic charge-up effect (Figure 3-30). The scanning angle is 7 degrees.

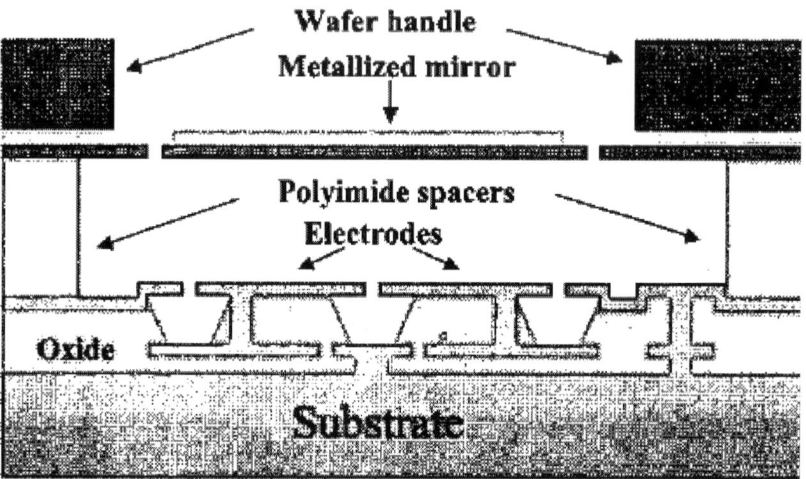

Fig. 3.30. Cross section of the SCS two-axis mirror developed by Lucent Technolgies (Picture courtesy of A. Gasparyan. Reprinted from [108] with permission)

NTT Corp. has reported a two-axis micromirror array driven by terraced electrodes. The mirrors and the electrodes are fabricated on separate chips and then bonded together. The use of terraced electrodes reduces the applied voltage by half, compared to regular parallel-plate-driven mirrors. The mirror is tilted by 5.4 degree at a maximum of 75 V. The resonant frequency of the fabricated mirror is approximately 1 kHz [98].

The two-axis scanner developed by Fujitsu Laboratories, Ltd. is based on vertical comb-drive actuators. SOI with 100-μm top and bottom (substrate) silicon layers has been used to fabricate the device [100]. The top silicon is for the moving comb fingers and mirror while the fixed fingers are made of the bottom silicon. V-shape torsion springs are adopted to improve the lateral stability which is a critical issue in comb-drive actuators. Rotation angle of +/- 5 degrees has been achieved with 60-V driving voltage.

Tellium, Inc. has demonstrated an electrostatic parallel-plate actuated two-axis scanner, featuring nonlinear servo closed-loop control [102]. The nonlinear servo closed-loop control enables the mirror to operate beyond the pull-in angle. Figure 3-31 shows the comparison of switching under open-loop and closed-loop operation. The closed-loop angular trajectory can exceed the pull-in (snap-down) angle and shows no overshoot. They have also developed a two-axis micromirror driven by both sidewall and bottom electrodes [104]. The addition of sidewall electrodes improves the linearity of the DC transfer characteristic. The mirror with sidewall electrodes also exhibits a larger scan angle than that driven by merely bottom electrodes (Figure 3-32).

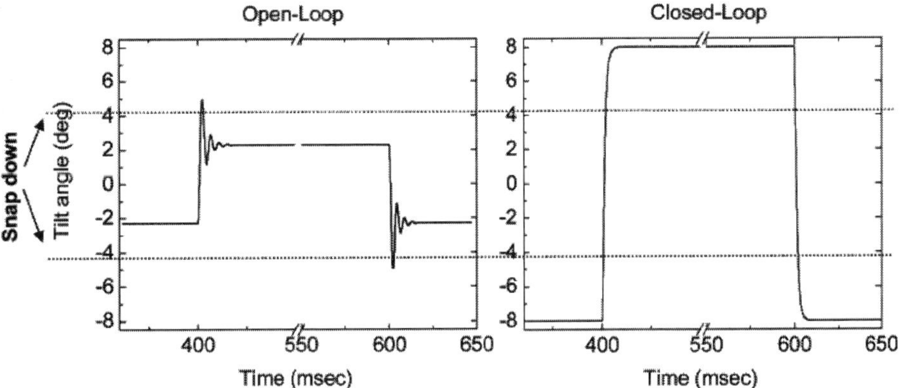

Fig. 3.31. Comparison of switching under open-loop and closed-loop operation (Picture Reprinted from [102] with permission)

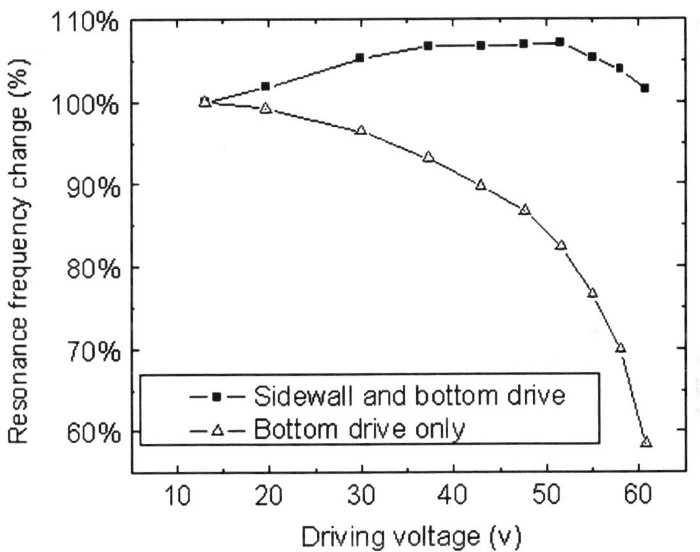

Fig. 3.32. DC transfer characteristics of the two-axis scanner of Tellium, Inc., with and without sidewall driving (Picture courtesy of C. Pu. Reprinted from [104] with permission)

4. Conclusion

In this Chapter, we present the history, common actuators, and fabrications of various Optical MEMS devices for display, imaging, and telecom applications. Each actuation mechanism has its own advantage that can be optimally deployed in each application. For instance, electrostatic actuation requires very low current and achieves short range in motion suiting for low power consumption application. In contrast, magnetic actuation needs high current (implying high power) and achieves long range in motion which is suitable for long range actuation application with less constraints on current driving limit.

The Optical MEMS technology promises to revolutionize nearly every product category with the ability to directly manipulate light or optical signal. With the integration of microelectronics and microoptical components, it is the possible realization of complete systems on a chip.

5. References

[1] Petersen, K. E., "Silicon torsional scanning mirror," IBM J. R&D, vol. 24, pp.631-7, 1980.

[2] M.C. Wu, "Micromachining for Optical and Optoelectronic Systems," Proc. IEEE, Vol. 85 (IEEE Press, Piscataway, N.J., 1997), pp. 1833–1856.

[3] R.S. Muller and K.Y. Lau, "Surface-micromachined microoptical elements and Systems," Proc. IEEE, Vol. 86, (IEEE Press, Piscataway, N.J., 1998), pp.1705-1720.

[4] Degani, O., Socher, E., Lipson, A., Leitner, T., Setter, D. J., Kaldor, S., and Nemirovsky, Y., "Pull-In Study of an Electrostatic Torsion Microactuator," IEEE J. Microelectromech. Syst., vol. 7, no. 4, pp. 373-379, 1998.

[5] Hah, D., Toshiyoshi, H., and Wu, M.C., "Design of Electrostatic Actuators for MOEMS," Proc. SPIE, Design, Test, Integration and Packaging of MEMS/MOEMS 2002, May 2002, Cannes, France

[6] R. A. Conant, J.T. Nee, K. Lau, R.S. Mueller "A Flat High-Frequency Scanning Micromirror", 2000 Solid-State Sensor and Actuator Workshop, Hilton Head, SC, pp. 6-9.

[7] P. R. Patterson, D. Hah, H. Chang, H. Toshiyoshi, M. C. Wu, "An Angular Vertical Comb Drive for Scanning Micromirrors", IEEE/LEOS International Conference on Optical MEMS, Sept. 25-28, 2001, Okinawa, Japan, p.25

[8] Schenk, H., Durr, P., Haase, T., Kunze, D., Sobe, U., Lakner, H., Kuck, H., "Large deflection micromechanical scanning mirrors for linear scans and pattern generation", Selected Topics in Quantum Electronics, IEEE Journal of, Sep/Oct 2000, vol. 6 Issue:5, pp. 715 – 722

[9] Sandner, T., Jung, D., Kallweit, D., Grasshoff, T., Schenk, H., "Microscanner with vertical out of plane combdrive", Optical MEMS and Nanophotonics (OMN), 2011 International Conference on Issue Date: 8-11 Aug. 2011, Istanbul, Turkey, pp. 33 - 34

[10] Judy, J.W. and Muller, R.S., "Magnetically Actuated, Addressable Microstructures," IEEE J. Microelectromech. Syst., vol. 6, no. 3, pp. 249-256, 1997.

[11] H. Miyajima et al., "A MEMS electromagnetic optical scanner for a commercial confocal laser scanning microscope," Journal of Microelectromechanical Systems, Vol. 12, No. 3, pp. 243-251, Jun. 2003.

[12] Il-Joo Cho, Euisik Yoon, "A low-voltage three-axis electromagnetically actuated micromirror for fine alignment among optical devices", Journal of Micromechanics and Microengineering vol. 19, no. 8, Aug. 2009.

[13] A. Jain, et al., "A two-axis electrothermal SCS micromirror for biomedical imaging", 2003 IEEE/LEOS International Conference on Optical MEMS 3, pp.14-15.

[14] Ankur Jain, Hongwei Qu, and Shane Todd, Gary K. Fedder, and Huikai Xie, "Electrothermal SCS micromirror with large-vertical-displacement actuation," 2004 Solid-State Sensor and Actuator Workshop Tech. Digest, June 2-6, Hilton Head, SC, pp.228-231.

[15] Lei Wu, Huikai Xie, "A large vertical displacement electrothermal bimorph microactuator with very small lateral shift" Sensors and Actuators A: Physical, vol 145-146, July-August 2008, pp. 37

[16] Lei Wu, Dooley, S., Watson, E.A., McManamon, P.F., Huikai Xie, "A Tip-Tilt-Piston Micromirror Array for Optical Phased Array Applications" Microelectromechanical Systems, Journal of, vol. 19 Issue:6, pp. 1450 - 1461, Dec. 2010

[17] Jingjing Sun, Shuguang Guo, Lei Wu, Lin Liu, Se-Woon Choe, Brian S. Sorg, and Huikai Xie, "3D In Vivo optical coherence tomography based on a low-voltage, large-scan-range 2D MEMS mirror", Optics Express, Vol. 18, Issue 12, pp. 12065-12075 (2010)

[18] Jingjing Sun, Huikai Xie, "MEMS-Based Endoscopic Optical Coherence Tomography" International Journal of Optics, vol 2011 (2011), Article ID 825629, 12 pages

[19] Zhen Qiu, Jeffrey S Pulskamp, Xianke Lin, Choong-Ho Rhee, Thomas Wang, Ronald G Polcawich and Kenn Oldham, "Large displacement vertical translational actuator based on piezoelectric thin films", 2010 J. Micromech. Microeng. 20 075016

[20] Holger Conrad, Jan Uwe Schmidt, Wolfram Pufe, Fabian Zimmer, Thilo Sandner, Harald Schenk, Hubert Lakner, "Aluminum nitride: a promising and full CMOS compatible piezoelectric material for MOEMS applications", Proc. SPIE 7362, 73620J (2009)

[21] Holger Conrad, Wolfram Pufe and Harald Schenk, "Aluminum Nitride Thin Film Development using Statistical Methods", 2011 International Students and Young Scientists Workshop „Photonics and Microsystems

[22] H.-J. Nam, Y.-S. Kim, S.-M. Cho, Y. Yee, and J.-U. Bu, "Low Voltage PZT Actuated Tilting Micromirror with Hinge Structure," 2002 IEEE/LEOS International Conference on Optical MEMS, Lugano, Switzerland, pp.89-90

[23] Bourouina,T., Lebrasseur,E., Reyne,G., Debray,A., Fujita, H., Ludwig, A., Quandt, E., Muro, H., Oki, T., and Asaoka, A., "Integration of Two Degree-of-Freedom Magnetostrictive Actuatio and Piezoresistive Detection: Application to a Two-Dimensional Optical Scanner," IEEE J. Microelectromech. Syst., vol. 11, no. 4, pp. 355-361, 2002.

[24] L. J. Hornbeck, "Digital Light Processing™ for High Brightness, High Resolution Applications," Proc. SPIE vol. 3013 (Electronic Imaging EI'97, Feb. 10-12, 1997, San Jose, CA).

[25] See for example, S. Senturia, Microsystem Design, Chapter 20, Kluwer Academic Publishers, 2001

[26] The Perspectra® product from Actuality Systems

[27] Z20/20™ product from VIZTA3D

[28] UV-Setter™ print-setting product from BasysPrint GmbH

[29] M. Liang, R. L. Stehr, A.W. Krause, "Confocal pattern period in multiple aperture confocal imaging systems with coherent illumination, Opt. Lett. 22, pp. 751-753, 1997

[30] C. MacAulay, A. Dlugan, Use of digital micro mirror devices in quantitative microscopy, Proc. SPIE Vol. 3260, 1998, pp. 201.

[31] A.L.P. Dlugan, C. E. MacAulay, and P.M> Lane, "Improvements to quantitative microscopy through the use of digital micromirror devices," Proc. SPIE 3221, pp. 6-11, 2000

[32] Q.S. Hanley, P.J. Verveer, M.J. Gemkow, D. Arndt-Jovin, T.M. Jovin, "An optical sectioning programmable array microscope implemented with a digital micromirror device," Journal of Microscopy, Vol. 196, Pt. 3 (1999), pp. 317-331

[33] E.P. Wagner II, B.W. Smith, S. Madden, J.D. Winefordner, M. Mignardi, "Construction and Evaluation of a Visible Spectrometer Using Digital Micromirror Spatial Light Modulation", Applied Spectroscopy, 49, 1715 (1995)

[34] D.M. Bloom, "The Grating Light Valve: revolutionizing display technology," Proc. International Society for Optical Engineering (SPIE), vol. 3013, Projection Displays III, pp. 165-71, 1997

[35] H. Urey, Retinal Scanning Displays, in Encyclopedia of Optical Engineering, to be published by Marcel-Dekker, 2003

[36] M. Freeman, "Miniature high-fidelity displays using a biaxial MEMS scanning mirror", MOEMS Display and Imaging Systems, Proc. SPIE Vol. 4985, San Jose, CA, Jan. 2003.

[37] H. Urey, "Torsional MEMS scanner design for high-resolution display systems," Proc. International Society for Optical Engineering (SPIE), vol. 4773, Optical Scanning II, pp. 27-37, 2002.

[38] Yan, S. Luanava, F.A. Dewitt IV, V. Cassanta, H. Urey, "Magnetic actuation for MEMS scanners for retinal scanning displays", Photonics West 2003, SPIE vol. 4985, pp.106-114, 2003

[39] Davis, W.O., Sprague, R., Miller, J., "MEMS-based pico projector display", Optical MEMs and Nanophotonics, 2008 IEEE/LEOS Internationall Conference on, Issue Date: 11-14 Aug. 2008, pp. 31 - 32

[40] Minsky, M. Microscopy Apparatus. US Patent # 3013467. 1957

[41] D.L. Dickensheets and G. S. Kino, "Silicon-micromachined scanning confocal optical microscope," Journal of Microelectromechanical Systems, Vol. 7, No. 1, pp. 38-47, March 1998.

[42] K. Murakami, A. Murata, T. Suga, H. Kitagawa, Y. Kamiya, M. Kubo, K. Matsumoto, H. Miyajima, and M. Katashiro, "A MINIATURE CONFOCAL OPTICAL MICROSCOPE WITH MEMS GIMBAL SCANNER", The 12th International Conference on Solid State Sensors, Actuators and Microsystems, Boston, June 8-12, 2003, pp. 587-590

[43] K.Murakami,et.al., "A MEMS gimbal scanner for a miniature confocal microscope", Optical-MEMS2002, 2002, TuA2 pp.9-10

[44] T. D. Wang, M. J. Mandella, C. H. Contag, and G. S. Kino, "Dual-axis confocal microscope for high-resolution in vivo imaging," Opt. Lett., 28(6), 414-416 (2003).

[45] J. T. C. Liu, M. J. Mandella, H. Ra, L. K. Wong, O. Solgaard, G. S. Kino, W. Piyawattanametha, C. H. Contag, and T. D. Wang, "Miniature near-infrared dual-axes confocal microscope utilizing a two-dimensional microelectromechanical systems scanner," Opt. Lett., 32(3), 256-258 (2007).

[46] H. Ra, W. Piyawattanametha, Y. Taguchi, D. Lee, M. J. Mandella, O. Solgaard, "Two-Dimensional MEMS Scanner for Dual-Axes Confocal Microscopy",Journal of Microelectromechanical Systems 16 (4), pp. 969-976 (2007).

[47] H. Ra, W. Piyawattanametha, M. J. Mandella, P.-L. Hsiung, J. Hardy, T. D. Wang, C. H. Contag, G. S. Kino, and O. Solgaard, "Three-dimensional in vivo imaging by a handheld dual-axes confocal microscope," Opt. Express, 16(10), 7224-7232 (2008).

[48] W. Piyawattanametha, H. Ra, M. J. Mandella, K. Loewke, T. D. Wang, G. S. Kino, O. Solgaard, and C. H. Contag, "3-d near-infrared fluorescence imaging using an mems-based miniature dual-axis confocal microscope," IEEE J. Sel. Top. Quantum Electron., 15(5), 1344-1350 (2009).

[49] Wibool Piyawattanametha, Hyejun Ra, Zhen Qiu, Shai Friedland, Jonathan T. C. Liu, Kevin Loewke, Gordon S. Kino, Olav Solgaard, Thomas D. Wang, Michael J. Mandella, and Christopher H. Contag, "In Vivo Near-infrared Confocal Microendoscopy in the Human Colorectal Tract", to be published in JBO 2012
[50] E.A. Swanson, D. Huang, M. R. Hee, J.G. Fujimoto, C.P. Lin, and C.A. Puliafito, "High-speed optical coherence domain reflectometry," Optics Letters, Vol. 17, Issue 2, pp 151, Jan. 1992
[51] D. Huang, E.A. Swanson, C.P. Lin, J.S. Schuman, W.G. Chang, M.R. Hee, T. Flotte, K. Gregory, C.A. Puliafito, and J.G. Fujimoto, "Optical Coherence Tomography", Science, 254:1178-1181, 1991
[52] Tearney, G.J., M.E. Brezinski, B.E. Bouma, S.A. Boppart, C. Pitvis, J.F. Southern, and J.G. Fujimoto, In vivo endoscopic optical biopsy with optical coherence tomography. Science, 1997. 276(5321): pp. 2037-9.
[53] A. M. Rollins, R. Ung-arunyawee, A. Chak, R. C. K. Wong, K. Kobayashi, M. V. Sivak, Jr., and J. A. Izatt, Opt. Lett. 24, 1358 (1999).
[54] Tuqiang Xie, Huikai Xie, Gary K. Fedder, and Yingtian Pan, "Endoscopic optical coherence tomography with a modified microelectromechanical systems mirror for detection of bladder cancers," APPLIED OPTICS, Vol. 42, No. 31, pp. 6422-6426, 1 November 2003.
[55] J. M. Zara, S. Yazdanfar, K. D. Rao, J. A. Izatt, and S. W. Smith, "Electrostatic micromachine scanning mirror for optical coherence tomography", OPTICS LETTERS, Vol. 28, No. 8, April 15, 2003, pp. 628-630
[56] A. D. Aguirre, P. R. Herz, Y. Chen, J. G. Fujimoto, W. Piyawattanametha, L. Fan, M. C. Wu, "Two-axis MEMS scanning catheter for ultrahigh resolution three-dimensional and en face imaging," Optics Express, Vol. 15, No. 5, pp. 2445-2453, March 2007.
[57] W. Piyawattanametha, P. Patterson, D. Hah, H. Toshiyoshi, and M. Wu, "A 2D Scanner by Surface and Bulk Micromachined Angular Vertical Comb Actuators," International Conference on Optical MEMS, August 18-21, Hawaii, USA, pp. 93-94
[58] W. Piyawattanametha, P. R. Patterson, D. Hah, H. Toshiyoshi, and M. C. Wu, "Surface- and Bulk- Micromachined Two-Dimensional Scanner Driven by Angular Vertical Comb Actuators," Journal of Microelectromechanical Systems, Vol. 14, No. 6, pp. 1329-1338, Dec. 2005.
[59] T. D. Kudrle, C. C. Wang, M. G. Bancu, J. C. Hsiao, A. Pareek, M. Waelti, G. A. Kirkos, T. Shone, C. D. Fung, and C. H. Mastrangelo, "Electrostatic Micromirror Arrays Fabricated with Bulk and Surface Micromachining Techniques," MEMS 2003, Kyoto, Japan, Jan. 2003, pp. 267-270
[60] R. K. Tyson and B. W. Frazier, "Microelectromechanical system programmable aberration generator for adaptive optics," Applied Optics Vol. 38 No. 1, pp 168-178, 1999
[61] C. Paterson, I. Munro, and J.C. Dainty, "A low cost adaptive optics system using a membrane mirror," Optics Express Vol. 6 No. 9 pp 175-185, 2000
[62] O. Albert, et al., "Smart microscope: an adaptive optics system using a membrane mirror," Optics Express Vol. 6 No. 9, pp 175-185, 2000
[63] G. Vdovin and V. Kiyko, "Intracavity control of a 200-W continuous-wave Nd:YAG laser by a micromachined deformable mirror," Optics Express Vol. 26 No. 11, pp. 798-800, 2001

[64] H. M. Dyson, R. M. Sharples, N. A. Dipper, G. V. Vdovin, " Cryogenic wavefront correction using membrane deformable mirrors, " Optics Express Vol. 8 No. 1, pp. 17-26, 2001
[65] L. Zhu et al., "Wave-front generation of Zernike polynomial modes with a micromachined membrane deformable mirror, " Applied Optics Vol. 38 No. 28, pp. 6019-6026, 1999
[66] Julie A. Perreault and Thomas G. Bifano, "HIGH-RESOLUTION WAVEFRONT CONTROL USING MICROMIRROR ARRAYS," The proceedings of Solid-State Sensor, Actuator and Microsystems Workshop Hilton Head Island, South Carolina, June 6-10, 2004, pp. 83-86
[67] Krishnamoorthy, R., Bifano, T. G., Vandelli, N., and Horenstein, M., "Development of MEMS deformable mirrors for phase modulation of light," Optical Engineering [36], pp. 542-548, 1997
[68] Currently MEMSCAP, Inc. Durham, NC
[69] Y. Hishinuma and E. H. Yang ,"Piezoelectric Unimorph Microactuator Arrays for Single-Crystal Silicon Continuous-Membrane Deformable Mirror" , JOURNAL OF MICROELECTROMECHANICAL SYSTEMS, VOL. 15, NO. 2, pp. 370-379, APRIL 2006.
[70] Michael A. Helmbrecht, Min He, Carl J. Kempf, Marc Besse, "MEMS DM development at Iris AO, Inc. ", Proc. SPIE 7931, 793108 (2011)
[71] M. Fujino, P. R. Patterson, H. Nguyen, W. Piyawattanametha, and M. C. Wu, "Monolithically Cascaded Micromirror Pair Driven by Angular Vertical Combs for Two-Axis Scanning," IEEE JOURNAL OF SELECTED TOPICS IN QUANTUM ELECTRONICS, VOL. 10, NO. 3, pp. 492-497, MAY/JUNE 2004
[72] H. Toshiyoshi and H. Fujita, "Electrostataic Micro Torsion Mirrors for an Optical Switch Matrix," IEEE J. Microelectromechanical Systems, Vol. 5, p. 231, 1996.
[73] L.Y. Lin, E.L. Goldstein, and R.W. Tkach, "Free-space micromachined optical switches with submillisecond switching time for large-scale optical crossconnects," IEEE Photonics Technology Letters, vol.10, p.525-7, 1998.
[74] B. Hehin, K.Y. Lau, and R.S. Muller, "Magnetically actuated micromirrors for fiber-optic switching," Solid-State Sensors and Actuator Workshop, Hilton Head Island, South Carolina, 1998.
[75] R. L. Wood, R. Mahadevan, and E. Hill, "MEMS 2-D matrix switch," 2002 Optical Fiber Communication (OFC) Conference, Paper TuO2, Anaheim California, 2002.
[76] Y. Yoon, K. Bae, and H. Choi, "An optical switch with newly designed electrostatic actuators for optical cross connects," 2002 IEEE/LEOS International Conference on Optical MEMS, Lugano, Switzerland, 2002.
[77] L.Y. Lin, E.L. Goldstein, and R.W. Tkach, "Angular-precision enhancement in free-space micromachined optical switches," IEEE Photonics Technology Letters, vol.11, p.1253-5, 1999.
[78] L.Y. Lin, E.L. Goldstein, and R.W. Tkach, "Free-space micromachined optical switches for optical networking," IEEE Journal of Selected Topics in Quantum Electronics, Vol. 5, P.4-9, 1999.
[79] T. Akiyama and H. Fujita, "A quantitative analysis of scratch drive actuator using buckling motion", Proc. 8th IEEE International MEMS Workshop, pp. 310 - 315, 1995.

[80] L.-Y. Lin, E.L. Goldstein, R.W. Tkach, "On the expandability of free-space micromachined optical cross connects," J. Lightwave Technology, Vol. 18, pp. 482 - 489, 2000.

[81] Li Fan, S. Gloeckner, P. D. Dobblelaere, S. Patra, D. Reiley, C. King, T. Yeh, J. Gritters, S. Gutierrez, Y. Loke, M. Harburn, R. Chen, E. Kruglick, M. Wu and A. Husain, "Digital MEMS switch for planar photonic crossconnects," 2002 Optical Fiber Communication (OFC) Conference, Paper TuO4, Anaheim, California, 2002

[82] R.T. Chen, H. Nguyen, M.C. Wu, "A high-speed low-voltage stress-induced micromachined 2x2 optical switch," IEEE Photonics Technol. Lett., Vol. 11, pp.1396-8, November 1999.

[83] Marxer, C.; de Rooij, N.F. "Micro-opto-mechanical 2x2 switch for single-mode fibers based on plasma-etched silicon mirror and electrostatic actuation," J. Lightwave Technology, vol. 17, no. 1, pp. 2-6, Jan. 1999.

[84] Noell, W.; Clerc, P.-A.; Dellmann, L.; Guldimann, B.; Herzig, H.-P.; Manzardo, O.; Marxer, C.R.; Weible, K.J.; Dandliker, R.; de Rooij, N. "Applications of SOI-based optical MEMS," IEEE J. Selected Topics in Quantum Electronics, vol. 8, no. 1, pp. 148-154, 2002.

[85] Marxer, C.; Thio, C.; Gretillat, M.-A.; de Rooij, N.F.; Battig, R.; Anthamatten, O.; Valk, B.; Vogel, P. "Vertical mirrors fabricated by deep reactive ion etching for fiber-optic switching applications," J. Microelectromechanical Systems, vol. 6, no. 3, pp.277-85, Sept. 1997.

[86] Neilson, D.T., et al.,"Fully Provisioned 112x112 Micro-Mechanical Optical Crossconnect With 35.8Tb/s Demonstrated Capacity," Optical Fiber Communication Conference, OFC 2000, March 7-10, Baltimore, MD, Vol. 4, pp.202-204.

[87] Lin, L.Y., Goldstein, E.L., "Opportunities and Challenges for MEMS in Lightwave Communications," IEEE J. Sel. Topics Quantum Elec., Vol. 8, No. 1, p.163, 2002.

[88] Syms, R.R.A., "Scaling Laws for MEMS Mirror-Rotation Optical Cross Connect Switches," IEEE J. Lightwave Tech., Vol. 20, No. 7, p. 1084, 2002.

[89] J. Kim et al., "1100x1100 port MEMS-based optical crossconnect with 4-db maximum loss," IEEE Photonics Technol. Lett., Vol. 15, pp.1537-9, November 2003.

[90] D. T. Neilson and R. Ryf, "Scalable micro mechanical optical crossconnects," 2000 IEEE/LEOS Annual Meeting, Paper ME2.

[91] V. A. Aksyuk et al., "238x238 micromechanical optical cross connect," IEEE Photonics Technol. Lett., Vol. 15, pp.587-9, April 2003.

[92] V. A. Aksyuk et al., "Beam-steering micromirrors for large optical cross-connects," IEEE J. Lightwave Tech., Vol. 21, No. 3, p. 634, 2003.

[93] M. Kozhevnikov et al., "Micromechanical optical crossconnect with 4-F relay imaging optics," IEEE Photonics Technol. Lett., Vol. 16, pp.275-7, Jan. 2004.

[94] R. Ryf et al., "1296-port MEMS transparent optical crossconnect with 2.07Petabit/s switch capacity," in Proceedings of 2001 OFC postdeadline paper, PD28.

[95] N. Yazdi, H. Sane, T. D. Kudrle, and C. H. Mastrangelo, "Robust sliding-mode control of electrostatic torsional micromirrors beyond the pull-in limit," 12th International Conference on Solid-State Sensors, Actuators and Microsystems (TRANSDUCERS 2003, 8-12 June 2003), vol. 2, pp.1450 - 1453.

[96] T. D. Kudrle, G. M. Shedd, C. C. Wang, J. C. Hsiao, M. G. Bancu, G. A. Kirkos, N. Yazdi, M. Waelti, H. Sane, and C. H. Mastrangelo, "Pull-in suppression and torque magnification in parallel plate electrostatic actuators with side electrodes," 12th

International Conference on Solid-State Sensors, Actuators and Microsystems (TRANSDUCERS 2003, 8-12 June 2003), vol. 1, pp.360 - 363.

[97] T. D. Kudrle, C. C. Wang, M. G. Bancu, J. C. Hsiao, A. Pareek, M. Waelti, G. A. Kirkos, T. Shone, C. D. Fung, and C. H. Mastrangelo, "Electrostatic micromirror arrays fabricated with bulk and surface micromachining techniques," IEEE The Sixteenth Annual International Conference on Micro Electro Mechanical Systems (MEMS 2003, 19-23 Jan. 2003), pp.267 - 270.

[98] Sawada, R.; Yamaguchi, J.; Higurashi, E.; Shimizu, A.; Yamamoto, T.; Takeuchi, N.; Uenishi, Y. "Single Si crystal 1024ch MEMS mirror based on terraced electrodes and a high-aspect ratio torsion spring for 3-D cross-connect switch," 2002 IEEE/LEOS International Conference on Optical MEMs, pp. 11 -12, 2002.

[99] Sawada, R.; Yamaguchi, J.; Higurashi, E.; Shimizu, A.; Yamamoto, T.; Takeuchi, N.; Uenishi, Y. "Improved single crystalline mirror actuated electrostatically by terraced electrodes with high aspect-ratio torsion spring," 2003 IEEE/LEOS International Conference on Optical MEMs, pp. 153 -154, 2003.

[100] Y. Mizuno et al., "A 2-axis comb-driven micromirror array for 3D MEMS switches," 2002 IEEE/LEOS International Conference on Optical MEMs, pp. 17 -18, 2002.

[101] N. Kouma et al., "A multi-step DRIE process for a 128x128 micromirror array," 2003 IEEE/LEOS International Conference on Optical MEMs, pp. 53 -54, 2003.

[102] P. B. Chu et al., "Design and Nonlinear Servo Control of MEMS Mirrors and Their Performance in a Large Port-Count Optical Switch," Journal of Microelectromechanical Systems, VOL. 14, NO. 2, pp. 261-273, APRIL 2005.

[103] J. I. Dadap et al., "Modular MEMS-based optical cross-connect with large port-count," IEEE Photonics Technol. Lett., Vol. 15, pp.1773-5, Dec. 2003.

[104] C. Pu et al., "Electrostatic Actuation of Three-Dimensional MEMS Mirrors Using Sidewall Electrodes, IEEE Journal of Selected Topics in Quantum Electronics," VOL. 10, NO. 3, pp. 472-477, MAY/JUNE 2004.

[105] X. Zheng et al., "Three-dimensional MEMS photonic cross-connect switch design and performance," IEEE J. Sel. Topics Quantum Elec., Vol. 9, No. 2, p.571, 2003.

[106] Aksyuk, V.A., Pardo, F., Bolle, C.A., Arney, S., Giles, C.R., Bishop, D.J., "Lucent Microstar micromirror array technology for large optical crossconnects," Proceedings of the SPIE, MOEMS and Miniaturized Systems, Sept. 2000, Santa Clara, CA, pp. 320-324.

[107] Aksyuk V.A., Simon, M.E., Pardo, F., Arney S., Lopez, D., and Villanueva, A., "Optical MEMS Design for Telecommunications Applications," 2002 Solid-State Sensor and Actuator Workshop Tech. Digest, June 2-6, Hilton Head, SC, pp.1-6.

[108] A. Gasparyan, et al., "Drift-free, 1000G mechanical shock tolerant single-crystal silicon two-axis MEMS tilting mirrors in a 1000x1000-port optical crossconnect", in Proceedings of 2003 OFC postdeadline paper, PD36.

[109] V.A. Aksyuk et al., "238×238 surface micromachined optical crossconnect with 2dB maximum loss," OFC (Optical Fiber Communication) 2002, Postdeadline paper, FB9-1 - FB 9-3.

13

Optical-Thermal Phenomena in Polycrystalline Silicon MEMS During Laser Irradiation

Justin R. Serrano and Leslie M. Phinney
Engineering Sciences Center, Sandia National Laboratories
Albuquerque, New Mexico,
USA

1. Introduction

Many microelectromechanical systems (MEMS) applications utilize laser irradiation as an integral part of the system functionality, including projection displays, optical switches, adaptive optics (Andrews et al., 2011; Andrews et al., 2008), optical cross-connects (Knoernschild et al., 2009), and laser powered thermal actuators (Serrano & Phinney, 2008; Serrano et al., 2005). When laser irradiation is incident on small-scale systems, such as these MEMS applications, the propensity for exceeding the thermal handling capability of the devices dramatically increases, often leading to overheating, and subsequent deformation and permanent damage to the devices. In most instances, this damage is a direct consequence of the device geometry and the material thermal properties, which hinder the transport of heat out of any locally heated area. Such thermally-driven failures are common in electrically-powered systems (Baker et al., 2004; Plass et al., 2004). However, for laser-irradiated MEMS, particularly those fabricated of surface-micromachined polycrystalline silicon (polysilicon), the optical properties can also affect the thermal response of the devices by altering how the laser energy is deposited within the material. Even more concerning in these types of devices is the fact that the thermal, optical, and mechanical response can be intimately coupled such that predicting device performance becomes difficult. In this chapter, we focus on understanding some of the basics of optical interactions in laser-irradiated MEMS. We will first look at how the optical properties of the materials affect the laser energy deposition within a device. We will then expand upon this by looking at the coupling that exists between the optical and thermal properties, paying particular attention to the implications that transient temperature changes have in the optical response, ultimately leading to device failure. Finally, we will look at various cases of laser-induced damage in polysilicon MEMS where the device geometry and design and optical-thermal coupling have led to device failure.

2. Optical interactions in MEMS

Understanding the coupling that exists between the thermal and optical behavior in laser-irradiated MEMS must begin by looking at the optical properties of the irradiated materials and at how the laser light interacts with each material. The primary factor that affects the magnitude of this interaction is the material's complex refractive index, $\hat{n} = n + ik$. A wave incident on the interface between two media of different refractive indices will undergo reflection and refraction, as shown in Fig. 1. The direction of the reflected and refracted beams follow the well-known laws of reflection and refraction:

$$\theta_r = \theta_i \text{ (law of reflection)}, \tag{1}$$

$$\hat{n}_2 \sin\theta_t = \hat{n}_1 \sin\theta_i \text{ (law of refraction; Snell's law)} \tag{2}$$

where θ_i, θ_r and θ_t are the angles of incidence, reflection, and refraction, respectively[1]. The following sections will discuss how the rules above are applied to laser-irradiated structures in order to obtain the magnitudes of the reflected, transmitted, and absorbed light, which ultimately dictate how the energy is deposited in an irradiated microsystem.

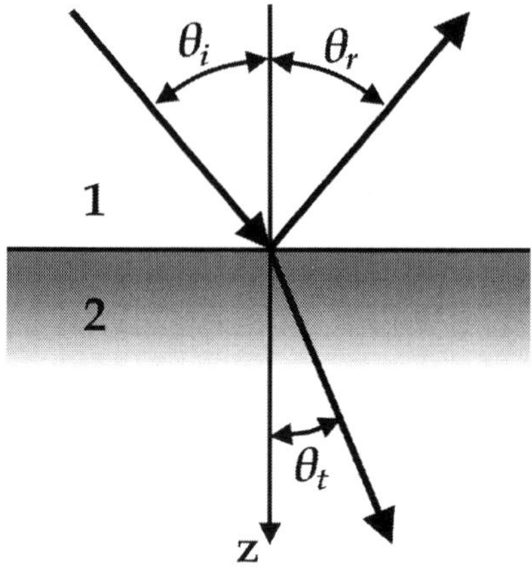

Fig. 1. Reflection and refraction of a plane wave incident on the interface between two media.

2.1 Optically thick systems

For monochromatic laser light, incident from vacuum ($\hat{n}_{vacuum} = 1.0$) at an angle θ_i upon a homogeneous, semi-infinite, non-magnetic medium of index $\hat{n}$, the reflectivity of the interface is dictated by the Fresnel coefficients (Born & Wolf, 1999):

$$r^s = \frac{\cos\theta_i - \sqrt{\hat{n}^2 - \sin^2\theta_i}}{\cos\theta_i + \sqrt{\hat{n}^2 - \sin^2\theta_i}}, \quad r^p = \frac{\hat{n}^2\cos\theta_i - \sqrt{\hat{n}^2 - \sin^2\theta_i}}{\hat{n}^2\cos\theta_i + \sqrt{\hat{n}^2 - \sin^2\theta_i}}, \tag{3}$$

where the law of refraction was used to rewrite the expressions in terms of the incident angle and medium refractive index only, and the subscripts *s* and *p* above refer to the *s-polarized* or *transverse-electric* (TE) and *p-polarized* or *transverse-magnetic* (TM) polarizations of

[1] For instances where the indices are complex, the quantity θ_t is also complex-valued and no longer has the same meaning as an angle of refraction.

the incident light, respectively. The surface reflectivity, that is the magnitude of the fraction of reflected energy, is given for either polarization[2], by:

$$R^s = \left|r^s\right|^2 \text{ and } R^p = \left|r^p\right|^2 . \tag{4}$$

For normal incidence, $\theta_i = 0$, the polarization dependence disappears and Eq. 3 reduces to the well-known expression for bulk surface reflectivity (Born & Wolf, 1999):

$$R = \left|\frac{\hat{n}-1}{\hat{n}+1}\right|^2 = \frac{(n-1)^2 + k^2}{(n+1)^2 + k^2} . \tag{5}$$

In any absorbing material (i.e., with $k > 0$) the light transmitted through the medium is attenuated in accordance with the Beer-Lambert law:

$$I(z) = I_o \, exp(-\alpha z) \tag{6}$$

where I_o is the intensity of light entering the surface, $\alpha = 4\pi k/\lambda$ is the linear attenuation coefficient of the medium at the wavelength λ, and z is the spatial coordinate with its origin at the surface. The inverse of the attenuation coefficient is known as the optical penetration depth

$$d_{opt} = \alpha^{-1} = \lambda/_{4\pi k} , \tag{7}$$

and it is the distance over which the light intensity is attenuated by $1/e$.

While the development above for the Fresnel coefficients assumes a semi-infinite medium (i.e., a single interface separating the two media), the significance of Eq. 6 is that any material whose of thickness $d >> d_{opt}$ can be considered optically thick, in the sense that it will behave the same as a semi-infinite medium. What constitutes an optically thick layer ultimately depends on the value of the complex part of the refractive index, k, as described in Eq. 7. For example, the penetration depth of silicon at $\lambda = 0.3$ µm is $d_{opt} \approx 5.8$ nm, very similar to that of aluminium at $\lambda = 0.4$ µm or gold at $\lambda = 0.7$ µm (Schulz, 1954); at longer wavelengths the penetration depth in silicon increases by over three orders of magnitude (on the order of several micrometers) due to the drastic decrease in the value of k. As we will show, the distinction between optically thick and optically thin films will have profound implications in the treatment of the optical thermal coupling that exists in laser heated MEMS.

2.2 Optically thin and multilayered systems

A different approach must be used in instances where the thickness of the irradiated film is comparable to the optical penetration depth. Such conditions are of significant relevance for surface micromachined polysilicon devices, which generally can have layers and gaps with thicknesses on the order of a few µm (Carter et al., 2005; MEMS Technologies Department,

[2] As a consequence of the two polarization conditions, there will be two independent values for reflectivity. For unpolarized irradiation, it is common practice to take the average of the two reflectivity values as the resultant reflectivity.

2008)—comparable to optical penetration depths at visible-to-near infrared wavelengths (Phinney & Serrano, 2007; Serrano & Phinney, 2009; Serrano et al., 2009). In theses cases, depicted in Fig. 3, light transmitted across the first interface will encounter a second interface and undergo reflection and refraction. The process of reflection and refraction at both interfaces can repeat itself numerous times, as shown in Fig. 3, and with each reflection, the wave can undergo a phase change of 180°. If the incident light is monochromatic, with sufficiently large coherence length (i.e., laser light), then the multiple reflections will interfere with each other constructively and destructively. This thin film interference will yield deviations from the values obtained with Eqs. 3 and 4 for the optical response of the irradiated surface.

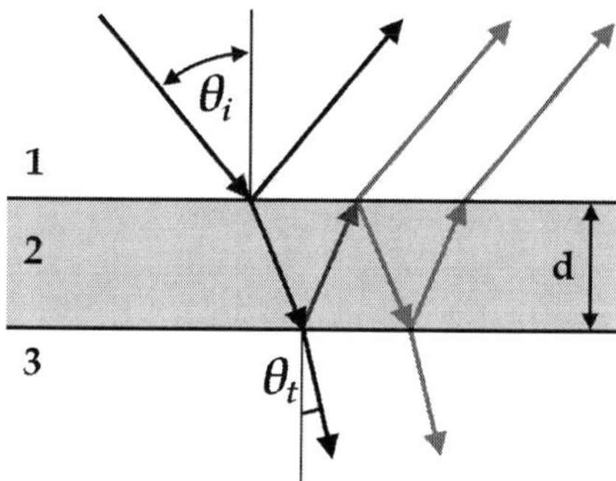

Fig. 3. Reflection and refraction in a multilayered system showing the multiple reflections from the two interfaces.

There are various ways to obtain a numerical description of the overall optical performance of such a multilayered system. The most common method is the transfer matrix method (Born & Wolf, 1999; Katsidis & Siapkas, 2002) whereby each individual layer is assigned a matrix of Fresnel coefficients, which capture the interaction of the incident wave with the layer. This method, while useful for obtaining the net response of the stratified structure, does not easily permit extracting information on how the energy is deposited within the layers, a detail of paramount importance when analyzing laser-irradiated MEMS. To obtain interlayer absorptance values, we turn to a similar analysis called the LTR method (Mazilu et al., 2001), which combines the layer responses in a modular form. This modularity then permits the extraction of the absorptances for the layers in the structure.

2.2.1 LTR method

The LTR method (Mazilu et al., 2001), which is stands for **L**eft-side reflectance, **T**ransmittance, and **R**ight-side reflectance, considers a stack of material irradiated from the left and right sides, as shown in Fig. 4. The technique leverages the fact that for an irradiated layered system only three terms are needed to fully describe its optical response—the reflectances of either side and a transmittance term. While most typically utilized for obtaining the net response of a stratified system, the modular nature of the LTR method facilitates the extraction of absorptance values for each individual layer, making it particularly useful for laser-irradiated MEMS (Serrano & Phinney, 2009; Serrano et al., 2009).

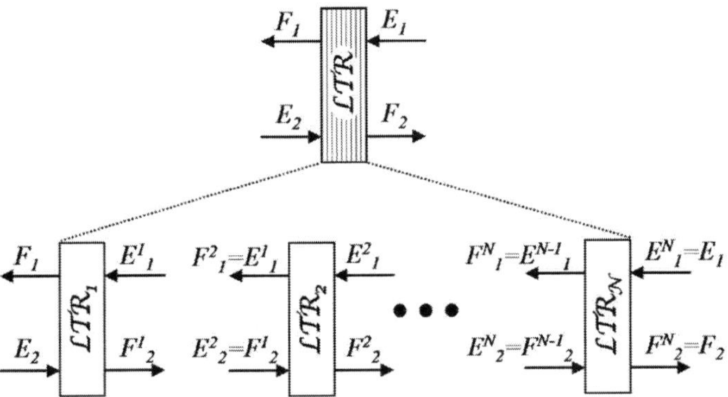

Fig. 5. Schematic representation of the LTR method. A multilayer stack is represented by an LTR element, where each layer is also made up of an LTR element.

This technique is similar to the transfer matrix method in that each layer is assigned a mathematical entity made up of the reflection and transmission coefficients for the layer. However, unlike the matrix method where the layer matrix depends on the properties of the media surrounding the layer, the coefficients are referenced with respect to vacuum (i.e., a wave is considered to be travelling into or from vacuum), simplifying the calculations and giving the technique its modularity. Thus, a three-element LTR vector containing the left- and right-side reflection coefficients, as well as the transmission coefficient, is defined as:

$$\mathbf{X} = \begin{pmatrix} L \\ T \\ R \end{pmatrix} = \begin{pmatrix} r\dfrac{1-p^2}{1-p^2r^2} \\ p\dfrac{1-r^2}{1-p^2r^2} \\ r\dfrac{1-p^2}{1-p^2r^2} \end{pmatrix}. \tag{8}$$

For a wave incident at an angle θ_i upon a layer of thickness d and refractive index $\hat{n}$, the coefficient p in Eq. 8 considers propagation in the medium and is defined as:

$$p = exp\left(i\tfrac{2\pi}{\lambda} d\sqrt{\hat{n}^2 - sin^2\theta_i} \right), \tag{9}$$

whereas the coefficient r considers reflection from the interfaces and for the two possible polarization conditions[3] is given by Eq. 3. The Fresnel coefficients above assume the wave

[3] As discussed in footnote 2, this method will yield polarization-dependent results for reflectance, transmittance, and absorptances. For unpolarized irradiation, the accepted value is the average of the two polarization cases.

travels from vacuum through the layer and out into vacuum once again. If the amplitudes of the fields incident on the layer from the right and the left are given, as shown in Fig. 5, by E_1 and E_2, respectively, the elements of $\mathbf{X}$ can be used to describe the amplitudes of the fields, F_1 and F_2, exiting the layer as:

$$F_1 = TE_1 + LE_2 \text{, and} \tag{10}$$

$$F_2 = RE_1 + TE_2 . \tag{11}$$

The LTR method additionally defines a vector for a single interface: one for an interface with a wave travelling from vacuum into a medium of index $\hat{n}$ ($\mathbf{S}_{01}$) and another for a wave travelling from the medium into vacuum ($\mathbf{S}_{10}$):

$$\mathbf{S}_{01}(\hat{n}) = \begin{pmatrix} L \\ T \\ R \end{pmatrix} = \begin{pmatrix} r \\ t_{01} \\ -r \end{pmatrix} \text{, and} \tag{12}$$

$$\mathbf{S}_{10}(\hat{n}) = \begin{pmatrix} L \\ T \\ R \end{pmatrix} = \begin{pmatrix} -r \\ t_{10} \\ r \end{pmatrix}, \tag{13}$$

where the coefficient r is given in Eq. 3 for the two polarization conditions, and

$$t_{01}^{s} = \frac{2\cos\theta_i}{\cos\theta_i + \sqrt{\hat{n}^2 - \sin^2\theta_i}}, \quad t_{01}^{p} = \frac{2\hat{n}\cos\theta_i}{\hat{n}^2\cos\theta_i + \sqrt{\hat{n}^2 - \sin^2\theta_i}}, \tag{14}$$

$$t_{10}^{s} = \frac{2\hat{n}\cos\theta_i}{\cos\theta_i + \sqrt{\hat{n}^2 - \sin^2\theta_i}}, \text{ and } t_{10}^{p} = \frac{2\hat{n}^2\cos\theta_i}{\hat{n}\cos\theta_i + \sqrt{\hat{n}^2 - \sin^2\theta_i}}. \tag{15}$$

Combination of multiple layers is implemented by the use of a composition rule, as shown below for two layers. Under the LTR scheme, each layer is considered a separate entity, separated from adjacent layers by a zero- thickness vacuum layer, such that the wave exits one layer into vacuum and enters the next layer from vacuum.

$$\mathbf{LTR} = \mathbf{X}_1 \oplus \mathbf{X}_2 = \begin{pmatrix} L_1 \\ T_1 \\ R_1 \end{pmatrix} \oplus \begin{pmatrix} L_2 \\ T_2 \\ R_2 \end{pmatrix} = \begin{pmatrix} L_1 + \dfrac{L_2 T_1^2}{1 - R_1 L_2} \\ \dfrac{T_1 T_2}{1 - R_1 L_2} \\ R_2 + \dfrac{R_1 T_2^2}{1 - R_1 L_2} \end{pmatrix} = \begin{pmatrix} \mathcal{L} \\ \mathcal{T} \\ \mathcal{R} \end{pmatrix}. \tag{16}$$

This rule enables modeling of a multilayer structure by sequential application of the composition rule to all the layers in the stack including the media on the left and right side of the multilayer structure.

$$\mathbf{LTR} = \mathbf{S}_{10}(\hat{n}_L) \oplus \mathbf{X}_1 \oplus \mathbf{X}_2 \oplus \cdots \oplus \mathbf{X}_{N-1} \oplus \mathbf{X}_N \oplus \mathbf{S}_{01}(\hat{n}_R) = \begin{pmatrix} \mathcal{L} \\ \mathcal{T} \\ \mathcal{R} \end{pmatrix} \tag{17}$$

Since the result of the composition is another LTR vector, if the fields E_1 and E_2 incident on the stack are known, then the remaining fields, F_1 and F_2, can be easily found using Eqs. 10 and 11.

While the LTR construct is useful for capturing the response of a multilayered structure irradiated from the front and the back (left and right in Fig. 3), only front-side illumination is considered here, as that is the most common configuration encountered in MEMS applications. For single-sided illumination the structure is assumed to be illuminated only from the left (i.e., $E_1 = 0$ in Fig. 5), and

$$\mathcal{L} = F_1/E_2 \text{ (left side reflection);} \tag{18}$$

$$\mathcal{T} = F_2/E_2 \text{ (transmission);} \tag{19}$$

$$\mathcal{R} = 0 \text{ (right side reflection).} \tag{20}$$

The total reflected, transmitted, and absorbed intensities are then:

$$R = |\mathcal{L}|^2, \tag{21}$$

$$T = |\mathcal{T}|^2 \frac{Re(n_R \cos\theta_R)}{Re(n_L \cos\theta_L)}, \text{ and} \tag{22}$$

$$A = 1 - R - T. \tag{23}$$

If the incident medium on the left is vacuum or air, Eq. 22 can be rewritten fully in terms of the angle of incidence and the substrate index, $\hat{n}_{sub}$, as

$$T = |\mathcal{T}|^2 \frac{Re\left(\sqrt{\hat{n}_{sub}^2 - \sin^2\theta_i}\right)}{\cos\theta_i}. \tag{24}$$

With the fields on the left- and right-most layers defined, the fields entering and exiting each layer can be obtained by recursively applying Eqs. 10 and 11 to each layer. Once these fields are defined, the individual layer absorptances can be easily obtained by noting that each layer is referenced to vacuum and the absorptance is simply the difference between the entering and exiting field magnitudes:

$$A_i = |E_1^i|^2 + |E_2^i|^2 - |F_1^i|^2 - |F_2^i|^2, \tag{25}$$

where the left-most fields of the first layer and the right-most fields for the last layer are obtained from Eqs. 18 and 19.

2.2.2 MEMS

As discussed in the previous section, the optical response of laser-irradiated materials depends strongly on various parameters. For optically thick materials, the refractive index of the irradiated medium determines the reflectivity of the surface and thus the fraction of the energy that is deposited in the material. When the optical penetration depth is comparable to film thickness, the geometry and composition of the structure becomes as important as refractive index in dictating the optical response. This becomes evident when analyzing the response of sacrificial micromachined MEMS fabricated from polysilicon.

In polysilicon-based MEMS the typical layer thickness is approximately 2 μm, with intermediate gaps of the same order (Carter et al., 2005; MEMS Technologies Department 2008). Such thicknesses are comparable to the penetration depth for both silicon and polysilicon for wavelengths above 550 nm (Jellison Jr & Modine, 1982a, 1982b; Lubberts et al., 1981; Xu & Grigoropoulos, 1993) and therefore the likelihood for thin film interference, as explained above, increases. Indeed, calculations carried out for air-spaced polysilicon structure fabricated from Sandia National Laboratories' SUMMiT-V™ process (MEMS Technologies Department, 2008), as shown in Fig. 6, show that the absorptance of the top-most layer can vary significantly as a function of the layer thickness. The multiple reflections from the various layers in the structure lead to conditions of local maxima and minima for different layer thicknesses. These extrema correspond to thicknesses where the interference between the multiply reflected waves is fully constructive or destructive as will be shown later.

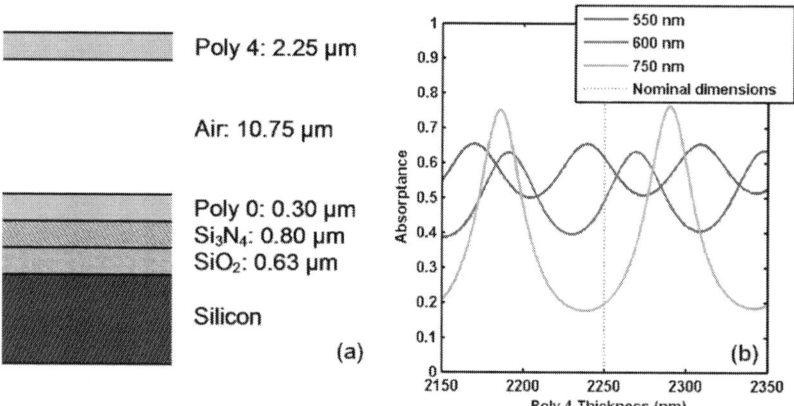

Fig. 6. (a) Schematic of a SUMMiT V™ polysilicon MEMS structure and (b) its optical response at different wavelengths as a function of the thickness of the top-most layer.

The variation in the amplitude and width of the absorptance peaks in this structure is related to the relative reflectivity of the two polysilicon surfaces at the particular wavelength much like a Fabry-Perot cavity (Born & Wolf, 1999) and will ultimately depend on the overall composition of the multilayered structure. For a coupled optical-thermal analysis, the existence of these periodic variations in the absorptance must be taken into account to predict the thermal behavior of laser-irradiated MEMS accurately.

3. Optical-thermal coupling in laser-irradiated MEMS

The previous section detailed the response of MEMS optical systems in strictly athermal terms. However, in laser-irradiated MEMS or MEMS exposed to extreme thermal environments the consequences of a changing thermal environment could be significant, especially in regards to the optical response. For simplicity, we shall consider cases where the incident laser energy is responsible for any temperature fluctuation in the irradiated structure, although the same principles are valid for structures subject to bulk external heating and laser irradiation (Burns & Bright, 1998).

Laser irradiation of an absorbing structure, such as micromachined polysilicon MEMS, will lead to a corresponding temperature increase. The magnitude of the induced temperature rise will depend on several factors, including the geometry, and thermal and optical properties of the irradiated materials. Because all of the parameters that play a role in determining the energy deposition exhibit some temperature dependence, the laser-induced heating of the structure will be dynamic in nature as the properties change during the heating event.

3.1 Temperature-induced geometry changes

We have already seen the potential effects of different layer thicknesses on the absorptance of an irradiated structure. However, while those fluctuations might arise out of manufacturing variability, the same effect can be observed during the heating of an as-built device. Geometrical and dimensional considerations during the heating result from any temperature-induced displacement and deformation of the MEMS when exposed to elevated temperatures (Knoernschild et al., 2010; Phinney et al., 2006). If the irradiating wavelength is in the optically thick regime for the irradiated material, the dimensional changes do not have a significant effect in the optical response of the structure since the incident energy is fully absorbed within the material. Nevertheless, depending on the structure, small deflections and deformations could have a significant effect on the heat transfer mechanisms on the heated device (Gallis et al., 2007; Wong & Graham, 2003).

When the conditions are such that thin film interference becomes important in the optical response, particularly for multilayered systems, the deformation will have a more dramatic effect. Depending on the design and geometry of the irradiated structure, the heating can alter both the thickness of the individual layers (via thermal expansion) and the spacing between them (via thermal expansion, buckling, etc.). Such deformations will produce changes in the absorptance of the laser irradiation, as shown in Fig. 7 for a Poly4 SUMMiT V™ structure similar to the one described by Phinney et al, (Phinney et al., 2006) and shown in Fig. 6a. The cantilevered structure in that reference suffered deflections of over 10 μm during laser irradiation. In Fig. 7, just a variation in the air gap height of ±500 nm suffices to demonstrate the type of deflection-induced changes in absorptance encountered in these tests. Assuming the deflection is caused by the temperature excursion of the structure, then a small change in gap height can lead to as much as a six-fold change in absorptance.

Additionally, due to the phase changes upon reflection, the trends in absorptance repeat for different values of thicknesses and gaps, as seen in Figs 6 and 7. The recurrence period can be estimated from Eq. 9 by finding the thickness increase Δd for which the path length difference is equivalent to an integer multiple of π :

$$\Delta d = \frac{m\lambda}{2\,Re\left(\sqrt{\hat{n}^2 - sin^2\theta_i}\right)},\ m = 1, 2, 3... \tag{26}$$

which, for the 800 nm example discussed, yields a recurrence period of 400 nm.

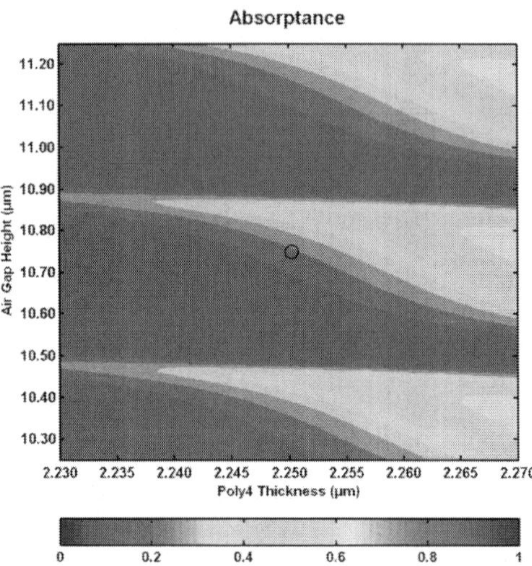

Fig. 7. Absorptance map for the geometry shown in Fig. 6a for λ = 808 nm as a function of layer and gap dimensions. The circle indicates the nominal dimensions for the geometry.

The outcome of such a variation in energy deposition can be detrimental—an increase in absorptance will cause additional heating and possibly lead to damage—or beneficial—a decrease in absorptance will permit the structure to withstand higher incident powers and avoid damage (Serrano & Phinney, 2009). Which situation is encountered with a particular device will depend on the irradiating conditions (wavelength and incidence angle), the optical properties, as well as the initial condition and the geometry of the device and the thermomechanical response of the structure. Because MEMS are primarily mechanical devices, these thermomechanical effects can typically be accounted and corrected for to reduce their contribution, much like it is done for electrically heated devices (Sassen et al., 2008).

3.2 Temperature-induced optical changes

In addition to purely mechanical effects caused by the heating, the temperature excursion will induce changes in the optical and thermal properties of the irradiated materials. While the variations in the thermal properties with temperature play a very important role in the thermal behavior of any laser-irradiated structure, their effects are generally noticeable for large temperature excursions. As we will show, the role of the temperature dependence of the material optical properties is, in some cases, more dominant and leads to marked changes in the thermal and optical performance of the irradiated structure over small temperature excursions.

For silicon-based materials, the complex index of refraction has been extensively studied as a function of temperature (Jellison Jr & Modine, 1982a, 1983; Sun et al., 1997; Xu & Grigoropoulos, 1993; Yavas et al., 1993). These works all show that the real part of the refractive index depends linearly with temperature:

$$n = n_o + \frac{dn}{dT}(T - T_o) \tag{27}$$

where n_o is the index at a reference temperature T_o and the slope $\frac{dn}{dT}$ typically has values on the order of 10^{-4} K^{-1} (Jellison Jr & Modine, 1982a, 1983; Sun et al., 1997; Xu & Grigoropoulos, 1993). The complex portion of the index, on the other hand, follows an exponential trend of the form:

$$k = k_o e^{\left(\frac{T-T_o}{T_R}\right)} \tag{28}$$

where k_o is complex index at T_o and the temperature T_R is an empirically determined reference temperature, which ranges in value from 498 K for bulk silicon (Jellison Jr & Modine, 1982a, 1983) to 680 K for different types of polysilicon (Sun et al., 1997; Xu & Grigoropoulos, 1993).

In optically thick systems, the change in complex refractive index will manifest itself as a change in surface reflectivity as a function of temperature. For silicon and polysilicon, this change is on the order of 10^{-5} K^{-1} (Jellison Jr & Modine, 1983) such that its impact on the thermal and mechanical response of irradiated devices is small. The same cannot be said for multilayered structures that are optically thin. In this case, the linear increase in the real part of the refractive index increases the effective path length difference between multiply reflected waves, changing the conditions for constructive and destructive interference from those present at the initial temperature. The exponential increase in the complex portion of the index, however, leads to a decrease in the optical penetration depth, reducing the effect of interference from deeper layers in the material. More importantly, the interplay between the two trends, when applied to the thin film interference equations discussed in the previous section, leads to temperature-dependent variations in the absorptance, as shown in Fig. 8 for the structure in Fig. 6a irradiated with 800 nm light. The most noticeable characteristic of the curves is the presence of temperature-periodic peaks. These result from the increase in the path length difference as the real portion of the index increases with temperature as given by Eq. 27. When the condition for fully destructive interference of the surface reflected waves is met, the absorptance of the layer increases. This condition is satisfied for

$$\tfrac{2\pi}{\lambda} d\, Re\left(\sqrt{(\hat{n}+\Delta\hat{n})^2 - sin^2\theta_i}\right) = \tfrac{2\pi}{\lambda} d\, Re\left(\sqrt{\hat{n}^2 - sin^2\theta_i}\right) + m\pi \text{ , } (m = 1, 2, 3, \ldots). \tag{29}$$

Solving the above relation for $\Delta\hat{n}$, and relating that to the temperature change through Eq. 27, we get:

$$\Delta T = Re\left(\frac{m\lambda\hat{n}}{2z\sqrt{\hat{n}^2 - sin^2\hat{\theta}_i}}\right)\left(\frac{dn}{dT}\right)^{-1}. \tag{30}$$

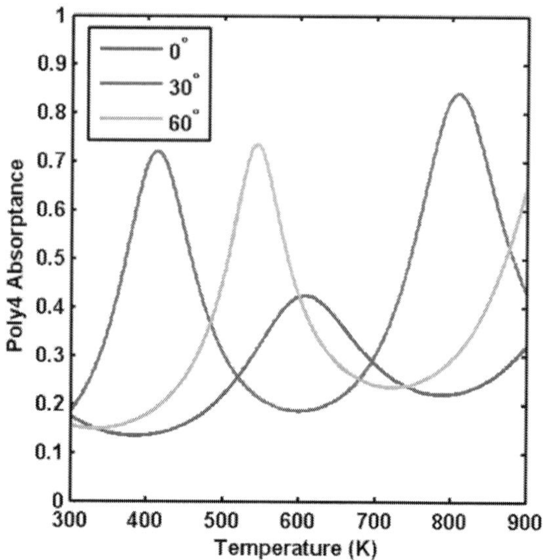

Fig. 8. Absorptance as a function of temperature and incidence angle for the geometry shown in Fig. 6a for λ = 800 nm.

The consequences of the peaks and valleys are significant for the behavior of laser-irradiated polysilicon MEMS. An incident laser on the surface will induce heating, leading to a change in absorptance, and corresponding changes in the sample temperature. The non-linear absorptance response thus creates stable and unstable conditions depending on the temperature of the sample. For temperatures in the range where the slope of the absorptance curves is negative, the system can achieve equilibrium since a temperature rise leads to decreased absorptance, reducing the energy deposition. For temperatures lying in the opposite side of the absorptance peak, the increase in temperature induces an increase in absorptance, leading to a significant increase in energy deposition and consequently an even greater temperature rise.

4. Laser-induced damage of polycrystalline silicon MEMS

The combination of multilayered design, coupled with temperature-induced changes in the optical properties ultimately leads to failure of laser-irradiated MEMS. From a design perspective, in addition to considering the primary mechanical function of the device—such as an actuator (Baglio et al., 2002; Oliver et al., 2003; Phinney et al., 2005; Phinney & Serrano, 2007; Serrano et al., 2005) or a shutter (Wong & Graham, 2003)—the design should also consider the optical and thermal behavior of the structure to reduce the likelihood of damage. To gain a better understanding of the design concerns associated with polysilicon optical MEMS, various experiments have been carried out that have provided insights into the importance of composition, optical energy deposition and thermal transport of heating (Baglio et al., 2002; Oliver et al., 2003; Phinney & Serrano, 2005; Serrano & Phinney, 2009;

Serrano et al., 2009). In this section, we will briefly look over some of the experimental results for laser-induced damage in the context of the optical and thermal analysis discussed in the previous sections.

For optically-powered MEMS thermal actuators (Baglio et al, 2002; Phinney & Serrano, 2007; Serrano & Phinney, 2009) most of the studies have mainly focused on empirically establishing the threshold power for damage. Typically, damage is defined as visible damage at the surface—in the form of a crater-like feature as shown Fig. 9—after initial irradiation of the surface. However, these studies also showed that damage could be initiated after prolonged exposure (on the order of minutes) to the laser irradiation, indicating the presence of a slow heating process. This behavior agrees qualitatively with the concepts discussed in the previous section. Thermal equilibrium for the irradiated structure cannot be achieved for the temperature where the absorptance exhibits a peak. Therefore, the system reaches a metastable equilibrium in the valleys of the absorptance curve as shown in Figs. 6 and 8. These valleys, however, do not represent a flat absorptance, but rather a slowly varying one. Thus, as the devices slowly heats up, the material's absorptance increases until the next absorptance peak is encountered and the deposited energy density is enough to cause damage of the device. The time-delayed damage observed is then evidence of the slow heating and approach of the temperature to the absorptance peak.

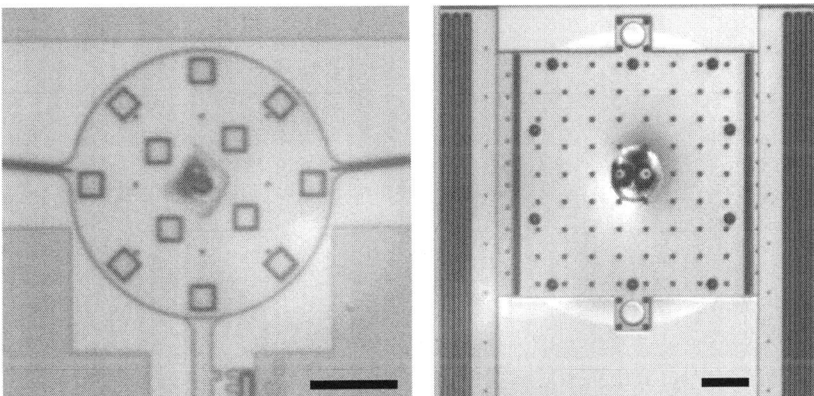

Fig. 9. Typical laser-induced damage on polysilicon MEMS structures. The scale bar on both is equivalent to 50 µm. (Phinney & Serrano, 2007; Serrano & Phinney, 2009)

The effect of the absorptance peaks can also explain the damage thresholds in laser-irradiated microsystems (Serrano & Phinney, 2009) that do not correlate with the number of layers present in the structure. The results show that a single-layer structure exhibited greater power handling capability than various multilayered ones. In said structures, the thin film interference phenomena leads to a minimum in absorptance, like the one shown Fig. 8 for normal incidence near room temperature. This minimum, coupled to improved

heat dissipation to the underlying substrate, permits the single layer structure to exhibit increased robustness to the laser irradiation compared to the multilayered structures.
The optical-thermal effects can also explain the temperature discontinuities observed in the temperature measurements of laser irradiated cantilevers and actuators (Serrano & Phinney, 2007; Serrano et al., 2009), shown in Fig. 9. As predicted above, the discontinuity corresponds to the presence of the peak in the absorptance curve. The surface temperature increases rapidly by 200 K as the peak is encountered. The temperature-power relationship regains a linear relation after the temperature reaches the opposite side of the absorptance peak. Numerical simulations of this experiment, utilizing the non-linear absorptance and known material and geometrical parameters for the irradiated structure, are in good agreement with the measured values, reproducing temperature discontinuity. This type of sudden increase in the temperature makes predicting a threshold power for laser damage in polysilicon structures extremely challenging without accurate knowledge of optical and dimensional properties.

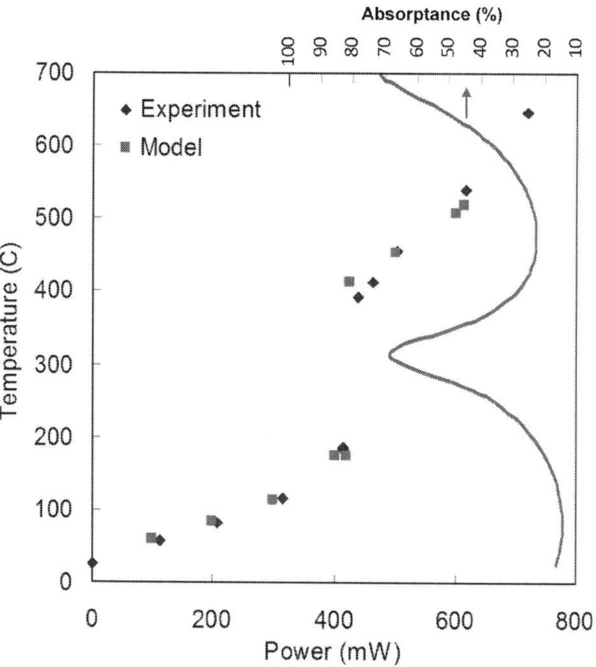

Fig. 10. Measured and modeled temperatures of a polysilicon MEMS structure measured irradiated with an λ = 808 nm laser. The discontinuity in the temperature results from a peak in the absorptance of the irradiated layer due to thin film interference effects (Serrano et al., 2009).

5. Conclusion

Understanding the thermal and optical response of laser-irradiated microsystems requires careful consideration of not only the individual thermal, optical, and mechanical parameters, but also the coupling that exists between them. Of particular importance is the impact that the change in the optical properties with temperature can have in the performance and reliability of these structures. To gain insight into the role that temperature and geometry play in the optical performance of these devices, one must utilize the basic optical relations in a way that is compatible with thermal analyses of a laser-heated structure. The LTR method has proven to be a very useful technique in these types of analyses since it can easily incorporate temperature dependant optical properties and readily provide the interlayer absorptances for the irradiated structures.

Once the temperature and optical fields are coupled in the analysis, a more accurate picture emerges of the thermal and optical behavior of the irradiated device. These coupled optical-thermal effects give rise to non-linear absorptance that can, in some instances, lead to increased resistance to laser damage by dynamically reducing the absorptance as the incident laser power is increased, while in other cases, the non-linear effects compound to enhance absorptance of the incident laser energy producing rapid temperature increases that eventually lead to device damage. A quantitative estimation of device robustness to determine in which regime of damage susceptibility a particular structure resides in therefore requires a complete description of the overall irradiating conditions as well as the device composition. For polysilicon-based devices, this type of analysis has shown reasonable agreement with the experimentally-observed thermal behavior, and can explain the observed damage trends of the laser-irradiated structures.

6. Acknowledgment

The authors would like to acknowledge the help and assistance of Allen Gorby, James Rogers, Wayne Trott, and Jaime Castaneda. Sandia National Laboratories is a multi-program laboratory managed and operated by Sandia Corporation, a wholly owned subsidiary of Lockheed Martin Corporation, for the U.S. Department of Energy's National Nuclear Security Administration under contract DE-AC04-94AL85000.

7. References

Andrews, J. R., Martinez, T., Teare, S. W., Restaino, S. R., Wilcox, C. C., Santiago, F. & Payne, D. M. (2011). A multi-conjugate adaptive optics testbed using two MEMS deformable mirrors. *Proceedings of MEMS Adaptive Optics V,* San Francisco, CA, USA

Andrews, J. R., Teare, S. W., Restaino, S. R., Martinez, T., Wilcox, C. C., Wick, D. V., Cowan, W. D., Spahn, O. B. & Bagwell, B. E. (2008). Performance of a MEMS reflective wavefront sensor. *Proceedings of SPIE,* Vol. 6888, pp. 68880C

Baglio, S., Castorina, S., Fortuna, L. & Savalli, N. (2002). Novel microactuators based on a photo-thermo-mechanical actuation strategy. *Proceedings of IEEE Sensors*, Orlando, FL, USA

Baker, M. S., Plass, R. A., Headley, T. J. & Walraven, J. A., (2004), *Final Report: Compliant Thermo-Mechanical MEMS Actuators LDRD #52553*, SAND2004-6635, Sandia National Laboratories, Albuquerque, New Mexico

Born, M. & Wolf, E. (1999). Principles of Optics: Electromagnetic Theory of Propagation, Interference and Diffraction of Light (7th edition), Cambridge University Press, Cambridge

Burns, D. M. & Bright, V. M. (1998). Optical power induced damage to microelectromechanical mirrors. *Sensors and Actuators A*, Vol. 70, No. 1-2, pp. 6-14

Carter, J., Cowen, A., Hardy, B., Mahadevan, R., Stonefield, M. & Wilcenski, S., (2005), *PolyMUMPs Design Handbook, Revision 11.0*, MEMSCAP, Inc.

Gallis, M. A., Torczynski, J. R. & Rader, D. J. (2007). A computational investigation of noncontinuum gas-phase heat transfer between a heated microbeam and the adjacent ambient substrate. *Sensors and Actuators A: Physical*, Vol. 134, No. 1, pp. 57-68

Jellison Jr, G. E. & Modine, F. A. (1982a). Optical absorption of silicon between 1.6 and 4.7 eV at elevated temperatures. *Applied Physics Letters*, Vol. 41, No. 2, pp. 180-182

Jellison Jr, G. E. & Modine, F. A. (1982b). Optical constants for silicon at 300 and 10 K determined from 1.64 to 4.73 eV by ellipsometry. *Journal of Applied Physics*, Vol. 53, No. 5, pp. 3745-3753

Jellison Jr, G. E. & Modine, F. A. (1983). Optical functions of silicon between 1.7 and 4.7 eV at elevated temperatures. *Physical Review B*, Vol. 27, No. 12, pp. 7466-7472

Katsidis, C. C. & Siapkas, D. I. (2002). General Transfer-Matrix Method for Optical Multilayer Systems with Coherent, Partially Coherent, and Incoherent Interference. *Applied Optics*, Vol. 41, No. 19, pp. 3978-3987

Knoernschild, C., Changsoon, K., Gregory, C. W., Lu, F. P. & Jungsang, K. (2010). Investigation of Optical Power Tolerance for MEMS Mirrors. *Journal of Microelectromechanical Systems*, Vol. 19, No. 3, pp. 640-646

Knoernschild, C., Kim, C., Lu, F. P. & Kim, J. (2009). Multiplexed broadband beam steering system utilizing high speed MEMS mirrors. *Optics Express*, Vol. 17, No. 9, pp. 7233-7244

Lubberts, G., Burkey, B. C., Moser, F. & Trabka, E. A. (1981). Optical properties of phosphorus-doped polycrystalline silicon layers. *Journal of Applied Physics*, Vol. 52, No. 11, pp. 6870-6878

Mazilu, M., Miller, A. & Donchev, V. T. (2001). Modular Method for Calculation of Transmission and Reflection in Multilayered Structures. *Applied Optics*, Vol. 40, No. 36, pp. 6670-6676

MEMS Technologies Department, (2008), *SUMMiT V™ Five Level Surface Micromachining Technology Design Manual: Version 3.1a*, Sandia Report No. SAND2008-0659P, Sandia National Laboratories, Albuquerque, NM

Oliver, A. D., Vigil, S. R. & Gianchandani, Y. B. (2003). Photothermal surface-micromachined actuators. *IEEE Transactions on Electron Devices,* Vol. 50, No. 4, pp. 1156-1157

Phinney, L. M., Klody, K. A., Sackos, J. T. & Walraven, J. A., (2005). Damage of MEMS thermal actuators heated by laser irradiation. *Proceedings of SPIE,* Vol. 5716, pp. 81-88

Phinney, L. M. & Serrano, J. R. (2007). Influence of target design on the damage threshold for optically powered MEMS thermal actuators. *Sensors and Actuators A,* Vol. 134, No. 2, pp. 538-543

Phinney, L. M., Spahn, O. B. & Wong, C. C. (2006). Experimental and computational study on laser heating of surface micromachined cantilevers. *Proceedings of SPIE,* Vol. 6111, pp. 611108

Plass, R. A., Baker, M. S. & Walraven, J. A. (2004). Electrothermal actuator reliability studies. *Proceedings of SPIE,* Vol. 5343, pp. 15-21

Sassen, W. P., Henneken, V. A., Tichem, M. & Sarro, P. M. (2008). Contoured thermal V-beam actuator with improved temperature uniformity. *Sensors and Actuators A: Physical,* Vol. 144, No. 2, pp. 341-347

Schulz, L. G. (1954). The Optical Constants of Silver, Gold, Copper, and Aluminum. I. The Absorption Coefficient k. *Journal of the Optical Society of America,* Vol. 44, No. 5, pp. 357-362

Serrano, J. R. & Phinney, L. M. (2008). Displacement and Thermal Performance of Laser-Heated Asymmetric MEMS Actuators. *Journal of Microelectromechanical Systems,* Vol. 17, No. 1, pp. 166-174

Serrano, J. R. & Phinney, L. M. (2009). Effects of layers and vias on continuous-wave laser heating and damage of surface-micromachined structures. *Journal of Micro/Nanolithography, MEMS and MOEMS,* Vol. 8, No. 4, pp. 043030

Serrano, J. R. & Phinney, L. M. (2007). Micro-Raman thermometry of laser heated surfaces. *Proceedings of ASME InterPACK 2007,* Vancouver, BC, Canada

Serrano, J. R., Phinney, L. M. & Brooks, C. F. (2005). Laser-Induced damage of polycrystalline silicon optically powered MEMS actuators. *Proceedings of ASME InterPACK 2005,* San Franscisco, CA, USA

Serrano, J. R., Phinney, L. M. & Rogers, J. W. (2009). Temperature amplification during laser heating of polycrystalline silicon microcantilevers due to temperature-dependent optical properties. *International Journal of Heat and Mass Transfer,* Vol. 52, No. 9-10, pp. 2255-2264

Sun, B. K., Zhang, X. & Grigoropoulos, C. P. (1997). Spectral optical functions of silicon in the range of 1.13-4.96 eV at elevated temperatures. *International Journal of Heat and Mass Transfer,* Vol. 40, No. 7, pp. 1591-1600

Wong, C. N. C. & Graham, S. (2003). Investigating the thermal response of a micro-optical shutter. *IEEE Transactions on Components and Packaging Technologies,* Vol. 26, No. 2, pp. 324-331

Xu, X. & Grigoropoulos, C. P. (1993). High temperature radiative properties of thin polysilicon films at the λ = 0.6328 μm wavelength. *International Journal of Heat and Mass Transfer,* Vol. 36, No. 17, pp. 4163-4172

Yavas, O., Do, N., Tam, A. C., Leung, P. T., Leung, W. P., Park, H. K., Grigoropoulos, C. P., Boneberg, J. & Leiderer, P. (1993). Temperature dependence of optical properties for amorphous silicon at wavelengths of 632.8 and 752 nm. *Optics Letters,* Vol. 18, No. 7, pp. 540-542

Part 4

MEMS Based Actuators

14

Piezoelectric Thick Films: Preparation and Characterization

J. Pérez de la Cruz
INESC Porto - Institute for Systems and Computer Engineering of Porto, Porto, Portugal

1. Introduction

Sol-gel technology allows the deposition of 1μm thick oxide films onto a variety of substrates at temperatures well below those conventionally used for bulk ceramic processing. Thin film processing temperatures as low as 500-700°C are typically used allowing ceramic materials to be incorporated into the silicon processing stages (Ohno et al., 2000; Wu et al., 1999). In addition, low processing temperatures reduce the inter-diffusion of atomic species between the different thin film layers and ionic vaporization, such as: lead in lead zirconate titanate oxide (PZT) films (Wu et al., 1999; Jeon et al., 2000). PZT thin films have been largely deposited in order to produce several types of devices, such as: membrane sensors, accelerometers and micromotors (Barrow et al., 1997). However, devices requiring larger actuation forces (i.e. high frequency transducers, vibration control devices) require thicker piezoelectric films (Tsurumi et al., 2000). In these cases, it is not practical to produce thick PZT films using standard sol-gel techniques, because of the increased cracking risk due to shrinkage nor is it desirable to produce thick films by a repetitive single layer deposition process due to the time required (Barrow et al., 1995; Zhou et al., 2000).
The interest in ferroelectric lead zirconate titanate thick films for device applications, including high-frequency ferroelectric sonar transducers (Bernstein et al., 1997), micro-electromechanical system devices (Polla & Schiller, 1995; Myers et al., 2003; (Akasheh et al., 2004), elastic surface wave devices (Cicco et al., 1996), hydrophones (Chan et al., 1999) and sensors (Xia et al., 2001), has increased in the last decades because PZT ferroelectric thick films possess the merits of both bulk and thin film materials (Barrow et al., 1997; Ledermann et al., 2003). PZT thick films devices not only work at low voltage and high frequency, as they are compatible with semiconductor integrated circuit, but also possess superior electric properties approaching near-bulk values. Naturally, processing of PZT thick films has also become an increasingly popular research field. Some approaches that have recently been studied to process thick films include electrophoretic deposition (Corni et al., 2008), pulsed laser deposition (Yang et al., 2003), screen printing (Walter et al., 2002) and sol-gel (Xia et al., 2001; Wang et al., 2003; He et al., 2003). Among them, the hybrid sol-gel technique has a special interest due to its low preparation cost and excellent stoichiometric control. Sol-gel also offers the capability to lay down thick layers anywhere between 0.1μm and 100μm, a thickness range that is difficult to achieve by other deposition techniques.

In this chapter an exhaustive review of the preparation of PZT thick films by infiltration method will be presented. Solution powder agglomeration, film densification, phase formation temperature, among others, will be some of the topics that will be analyzed. Finally, the structural and electrical properties of the PZT thick films as a function of the number of solution infiltrations will be highlighted.

2. Formation and structural characterization of the thick PZT films

2.1 Flow diagram of hybrid PZT films

It is well-accepted that hybrid powder sol gel coating technology is an excellent technique to develop high quality thick ceramic films of more than 1μm, while all the benefits of sol gel, i.e. ease of fabrication, ability to coat complex geometries and relative cost effectiveness, remains intact (Barrow et al., 1997; Barrow et al., 1995; Dorey et al., 2002, Pérez et al., 2007). It is also known that high-quality lead zirconate titanate (PZT), yttria- and ceria-stabilized zirconia, titania, silica and alumina thick films with more than 100 μm could be fabricated by this method (Barrow et al., 1995). However, questions remain; for instance, which are the advantages and disadvantages of this method?

It is clear that the advantages of the method are associated with the possibility to obtain high-performance thick films with up to 100μm on a variety of substrate material and shapes, but what about the disadvantages? The main problem of this method is associated with the difficulty to obtain dense films that reproduce the ceramics properties. This problem affects all type of thick films; however, it is more pronounced in ferroelectric materials like PZT, where the dielectric, ferroelectric and piezoelectric properties are highly influenced by the film densification.

The preparation process of thick films, specifically PZT, is divided into two main steps: i) the selection of the sol-gel matrix and ii) the dispersion of the PZT precursor powder in the sol-gel matrix. The sol-gel matrix is selected taking into account its composition, viscosity, and endurance to aging, etc. Once the selection of the sol-gel matrix is completed, the second step is the dispersion of the PZT powder into PZT sol-gel matrix. The powder dispersion process guarantees the homogeneity of the suspension and eliminates the possibility of agglomerate formation.

Several authors have considered that this step is fundamental in the preparation of high quality thick films (Barrow et al., 1997; Barrow et al., 1995; Pérez et al., 2007; Kholkin et al., 2001; Simon et al., 2001], because it determines the powder agglomeration degree and subsequent densification of the films. Few authors have showed that the level of powder agglomeration inside the sol gel matrix depends on the shape and size and the method used to obtain these powders (Simon et al., 2001). Moreover, it is consensual that agglomeration increases with time, due to the surface tension force and the sedimentation process.

To avoid the agglomeration process that take places before and during the thick film preparation, some authors have used organic dispersants (i.e., buthoxyethoxy-ethyl acetate (BEEA)), high molecular weight solvents (i.e., α-terpineol), binders (i.e., polyvinyl butyral (PVB)) and plasticizers (i.e., polyethylene glycol (PEG)). Figure 1 shows the grain size distribution of PZT and PT powders prepared by different synthesis routes, which were dispersed using the above-mentioned compounds (Simon et al., 2001). It is clear that powder preparation processes and consequently grain size and shape have a fundamental role in the powder agglomerate formation. For instances, PZT powder prepared by conventional solid state oxide methods (dry method or DM) shows a mean grain size of 2.2 μm, while PZT powder prepared by a coprecipitation process (web method or WM) shows a mean grain size of 1.9 μm. Some readers could assume that a progressive decrease in powder size could contribute to a smaller powder agglomeration inside the sol-gel matrix, improving the films densification. However, this is not completely true, as when the power size decreases, the effective surface of the powder increases. This increase in the powder effective surface results in an increase of the surface tension force, which strongly contribute to agglomerate formation. An evident example can be observed in Fig. 1, the coprecipitation process produces smaller grain size PZT powders; however, during the sol-gel PZT power mixture a ~10 μm agglomerated is observed.

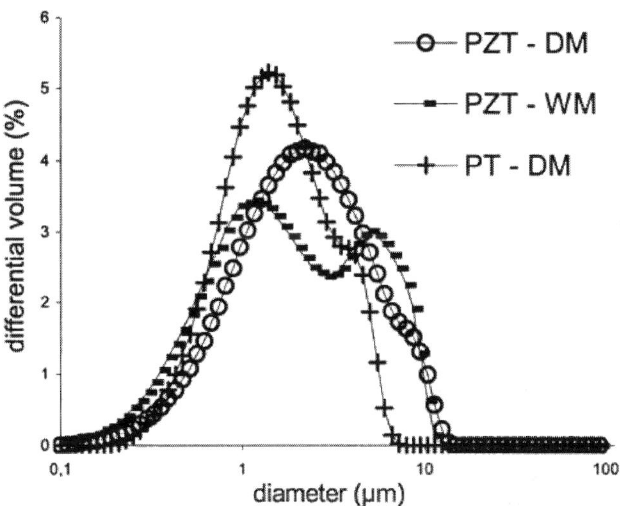

Fig. 1. Grain size distribution of PZT and PT powders prepared by different synthesis routes, and dispersed using α-terpineol solvent, BEEA dispersant, PVB binder and PEG plasticizer (Simon et al., 2001). (Copyright Elsevier)

Similar agglomeration behavior is also observed in high molecular weight free sol gel matrixes. In this case authors have used an ultrasonic dispersion process combined with a relatively high viscous sol-gel matrix in order to eliminate the PZT powder agglomeration (Pérez et al., 2007). As the ultrasonic dispersion time increases, agglomerate formation decreases, as shown in Fig.2. It is clear that the ultrasonic process disperses the agglomerates resulting in an increase of low diameter differential volume. The reduction of the PZT powder agglomeration and the increase in the low diameter differential volume improve the densification of the film during the preparation and reduce the formation of *closed pores*, which cannot be filled by any post-deposition process, such as: infiltration.

It is notable that the ultrasonic process eliminates the agglomerate formation; moreover, it also negates the use of the high molecular weight solvents, dispersants, binders and plasticizer, resulting in shorter drying and pre-annealing processes during the thick film preparation process. The elimination of high molecular weight compounds from the sol-gel matrix reduces in large scale pore formation, improving the PZT thick films densification.

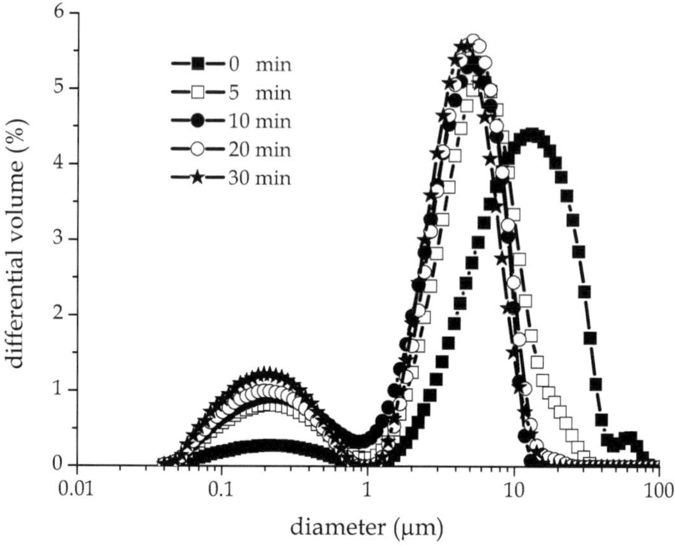

Fig. 2. Distributions of the TRS600 PZT powder particle under different ultrasonic mixing times (Pérez et al., 2007). (Copyright Elsevier)

Hybrid sol-gel/powder formation is followed by the film preparation process. The PZT thick film preparation process practically does not differ to the standard sol-gel thin film preparation (Barrow et al., 1997; Barrow et al., 1995; He et al., 2003; Dorey et al., 2002; Pérez et al., 2007; Kholkin et al., 2001). It is known that PZT thick films could be deposited in several types of substrates; however, it is common that PZT thick films are deposited onto platinized silicon wafers ($Pt/Ti/SiO_2/Si$), due to the higher conductivity of the platinum (Barrow et al., 1995). Prior to the coating step the substrate should be cleaned to remove the substrate dirt. It is usual that the removing of any residual organics of the substrate surface involves low molecular weight volatile compounds, such us: methanol, ethanol, isopropanol, and acetone, among others. However, other techniques like ultrasonic bath and plasma etching are regularly used during the substrate cleaning process (Dorey et al., 2002; Pérez et al., 2007). Afterward, PZT thick films are grown up by depositing a consecutive number of layers. Each layer consists of an initial composite layer, which is deposited by covering the entire wafer surface with the composite slurry and then spinning for 30 s to 60 s in the 2000-4000 rpm spinning frequency range. The spinning frequency and time are normally optimized in order to obtain the desire thickness of each individual layer, which depends on the powder mean size and also on the viscosity of the sol-gel matrix.

Each individual layer is then subjected to a heat treatment process at an intermediate temperature designed to remove the organic component and to pyrolise the PZT sol-gel. This heat treatment process is divided in two step: 1) the drying process, which is carried out from 100°C up to 300°C during ~60 s and 2) the pre-annealing process (or calcination) that is carried in the 300°C-500°C temperature range form few seconds up to various minutes (Barrow et al., 1997; Barrow et al., 1995; He et al., 2003; Dorey et al., 2002; Pérez et al., 2007; Kholkin et al., 2001). The drying and pre-annealing temperature used in the heat treatment of each layer should take into account the type of solvent and the organic compounds utilized in the preparation of the PZT sol-gel matrix and also the stoichiometry of the material that is to be prepared. For that reason, prior to the preparation process it is convenient to carry out the decomposition analysis of the sol-gel matrix by using thermogravimetric and differential thermal analysis techniques. These techniques supply the characteristic decomposition temperatures of the sol-gel matrix. Drying and/or pre-annealing the PZT *green* layers above the characteristic temperatures should be carried out gradually, because it could result in the formation of cracks.

On the other hand, PZT thick films prepared by a sol gel matrix rich in high molecular weight compounds, like dispersants, binders and plasticizers, show long drying and pre-annealing processes that range from 1 hour up to 24 hours. The long time drying and pre-annealing processes are necessary in order to evaporate the solvents and burn all the high molecular weight organic compounds used in the PZT sol-gel matrix preparation avoiding, therefore, the formation of crack in the films.

Once the desire thickness is obtained the whole film is subjected to a crystallization stage where the pre-annealing sample is sintered at higher temperatures designed to develop the perosvkite phase. In the case of PZT thick films the sintering temperature (also known as annealing temperature) can vary from 400°C up to 800°C as a function of the zirconium/titanium ratio, while the annealing time could go from 30 up to 60 minutes. Using this technique, several authors have prepared thick, crack free PZT films (Barrow et al., 1995; Dorey et al., 2002; Pérez et al., 2007). However, Barrow *et. al.,* (Barrow et al., 1995) were the first to attribute the crack free nature of these films to i) the presence of large amounts of powder that results in a decrease of the level of sol-gel present and hence lower shrinkage; (ii) strong bonding between the sol-gel and the PZT particles making cracking less likely.

On the other hand, Wu *et. al.,* (Wu et al., 1999) proved that the incorporation of PZT powder into the PZT sol-gel solution shows additional benefits. The addition of ~1 wt.% of PZT powder decreases the perovskite formation temperature by 50 °C, increasing substantially the dielectric and ferroelectric properties of these films (Wu et al.,1999). It is believed that the incorporation of PZT micro-powders in a PZT sol-gel matrix promotes heterogeneous nucleation of the perovskite phase coming from the sol-gel, resulting in a randomly orientated PZT film.

Dorey *et. al.,* (Dorey et al., 2002) and Pérez *et. al.,* (Pérez, 2004) observed that a graded structure is obtained when the PZT thick film is not infiltrated with the precursor PZT sol-gel solution, during the preparation process. This graded structure results because the sol-gel is drawn from the slurry into the underlying porous composite layer (Dorey et al., 2002). It is believed that as the number of layers increased, the lower composite layers become further enriched with sol-gel. Thus, the bottom composite layer of a four layer structure will be effectively infiltrated three times (Dorey et al., 2002). It is evident that the infiltration of the bottom layer is conditioned by the porous infiltration saturation. Pérez *et. al.,* observed that there is a progressive infiltration of PZT graded structure up to four infiltrations (Pérez, 2004). However, at higher infiltrations the porosity of the bottom layer practically does not change, which can be further explained based on Darcy´s law (Scheidegger, 1974).

It is consensual that the formation of a graded structure is detrimental to the dielectric, ferroelectric and piezoelectric properties of the PZT thick films. Hence, an intermediate infiltration of each individual composite layer and a final infiltration of the whole film are necessary, leading to a homogenization of the film structure, fixing the graded structural problem, as shown by Perez and co-workers (Pérez et al., 2007; Pérez, 2004). The infiltration of each individual layer with sol-gel prior to the deposition of the next composite layer results in a relative densification of the layer and a strengthening of the powder compact without shrinkage, while the final infiltration is used to improve the PZT thick film surface (Tu et al., 1995). Thus, with this method it is possible to produce high-quality PZT thick composite films with uniform densities and good surfaces.

2.2 Infiltration of the PZT thick films

It was mentioned above that the infiltration process of the PZT thick film could be explained based on Darcy´s law (Scheidegger, 1974). The experiments of Darcy were focused on the volumetric flow rate (Q) of the fluid through a sand column, which is similar to the infiltration of a sol-gel solution through a porous film. It can be observed that flow occurs only in the pore space. Thus, the effective area of flow is not the entire column cross section (A), but this area multiplied by the porosity ($\mu \cdot A$). Note that although the porosity μ is a volume fraction, it is also useful in determining an average effective area of flow. Thus the average speed of the macroscopically one-dimensional flow in a cross section of Darcy's column relative to the solid grains is:

$$\left| v^l - v^s \right| = \frac{Q}{\mu A} \tag{1}$$

where v^l is the velocity of the liquid in a porous media and v^s is the velocity of the media. It is clear that in the thick films case the velocity of the media is ~0.

Taking into account that Darcy demonstrated experimentally that the volumetric flow rate of a liquid (i.e., water) down through the porous medium is proportional to the head difference across the sand column ($h_2 - h_1$), and the cross sectional flow area and is inversely proportional to the packed height of the column (L), such that:

$$Q = KA \left| \frac{h_2 - h_1}{L} \right| \tag{2}$$

where K is referred to as hydraulic conductivity (or permeability of the porous body) and it is a function of both the porous medium and the fluid properties, being practically constant for a particular packing even when the flow rate in the column changes. One can obtain the differential form of Darcy´s law, combining the equations 1 and 2 when the limit of the column length is reduce (L→0), as shown:

$$\mu(v^l - v^s) = K^l \frac{dh}{dz} \tag{3}$$

where the hydraulic conductivity (K^l) has been labelled with a superscript (l) to emphasize that in a particular medium, its value will depend on the properties of the fluid phase.

$q^l = \varepsilon(v^l - v^s)$ In a general case the Darcy velocity can be represented in terms of the gradient of pressure elevation heads as:

$$q^{l} = \frac{K^{l}}{\rho g}(\nabla P - \rho g) \tag{4}$$

where ρ is the density of the liquid and P the pressure experimented by the liquid. This equation is accurate when the pressure gradient in the liquid column is balanced by gravitational forces. For this equilibrium state, the Darcy velocity is zero. The equations propose that for a "small" imbalance in the gradients that drive the flow, the Darcy expressions reasonably describe the velocity (or infiltration velocity). Confirming experimental evidence indicates that Darcy's correlation is a useful expression to describe flow through porous medium (Scheidegger, 1974). It is clear that the Darcy velocity is not a true velocity of the fluid but represents an effective flow rate through the porous medium.

Several pressures could influence the infiltration process. However, it can be resumed to three main pressures, the applied pressure *Pa* (which includes the hydrostatic pressure), the capillarity force *Pc* and the internal opposing pressure of the compressed gas *Pi*, as shown by Kholkin et. al., (Kholkin et al., 2001). During a "static" infiltration, such as used during a dip-coating process, the infiltration is mainly dominated by the hydrostatic pressure, the capillarity pressure and the internal opposing pressure of the compressed gas; however, during a "dynamic" infiltration, such as used during a spin-coating process, the infiltration is mainly dominated by the applied pressure due to the centrifugation process.

Kholkin et al., (Kholkin et al., 2001) has predicted based on the modified Darcy´s law reported by Scheidegger et. al., (Scheidegger, 1974) (see Equation 5) that for a static infiltration process, where the flow of the liquid into the porous media in controlled by the capillarity pressure, the distance of liquid introduced in the PZT thick film (z) during the time (t=30s) is ≈ 2mm.

$$z = \left(\frac{2KP}{\eta}\right)^{\frac{1}{2}} t^{\frac{1}{2}} \tag{5}$$

This value is 200 times greater than the thickness of the PZT thick film deposited in this study (~10nm). Authors suggest that this value was reached because they did not take into account the increase of the internal opposing pressure of the compressed gas due to the no evacuation of the replaced gas (Kholkin et al., 2001). However, there are other factors like the viscosity of the solution and the permeability of the porous body that should be taken into account. It is known that as the number of infiltrations increase, the viscosity of the solution increases, resulting in a decrease of the infiltration depth (Dorey et al., 2002). On the other hand, the permeability of a powder compact is expressed by the Kozeny-Carman expression:

$$K = \frac{D^{2}(1-\rho)^{3}}{36C\rho^{2}} \tag{6}$$

where D is the average grain size and C de is a parameter that define the shape and tortuosity of the porous channel (Scheidegger, 1974). The C value for a powder compact is estimated as ≈5; however, it increases exponentially as the number of infiltrations increase, mainly because the porous size is reduced. It is believed that this is the major factor that reduces the infiltration depth in powder compact in general and in PZT thick films in particular.

In the next section, one analyzes the effect of the number of top infiltrations in structural, dielectric and piezoelectric properties of a intermediate infiltrated PZT thick film. Moreover, the dielectric properties of the infiltrated PZT thick films will be simulated based on 0-3 and cube ceramic/ceramic composite models, where the numbers 0 and 3 describe the connectivity of the two phases of the material (i.e., the sol gel matrix interconnected in the three directions (3) whereas PZT powder particles are not connected in any direction (0)) (Newnham et al., 1978). Finally, the structural and electrical results will be compared with the ones reported by Dorey *et. al.*, and Ohno *et al.*, in PZT thick film infiltrated with a high molecular weight prepared PZT solution (Ohno et al., 2000; Pérez et al., 2007).

3. Structural, electric and piezoelectric characterization of the PZT thick films

3.1 Structural and microstructural characterization of the infiltrate PZT thick films

There are practically no differences between the structural and microstructural characterizations of thin and thick PZT films. However, it should be taken into account that PZT thick film phase formation, grain growth, crack formation, etc., are highly conditioned by the precursor powder and the powder agglomerate formation, while in PZT thin films they are mainly conditioned by the substrate structure, the characteristic of the PZT precursor solution and the preparation conditions.

3.1.1 Structural characterization of the precursor powder and the infiltrated PZT thick films (X-ray analysis)

Prior to the structural characterization of the PZT thick films it is important to carry out the precursor powder characterization. Figure 3 a) shows the X-ray diffraction pattern of the PZT TRS600 precursor powder used in the preparation of infiltrated PZT thick films (Pérez et al., 2007, Pérez, 2004). The diffraction pattern of this PZT powder shows the presence of a pure perovskite PZT phase and two marginal extra diffraction peaks (26.8° and 33.12°) that are associated with the lead excess and other additives (such as Nb_2O_5) used in the preparation of the commercial PZT powders (TRS, 1998, Kholkin et al., 2000). Normally, the additive compounds used in the preparation of commercial PZT powder have the objective of improving its dielectric and piezoelectric properties. However, they also show extra attributes when the PZT powders are used in the preparation of PZT thick films. For instances, the lead oxide excess present in the PZT precursor powder guarantee that the PZT thick films remain stoichiometric after the annealing process.

Figure 3b) shows the X-ray diffraction patterns of PZT thick films prepared with different number of top infiltrations. The analysis of the (110) PZT diffraction peak shows a decrease of the width and a small shift of the peak position, relatively to the TRS600 PZT commercial powder, as the number of infiltration steps increases. It is evident that the infiltration process reduces the PZT thick film surface roughness, which results in a better relation between incident and diffraction angle of the films. For that reason, the decrease in the width of the (110) PZT diffraction peak is mainly attributed to the decrease of the surface roughness.

In contrast, the shift in the peak position of the maximum as the number of top infiltration layers increase, has been associated with two factors: i) a small difference between PZT powder and PZT solution compositions and ii) a change in the stress of the films provoked by the infiltration cycle, which somehow compacts the film structure.

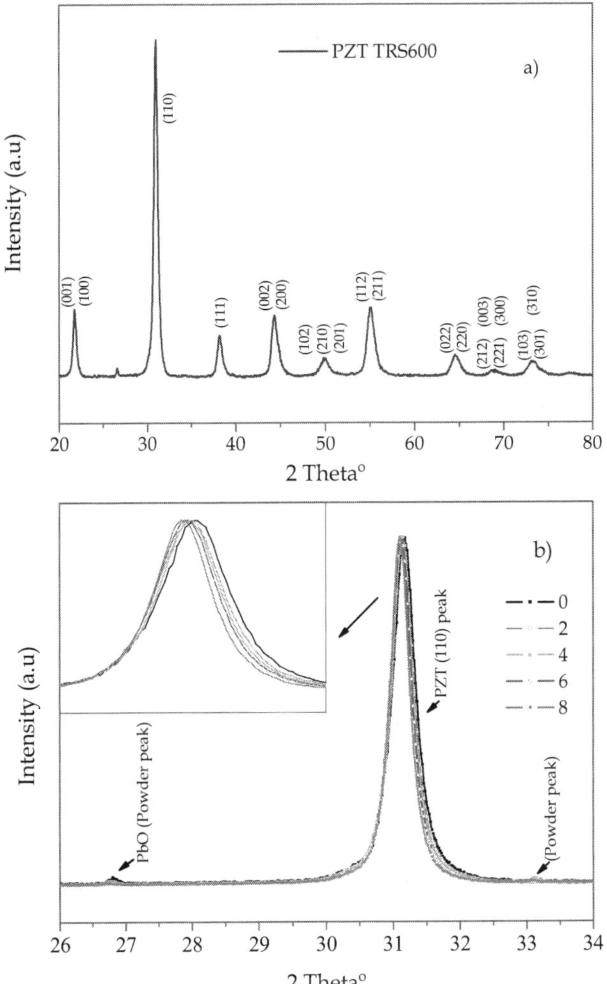

Fig. 3. X-ray diffraction patterns of a) TRS600 PZT precursor powder and b) PZT thick films prepared with different number of top infiltrations (0, 2, 4, 6, and 8) (Pérez et al., 2007). (Copyright Elsevier)

Finally, a decrease in the intensity of the extra TRS600 X-ray diffraction peaks (visible at 26.8° and 33.12°) is observed, showing that as the number of infiltrations increase the factional volume of the formed sol-gel PZT phase is more palpable. This fact emphasizes the idea that as the number of infiltrations increases the number of pores decreases due to a complete coverage by the sol-gel solution. Moreover, the decrease in the 26.8° X-ray diffraction peak could be also associated with the possible evaporation of the lead oxide during the PZT thick film heat treatment process.

3.1.2 Microstructural characterization of the infiltrated PZT thick films (SEM analysis)

Following the crystallographic structural analysis, the microstructural analysis of standard and infiltrated PZT thick films reveals important aspects like: surface roughness, surface cracks, powder agglomeration, film thickness, porosity, among others. When the film is not infiltrated it is easy to observe cracks and small powder agglomerates in the surface of the films, while after few infiltrations the PZT thick films show a smoother and crack-free surface, as shown in Fig.4. It is visible that the infiltration process results in a decrease of the film porosity, increasing the connectivity of the composite films. This fact is extremely important in order to improve the dielectric and ferroelectric properties of the PZT thick films. Moreover, the infiltration process also reduces the film surface roughness, which is helpful for a good adhesion between the film and the top electrode and for a correct estimation of the thickness of the films that will be used in the further calculation of the dielectric and ferroelectric properties.

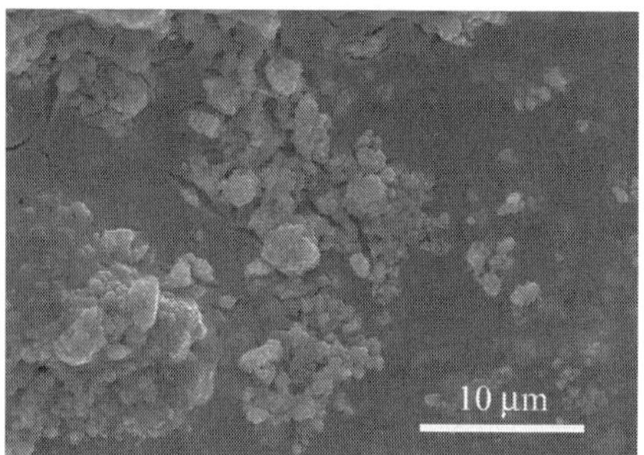

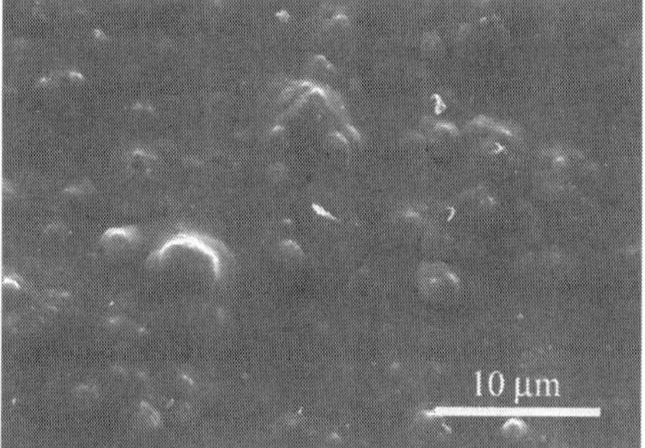

Fig. 4. SEM plan-view images of the PZT thick films with 0 (left) and 8 (right) top infiltrations (Pérez et al., 2007). (Copyright Elsevier)

Figure 5 shows the SEM cross-section microstructure images of the PZT thick films prepared with 0 and 8 top infiltrations. Relatively dense microstructure is observed in the film with 8

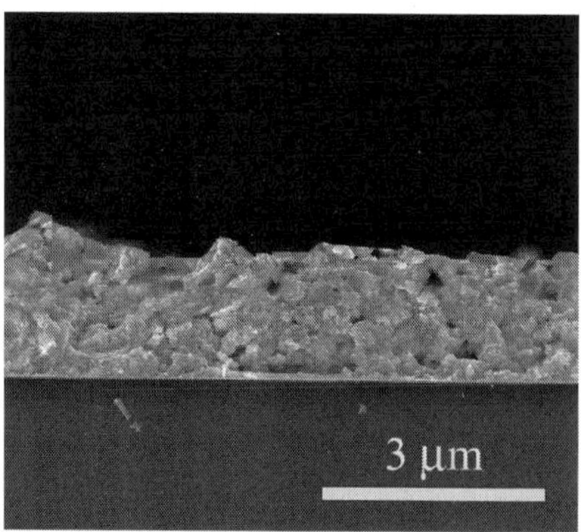

Fig. 5. SEM cross section images of the PZT thick films after 0 (left) and 8 (right) top infiltrations (Pérez et al., 2007). (Copyright Elsevier)

infiltrations, while film without infiltration shows a porous structure mainly in the last deposited layer. It is also visible in the cross section images that the infiltration process results in a decrease of the film porosity; however, there is some remaining porosity that cannot be eliminated. It is called *internal or closed porosity* and results from a premature closing up of the pores channels.

Dorey et al., report a similar behavior in infiltrated PZT thick films using a high molecular weight precursor solution, as shown in Fig. 6 (Dorey et al., 2002). We can see that in films with four infiltrations of this solution, the internal porosity is higher than those reported by Pérez et. al., (Pérez, 2004). It is believed that the higher closed porosity observed in these films result from the increase of the density and viscosity of the precursor solution. Nevertheless, other factors like average grain size and shape and tortuosity of the pore channel should be also taken into account.

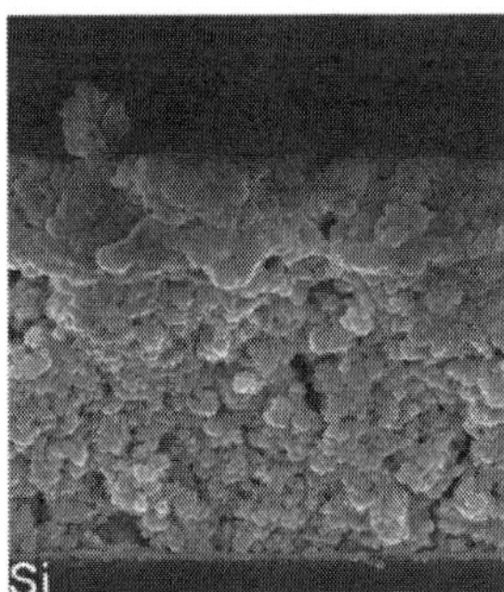

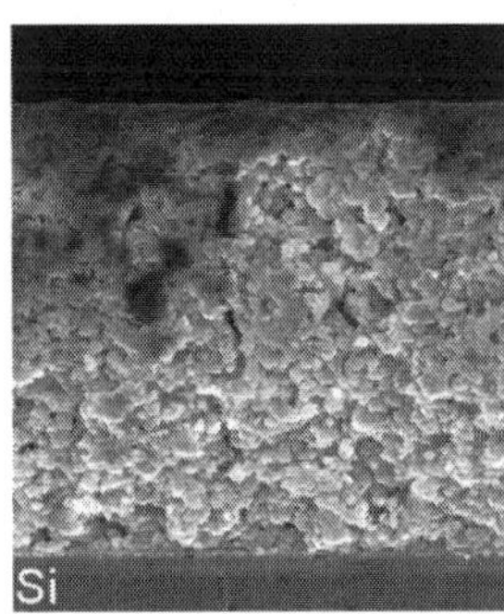

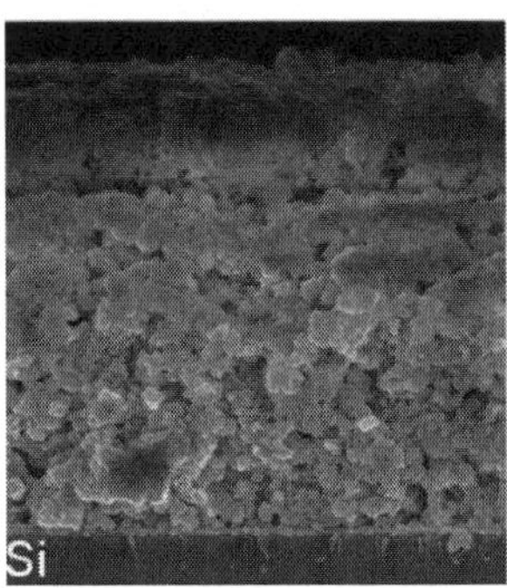

Fig. 6. SEM cross section images of the PZT thick films after 0 (left), 2 center and 4 (right) infiltrations using a high molecular weight precursor solution (Dorey et al., 2002). (Copyright Elsevier)

Surface roughness and porosity can be responsible for the possible deterioration of the dielectric constant and polarization values and the increase dielectric losses in PZT thick film. As mentioned, they are highly affected by the infiltration process. For this reason, in the next section one presents the dielectric, ferroelectric and piezoelectric behavior of the PZT thick films as a function of the number of infiltrations.

4. Electrical properties of the PZT thick films as a function of the number of infiltrations

We have already seen that in PZT thick films the structural and microstructural properties are highly dependent on the number of infiltrations, but what about the dielectric, ferroelectric and piezoelectric properties? It is well-known that in bulk materials the dielectric, ferroelectric and piezoelectric properties are highly dependent on grain size, porosity, phase formation, stoichiometric, crystallographic orientation, amongst others. However, in PZT thick films this dependence may be notably different because the electrical properties are also affected by substrate clamping, surface powder agglomeration and the mixture of phases coming from the precursors PZT powder and PZT sol-gel solution, etc.

4.1 Dielectric behavior of the PZT thick films as a function of the number of infiltrations

Figure 7 shows the dielectric constant and dielectric losses behavior reported by Pérez and co-workers for infiltrated PZT thick films (Pérez et al., 2007). It is observable that the dielectric constant increases as the number of infiltrations increase, while the dielectric loss remains around 0.05. This dielectric constant behavior is consistent with the decrease in porosity as the film is infiltrated, reaching its maximum around 2320. This value is far from the 3420 reported on the TRS600 PZT powder (Kholkin et al., 2001, TRS, 1998). However, it should be taken into account that a phase mixture is presented in the PZT thick film, being the dielectric constant of the PZT layer obtained from the sol-gel solution ~1900 (Pérez et al., 2004), and, furthermore, that there is some closed porosity .
The increases in the dielectric constant as the number of infiltrations increase could be easily explained based on connectivity models (Barrow et al., 1997; Kholkin et al., 2001). These models describe the connectivity degree between the PZT powders particles inside the PZT films matrix formed after the deposition process.

4.1.1 Dielectric constant models for connected composite material

Based on the deposition process of the PZT infiltrated thick films reported by Pérez and co-workers, which includes intermediate infiltrations, the dielectric constant might be modeled as a 0-3 composite material, where the sol gel matrix is fully connected in three directions and the ceramic particles (powder) are not connected in any direction. Assuming that the particles are uniformly dispersed in the sol gel matrix, then the resulting dielectric constant can be given by (Moulson & Herbert, 2004):

$$\varepsilon_m = \varepsilon_2 \left[1 + \frac{3 * V_1(\varepsilon_1 - \varepsilon_2)}{\varepsilon_1 + 2 * \varepsilon_2 - V_1(\varepsilon_1 - \varepsilon_2)} \right] \tag{7}$$

where ε_1 is the dielectric constant of the PZT powder, ε_2 is the dielectric constant of the sol gel matrix, and V_1 is the volume fraction of the ceramic powder.
Practically, this equation is only valid for low powder volume fraction values $V_1<0.1$ (Barrow et al., 1997), because at higher concentrations, the dispersed phase starts to form continuous structures throughout the bulk that have nonzero connectivity. However, due to the intermediate infiltration process carried out in the preparation of the PZT thick films the 0-3 connectivity is practically maintained. Thus, this model can be applied for relatively higher powder volume fraction values.
Although the 0-3 connectivity model is not faraway from reality, a more likely scenario is that proposed by Pauer (Barrow et al., 1997; Pauer, 1973], where the material exists in a combination of series and parallel phases (cube model). In this case the effective dielectric constant can be calculated by:

$$\varepsilon_m = \frac{\varepsilon_1 \varepsilon_2}{(\varepsilon_2 - \varepsilon_1) V_1^{-1/3} + \varepsilon_1 V_1^{-2/3}} + \varepsilon_2 (1 - V_1^{2/3}) \tag{8}$$

Assuming that in a highly infiltrated PZT films the powder/solution volume fraction is 0.66/0.33, the dielectric value calculated by both models is ~2580, a little higher than the one obtained in the experimental results (see Figure 8). Note both models did not take into consideration the porosity of the films. For that reason, we could expect lower effective dielectric values than 2580.

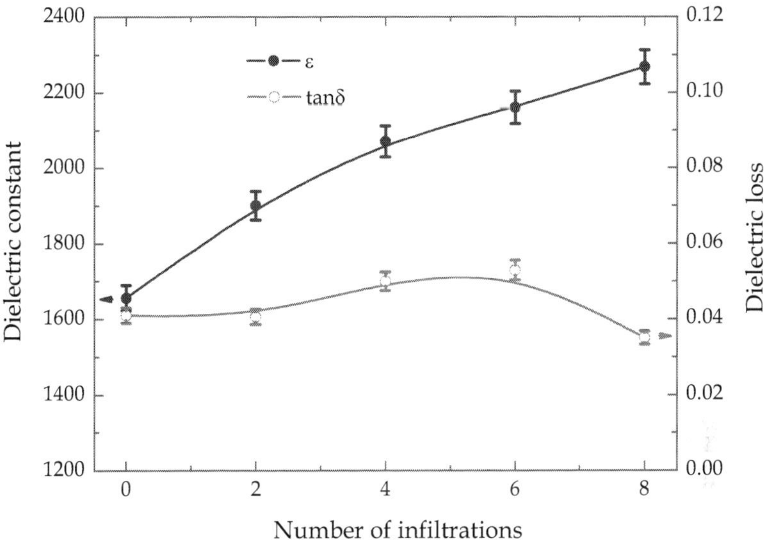

Fig. 7. Dielectric behavior of the PZT thick films, prepared based on a low molecular weight precursor solution, as a function of the number of infiltrations (Pérez et al., 2007). (Copyright Elsevier)

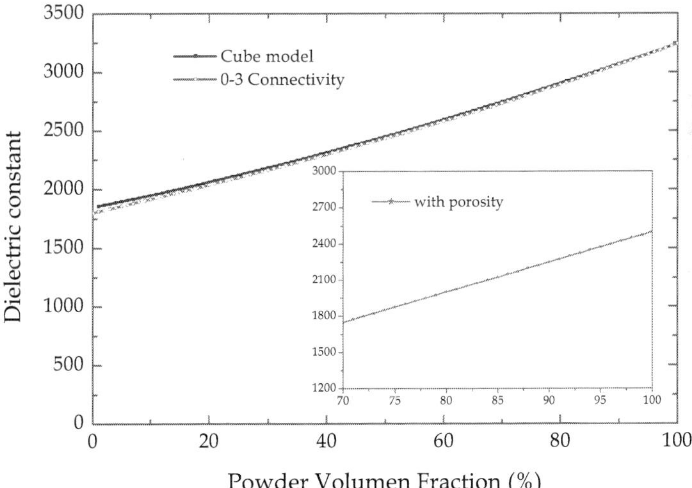

Fig. 8. Dielectric constant of PZT composite predicted by 0-3 composite and cube models. Inset plot shows the dielectric constant values calculated taking into account a 30% to 0% variable porosity in parallel with the PZT composite (0.66/0.33 powder/solution ratio).

In the hypothetic scenario that 8% of the films is free space or porosity filled by air and also that they are connected in parallel with the composite material (powder/solution), the dielectric constant values match with the experimental results (see inset plot Figure 8). At the extremes, one observes deviations from the experimental values due to the irremovable porosity at higher dielectric volume fraction and the actuation of a serial porosity component at a "lower" composite volume fraction. The irremovable porosity is enclosed in the powder agglomeration, which hinder its elimination while the serial porosity contribution appears when the composite films are not infiltrated. It is clear that in non-infiltrated PZT thick films the porosity contributes to both serial and parallel capacitances of the system, degrading the dielectric constant of these films.

4.2 Ferroelectric behavior of the PZT thick films as a function of the number of infiltrations

Figure 9 shows the hysteresis loop of PZT thick films prepared using different number of infiltrations. It is visible that the remnant polarization values increases with the increase of the number of infiltrations. A large remnant polarization in the orders of *Pr*=35 μC/cm² and a small coercive field *Ec*=59 kV/cm are obtained in PZT thick film prepared with 8 top infiltrations, as shown in the inset of Figure 9. The remnant polarization value is similar to the one reported for 1μm coating PZT films prepared using a modified solution (Pérez et al., 2004). However, the coercive field is lower suggesting that it is easier to switch the polarization in infiltrated PZT thick films. In this study, the polarization trend results for a decrease of the porosity and improvement of the films surface as the number of infiltration

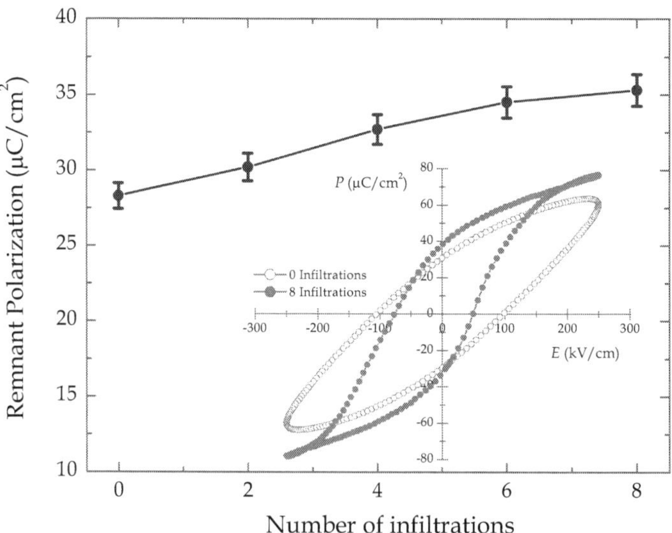

Fig. 9. Remnant polarization of the PZT thick films as a function of the number of top infiltration. Inset plot show the hysteresis loop of the films with 0 and 8 infiltrations (Pérez et al., 2007).

increase. It is comprehensible that as the channel of the pore and the pore size itself are reduced; the infiltration becomes more and more difficult, resulting in a blocking of the infiltration process. The saturation in dielectric constant and remnant polarization also indicates that infiltration and pore size reduction is almost stopped for higher number of infiltrations.

4.3 Piezoelectric behavior of the PZT thick films as a function of the number of infiltrations

Figure 10 illustrates the piezoelectric coefficient as a function of the number of infiltration and the piezoelectric loop of the PZT thick film prepared with 8 top infiltrations. The trend in the piezoelectric coefficient is similar to those reported for the polarization, showing a piezoelectric coefficient (d_{33}) of ~65 pm/V in the films with 8 top infiltrations. It is notable that the piezoelectric coefficients observed in the 8 times infiltrated PZT thick film is of the same order of magnitude as those reported for a 1μm PZT coating films prepared using a modified solution (Pérez et al., 2004). On the other hand, in the non-infiltrated film the piezoelectric coefficient is not reported due to the high surface roughness, which does not allow the piezoelectric measurement of this film.

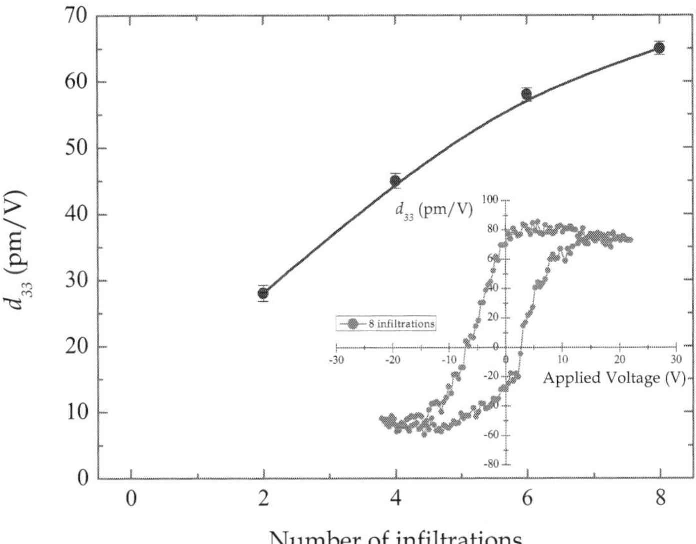

Fig. 10. d_{33} piezoelectric thick film behavior as a function of the number of infiltrations. Inset plot show the piezoelectric loop of the PZT thick films prepared using 8 top infiltrations.

Dorey *et. al.*, reported the same piezoelectric coefficient (~65 pm/V) for PZT thick film infiltrated with a high molecular weight solution (Dorey et al., 2002). The d_{33} coefficients obtained by Dorey and Pérez suggests that the piezoelectric response of the infiltrated PZT thick films is highly conditioned by the piezoelectric response of the PZT phase formed from the sol-gel solution (Dorey et al., 2002; Pérez et al., 2007).

Finally, it should be noted that although the piezoelectric coefficient of highly infiltrated PZT thick film does not reach the piezoelectric coefficient of the PZT ceramics, it is good enough for micromechanical applications. Note it is much higher than for non-ferroelectric piezoelectrics clamping by rigid substrate (ZnO, AlN) (Trolier-McKinstry & Muralt, 1973) and in the same order as that of PZT thin film deposited onto a platinized substrate (Pérez et al., 2004).

5. Conclusion

In this work an exhaustive review on the preparation of PZT thick films have been carried out, taking specific focus in the effect of the infiltration in the preparation of high-quality films. Solution powder agglomeration, films densification, phase formation temperature, among others, have been analyzed. Moreover, the infiltration process is discussed based on Darcy´s law, while the improvement of structural and microstructural properties has been analyzed as a function of the infiltration process. The dielectric properties are characterized as function of the number of infiltrations and the results compared with those obtained by the 0-3 composite connectivity and the cube models. Finally, ferroelectric and piezoelectric properties are also discussed, as function of the number of infiltrations.

6. References

Akasheh, F.; Myers, T.; Fraser, J.; Bose, S. & Bandyopadhyay, A. (2004). Development of Piezoelectric Micromachined Ultrasonic Transducers, *Sensors and Actuators A-Physical*, Vol.111, No.2-3, pp. 275–287

Barrow, D. A., Petroff, T. E. and Sayer M. (1995). Thick Ceramic Coating Using a Sol Gel Based Ceramic-Ceramic 0-3 Composite, *Surface and Coating Technology*, Vol. 76-77, No. 2-3, pp. 113–118

Barrow, D. A.; Petroff, T. E.; Tandon, R. P. & Sayer M. (1997). Characterization of Thick Lead Zirconate Titanate Films Fabricated Using a New Sol-Gel Based Process, *Journal of Applied Physics*, Vol.81, No.2, pp. 876–881

Bernstein J. J., Finberg S. L., Houston K., Niles L. C., Chen H. D., Cross L. E., Li K. K., and Udayakumar K. (1997). Micromachined High Frequency Ferroelectric Sonar Transducers, *IEEE Transactions on Ultrasonics Ferroelectric and Frequency Control*, Vol. 44, No.5, pp. 960–969

Chan, H. L. W.; Lau, S. T. & Kwok, K. W. (1999). Nanocomposite Ultrasonic Hydrophones, *Sensors and Actuators A-Physical*, Vol.75, No.3, pp.252–256

Cicco, G. D.; Morten, B. & Prudenziati, M. (1996). Elastic Surface Wave Devices Based on Piezoelectric Thick-Films, *IEEE Transactions on Ultrasonics Ferroelectric and Frequency Control*, Vol.43, No.1, pp. 73–77

Corni, I.; Ryan, M. P. & Boccaccini, A. R. (2008). Electrophoretic Deposition: From Traditional Ceramics to Nanotechnology, *Journal of the European Ceramic Society*, Vol.28, No.7, pp. 1353–1367

Dorey, R. A.; Stringfellow, S. B. & Whatmore, R. W. (2002). Effect of Sintering Aid and Repeated Sot Infiltrations on the Dielectric and Piezoelectric Properties of a PZT Composite Thick Film, *Journal of the European Ceramic Society*, Vol.22, No.16, pp.2921–2926

He, X. Y.; Ding, A. L.; Zheng, X. S.; Qiu, P. S.; & Luo, W. G. (2003). Preparation of PZT(53/47) Thick Films Deposited by a Dip-Coating Process, *Microelectronic Engineering*, Vol.66, No.1-4, pp.865–871

Jeon Y., Chung J., and No K. (2000). Fabrication of PZT Thick Films on Silicon Substrates for Piezoelectric Actuator, *Journal of Electroceramics*, Vol.4, No.1, pp. 195–199

Kholkin, A. L.; Yarmarkin, V. K.; Wu, A.; Vilarinho P. M. & Baptista J. L. (2000). Thick Piezoelectric Coatings via Modified Sol-Gel Technique, *Integrated. Ferroelectrics*, Vol.30, No.1-4, pp.245–252

Kholkin, A.L.; Yarmarkin V.K.; Wu, A.; Avdeev, M.; Vilarinho, P. M. & Baptista, J. L. (2001). PZT-Based Piezoelectric Composites via a Modified Sol-Gel Route, *Journal of the European Ceramic Society*, Vol.21, No.10-11, pp.1535–1538

Ledermann, N.; Muralt, P.; Baborowski, J.; Gentil, S.; Mukati, K.; Cantoni, M.; Seifert, A. & Setter, N. (2003). {100}-Textured, Piezoelectric $Pb(Zr_xTi_{1-x})O_3$ Thin Films for MEMS: Integration, Deposition and Properties, *Sensors and Actuators A-Physical*, Vol.105, No.2, pp. 162–170

Moulson, A. J. & Herbert, J. M. (1990). *Electroceramics*, University Press, Cambridge, UK

Myers, T. B.; Bose, S.; Fraser, J. D. & Bandyopadhyay, A. (2003). Micro-Machining Of PZT-Based MEMS, *American Ceramic Society Bulletin*, Vol.82, No.1, pp. 30–34

Newnham, R. E.; Skinner, D. P. & Cross, L. E. (1978). Connectivity and Piezoelectric-Pyroelectric Composites, *Material Research Bulletin*, 13, No.5 (1978) pp.525–536

Ohno, T.; Kunieda M.; Suzuki, H. & Hayashi, T. (2000). Low-Temperature Processing of $(PbZr_{0.53}Ti_{0.47}O_3)$ Thin Films by Sol-Gel Method, *Japanese Journal of Applied Physics*, Vol. 39, No.9B, pp. 5429–5433

Pauer, L.A. (1973). Flexible piezoelectric materials, *IEEE International Convention Record*, Vol.21, pp.1–5

Pérez, J. (2004). Preparation and Characterization of Ferroelectric PZT Films for Electromechanical and Memory Applications", *Ph.D. Thesis Dissertation*, University of Aveiro, Portugal

Perez, J.; Vilarinho, P. M. & Kholkin A. L. (2004). High-Quality $PbZr_{0.52}Ti_{0.48}O_3$ Films Prepared by Modified Sol–Gel Route at Low Temperature, *Thin Solid Films*, Vol.449, No. 1-2, pp. 20–24

Pérez, J.; Vyshatko, N. P.; Vilarinho, P. M. & Kholkin, A. L. (2007). Electrical Properties of Lead Zirconate Titanate Thick Films Prepared by Hybrid Sol-Gel Method with Multiple Infiltration Steps, *Materials Chemistry and Physics*, Vol.101, No.2-3, pp.280–284

Polla, D. L. & Schiller, P. J. (1995). Integrated Ferroelectric Microelectromechanical Systems (MEMS), *Integrated Ferroelectrics*, Vol.7, No.3-4, pp. 359–370

Scheidegger, A. E. (1974), The Physics of Flow Through Porous Media, (3rd Ed.), ISBN 10-0802018491, University of Toronto Press, Toronto, Canada

Simon, L.; Le Dren, S.; & Gonnard, P. (2001). PZT and PT Screen-Printed Thick Films, *Journal of the European Ceramic Society*, Vol.21, No.10-11, pp.1441–1538

Trolier-McKinstry, S. & Muralt, P. (2004). Thin Film Piezoelectrics for MEMS, *Journal of Electroceramics*, Vol.12, No.1-2, pp.7–17

TRS ceramic Data Sheet, "*Properties of TRS Soft Piezoceramics*", 8-3-98, (1998)

Tsurumi, T.; Ozawa, S.; Abe, G.; Ohashi, N.; Wada, S. & Yamane, M. (2000). Preparation of $PbZr_{0.53}Ti_{0.47}O_3$ Thick Films by an Interfacial Polymerization Method on Silicon

Substrates and Their Electric and Piezoelectric Properties, *Japanese Journal of Applied Physics*, Vol.39, No.9B. pp. 5604–5608

Tu, W. C. & Lange, F. F. (1995). Liquid Precursor Infiltration Processing of Powder Compacts 1. Kinetic Studies and Microstructure Development, *Journal of the American Ceramic Society*, Vol.78, No.12, pp. 3277–3282

Walter, V.; Delobelle, P.; Le Moal, P.; Joseph, E.; & Collet, M. (2002). A Piezo-Mechanical Characterization of PZT Thick Films Screen-Printed on Alumina Substrate, *Sensor and Actuators A-Physical*, Vol.96, No2-3, pp.157–166

Wang, Z.; Zhu, W.; Zhao, C. & Tan, O. K. (2003). Dense PZT Thick Films Derived from Sol-Gel Based Nanocomposite Process, *Material Science Engineering B*, Vol.99, No.1-3, pp.56–62

Wu, A.; Vilarinho, P. M.; Miranda Salvado, I. M.; Baptista, J. L.; de Jesus, C. M. & da Silva, M. F. (1999). Characterization of Seeded Sol-Gel Lead Zirconate Titanate Thin Films, *Journal of the European Ceramic Soc*iety, Vol.19, No.6-7, pp. 1403–1407

Xia, D.; Liu, M.; Zeng Y. & Li, C. (2001). Fabrication and Electrical Properties of Lead Zirconate Titanate Thick Films by the New Sol–Gel Method, *Material Science and Engineering B*, Vol. 87, No.2, pp. 160–163

Yang, C.; Liu, J.; Zhang, S. & Chen, Z. (2003). Characterization of $Pb(Zr,Ti)O_3$ Thin Film Prepared by Pulsed Laser Deposition, *Material Science Engineering B*, Vol.99, No.1-3, pp.356–359

Zhou, Q. F.; Chan, H. L. W. & Choy, C. L. (2000). PZT Ceramic-Ceramic 0-3 Nanocomposite Films for Ultrasonic Transducer Applications, *Thin Solid Films*, Vol. 375, No.1-2, pp. 95–99

15

Possibilities for Flexible MEMS: Take Display Systems as Examples

Cheng-Yao Lo
National Tsing Hua University
Taiwan (Republic of China)

1. Introduction

After the development of cathode ray tube (CRT) in the 19th century and the commercialization of television (TV) in the 1930s display devices are always one of the dream products in daily life. The revolutionary innovation from black-and-white to color display, from small size to large area, from curved surface to flat panel, and from space-consuming tube to short tube, all proved that the demands and distribution of display device was growing and played a critical role on civilization and industrialization.

Along the popularization of personal computer, the demand of CRT put the display technology and its industry to a highly growth field for the past few decades until the late 20th century. On the other hand, with the leaping progress of research and development on liquid crystal, liquid crystal display (LCD) also attracted customers' attention and overwhelmingly replaced CRT and suppressed CRT industry because of its light weight and thin body. Even though LCD's starting price was high, most of customers were still switching their display device from CRT since the bulky CRTs cost more when talking about office or house rent. Similarly to CRT, researchers also spent time developing thinner, lighter, wider viewing angle, shorter response time, and larger size LCDs. As a result, LCD became a main stream not only for computer displays, but also for recreation displays such as TV. In the same time, plasma display panel (PDP), and organic light emitting diode (OLED) also found their application fields as a flat panel display device to replace the role of CRT. Unfortunately, PDP's high resolution and fast response time come with low life time and high power dissipation. The original advantage of large size display was also gradually replaced by the up-to-date LCD produced from the 6th-8th generation glass substrate. Thus PDP market is now suppressed by LCD and PDP manufacturers are also reducing their production. Although OLED was proved to be display-capable, recently commercialized OLED's blue color degradation still limits its application on information display. Since the demand for larger display size, wider viewing angle, and smaller body size are still growing, projection display such as back projection TVs and overhead projectors are built in parallel. Back projection TV, owing to its bulky size and low resolution, had limited applications and disappeared from market rapidly; overhead projectors, even though still find their way in the early 21th century, it was neither portable nor long lasting. Thus, how to realize a light weight, small size, and large display area comes to one end - flexible display. When display device becomes flexible, it must be light weight and portable. It must also be with compact size when folded or rolled, and must be supporting large area when unfolded or unrolled. Hence the first idea within its applications is focused on electronic paper (e-paper). With the arising of the attitude towards ecological friendliness, the demands of paper reduction also transfer to and stress on the development of flexible display as an e-paper. A successfully flexible display is thus expected to be bendable, portable, light weigh, low cost, durable, and even disposable. On the other hand, from the technology point of view, to manufacture a flexible display typically requires a flexible substrate such as polymer plastic material, which adds more complicated factors into traditional electronic device's production line.

From the process flow we understand that there are several disadvantages when use polymer plastic material as flexible substrate. For example, high temperature thin film deposition will cause material deformation since most polymers' glass transition temperatures (T_g); points or even melting point are below 200°C. Some polymer materials are also weak to chemical treatment or plasma bombardment which usually carries ultraviolet (UV) light. Thus, in order to successfully handle flexible substrate during production, a new process system should be set up. This process system is expected to take advantage of substrate's flexibility for variety and to avoid previously mentioned drawbacks on polymer materials. As discussed before, the flexible display has to support large display area thus a large substrate also introduces process difficulties in a production line. Take current semiconductor apparatus as an example, up to 12-inch circular substrate can be supported with lowest unit cost. However, a 12-inch diameter area cannot be called large area display. The latest 8th generation LCD apparatus which supports up to 2160×2460mm substrate size will be very helpful for large area display but its transportation is always a problem. To overcome these issues, traditional printing processes are taken into consideration because flexible polymer substrate is just like paper, which can be bent during process and can be produced continuously. Both the material flexibility and process continuity are positive factors for large area device realization.

1.1 Flexible display systems

1.1.1 Gyricon

Historically, the very first flexible display system was realized with "Gyricon" media by Sheridon and Berlovitz in 1977 and its conceptual cross-sectional view is illustrated in Figure 1. This kind of material contains black and white colored beads (balls) with averaged size of 100μm. Since different colors on the bead carry different charge polarity and the beads are enclosed in liquid, specific color can be directed to specific direction by outside electric field. This kind of system can perform either binary (black and white) data or gray scale reflective light. This system also mimics the printing process on papers by putting this Gyricon sheet into programmable electric field "printer". Unfortunately this project was closed in 2005 for not profitable reasons by Xerox. Some main advantages and disadvantages of electrophoretic system are listed below:

Advantage

1. The first demonstration of e-paper with little flexibility,
2. Supports A4 size,
3. Pattern is programmable by controller (computer),
4. Supports gray scale display.
5. Zero power consumption for static display.

Disadvantage

1. Coarse resolution,
2. Slow response time,
3. No full color,
4. Impure color,
5. High operation voltage.

Fig. 1. Schematic cross-sectional view of Gyricon system.

1.1.2 Electrophoretic display

Electrophoretic system inherited some basic concepts from Gyricon. As depicted in Figure 2, electrophoretic system further reduced the beads' sizes and increased their density. The main difference between electrophoretic beads and Gyricon beads is there's only one color on one electrophoretic bead. This change makes electrophoretic system possible to show higher resolution and contrast. Furthermore, since beads' sizes (approximately 1μm) are smaller than Gyricon's, gray scale display can be divided into finer details. Similar to Gyricon, an outside electric field plays the critical role on controlling the charged beads' movement in a liquid environment. But different from Gyricon, where beads have to rotate inside the liquid and stabilize after long time; the electrophoretic beads don't have to wait for stabilization since the beads are with only one color. This makes the electrophoretic system response faster than Gyricon. Likewise, electrophoretic system does not require any maintenance voltage after operation. Some main advantages and disadvantages of electrophoretic system are listed below:

Advantage

1. Finer resolution (compared to Gyricon),
2. Faster response time (compared to Gyricon),
3. More gray scale options,
4. Capable for mass production,

Disadvantage

1. Still slow response time (for human perception),
2. Too high operation voltage.

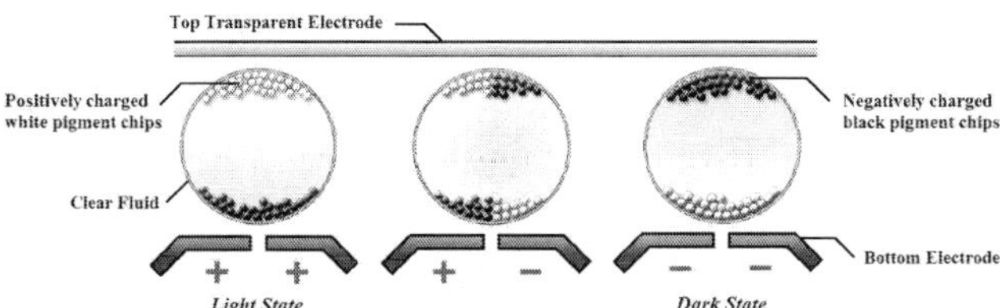

Fig. 2. Concept of electrophoretic system.

1.1.3 Electrowetting display

Compared to particle based Gyricon and electrophoretic display systems, electrowetting system uses all liquid material for color modification. This kind of concept takes advantage of Young-Lippmann's equation (Equation 1) to modify different surface energies between the droplet and the substrate underneath, which in turn changes the contact angle between the droplet and the substrate. Here, γ_{LG} is the surface tension between liquid and gas, γ_{SG} is the surface tension between solid and gas, γ_{SL} is the surface tension between solid and liquid, θ is the contact angle, V is the applied voltage, and C is the electric capacitance per unit area in the region of contact between a metal surface and the electrolyte drop.

$$\gamma_{LG} \cos\theta = \gamma_{SG} - \gamma_{SL} + \frac{CV^2}{2} \quad (1)$$

According to Equation 1 and Figure 3, the applied voltage will change the droplet's contact angle. A smaller contact angle represents a larger droplet diameter while a larger contact angle represents a smaller droplet diameter. By following this concept, an electrowetting display system was designed: The intermediate liquid was dyed for different colors while the water was kept transparent. Under normal (OFF) condition, the intermediate color liquid is laying under transparent water thus the reflective light shows intermediate liquid's color. Under operation (ON) condition, the intermediate color liquid is pressed into a specific corner of a pixel, leaving the rest area only with transparent water. Thus the reflection light with background color appears. By switching this system ON and OFF, two different colors can be switched for display purpose as shown in Figure 4. Since electrowetting system is using all liquid material for color modification, it is supposed to be flexible for display application. Some main advantages and disadvantages of electrowetting system are listed below:

Advantage

1. Gray scale is ideally controllable by applied voltage (contact angle),
2. Low cost,
3. Good color purity (not by mixing colors),
4. Can be designed for transmission or reflection type.

Disadvantage

1. Low resolution (limited by liquid suppressed at corners),
2. Low stability (gravity influence on liquid),
3. High power consumption (continuous power supply when ON).

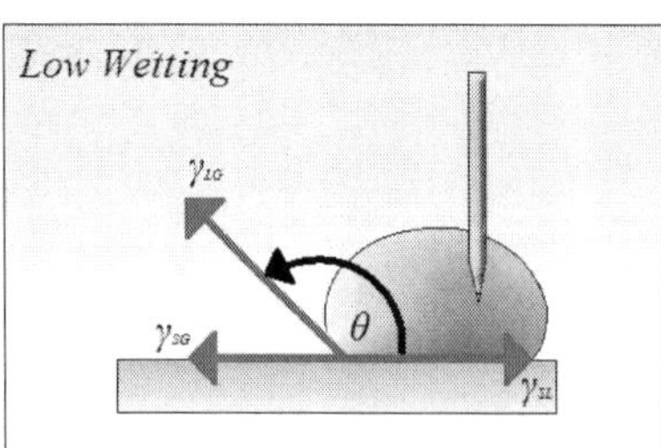

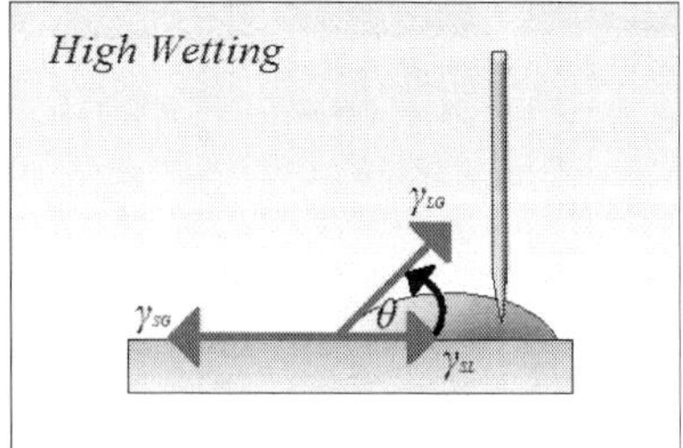

Fig. 3. Contact angle change by applied voltage is the basic of electrowetting.

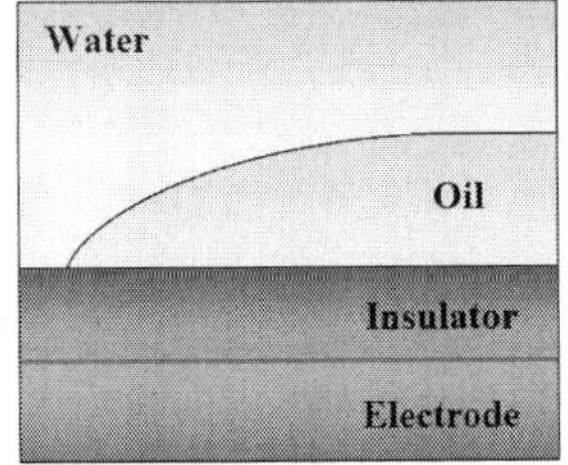

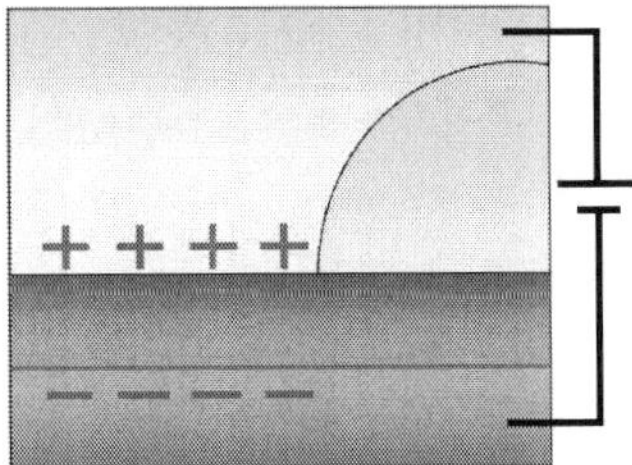

Fig. 4. Switching colors by controlling liquid's electrowetting performance.

1.1.4 Electrochromic display

Gyricon and electrophoretic systems are using physical control by electric field outside on colored particles and electrowetting system is also using physical control but the main material contains only liquid. Here, the electrochromic system is using chemical concept to change material's charging condition in dielectric electrolyte in order to change its light absorption band as shown in Figure 5. The electrochromic material can be either dissolved in the electrolyte or coated on the electrode substrate. Typical materials in electrochromic system are WO_3 and TiO_2. WO_3 had been reported with capability to change its colors between transparent under oxidation state and blue under reduction state; TiO_2 had been reported with capability to change its colors between transparent under reduction state and white under oxidation state. Thus, using only one electrochromic material can realize a two-color system and a combination of two electrochromic materials can realize a multiple color system. When the electrochromic material is thinly coated on the electrode, the whole structure will be bendable; when the electrochromic material is dissolved in electrolyte, the whole structure will also be bendable. Thus the electrochromic system can be used as a flexible color filtering device. Some main advantages and disadvantages of electrowetting system are listed below:

Advantage

1. Possible combinations for various colors,
2. Low operation voltage,
3. Can be designed for either transmissive or reflective.

Disadvantage

1. Low response time,
2. Color purity (depends on material's natural characteristic),
3. No black color.

All previous described technologies are supposed to be applicable on flexible substrate since all Gyricon, electrophoretic, electrowetting, and electrochromic (if material dissolves in electrolyte) are using liquid as intermediate or main material. But this also implies a reliability concern under critical operation conditions such as: high/low temperature, vibration or shock, and gravity influence, not mention to the jeopardy when the whole system is breaking and the chemical or electrolyte is leaking.

After reviewing these technologies from their basic operations, a summary can be made: Both physical and chemical concepts make the system slim and simple, thus the whole system can be fabricated on a thin substrate for flexible applications while reliability is a special concern. In contrary, stable, predictable, reliable, and reproducible mechanical system is considered for flexible application. In which, electrostatic force controlled micro scale system with mechanical movement for color filtering is set to solve the reliability problem. With these settings, the structure becomes a micro electro mechanical system (MEMS).

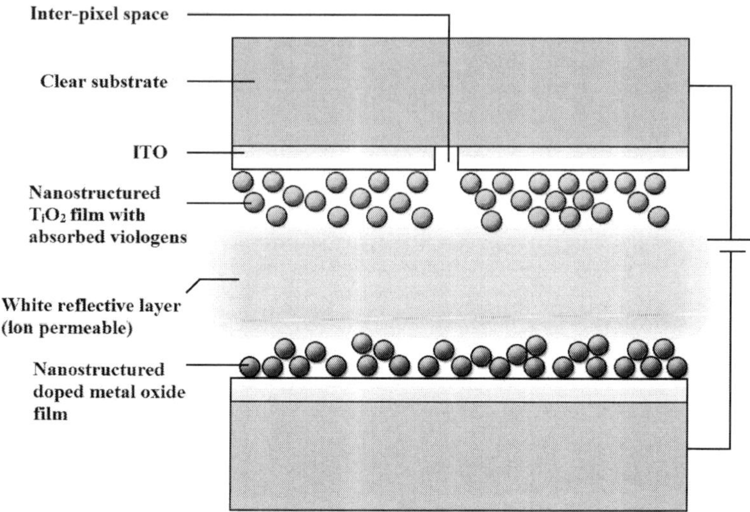

Fig. 5. The basic composition of an electrochromic system.

1.2 MEMS controlled display system

MEMS device is usually fabricated on solid substrate with batch photolithography process. In this section, some commercially realized MEMS display system will be discussed and reviewed by its color modification classifications.

1.2.1 Reflection

The most famous commercial reflective MEMS display system is Texas Instruments (TI) digital micromirror device (DMD). This device is fabricated on silicon (Si) substrate with complicated mechanical movement design. The key module - micromirror - is mounted on the center part of a torsion beam supported platform and the tilt angle of the platform can be controlled by electrostatic force to ±12°. With this setting, the reflection light from light source can be directed to display location (ON state, with color) or a shutter (OFF state, black). This is the realization of basic optical MEMS mirror design and people can generate three primary colors by implementing three DMD modules together or by using a color filter on a single DMD module.

1.2.2 Diffraction

When optical slits' sizes are well designed, optical diffraction takes place when light goes through the slits. Take white light as an example, different wavelength components diffract into different direction. Thus when viewing angle is fixed, different diffraction grating designs generate different primary colors. Sony's (originally developed by Silicon Light Machines) grating light valve (GLV) is one of the applications. Its MEMS part lies on the control and movement of its thin periodical metal ribbons. The ribbons reflect incident light under OFF state and specific wavelength is diffracted into designed direction when electrostatic force is applied. The individual control off each ribbon makes the system with different diffraction spatial frequencies for different colors. This device is usually made with complementary metal oxide semiconductor (CMOS) process.

1.2.3 Switching

Shutter is one of the applications in MEMS field and most uses of shutters are on optical or display categories. Pixtronix's digital micro shutter (DMS) is the representative device. DMS is fabricated by photolithography process on solid substrate with a suspension beam on opposite sides. The mechanical movement of the shutter layer opens and closes the output light from below generated by white light source. Its full color presentation comes from ultra fast switch rate which allows >1000 colors per second and avoids video fragments and color breakups.

1.2.4 Interference

Color interference takes place when a light beam is interfered by itself. To achieve this, one can put a dielectric material in another intermediate as shown in Figure 6. A key point to generate self interference is to have both reflection and transmission light at each interface. For example, light goes through Interface 1 or Interface 2 will generate two waves: one transmission light and one reflection light. Interference happens when transmission light from Interface 1 (t_1) encounters reflection light from Interface 2 (r_2). As a result, both constructive and destructive interference lights are formed as output. The Fabry-Perot interference condition (Equation 2) describes the constructive (visible) light under certain criteria. This Equation implies that when the incident angle (θ) and index of refraction (n) are understood, the output interference wavelength (λ) can be determined by dielectric material's thickness (d). The index m in the equation means any positive integer. A multiple layer stack for Fabry-Perot interference is normally the basic design concept of wavelength filters for visible colors and invisible transmission applications . Qualcomm's interferometric modulator (iMOD) took this advantage and commercialized small scale, low power consumption, high contrast device for display application. As shown in Figure 7, the reflection light's wavelength is determined by the gap distance between the solid substrate and a deformable metal membrane. Its OFF states reflect three primary colors and its ON states interfere the output lights to invisible region to compose a full color display.

$$2nd = m\lambda \cos\theta_1 \tag{2}$$

Although most of these MEMS ideas require solid substrate and CMOS photolithography process, some of them already showed flexible system with soft substrate when explaining MEMS with generalized terms: A mechanical movement system controlled by electrostatic force in micrometer scale. The rest issues lie on how to process or manufacture such flexible system.

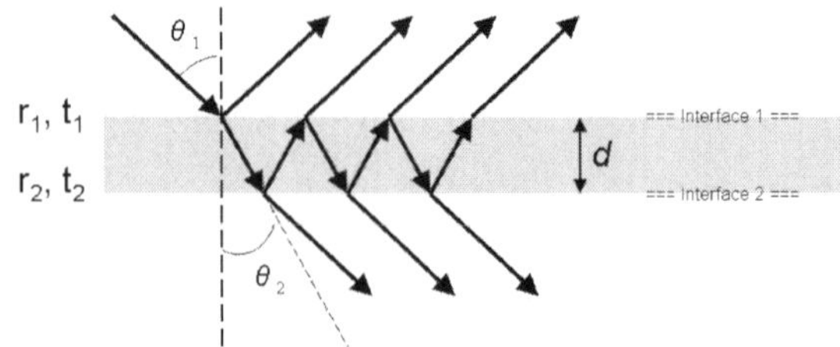

Fig. 6. A three-layer (two-interface) optical interferometer.

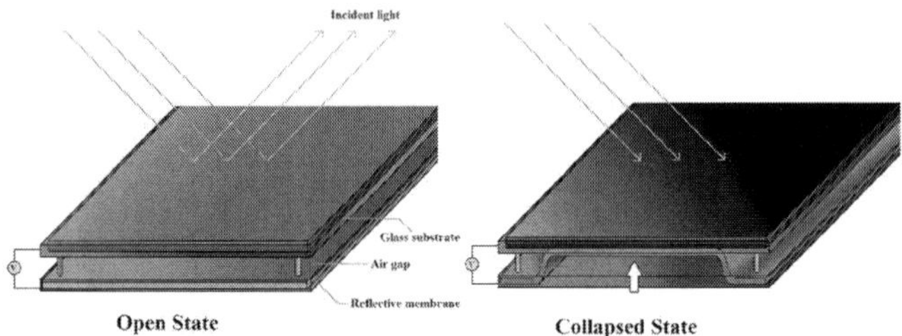

Fig. 7. A reflective Fabry-Perot interferometer iMOD.

1.3 Novel printing process system

Manufacturing technique for flexible electronic devices, especially for display devices, is the basic but also a crucial factor. As described and summarized previously, conventional CMOS photolithography techniques are not 100% applicable on polymer flexible substrates owing to heat, UV exposure, chemical treatment, and plasma bombardment. Thus, new process which is suitable for polymer substrate should be firstly developed to support the MEMS display system design. With polymer material's natural paper-like characteristic – stocks in a roll, conventional paper printing process seems workable for patterning the system's circuits as well as its structure. Here, some newly developed printing process will be introduced and evaluated to see how it can be modified and applied on a flexible polymer substrate based display system.

1.3.1 Gravure printing

After reviewing some trench patterning techniques, it is necessary to consider flat, continuous, and uniform layer stack. Within printing techniques, gravure printing is one of the most famous systems for ink printing on materials such as paper, plastic, and clothes. Its advantages are low cost, addable multiple inks, and gray scale. Its characteristic of low cost comes from the continuous mass production; its addable multiple colors comes from the combination of individual colors prepared by different cylinder to form a color mixture; its gray scale comes from the different designs of cell depth, cell density, cell angle, cell size, and cell shape. The cells are made by laser engraving and are recessed from the cylinder surface. A schematic plot of gravure printing in working is illustrate in Figure 8 and is usually used for continuous process. As shown in this figure, the ink cell on the cylinder represents how dense, how large, how high the printed patterns will be. Thus, gravure printing is usually used for thick layer transfer. When ink's solid content is low (less sticky), printed high pattern density ink will spread and then merge together. This behavior provides a solution for continuous thick layer preparation.

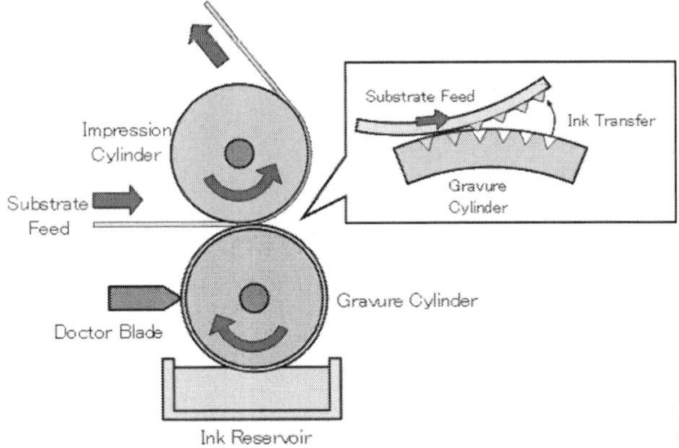

Fig. 8. An example of gravure printing system.

1.3.2 Flexography printing

Gravure printing is not possible to support thin layer deposition because in order to let printed inks merge, dense, deep, large ink cells are expected. Thus the transferred ink layer are usually ranging from 5-10μm. Flexography, as illustrated in Figure 9, uses a pattern plate to introduce ink from the anilox roller to the substrate. Since the patterns on the plate are

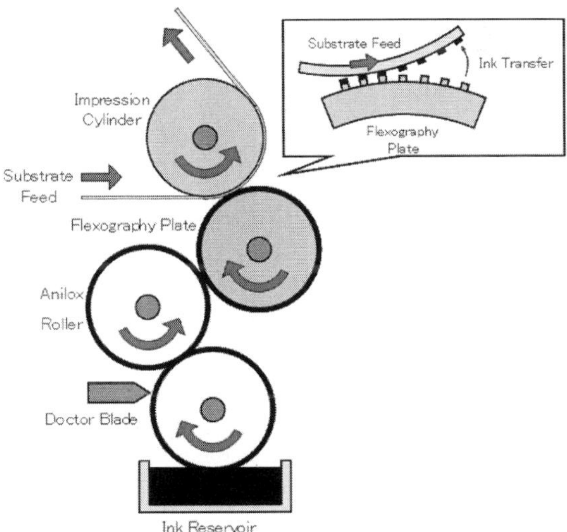

Fig. 9. An example of flexography printing system.

raised from the plate surface, the transferred ink amount depends only on the substrate's surface condition and ink's characteristic. However, flexography is suitable for thin layer (<2μm) deposition for good uniformity. There's also no ink rheology requirement generated from gravure printing's ink merge process. For pattern isolation, both hot embossing and laser ablation introduce fast and simple solutions but also bring some uniformity concerns. For layer deposition, even though gravure is used for thick layer and flexography is used for thin layer, how precise the thicknesses are seriously influence the optical design in a display device. But apparently, to perform integration process on a flexible substrate, the printing techniques described in this section are compulsory. Other printing techniques such as screen, inkjet, and offset are not suitable for this study but are widely discussed for flexible electronic devices' applications.

1.4 Target application

Section 1.1 already detail described why flexible display is necessary in the future and how those promising technologies are being realized and commercialized nowadays. The technologies introduced in section 1.1 have different target applications and markets and thus are with different concepts. It is obvious that no single technology can satisfy all requirements with all advantages such as low power consumption, high brightness, and fast response time. As a result, this chapter wants to cover and target at the large scale flexible display area for signage, advertisement, and decoration purpose. This means that this chapter is not necessarily pursuing a fine resolution, vivid true color, and fast response which are fundamental factors for TVs and monitors. Nevertheless, this study still looks for and tries to realize these good characteristics as reasonable as possible under some natural limitation such as availabilities of materials and configuration of apparatus. One of the most interesting examples of its applications for this study is to replace the mosaic windows which are usually decorated in churches as a motive. When the above targets are realized, the large scale flexible display sheet will be very distinguishable from previously mentioned flexible display systems and also those MEMS devices listed in section 1.2. Finally, this device will not only support uneven surfaces but will also be programmable to change the mosaic patterns without artificial backlight.

2. System design, material evaluation, mechanism, and simulation

Some flexible display ideas and control mechanism have been introduced. Within them, the Fabry-Perot was evaluated as the most promising system to be controlled by MEMS. This chapter then chose the MEMS as the flexible display's control system with Fabry-Perot color interference concept.

2.1 The Fabry-Perot interferometer

2.1.1 Mono layer model

The model in Figure 10 shows the basic composition of a Fabry-Perot interferometer. In the figure there are two intermediates which form two interfaces. When a light goes through the Intermediate 1 and reaches the Interface 1, a reflective light and a transmissive light will be generated at Interface 1. The transmissive part will again be divided into a reflective light and a transmissive light at Interface 2 when it goes through the Intermediate 2 and reaches the Interface 2. Likewise, a reflective light and a transmissive light will be generated at any interface even though both the reflective light and the transmissive light decay in the intermediates.

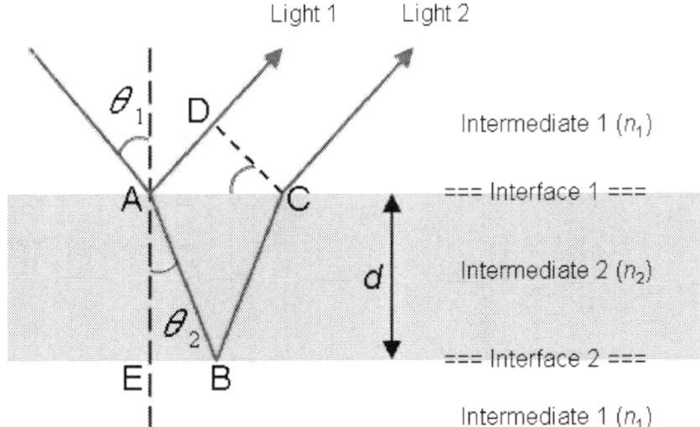

Fig. 10. A basic Fabry-Perot interferometer.

Color interference takes place when there is optical path length difference (Γ) between two or more light components traveling together. In this figure, the reflective Light 1 and the reflective Light 2 have optical path length difference Γ:

$$\Gamma = n_2(\overline{AB} + \overline{BC}) - n_1\overline{AD} \tag{3}$$

Here, n_1 and n_2 is the index of refraction of Intermediate 1 and Intermediate 2, respectively. Since the angle of incidence equals to the angle of reflectance, which is θ_2 in Figure 10, thus,

$$\overline{AB} = \overline{BC} = \frac{d}{\cos\theta_2} \tag{4}$$

$$\begin{aligned}\overline{AD} &= \overline{AC}\sin\theta_1 \\ &= 2\overline{EB}\sin\theta_1 \\ &= 2(\overline{AE}\tan\theta_2)\sin\theta_1 \\ &= 2d\tan\theta_2\sin\theta_1\end{aligned} \tag{5}$$

According to Snell's law:

$$n_1\sin\theta_1 = n_2\sin\theta_2 \tag{6}$$

The distance $\overline{AD}$ becomes

$$\overline{AD} = 2d\tan\theta_2(\frac{n_2}{n_1}\times\sin\theta_2) \tag{7}$$

Replace $\overline{AB}$, $\overline{BC}$, and $\overline{AD}$ by Equation 4 and Equation 5 into Equation 3,

$$\Gamma = 2n_2 d \frac{1 - \frac{1}{n_1} \times \sin^2 \theta_2}{\cos \theta_2} \quad (8)$$

With trigonometric function:

$$\cos \theta_2 \tan \theta_2 = \sin \theta_2 \quad (9)$$

$$\cos^2 \theta_2 + \sin^2 \theta_2 = 1 \quad (10)$$

and take atmospheric air as Intermediate 1 with n_1=1, the optical path length difference becomes

$$\Gamma = 2n_2 d \cos \theta_2 \quad (11)$$

A constructive interference takes place when the two reflective lights are in-phase and a destructive interference takes place when the two reflective lights are out-of-phase. A maximum constructive interference happens when the two lights are with 0° phase difference or zero (or 2п) phase change:

$$2n_2 d \cos \theta_2 = m\lambda \quad (12)$$

Similarly, a minimum destructive interference happens when the two light are with 180° phase difference or п phase change:

$$2n_2 d \cos \theta_2 = (m - \frac{1}{2})\lambda \quad (13)$$

Here m is an integer and λ is the wavelength for both cases. The interference from the transmissive side can be also evaluated from the Interface 2.

2.1.2 Multiple layer model

Since the system is designed for information display, a maximum constructive interference is expected. According to Equation 12, one can easily design specific output light (wavelength) with specific intermediate (n_2, d) under fixed angle of incidence θ_2. It is also possible to calculate a multilayer system according to Equation 8 when the intermediate material is not air. Based on Figure 10, a multilayer structure shown in Figure 11 was chosen for color filtering. A premise is also made here: The color (color 1) filtered by the structure in Figure 11(a) is different from the color (color 2) filtered by the structure in Figure 11(b). The change of the multilayer structure hence lies on the mechanical control by MEMS. Here a special note should be put that a color of either color 1 or color 2 is not necessary to be both destructive interferences, rather, a color will be good enough to distinguish from another even though it is not formed by interference. The multilayer system in Figure 11 is switching between six layers and five layers (excluding ambient air layers: Intermediate 1). Thus, when talking about a multilayer system with more than three intermediates (four interfaces), the optical path length difference becomes relatively complicated which will be discussed and simulated by commercial software later.

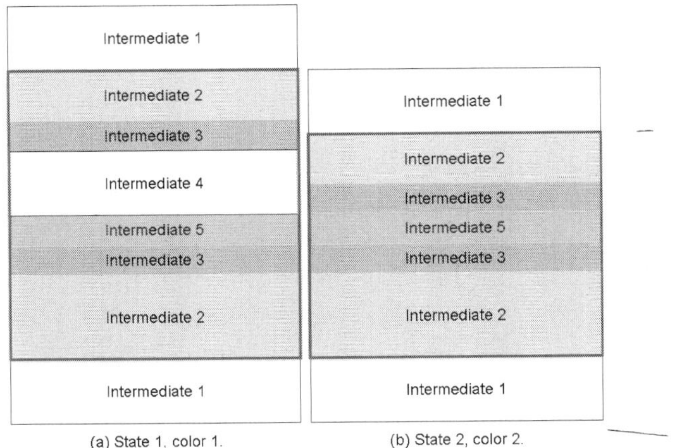

Fig. 11. The system design for two colors.

2.2 Color design and simulation

2.2.1 Color purity consideration

With the layer definition in Figure 11 and the Fabry-Perot interferometer concept, color designs in this section will help to decide how thick those layers should be and what kind of optical characteristics should they have for the three primary colors: red, green, and blue for full color applications. Previous report showed distinguishable yet poor colors especially for red. The root cause is the extra peak in blue region for red color, as a result a pink or purple color was shown. The author also implied a solution with new layer design which reduced the isolation layer's (Intermediate 5 in Figure 11) thickness from 370nm to 185nm. However, 185nm thickness was neither achieved nor disclosed. Besides the color purity issue, its transmittance for red, green, and blue was low and also not balanced. The unbalanced transmittance increased the design difficulty for backlight.

Within normally used metals, aluminum (Al) and copper (Cu) are widely used for their low cost and good conductivity while silver (Ag) and gold (Au) are also good but expensive. From the index of refraction (n) point of view, all Au, Al, and Cu show large difference under different wavelength. From the light wave phase (Φ) point of view,

$$\begin{aligned}\Phi &= \alpha \times x \\ &= \frac{2\pi}{\lambda} \times x \\ &= \frac{2\pi f}{n \times v_p} \times x\end{aligned} \tag{14}$$

where α is the wavenumber, x is the propagation distance, λ is the wavelength, n is the index of refraction, f is the frequence and v_p is the phase velocity. Thus the same input light with the same phase will generate two different phase output light owing to different index of refraction. According to this, a relatively uniform n value distribution for visible region (400-700nm) of a material is highly expected to solve the unbalanced intensity issue. Ag showed small n value difference across the visible region which suggests a structure change solution: To replace Al by Ag for Intermediate 3 as electrodes in Figure 11. However, since the n value change and according to Equation 12, a suitable intermediate thickness (d) should also be evaluated.

2.2.2 Transmittance consideration

On the other hand, when trace the low transmittance with Al's optical parameters, a hint from its absorption coefficient (k) also emerges. Since k value means the decay behavior as well as how the light is absorbed in the intermediate, the higher k value is the lower the transmittance is. Al's k value ranges from 3.9 to 7 in visible region which is relatively higher when compared to Au, Ag, and Cu. We also understand that even Ag has higher k value in red and green region compared to Au and Cu, it is very suitable to replace Al to increase the transmittance for all red, green, and blue colors. Figure 12(a)-(c) and Figure 12(d)-(f) are the simulation results done by commercial software Optas-Film with structures of Figure 11(a) and Figure 11(b), respectively. The only variable in Figure 12(a)-(c) was the thickness of Intermediate 4 and the only variable in Figure 12(d)-(f) was the thickness of Intermediate 5 in Figure 11. The best condition for Intermediate 4's thickness was 600nm which shows a very distinguishable white color. With this setting, the best conditions of Intermediate 5's thickness are 160, 325, and 240nm for red, green, and blue, respectively. The transmittance in Figure 2-4 followed the optical design and thus proved higher transmittance with balanced output for all primary colors.

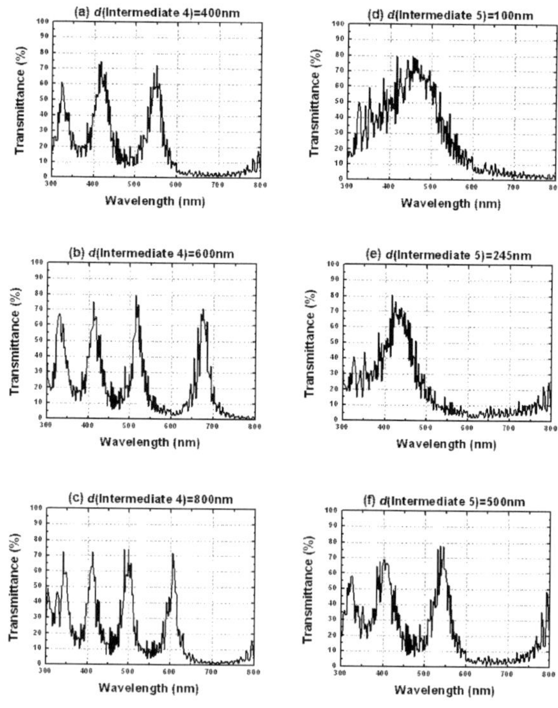

Fig. 12. Simulated transmittance with structures in Figure 11.

2.2.3 Color purity deviation

Without putting the output wavelength on a color chart it is very difficult to judge whether the output colors are vivid and pure or not; without quantifying the improvement it is very difficult to tell how good the new design is. Normally, the color purity is measured and illustrated on the CIE chromaticity diagram published in 1931 with (x, y) coordinate system or in 1976 with (u', v') coordinate system. The translation between these two systems follows the following formulas:

$$u' = \frac{4x}{-2x+12y+3} \qquad v' = \frac{9y}{-2x+12y+3} \tag{15}$$

Here, only the CIE 1931 chromaticity system is used. Figure 13 is the color purity comparison of previous work, this design, and a CRT display. One can easily find that this work greatly pushed the green color to a better purity place while keeping the red and blue colors' purity similar. In order to quantify how much the improvement is, a color purity deviation (*CPD*) is firstly defined as the shortest distance between two points on the diagram with the following equation:

$$CPD = \sqrt{(x_T - x_{SE})^2 + (y_T - y_{SE})^2} \tag{16}$$

Here the subscript T means the target coordinate and the subscript SE means the simulated or experiment data coordinate. The best and the smallest *CPD* value are both 0 (zero). Based on the optical parameters analysis, the new design which inaugurated Ag as electrode material improved all the transmittance, balance between all colors, and the color purities. The incoming challenge thus lies on how to switch this Fabry-Perot interferometer system between the two states illustrated in Figure 11.

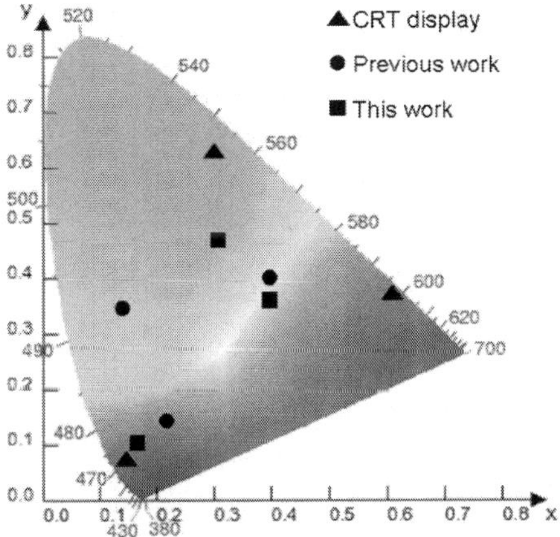

Fig. 13. Color purity distribution and the improvement comparison.

2.3 The MEMS model and simulation

A MEMS controlled system to switch a multilayer structure (Figure 11) of Fabry-Perot interferometer for different colors has been decided in previous section. Since the color design in section 2.2 already showed a very promising two-state color system, this section will handle how to design and prepare the structure in Figure 11 as a MEMS. Within all MEMS driving methods, the electrostatic way is believed to be suitable for structure in Figure 11 owing to the straightforward vertical movement. The structure in Figure 11 is also designed for electrostatic driving since the two Intermediate 3 materials can serve as the two parallel electrodes and the Intermediate 5 material can serve as the isolation layer when the two parallel parts are in contact.

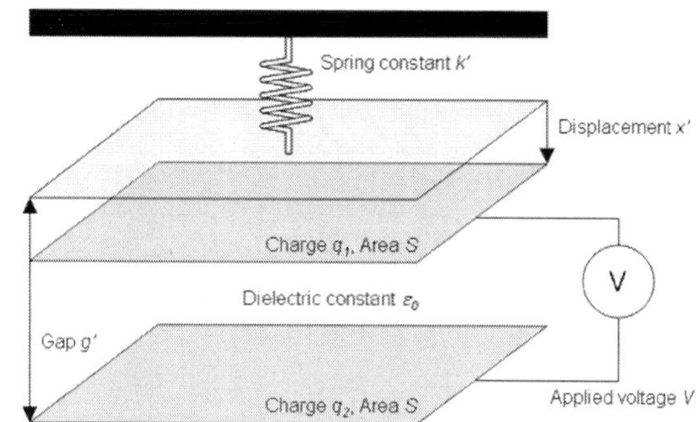

Fig.14. A parallel plate actuator system.

Suppose there are two particles in a space separated by distance g, a Coulomb (electrostatic) force F exists between these two charges:

$$F = k_C \frac{q_1 q_2}{g^2} \tag{17}$$

Here k_c is the Coulomb force constant whose value is $8.99\times10^9 Nm^2C^{-2}$, and q_1, q_2 are the particle charges. A parallel plate system shown in Figure 14 consists two conductive layers and those layers are capable to stock charges. Charges can be supplied by outside source and the parallel plates start to attract with each other when the Coulomb force is strong enough. Thus, when two large plates are separated by spacer structure at the edges, its center part can be treated as a parallel plate system. A schematic plot is illustrated in Figure 15.

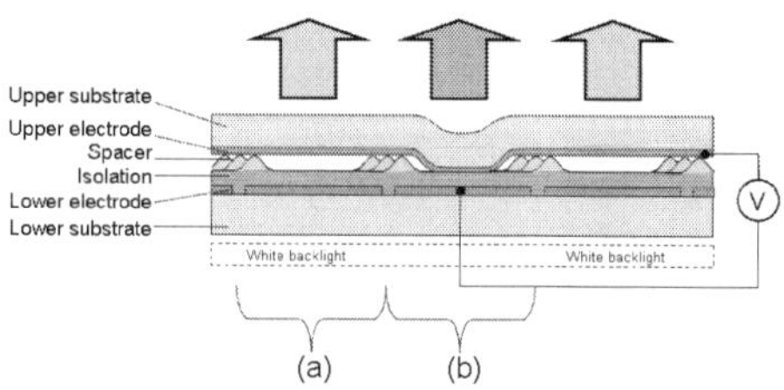

Fig. 15. The MEMS structure in (a) OFF state and (b) ON state.

2.3.1 Mathematical model

Even though the main part of Figure 15 can be expressed by a parallel plate, the whole structure still contains fixed ends close to spacer structures. A complex model combines these two parts is then necessary. The complex model of a single pixel (one parallel plate set) was divided into two parts horizontally and was further divided into two parts of a single-end fixed cantilever and a parallel plate as shown in Figure 16. The left part of Figure 16 is a cantilever which is fixed at one end and bending at another. The right part of Figure 16 is a parallel plate system which moves up (OFF state) and down (ON state) when applied with electrostatic force. Since the cantilever part is connected to the parallel plate part, when the plate moves down the cantilever is pulled down. When ON, the parallel plate system suffers an electrostatic force and tends to stay in contact while the cantilever suffers a reaction force and tends to return to the original (upper) position. In this model, the displacement x' in Figure 14 moved the entire gap g.

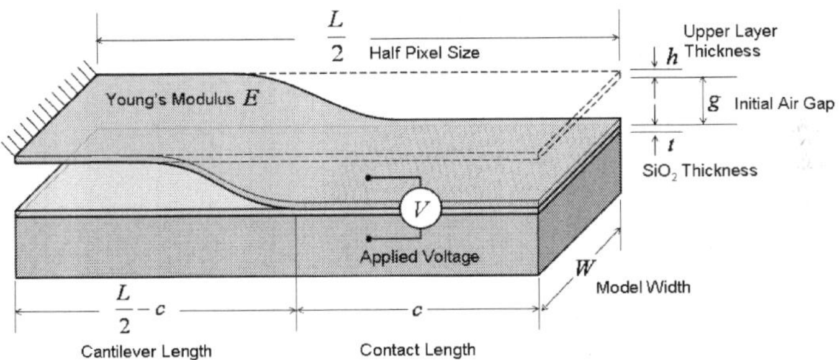

Fig. 16. The MEMS model from a half pixel of Figure 15(b).

The electrostatic force on the parallel plate system is:

$$F_p = \frac{1}{2}\varepsilon_0\varepsilon_r \frac{W \times c}{[(g+t)-g]^2} V^2 = \frac{1}{2}\varepsilon_0\varepsilon_r \frac{W \times c}{t^2} V^2 \tag{18}$$

The reaction force on the cantilever part is:

$$F_c = k' \times g = \frac{EWh^3}{4(\frac{L}{2}-c)^3} \times g \tag{19}$$

An overall stable system will be formed when these two force are in equal:

$$\frac{1}{2}\varepsilon_0\varepsilon_r \frac{W \times c}{[(g+t)-g]^2} V^2 = \frac{EWh^3}{4(\frac{L}{2}-c)^3} \times g \tag{20}$$

2.3.2 Simulation prediction

By rearranging Equation 20, a relationship between the contact length (c) and the applied voltage (V) can be set up. With this relationship, one can estimate the contact area and its percentage under certain applied voltage. Similarly, one can also expect the operation voltage for specific contact area. The following simulations were performed under the following parameter settings: ε_0 = 8.85×10^{-12} A^{2}s^{4}kg^{-1}m^{-3}, ε_r = 3, h = 16μm, L = 2000μm, t = 0.3μm, E = 6.1GPa, and g = 0.6μm. Figure 17 is the simulation result with different pixel size (L). From these figures we understand that under the same applied voltage, a larger pixel will result in a larger contact area. From these figures we also understand that when one wants to achieve, for example, 90% contact area, a great operation voltage difference (55V for 1000μm pixel and 15V for 2000μm pixel) appears in Figure 17(a)-(b). Figure 17(c) is the simulation results with different spacer thickness (g). We understand that the operation voltage can be further reduced from 15V to 10V when change the spacer thickness from 600nm to 300nm. An examination in Figure 17(d) also indicated that when change the upper layer's thickness (h) from 16μm to 8μm, the operation voltage can be further reduced from 10V to 5V. Thus, a combination of these improvement designs, one can expect and design a low operation voltage device with this MEMS model. Other parameters concerning material characteristics such as ε_r and E, can also help on the operation voltage lowering but will not be considered here.

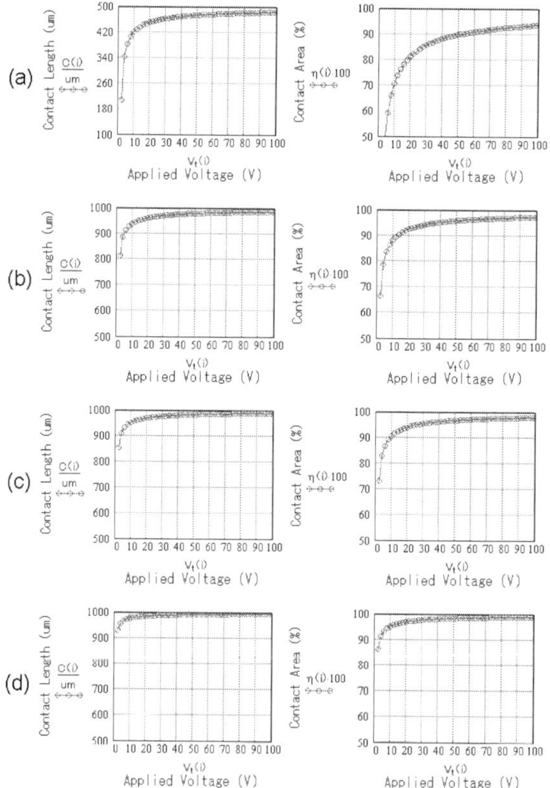

Fig. 17. Simulation results with different parameters from the MEMS model.

2.4 The horizontal structure design

2.4.1 Air pressure consideration

Until now, the device structure is designed and discussed vertically in detail thus its horizontal dimension and structure should also be considered. Previous report indicates high operation voltage with an enclosed Intermediate 4 in Figure 11 which is shown here in Figure 18(a) from its top. The author suggested some solutions to lower down the operation voltage such as to use thinner upper layer, to use thinner Intermediate 4, and to design a larger system. However, if the Intermediate 4 is trapped inside the system when ON, it will become a movement barrier or cause reliability issue unexpectedly. Since atmospheric air is designed for Intermediate 4, it is very possible to reduce the air pressure trapped inside the enclosed spacer area to alleviate the operation voltage. Figure 18(b) is the top view of a newly designed structure. Compared to Figure 18(a), the new design has some openings (air channel) on specific locations. These air channels serve as air evacuation paths when the device is ON. Figure 19 to Figure 21 are the simulations done with commercial software MEMSOne to explain how flexible can the air channel be designed and how the corresponding structure moves. Note that since there are design limitations on the structure by this software, the color legend means the areas in moving instead of its absolute displacement value. The value of opening ratio means the air evacuation efficiency while different designs imply different display shape because the air channel area can also be turned on if the opening ratio is large. Compared to the baseline design, we understand that increasing the opening ratio helps on enlarging the MEMS movement area. In the design of Figure 19, which represents the basic design in Figure 18 where the air channel was put in the center part of one spacer side, the displacement area increases when the opening ratio increases. The extreme model (Figure 19(c)) indicates that the displacement switched to the air channel area rather than the pixel area.

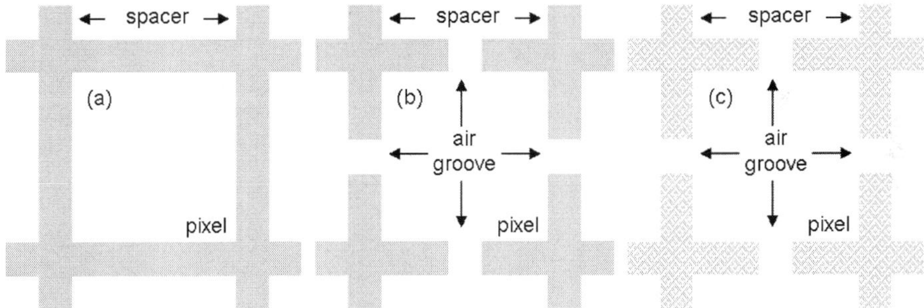

Fig. 18. Renewed spacer layer design's top views.

Similar effects also appeared on the Figure 20 designs, in which the air channel was divided into two sub-channels and were put at the both ends of one spacer side. One can find that up to 40% opening ratio, the displacement area follows the Figure 19 designs but the opening ratio of 60% (not shown here) and 80% (Figure 20(c)) ones helped the displacement continued to expand inside the pixel area. With this design improvement, we can positively change the unexpected displacement area caused by air channel to a reasonable and expectable area within the pixel. Based on Figure 20, the sub-channels were moved to the two ends of one spacer side and the spacer corners were also removed. The opening ratio of one spacer side was thus still the same. The simulation result in Figure 21 did not change when the opening ratio is smaller than 20%, but interesting displacement took place at the spacer corners for opening ratio larger than 40%. This kind of motivation came from Figure 19(c), in which the displacement expanded to the direction with spacer's opening ratio large enough. Similarly, when the opening ratio in Figure 21(c) was large enough, the displacement area expanded to the corners when keeping its shape the same of a square. The structure design and simulation showed very promising results to realize the design in Figure 11.

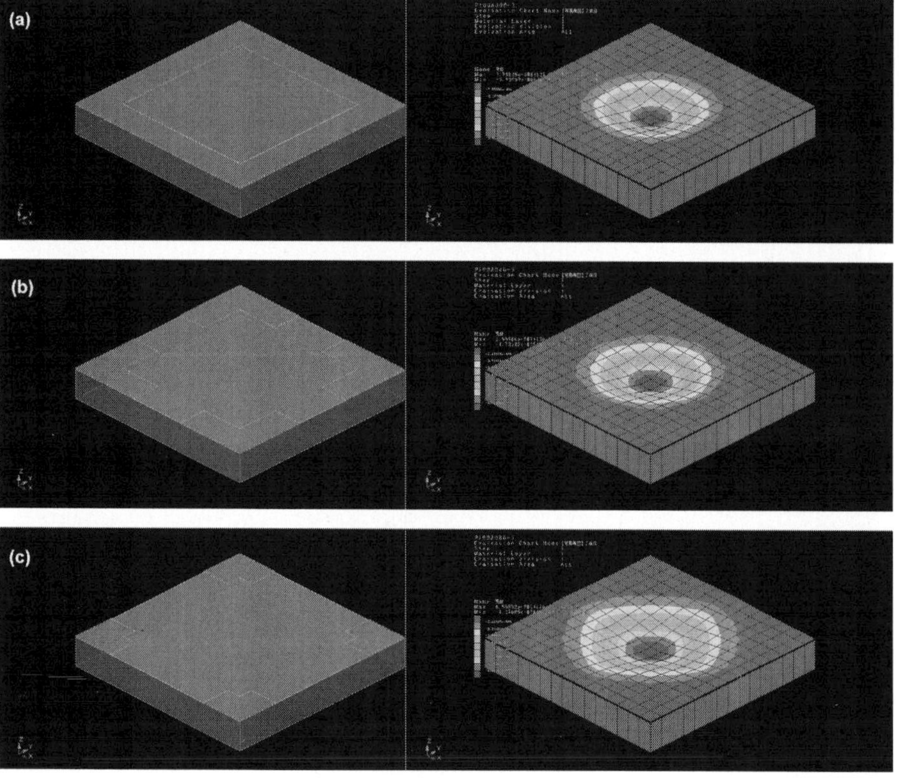

Fig. 19. Simulation results of opening ratio of (a) 0%, (b) 40%, (c) 80% for design 1.

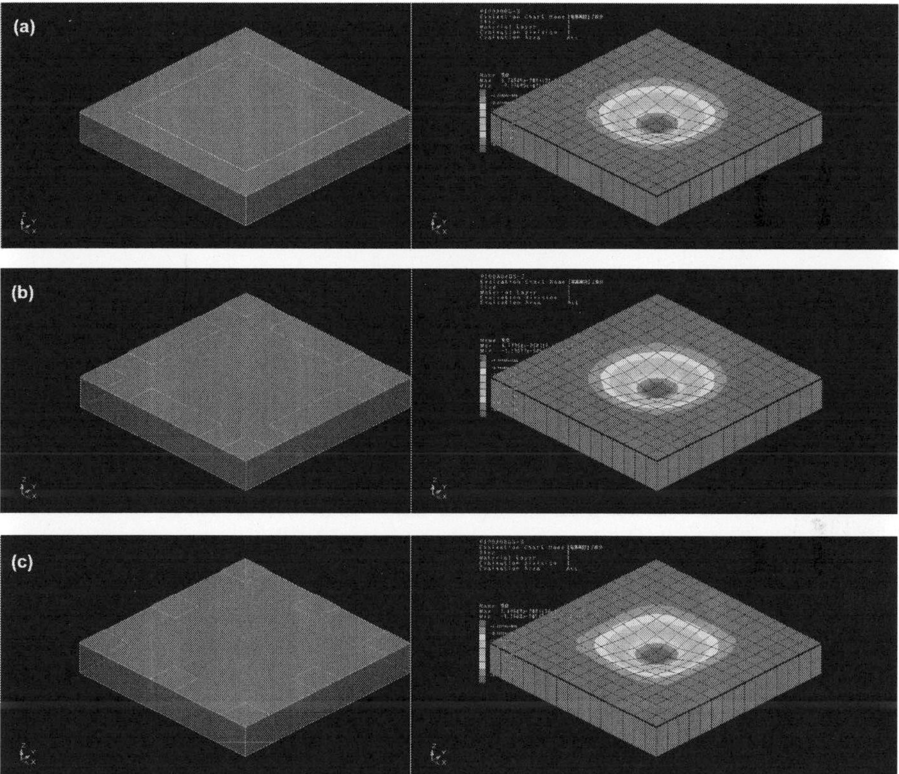

Fig. 20. Simulation results of opening ratio of (a) 0%, (b) 40%, (c) 80% for design 2.

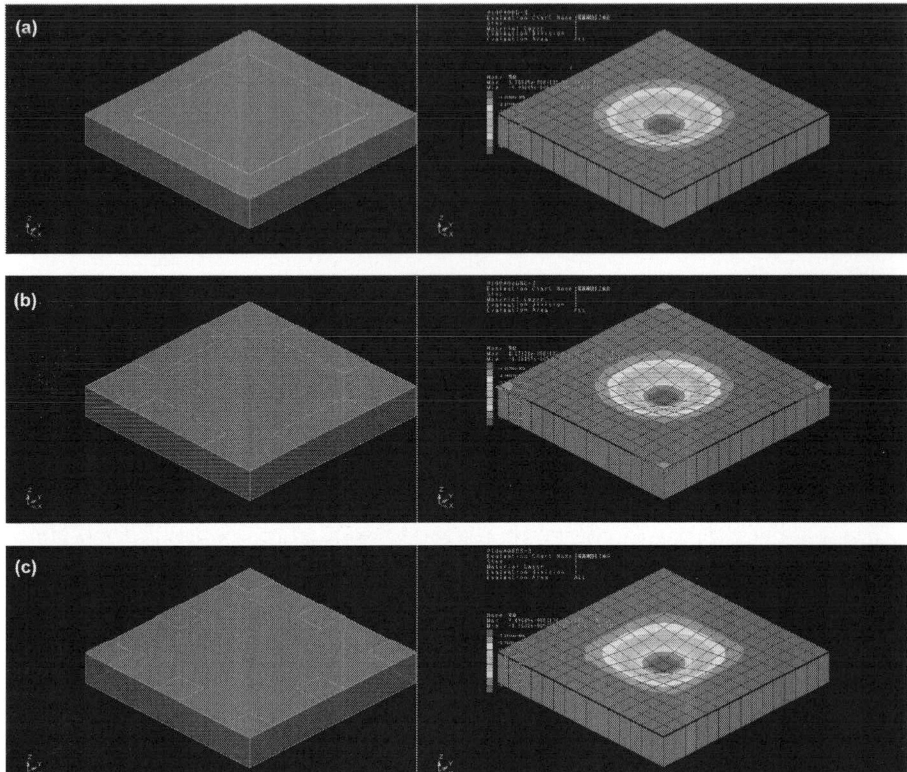

Fig. 21. Simulation results of opening ratio of (a) 0%, (b) 40%, (c) 80% for design 3.

3. Fabrication based on printing techniques

As introduced before, a novel process should be used for the special requirement on not only the structure's flexibility but also on the dot spacer layer design. Several promising solutions were examined in section 1.3 and with the structure set up before; material, process, and concerns will be discussed layer by layer here.

3.1 Material selection and system setup

3.1.1 Substrate

According to Figure 11, the whole structure will be made and the device will be operated in atmospheric ambient. Besides the Intermediate 1 in Figure 11, the multilayer structure contains six layers. The substrate material plays a crucial role for flexibility since it is the thickest part of the structure. When its thickness is below 500μm, it is with sufficient flexibility for display applications. However, when a large curvature is expected, only thinner substrates can satisfy this requirement. The latest glass manufacturing techniques support 30μm thick commercial products for large scale (300×400mm). Even though thin glass substrates provide very promising options for curved or flexible applications with large curvature, its fragility still limits its realization on flexible electronic devices especially for portable products. The potential safety and reliability concerns also put a barrier between its benefit and realization. In contrary to the fragile glass, elastic polymer material (plastic) is a very good option for the substrate. Because the plastic substrate will be used for the flexible display system, some special requirements including:

1. Flexibility - Low Young's modulus (*E*) is highly expected,
2. Transparency - High transmittance in visible region (400-700nm) is necessary,
3. Cutoff - Unexpected wavelength (<400nm and >700nm) should be screened out,
4. Stability - Should be thermally and electrically stable,
5. Reliability - Have to be highly moisture, gas, and chemical resistive,

are primary material selection principles. Within polymer materials, one can screen out polyvinyl chloride (PVC), polycarbonate (PC), polyethylene (PE), and polyimide (PI) from stability, reliability, deformation, and transparency point of view, respectively. With these concerns, polyethylene teraphthalate (PET) and polyethylene naphthalate (PEN) are relatively suitable for this flexible MEMS design. PET is also famous for its low cost and high transmittance in visible region while PEN is famous for its high temperature stability and sharp cutoff performance for UV light. According to these characteristics, PEN was chosen as the substrate material.

3.1.2 Process environment for plastic substrates

Besides embossing and laser ablation, which are patterning techniques for isolation instead of layer stack, the other printing methods are all printing process related ideas. However, within the printing process ideas, the screen printing and ink jet printing are batch processes which do not provide any help on improving the low throughput in photolithography. A compromise between resolution and throughput results in the flexography and gravure printing. Their working concepts have been explained in section 1.3 and the detail process parameters and system specifications will be discussed in the following sections.

3.1.3 Ink

In printing process, ink plays a very important role. Refer to Figure 11, four layers should be processed besides plastic substrates. Within these four layers, two electrodes are Ag; and the isolation is SiO_2. Since the two substrates have to be laminated after process, the spacer should also cover the lamination job. A commercial standard spin-on SiO_2 (TOK, OCD T7-12000-T) was chosen for isolation. This material is composed of $RnSi(OH)_{4-n}$ and additives (diffusion dopants, glass matter forming agent, and organic binder) dissolved in organic solvents (ester, ketone, and mainly consisting of alcohol) in liquid form and thus is suitable for printing process. Its SiO_2 solid content is 12wt% and its thickness can be controlled by curing temperature, time, and spin speed if prepared by spin-on process. Because the roll-to-roll (reel-to-reel, R2R) system uses gravure printing, whose printing thickness can be adjusted by cylinder cell design, only the curing temperature and time were studied for the thickness control. Figure 22 is the thickness change after thermal and UV treatment which are two optional steps in the process system. Curing temperature was controlled between 100-150°C for less than 30min in this study. After the thermal treatment a 2min 12.5mW/cm^2 UV exposure was applied. The thickness change was mainly because the evaporation of solvent and the thickness is basically inversely related to temperature and time. Since the R2R system is designed for high volume production and the plastic substrate is with low operation temperature limit, the data point of 140°C for 1min was set as the process variation boundary. On the other hand, since the spacer has to be sticky but does not have to be precise for physical and optical requirements, commercial adhesive glue (Herberts, EPS71) mixed with 35wt% hardener (Herberts, KN75) was chosen as the spacer material.

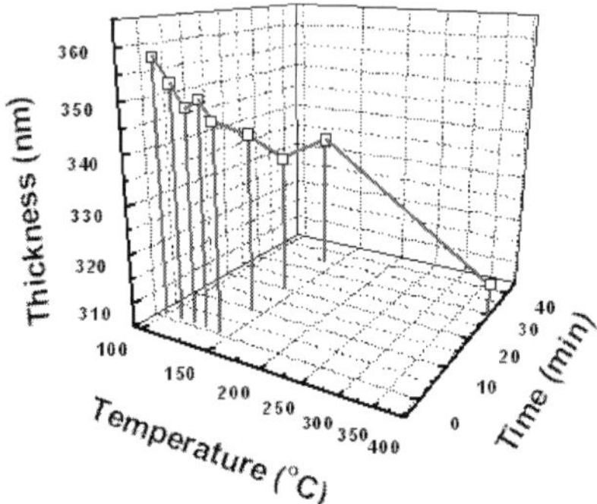

Fig. 22. Spin-on SiO_2 thickness variation after thermal treatment.

3.1.4 The roll-to-roll system

The final production system setup is illustrated in Figure 23. The flexible substrate is unwound from a roll and is then transferred into several different process steps. These

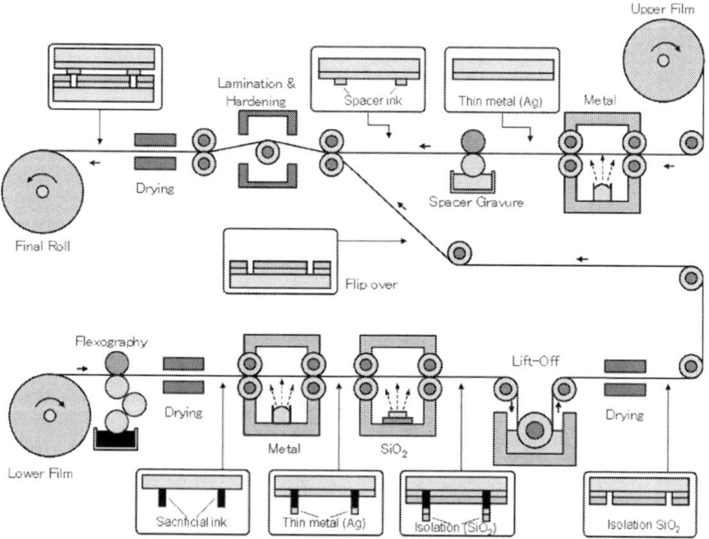

Fig. 23. Setup of the roll-to-roll process system.

process steps contain not only previously mentioned flexography and gravure printing but also conventional sputtering, cleaning, and drying/heating units. In the same time another or more substrate rolls can also be processed with similar sequences to form required patterns and layer stacks. These processed substrates might be finally laminated together and rewound back into a roll to complete this "roll-to-roll" process concept. Of course the final products can also be cut into sheets as a "roll-to-sheet" system. Note that even there is no multilayer combination and lamination process, the single substrate process is still a roll-to-roll system with roll-to-sheet capability. Before the whole continuous system is set up, individual parts in Figure 23 will be evaluated and single process step will be developed firstly with discrete units in the following sections.

3.2 Flexography printing

The concept of flexography printing was described in section 1.3. As mentioned previously, the flexography plate contains patterns raised from the surface and introduces very little ink from an anilox instead of the ink tank. Besides the adhesion force between flexography plate and ink, the plate merely plays a crucial role for ink transferring. Unfortunately, no commercial Ag ink can satisfy the specs and thus the direct pattern printing by flexography became inapplicable. A drawback of flexography printing laid on the uneven printed surface. The printed surface uniformity depends on a compromise between pattern integrity (contrast) and flatness: high pattern integrity requires thick (high solid content) ink, which in turn leaves peaks and valleys on the printed surface. Another drawback is that the printing process (not only limited to flexography) is highly direction oriented system: the pattern integrity is better along the printing direction (machine direction, mechanical direction, MD). Figure 24 is a printed example that along the printing direction more complete test lines are available. One can make modifications on ink quality and printing speed but can relatively induce worse surface uniformity and lower throughput unexpectedly. This printing behavior implies that one should design finer resolution along the printing direction to avoid broken lines. The process speed of Figure 24 was set to be 5m/min and its technique of lift-off will be explained in the next section. However, the drawback of surface uniformity is also just the merit of PR in photolithography: the rough surface provides good chemical resolving path during PR removal. With this advantage, a photolithography-like lift-off process was developed based on flexography printing.

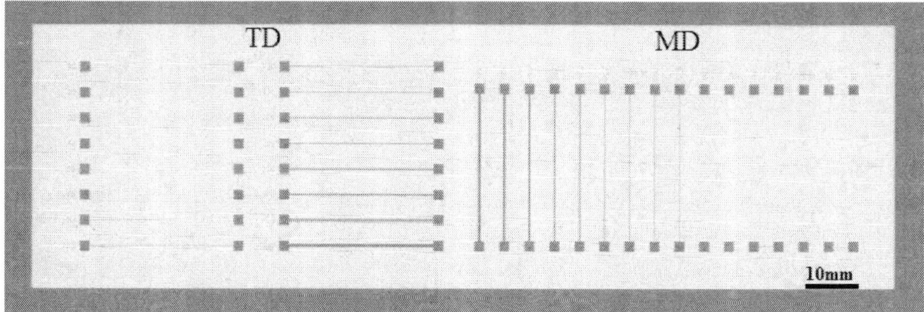

Fig. 24. Printing direction is with better resolution thus horizontal lines are thinner.

3.3 Lift-off

PR is recognized to be a patterning mask which influences the final structure on the substrates. With the uneven property of flexography printed surface, the ink was also treated as a sacrificial layer which provides protections on the substrate for layer stacking in the same figure. Here we modified Figure 23 from destructive process to constructive process for the same comparison basis. One can easily find that a process step of PR (ink) coating was omitted - the sacrificial pattern was directly printed from flexography plate while the photolithography exposed UV light through a photo mask. Similar to photolithography, this lift-off process is also a multilayer capable step as long as the sacrificial layer is thicker than the total thickness of multilayer. Figure 25 is a step-by-step example of lift-off process substrate. The thickness of sacrificial layer was ranging from 1-2μm which was higher and sufficient for multilayer stack with total thickness less than 1μm. The Ag sputtering was controlled with 20nm according to the color interference design in section 2.2. The third and the last step for the lift-off process is the sacrificial layer removal. In corresponding to the ink composition, basic chemicals such as acetone and ethyl acetate are suitable to dissolve it. It was also obvious that the substrate was flexible and was capable for large curvature process at this step. With ultra sonic vibration's help, the sacrificial ink dissolved in acetone in only seconds. Another important factor of heat influence on the removal efficiency was also carried out because of the drying unit after the sacrificial ink flexography printing. Here the sputter process influence was omitted because it was controlled under 60°C.

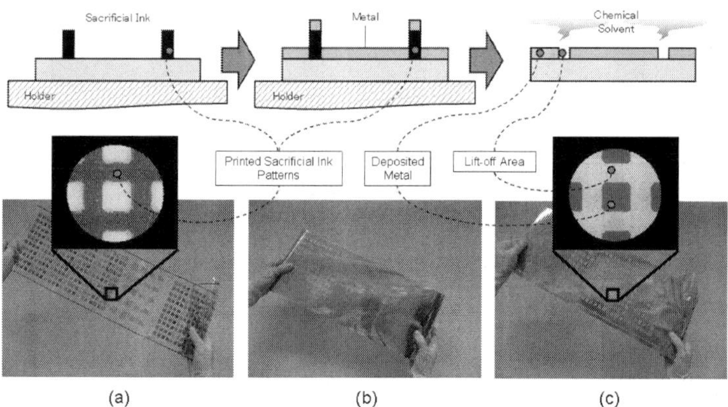

Fig. 25. Printing process is simpler than photolithography on step numbers: (a) sacrificial ink printing, (b) metal layer coating, and (c) sacrificial layer removal. The dimension of the square pattern in (a) and (b) is designed as 2000μm.

3.4 Gravure printing

The main difference between flexography printing and gravure printing lies on the controllability of printed layer. As shown in Figure 8 to Figure 9 and experimental results in section 3.2, it is apparent that the printed layer thickness and surface uniformity by flexography is merely controllable. The only solution for thick layer patterning is to choose from screen printing, ink jet printing, and gravure printing. The screen printing was screened out in section 3.1 for its low process speed and batch production characteristic. From the ink engineering point of view, ink jet printing and gravure printing provide similar ink droplet behavior on the substrate but the control variety of ink jet printing is less yet it is also a kind of batch process. When thoroughly review the cell design of a gravure cylinder in Figure 26, one can easily find its control varieties such as: cell width, wall width, channel width, channel depth, cell density, screen angle, depth, and stylus angle. All these factors' combination results in a final parameter of volume. Table 1 is a list of designed patterns for different printed thickness as well as wetting performance. The adjustable parameters are the printing speed, the pressure force, and the contact angle between the doctor blade and the cylinder.

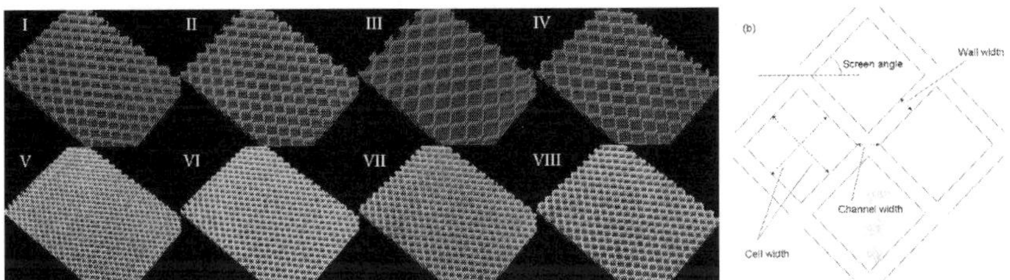

Fig. 26. A cell design set on the gravure cylinder and parameter definitions.

Design	I	II	III	IV	V	VI	VII	VIII
Mesh (line/cm)	60.8	53.8	41	47.8	116.4	104.3	89.9	80.9
Cell width (μm)	152	167.5	232.1	197.6	78.8	91.1	103.4	114.8
Wall width (μm)	12.6	18.3	11.7	11.7	7.1	4.8	7.8	8.8
Channel width (μm)	24	41	47	35	9	19	24	28
Channel depth (μm)	3.5	7.4	8.3	9.5	2	0	4.5	4.8
Cell density (%)	85.3	81.3	90.6	89.1	84.2	90.2	86.5	86.3
Cell depth (μm)	53.1	51.7	42.5	53	52.6	53.6	52.8	53.1
Screen angle(°)	56.9	60.8	58.6	59.4	31	34.4	39.1	43.5
Stylus angle (°)	121.5	119.8	129.8	130.1	119.1	118	118.7	118.3

Table 1. Detail list of cell designs.

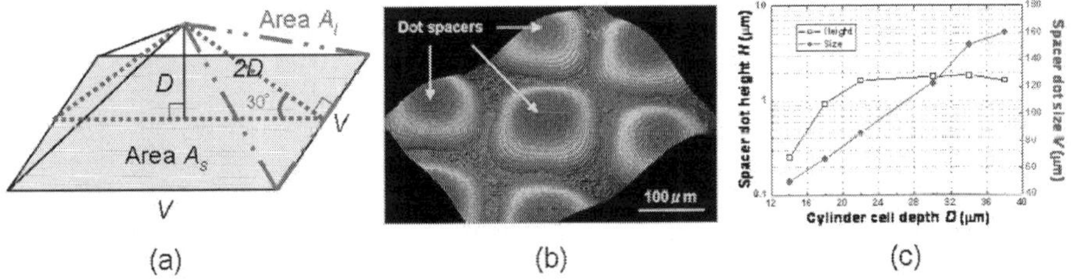

Fig. 27. (a) Cell model, (b) printed structures, and (c) relationship between (a) and (b) of gravure printing.

Figure 27(a) is a schematic plot for a single cell on the cylinder, the pyramid structure was laser engraved from the stainless steel cylinder. Figure 27(b) is the three dimensional printed structures measured optically and Figure 27(c) is an experimental plot for the printed pattern's size and its height. The interesting experiment result falls on that the printed pattern's size is linearly positively related to its designed dimension but the printed pattern's height is only in small range positively related to its designed dimension and finally trends to a saturation behavior. To explain this, the pyramid model in Figure 27(a) is used: The ink transfer process is a balance of force competition between the interface between the substrate (F_{is}) and the ink and the interface between the ink and the cell wall (F_{ic}). When F_{is} is larger than F_{ic}, the ink will tend to adhere to the substrate based on a premise that the ink quality is uniform within the whole droplet. Inversely, the ink will tend to stay inside the cell. With the relationship between the cell width (V) and the cell depth (D), angle γ can be calculated:

$$\gamma = \tan^{-1}(\frac{2D}{V}) \tag{21}$$

The stylus angle is approximately 120° and the total area (A_c) of four cell walls (A_t) is:

$$\begin{aligned} A_C &= 4A_t \\ &= 4VD \end{aligned} \tag{22}$$

Since the angle γ is 30°, the relationship between V and D is:

$$V = 2\sqrt{3}D \tag{23}$$

When replace V in Equation 22 with Equation 23:

$$\begin{aligned} A_C &= 4A_t \\ &= 8\sqrt{3}D^2 \end{aligned} \tag{24}$$

It is also obvious that the area of the opening area of the cell (A_s) is:

$$\begin{aligned} A_S &= V^2 \\ &= 12D^2 \end{aligned} \tag{25}$$

A comparison of A_c and A_s indicates that the difference of adhesion force becomes larger and larger when increasing the cell volume. As a result, the transfer of ink from the cell to the substrate becomes more and more difficult and final reaches its limitation. This special behavior will be alleviated if design the stylus angle to a larger value and the linear region can be extended. In order to keep the spacer dot height (H) – cylinder cell depth (D) curve linear, a critical angle γ_c, from Equation 21, can be calculated based on A_c=A_s or V=4D:

$$\begin{aligned} \gamma_C &= \tan^{-1}(\frac{2D}{V}) \\ &= \tan^{-1}(\frac{1}{2}) \\ &= 26.6^{\circ} \end{aligned} \tag{26}$$

When the angle γ is controlled smaller than γ_c, the adhesion force F_{is} and F_{ic} are at least balanced thus the transferred ink amount will be in linear relationship with the cell volume. For the structure design, the desired thicknesses fall in the linear region in Figure 27(c), thus the stylus angle of 120° was used. Experimented results shows that bigger and denser cell designs helped the printed ink to spread and merge (wetting). Further study and optimization should be done after this work for flatter surface. The final gravure printing process parameters for the isolation layer is summarized in Table 2. The best resolution of gravure printing was 200μm.

Target	Cylinder design (Table 1)	Pressure (N)	Speed (m/min)	Contact angle between doctor blade and cylinder (°)
Red (160nm)	V	550	32	60
Green (325nm)	VII			
Blue (245nm)	VI			

Table 2. The process parameters for different SiO_2 targets.

3.5 Lamination and finishing

After the process for both substrates in Figure 23 separately but before their lamination, one more study was performed to improve the MEMS flexible display device's contrast. Previous study only targeted on the demonstration of a single display pixel and did not address much on the overall performance of combined performance. A final compromise was made for the adhesive ink: a 35wt% glue solid content with 30wt% tiny black pigment (Sun Chemical, 049-72784) to block the transmission light from spacer areas. The lamination process was performed manually right after the spacer printing process. A final drying process was also applied right after the lamination step under 120°C for 1h to remove the extra solvent from the inks to reduce the reliability risks. The continuous substrate processed by roll-to-roll facility will be rewound back into a roll for stock and transfer before cut out into pieces for applications but the small area demonstrators fabricated by discrete tools discussed in these sections will be used for test in the next part.

4. Characterization and analysis

The discrete roll-to-roll printing system was used to check the printing capability, to characterize ink properties, and to study the system design. From system's point of view the continuous roll-to-roll printing processes handled the large area substrate before it was cut into small pieces for discrete roll-to-roll printing processes. From automation's point of view both the continuous and discrete roll-to-roll printing processes were automatically performed but the lamination process was semi-auto for alignment. The final production result of this study was not influenced by any factor of the system or the automation settings. Figure 28 is the pictures for the final demonstrators of (a) 3×3 active matrix array and (b) 21×39 passive matrix array. The 3×3 active matrix array device was examined and evaluated for various characteristics.

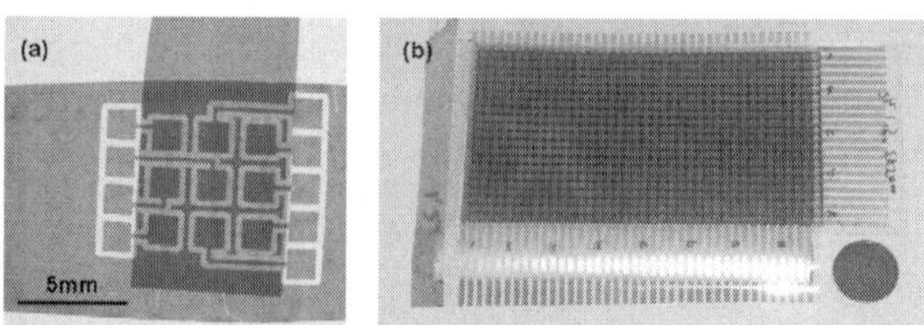

Fig. 28. (a) 3×3 active-matrix and (b) 21×39 passive-matrix demonstrators.

4.1 Optical performance

4.1.1 Color purity

The color purity simulation was done in section 2.2. The main influencing factor for color purity was explained in the same section by the control of metal electrodes. The best performance was concluded with 20nm Ag for its relatively balanced n value and relatively smaller k value. The color purity in CIE 1931 chromaticity diagram was measured by color tester (Yokogawa Denki, 3298F) with (x, y) axis system. A luminescent light (5500k) was used as the backlight. Figure 29(d) is the real performance of Ag electrode devices. The results typically followed the simulations and purer colors can be expected. From the datasheet PEN shows the best cutoff in UV region. Since strong UV light is harmful to human eyes and this MEMS display device claims sunlight as the backlight, a suitable substrate like PEN helps a lot on the UV cut out performance. When considering the visible region (400-700nm) in Figure 29(d), one can easily understand that only a single main peak appears in one color design which advanced each color's purity.

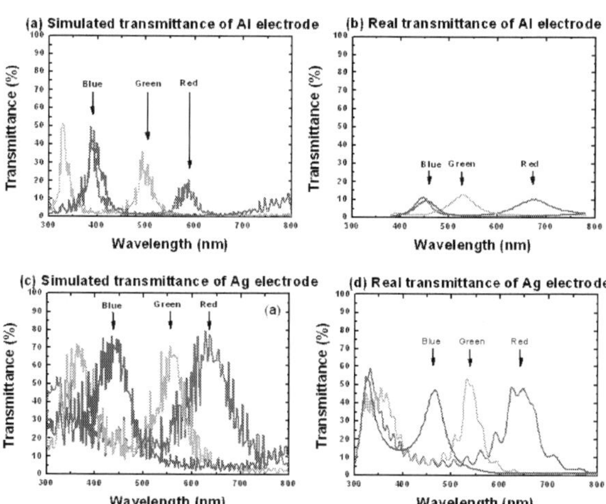

Fig. 29. Optical transmittance of different samples.

4.1.2 Color purity deviation

A color purity deviation (*CPD*) was defined in Equation 16 and was used to judge the color purity improvement in Figure 13. Figure 30 is the CIE chromaticity diagram for real devices made with Al and Ag. As predicted by simulation in section 2.2, Ag samples showed better distribution for the three primary colors than Al samples did. Ag samples actually showed distinguishable red improvement, better blue, and comparable green. An averaged larger than 50% *CPD* refinement can be seen in red and blue colors. The *CPD* also implies that green color almost reached the target; blue color has little room to be improved; and red color still has great space to be refined.

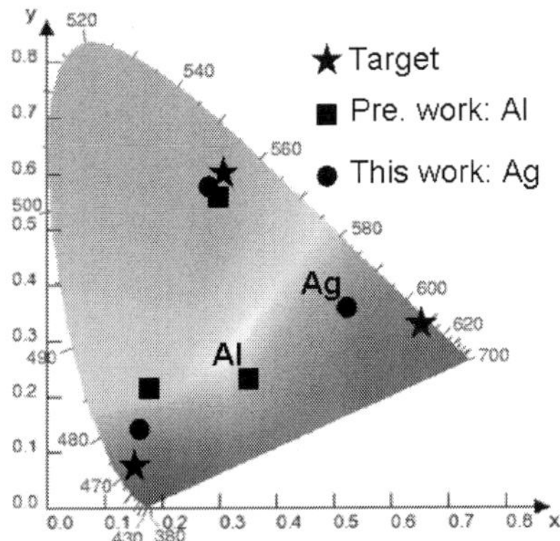

Fig. 30. Color purity comparison of previous and this works.

4.1.3 Transmittance

From Figure 29, around 20% transmittance difference was visible in both Al and Ag samples between simulation and real devices. This is believed to be the offset of simulation and real device. Besides the offset, as high as 40% unbalanced transmittance difference was in all Al colors but there is only less than 10% unbalanced transmittance difference in all Ag colors. Another important factor when judging the transmittance is its intensity. As shown in Figure 29(a)-(b), the intensities of all colors made by Al electrode are weak. This will become a serious perceptual issue if the substrate or extra protection layer absorbs some more intensity or when the backlight is weak. The common goal for all kinds of transmissive display device or color filter is to increase its transmittance intensity. In contrary, Ag samples showed higher intensity from both simulation and real device results.

4.1.4 Process variation induced color shift

The description in section 2.2 and the structure design implies that any variation of optical property and thickness of each layer will influence the output transmittance and color purity, it is important to understand how serious can these variation be. Table 3 is the simulation results for different Ag thicknesses. Since Ag was originally designed for 20nm and precise control was difficult, here a 25% (5nm) is set for the simulation. From the *CPD* point of view one may guess that 25nm should be the best setting for the smallest *CPD* for all colors. However, 5nm thickness lowered the highest transmittance peaks in red (618nm), green (562nm), and blue (426nm) for 7%, 15%, and 9%, respectively. Thus a 20nm Ag was finally decided as the process target. Table 4 is the simulation results for different isolation thicknesses, the variation is set to be ±40nm from each target. From the tables one can find serious color shifts along all settings, which means that slight process difference will result in great color change. Since the isolation layer does not have transmittance issue, the thickness selection was made at the best locations for each color so that the smallest *CPD* also took place in these designs.

	Ag=15nm	Ag=20nm	Ag=25nm	Ag=30nm
Red SiO_2=160nm	0.26	0.21	0.16	0.13
Green SiO_2=325nm	0.22	0.17	0.12	0.09
Blue SiO_2=245nm	0.10	0.04	0.00	0.03

Table 3. *CPD* values with electrode layer (Ag) thickness skew.

Red	SiO_2=120nm	SiO_2=160nm	SiO_2=200nm
	0.39	0.21	0.40
Green	SiO_2=285nm	SiO_2=325nm	SiO_2=365nm
	0.34	0.17	0.35
Blue	SiO_2=200nm	SiO_2=245nm	SiO_2=285nm
	0.16	0.04	0.22

Table 4. *CPD* values with isolation layer (SiO_2) thickness skew.

The thickness variation of spacer layer was also performed and summarized in Table 5. Since there is no target value, no *CPD* calculation had been made. However, the data points represent the OFF state color output, thus how to keep these data points as close as possible to let OFF state color the same is very important. One simple way to check how close those data points are is to calculate the triangle area enclosed by those data points. The Heron's formula (or Hero's formula) describes the triangle's area (A_{OFF}) by its three lengths:

$$A_{OFF} = \sqrt{s(s-L_a)(s-L_b)(s-L_c)} \tag{27}$$

where

$$s = \frac{L_a + L_a + L_c}{2} \tag{28}$$

and

$$\begin{aligned} L_a &= \sqrt{(x_R - x_G)^2 + (y_R - y_G)^2} \\ L_b &= \sqrt{(x_G - x_B)^2 + (y_G - y_B)^2} \\ L_c &= \sqrt{(x_B - x_R)^2 + (y_B - y_R)^2} \end{aligned} \tag{29}$$

Here L_a, L_b, and L_c are the three lengths of a triangle; (x_R, y_R), (x_G, y_G), and (x_B, y_B) are the coordinates of each point. The smallest value of A_{OFF} happens on the 750nm spacer case within the three trials. Nevertheless, this spacer layer was gravure printing process prepared and its rheology characteristics have been plotted in Figure 27(c) which showed very limited linear region. In order to operate the gravure printing with sufficient process window, a compromised design of 600nm which located in the center of the linear rheology region was chosen.

	Spacer=450nm	Spacer=600nm	Spacer=750nm
Red SiO_2=160nm	(0.25, 0.22)	(0.30, 0.24)	(0.23, 0.25)
Green SiO_2=325nm	(0.30, 0.25)	(0.25, 0.26)	(0.22, 0.25)
Blue SiO_2=245nm	(0.22, 0.27)	(0.22, 0.26)	(0.28, 0.24)
A_{OFF}	0.001943	0.000300	0.000198

Table 5. CIE coordinates with spacer height skew.

4.2 Structural performance

4.2.1 Air channel

Even though the MEMS model also predicted several different solutions to lower the operation voltage with Equation 20, a better solution which avoids changing the device's vertical design was proposed in section 2.4. The introduction of air channel is believe to be helpful to reduce the air pressure trapped inside a single pixel when ON. As shown in Figure 31, the Newton's ring means the colorful interference part at the edge of an interferometer. The root cause of Newton's ring was the different optical interference path lengths (Γ). These different optical path lengths in turn represented different output interfered colors. An interesting behavior in the figure is that the Newton's ring's size neither increases linearly nor increases infinitely.

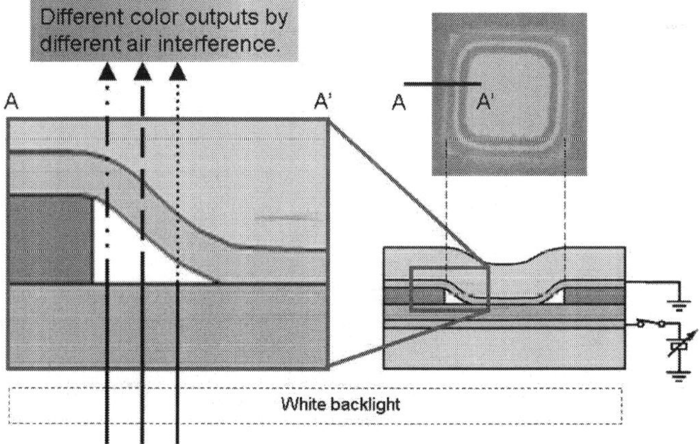

Fig. 31. A schematic plot to explain how Newton's ring took place.

Rather, the increment decreased along the increasing pixel size. A special saturation behavior of the Newton's ring's size appeared in Figure 32(a). This means that the increased pixel size will only make the display aperture looks larger instead of reduce the Newton's ring. This conclusion strongly supports the necessity of a revolutionary structure change. With different spacer coverage designs - 100% (no air channel), 90%, 80%, and 60% - the experimental data showed great amount of improvement. The width of Newton's ring reduced with all the coverage designs in Figure 32(b).

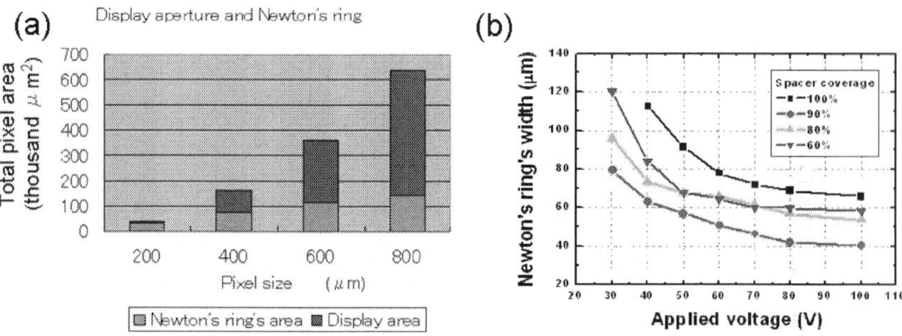

Fig. 32. Different spacer coverage alleviated the operation voltage.

Figure 33 is the picture which explains why saturation took place even with air channel design and why larger air channel did not yield in smaller Newton's ring: When the air channel is spacious enough, high applied voltage will let the lower layer attract the upper layer in the air channel area. Since the upper layer was put on the spacer and both sides (pixel area and air channel area) were competing each other, a see-saw performance showed - The more the air channel area in contact, the less the pixel area in contact. Thus a proper instead of a wide air channel is preferred. In this experiment a 90% coverage showed the smallest Newton's ring. This behavior was also obvious during simulation in Figure 20 and Figure 21: The display area tended to expand to the central part of air channel when the channel was wide enough but the display area tended to expand to the four corners when the channel with the same coverage was divided into two parts and were put aside. A combination of the experimental and simulation data suggested narrow and separate air channels are better. However, consider the resolution of printing process and the function of spacer layer for lamination, a single and large air channel was decided for the final structure.

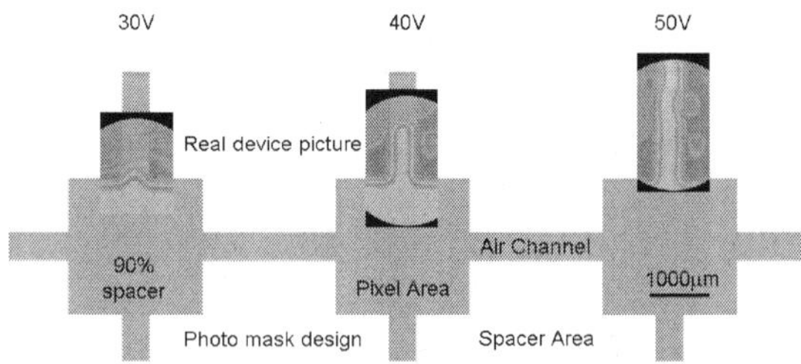

Fig. 33. Contact areas protruded into spacer areas when design was not optimized.

4.3 Electrical performance

4.3.1 Factors influencing electrical performance

The main target on the electrical performance of this MEMS display device is to reduce its operation voltage as described in the design part in section 2.3. However, to reduce the upper layer thickness together incorporate with the handling issue in which the electrostatic force is too strong on the <20μm PEN. The thin PEN will be easily attracted to the rubber pad and other equipment parts during printing and lamination processes by electrostatic force. The ultra thin upper layer will also induce special concerns on reliability. If extra layer should be added unto the whole structure, the transmittance and the optical performance should also be re-designed. Thus to reduce the thickness of the upper layer is not adequate.

	Spacer=400nm	Spacer=500nm	Spacer=600nm
Red	(0.31, 0.21)	(0.19, 0.33)	(0.30, 0.24)
SiO_2=160nm	0.10	0.12	0.07
Green	(0.25, 0.24)	(0.25, 0.26)	(0.25, 0.26)
SiO_2=325nm	0.09	0.08	0.08
Blue	(0.19, 0.31)	(0.28, 0.24)	(0.22, 0.26)
SiO_2=245nm	0.12	0.08	0.10

Table 6. *CPD* and its CIE coordinate under different spacer height settings.

To reduce the spacer height is not also a proper solution because the spacer height in OFF state also influences the output color as described in section 4.1. Table 6 is the list of simulated *CPD* under OFF state with white target of (0.31, 0.31) on CIE 1931 chromaticity diagram. The design goal not only fell on the small *CPD* but also required a small *CPD* difference between different colors. Both 400nm and 600nm spacer height designs are with smallest *CPD* differences $(0.10-0.07=0.12-0.09=0.03)$ but from the gravure printing characteristic point of view in Figure 27(c), the original 600nm design falls on the center part of the linear region thus provides more confidence on process control. Since the isolation layer thickness is the key for color interference, to reduce its thickness while keeping the same color design is then inaccessible. The final possibility fell on the pixel size and since this study aims on a large area display device for decoration, a 2000μm pixel size was set in section 2.3. Note that even though a 15V operation voltage was simulated in the same section, actual driving voltage was far higher than expectation in previous publication as listed in Table 7.

Pixel Size	Item	Red, SiO_2=370nm		Green, SiO_2=310nm		Blue, SiO_2=240nm	
		Sim.	Real	Sim.	Real	Sim.	Real
200μm	Voltage	65V	153V	55V	118V	40V	101V
	aperture	19%		19%		22%	
400μm	Voltage	40V	153V	30V	118V	28V	101V
	aperture	54%		54%		58%	
600μm	Voltage	34V	153V	25V	118V	22V	101V
	aperture	70%		68%		70%	
800μm	Voltage	28V	153V	25V	118V	18V	101V
	aperture	76%		78%		78%	

Table 7. Operation voltage difference between simulation and real device [35].

4.3.2 Combinational performance

Figure 34 is the three primary color pixels of the 3×3 MEMS flexible display demonstrator made partially by continuous and partially by discrete roll-to-roll printing processes operated under <20V. According to simulations, the contact area should be 93%(15V), 92%(20V), and 94%(20V) for red, green, and blue, respectively. These data typically match the expectations and the trend in Table 7. The big difference between these data and previous study is the Newton's ring's size. In the previous study the Newton's ring can be found under both small and large display apertures but the Newton's ring can merely be found in the green and blue pixel of this study because of the air channel design. This kind of improvements can be attributed to:

1. Low operation voltage – By using the air channel design to evacuate air pressure when ON;
2. Small Newton's ring – By replacing the steep spacer made by photolithography by the oblique ink spacer made by gravure printing.

The first merit was fully explained in section 4.2 and the second merit can be explained with Figure 35. Since the spacer structures were printed with pyramid shapes as shown in Figure 27(b), their oblique surfaces provide supports for the upper layer when ON. Also because of the black pigment doping in the spacer ink, the Newton's ring's color was blocked by the spacer structure.

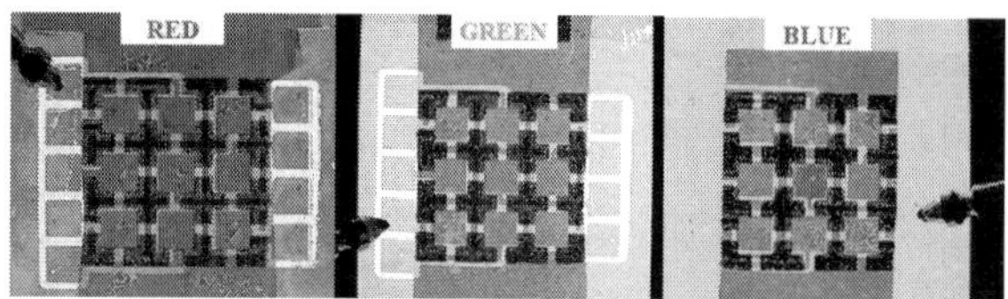

Fig. 34. The 3×3 test sample under ON state. A single pixel (square) is designed as 2000μm.

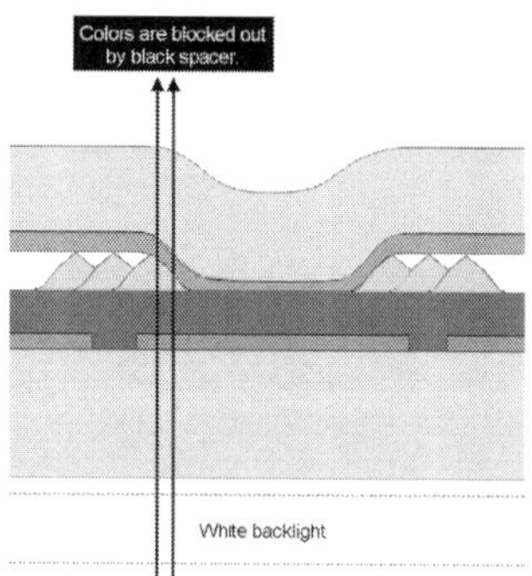

Fig. 35. A schematic plot to explain how Newton's ring was suppressed.

4.4 Yield performance

Section 4.1 to section 4.4 reviewed the MEMS flexible display device's characteristics. This section will review the roll-to-roll process's integrity from the mass production point of view.

4.4.1 Sheet to sheet uniformity

Figure 36 is the cumulative plot for electrode's sheet resistance (R_s). Since R_s excludes the influence by thickness, it represents a normalized impedance to its area with the following equation:

$$R_S = \frac{R \times w}{l} \tag{27}$$

where R is the sheet resitance. Note that the real line length (l) and real line width (w) will differ from the designed value, only the real value should be used to correlate with area size. Even though section 3.1 suggested a better flexography printing resolution along MD direction, electrical test revealed that the finest resolution was about 40μm for both transverse direction (TD) and MD. The data in this figure came from continuous 10m substrate with repeated 18 patterns for 8 times (sheets). The failure rate was 2.78% (4 out of 144) which is very compatible with current commercial semiconductor process lines. The whole patterning process done by the continuous roll-to-roll system including sacrificial ink printing, metal sputtering, and ultrasonic assisted lift-off was successfully developed and proved. From the figures we also understood that narrower lines were with larger standard deviations which implied poorer resolution controls. From the results, the smaller standard variation value of vertical patterns (0.54ohm/sq) also suggested better printing integrity along the MD direction. The TD patterns (whose standard variation is 0.85ohm/sq) showed finer lines but was by chance. Thus when one wants to try to obtain fine lines, it is suggested to design patterns normal (90°) to the printing direction but when one wants to obtain stable performance, it is suggested to design patterns along the printing direction. With these data, the developed lift-off process is suggested for the wider than 55μm line width applications.

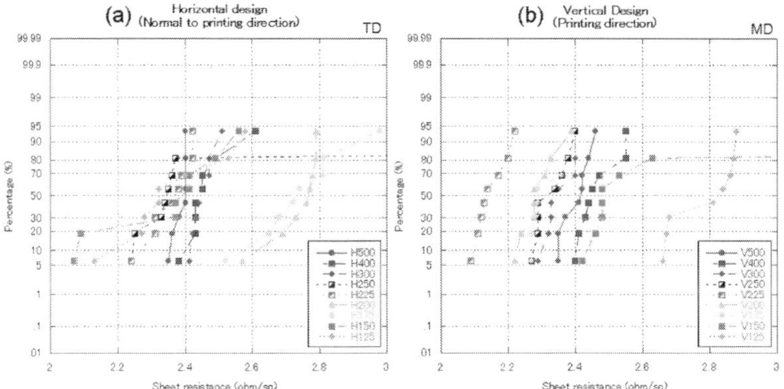

Fig. 36. Sheet resistance yield plots of electrode layer with (a) TD and (b) MD pattern.

4.4.2 Within sheet uniformity

Another test line set which occupies the whole sheet was used to check the within sheet uniformity. These test lines were designed only along the MD direction. Figure 37 is the cumulative plot for a 2mm long line which was used to fabricate passive matrix samples. Compared to Figure 36, these data were perfectly distributed with 100% as a sharp line since the line width were relatively wider than the lines in the test pattern set, these 2mm lines were thus with less ink wetting induced variations from gravure printing. Even though the study goal is a large area MEMS controlled flexible display device, to develop a process which can support the requirement of the display system emerged parasitically. Thus this characterization section reviewed not only the device itself, but also the yield of the production line.

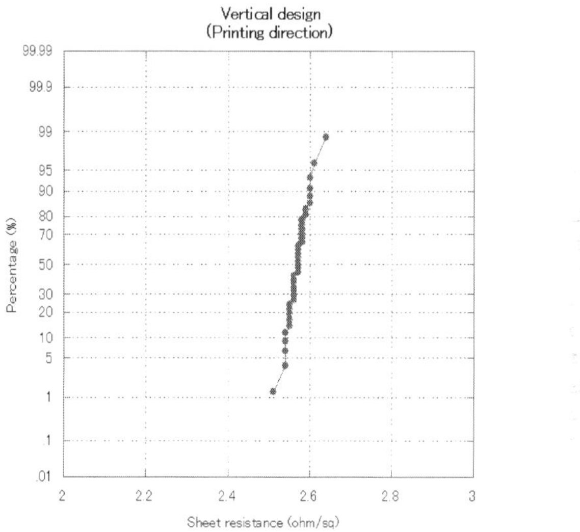

Fig. 37. Sheet resistance yield plots of electrode layer test patterns on the same sheet.

5. Summary and discussion

After review the micro MEMS display device's electrical, mechanical, and optical behaviors in previous section, this section will deal with some special considerations. These considerations came with the original design and sometimes worsened along the long term operation or the mass production. Thus the discussions on these considerations help on verifying some root causes of issues and also help on improving the device into a more complete design.

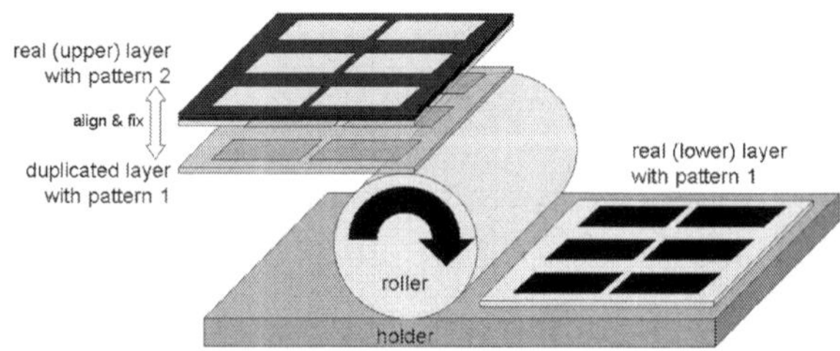

Fig. 38. The schematic plot for manual alignment and semi-auto lamination.

5.1 Alignment accuracy

This MEMS flexible display device was made partially by automatic continuous roll-to-roll system and partially by semi-auto discrete processes. Since the alignment apparatus was not yet installed in the roll-to-roll system shown in Figure 23, the alignment process during lamination of the two layers was performed manually. As shown in Figure 38, the lamination was performed with the test printer by the following procedures:

1. Prepare a duplicated lower layer pattern on a thin substrate,
2. Align the real upper layer with the duplicated layer and put them on the roller,
3. Put the real lower layer on the plate,
4. Activate the roller to laminate the two real layers,
5. Remove the dummy layer to obtain the laminated device.

The solution for misalignment is to install the lamination process into the continuous production line and control the same misalignment amount over a long distance. Figure 39 is the schematic plot for this idea. Let $L_1 >> L_2$ and a is the smallest misalignment done by semi-auto system with manually alignment. Since the a value is fixed no matter how long the substrate is, when the process was aligned with a long substrate (L_1) and cut into smaller sheets (L_2) the misalignment amount b will be smaller than a. This kind of comparison was made base on the concept of unit length (here, the L_2). For example, the misalignment amount roughly reduced to 10% on the small area when the process distance was 10 times longer; the misalignment amount roughly reduced to 1% on the small area when the process distance was 100 times longer. When good alignment is expected on small areas, alignment mark can be added on both layers and registered and adjusted optically. Moreover, a feedback system which is capable to adjust the cylinder's location will also be helpful to adjust the alignment.

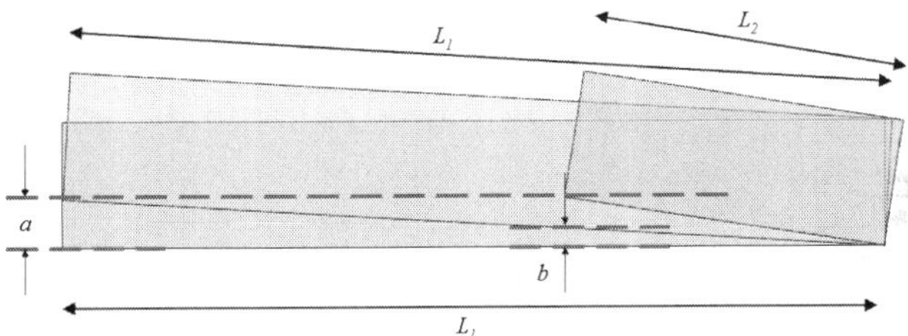

Fig. 39. The longer substrate helps on reducing the manual misalignment per unit length.

5.2 Color degradation

During the experiments and evaluations, a color degradation issue was found. The display color degraded from the original color after long term, high stress (voltage) operation. Note that the reliability test was cumulatively stressed from the low voltage → short term → long term → high voltage → short term → long term. Previous study attributed similar behavior to the reliability of thin electrode layer (12nm aluminum) and the strong electrostatic adhesion between upper electrode layer and the isolation layer. When they are in contact under stress, the upper electrode pealed off from the upper substrate and became incapable for color interference anymore. There was a light trend of display area with test sequence. This was because the isolation layer thickness difference and the charges started to accumulate from the thinnest areas. However, the charges were not smoothly removed because the electric power was removed suddenly thus the display area did not 100% return to its original state. A study done with a periodical electric power supply which continuously switched between two polarities and thus the charges could be removed and the upper layer could return back to the original state. The electrodes with accumulated charges were further supported with more and more charges under the cumulative stress test, thus the display area expanded larger and larger. Some solutions for this reliability issue are thus proposed in three ways:

A. Solid isolation material

In order to obtain high and solid insulating material in a roll-to-roll process system, a sputter process is necessary and the sputter still matches the mass production requirement as the metal sputter process does. To continuously include the isolation layer sputter in the same apparatus of metal target can further reduce the process time (vacuum pumping) and the equipment space (add one target instead of add a machine).

B. Complete drying and curing process

Since the highest working temperature of plastic material is normally lower than 200°C, the over 350°C curing temperature for SiO_2 curing is not acceptable. This issue can be solved by replacing the plastic substrate by glass. Even though the glass substrate is fragile for shear stress, it can be made thin and thus flexible. Section 3.1 gave one example of how thin, how large, and how flexible can the thin glass be.

C. Doubled isolation layers

Either the sputtered SiO_2 or the completely cured SiO_2 ink will not solve the profound impact of electrostatic adhesion between upper electrode layer and the isolation layer. As a result, the sputtered SiO_2 or the completely cured SiO_2 ink will still degrade owing to the detachment of electrode material from the upper substrate. To prevent the electrode material from detaching from its substrate, a good solution is to cover the electrode with the isolation material. When this device is ON, the combinational thickness matches the color design value in the second section and the output color is expected to be the same. However, as summarized in Table 8, the OFF state non interfered color needs to be specially designed. The thickness in this table was easily divided the isolation layers into equivalent two parts from its originally designed value. Compared to the data for single layer, the data for double layer will form a larger triangle which is a drawback explained in section 4.1.

Structure	Isolation layer thickness (nm)	CIE coordinate when OFF	Difference
Single layer	Red=160	(0.30, 0.24)	---
	Green=325	(0.25, 0.26)	
	Blue=245	(0.22, 0.26)	
Double layer	Red=80×2	(0.32, 0.24)	acceptable
	Green=162×2	(0.29, 0.21)	acceptable
	Blue=122×2	(0.23, 0.44)	large

Table 8. OFF color difference from single and double layer designs.

5.3 Surface condition

Previously study concluded that the surface unevenness was the root cause for the poor color purity. The reference also concluded that the same issue caused the isolation breakdown. Furthermore, the surface unevenness structures distributed on the substrate surface mentioned in the reference were dense and uniformly spread on the substrate, the color interference of the defects then should not be as uniform as the data disclosed in the same reference. But even though the hypothesis was not correct, this surface defect should be eliminated. Figure 40 is the surface profile analyzed by atomic force microscope (AFM, SII NanoTechnology Inc., Nanopics). Here R_q is the root mean square value of measured data. The surface condition was monitored (a) before metal electrode sputtering, (b) after metal electrode sputtering, (c) after pattern lift-off, and (d) after isolation layer printing. There was no significant difference for the first three steps which suggested the substrate's profile was always inherited and followed from layer to layer until the SiO_2 was capped. Compared to the sputter and lift-off process, the gravure SiO_2 printing and the following drying and curing processes covered all these surface defects. The SiO_2 droplets transferred out from the gravure cylinder reflowed, spread, and finally merged before drying process provided a more uniform surface than the original substrate did. Thus the printing rheology is a key point for study.

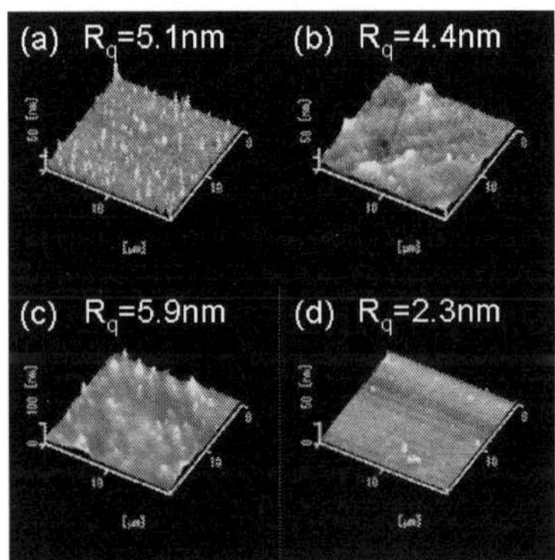

Fig. 40. Surface profile after different processes.

5.4 True colors

The purer red provided by Ag appeared owing to Ag's suitable optical parameters. However, the simulated and experimental data were not close to target red identified in Figure 13. Other metals such Au and Cu were examined to see how they provide interfered colors. Figure 41 is the simulations done with Au and Cu electrodes. As indicated by its color purity deviation (*CPD*), the red color by Au and green color by Cu provided smallest values. However, these values were not small enough to be true colors. Thus true colors of red and green require either new electrode materials or a multiple layer which serves as a single electrode layer to perform color interference. Other popular metals such as chrome (Cr) and cobalt (Co) were exempt from the simulation database because of their poor color and transmittance performance during simulation.

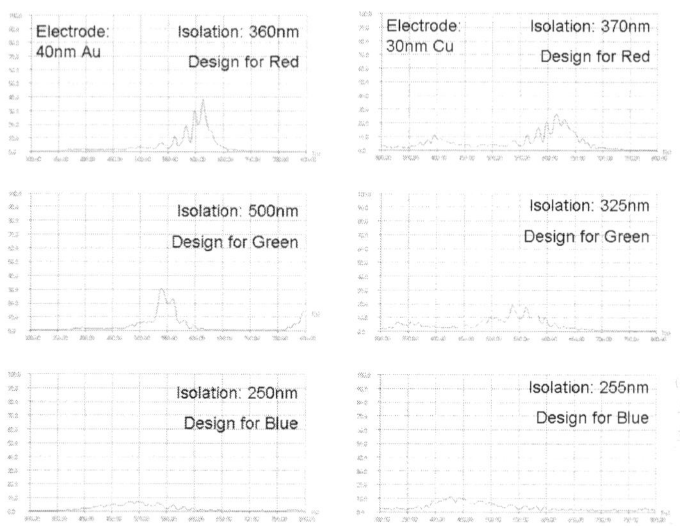

Fig. 41. Simulated best design with Au electrodes.

5.5 Structures

Even though the combinational full color display is not a must of the goal for decoration applications, this study implied a full color possibility by combining all three primary colors. However, the original design, which used a unified spacer thickness of 600nm, showed different combinational layer thickness. When combine the structures in Figure 11, a total thickness difference will appear since the isolation (Intermediate 5) thicknesses are different. This means that when using a unified lower substrate, the upper substrate will not be flat as shown in Figure 42. The uneven layer will not only cause process difficulties but will also result in reliability concerns. To overcome this, a structure with unified upper layer which requires different spacer heights and different isolation thickness for different colors should be designed. Some candidate solutions suggested in Table 9 provide uniform OFF state colors (same CIE coordinate). The process detail will be complicated but by doing this, the process issue disappears and the spacer heights can be controlled by gravure printing's cylinder and ink engineering discussed in section 3.4.

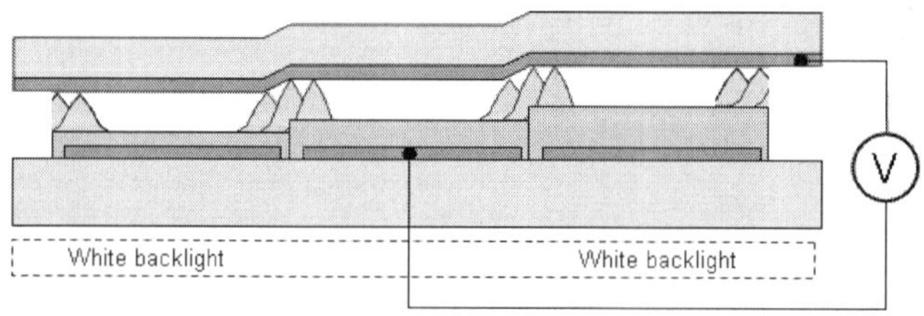

Fig. 42. Unified spacer height will leave an uneven upper layer.

	Solution 1		Solution 2		Solution 3	
	Spacer height (nm)	CIE coordinate when OFF	Spacer height (nm)	CIE coordinate when OFF	Spacer height (nm)	CIE coordinate when OFF
Red	765	(0.25, 0.26)	685	(0.22, 0.26)	600	(0.30, 0.24)
Green	600		520		445	
Blue	680		600		515	

Table 9. Color behavior under OFF states with uneven spacer heights

5.6 Processes

Recent research and development in the printed electronics field indicated usable Ag nano particle ink, which can be a good candidate for printed electrode. But designer also has to consider the influence of high resistivity of this ink because even though the resistivity does not influence the electrostatic behavior, the device requires a high electrical yield for connection lines. After these discussion focused on specific issues from process, material, and operation point of view with possible solutions, one can further expect a perfect final display device which is controlled by MEMS and manufactured by printing processes with flexibility.

6. Summary

The flexible MEMS by roll-to-roll printing system is an epoch-making concept that the system require neither solid substrate nor photolithography process.

A complex model which is a superposition of a single-end fixed cantilever and a parallel plate was proposed and proved by simulation and experiments. The model clearly indicated key parameters which play crucial roles on operation voltage which is also a key for portable, light weight, low cost, and maybe disaposable appliation. The Fabry-Perot color interferometer concept was taken as a demonstrator which was controlled by the MEMS. With this complex model, similar MEMS device can easily achieve low operation voltage which is a must for commercialization and safety concern.

The roll-to-roll process system showed a budget-friendly, high-efficiency, and large area supportive production and suggested high potential on replacing current photolithography technique for flexible applications.

With the successful electrical, mechanical, and optical demonstration, the large area flexible MEMS as well as the roll-to-roll printing process system opened great applications on portable and disposable electronic devices thus provided significant values in design and production fields.

7. Acknowledgment

The author wants to thank Prof. Dr. Hiroyuki Fujita and Prof. Dr. Hiroshi Toshiyoshi (both are with the University of Tokyo, Japan) for their advisory helps. The author also wants to thank Kuan-Hsun Liao, Sheng-An Kuo, and Chung-Yuan Yang for their lab works. The related works were partially supported by the following projects:

National Science Council (NSC) of Taiwan (Republic of China)

Research project (ID: 99-2218-E-007-018-MY2),
New Energy and Industrial Technology Development Organization (NEDO) of Japan
Industrial Technology Research Grant Program (ID: 06D48522d),
Japan Society for the Promotion of Science (JSPS)
International Training Program (ITP),
Finnish Funding Agency for Technology and Innovations (TEKES)
Funding decision 40104/07.

8. References

Abe, T., Yamashita, J., Shibata, H., Kato, Y., Matsumoto, H., & Iijama, T. (2008). *High-Accuracy Correction of Critical Dimension Errors Taking Sequence of Large-Scale Integrated Circuits Fabrication Processes into Account*, Journal of Micro-nanolithography MEMS and MOEMS, Vol. 7, Iss. 4, (October-December 2008), pp. 043008, ISSN 1932-5150

Author, G., & Martin, B. (1996). *Investigation of Photoresist-Specific Optical Proximity Effect*, Microelectronic Engineering, Vol. 30, Iss. 1-4, pp. 133-136, ISSN 0167-9317

Blanchet, G, B., Loo, Y., Rogers, J, A., Gao, F., & Fincher, C. R. (2003). *Large Area, High Resolution, Dry Printing of Conducting Polymers for Organic Electronics*, Applied Physics Letters, Vol. 82, No. 3, (January 2003), pp. 463-465, ISSN 0003-6951

Blanchet, G., & Rogers, J. (2003). *Printing Techniques for Plastic Electronics", Journal of Imaging Science and Technology*, Vol. 47, No. 4, (July-August 2003), pp. 296-303, ISSN 1062-3701

Boer, W, den. (2005). *Active Matrix Liquid Crystal Displays*, Elsevier Inc., ISBN 978-0-7506-7813-1, Burlington, USA

Bogacz, S., & Trafton, J, G. (2005). *Understanding Dynamic and Static Displays*, Cognitive Systems Research, Vol. 6, Iss. 4, (January 2005), pp. 312-319, ISSN 1389-0417

Corr, D., Bach, U., Fay, D., Kinsella, M., McAtamney, C., O'Reilly, F., Rao, S.N., & Stobie, N. (2003). *Coloured Electrochromic "Paper-quality" Displays Based on Modified Mesoporous Electrodes*, Solid State Ionics, Vol. 165, pp. 315-321, ISSN 0167-2738

Crawford, G, P. (2005). *Flexible Flat Panel Displays*, John Wiley & Sons, Ltd., ISBN 978-0-470-87048-8, Chichester, England

Crowley, J, M., Sheridon, N, K., & Romano, L. (2002). *Dipole Moments of Gyricon Balls*, Journal of Electrostatics, Vol. 55, pp. 247-259, ISSN 0304-3886

Cummins, D., Boschloo, G., Ryan, M., Corr, D., Rao, S, N., & Fitzmaurice, D. (2000). *Ultrafast Electrochromic Windows Based on Redox-chromophore Modified Nanostructured*, Journal of Physical Chemistry B, Vol. 104, pp. 11449-11459, ISSN 1520-6106

Hernandez, G. (1988). *Fabry-perot Interferometer*, Cambridge University Press, ISBN 0-521-36812-X, New York, USA

Ida, N. (2004). *Engineering Eletromagnetics*, Springer-verlag New York, LLC., ISBN 0-387-20156-4, New York, USA

Johnson, M, T., Zhou, G., Zehner, R., Amundson, K., Henzen, A., & Van de Kamer, J. (2006). *High-Quality Images on Electrophoretic Displays*, Journal of Society for Information Display, Vol. 14, Iss. 2, pp. 175-180, ISSN 1071-0922

Kim, J., Yang, K., Hong, S., & Lee, H. (2008). *Formation of Au Nano-patterns on Various Substrates Using Simplified Nano-Transfer Printing Method*, Applied Surface Science, Vol. 254, pp. 5607-5611, ISSN 0169-4332

Leech, P, W., & Lee, R. A. (2007). *Hot Embossing of Diffractive Optically Variable Images in Biaxially-Oriented Polypropylene,* Microelectronic Engineering, Vol. 84, pp. 25-30, ISSN 0167-9317

Lin, S, C., Lee, S, L., & Yang, C, L. (2009). *Spectral Filtering of Multiple Directly Modulated Channels for WDM Access Networks by Using an FP Etalon,* Journal of Optical Networking, Vol. 8, Iss. 3, (March 2009), pp. 308-316, ISSN 1943-0620

Lo, C., Huttunen, Hiitola-Keinänen, O. -H., Petäjä, J., Hast, J., Maaninen, A., Kopola, H., Fujita, H., & Toshiyoshi, H. (2008). *Active Matrix Flexible Display Array Fabricated by MEMS Printing Techniques,* Proceedings of The 15th Int. Display Workshop, pp. 1353-1356, Niigata, Japan, December 3-5, 2008

Lo, C., Huttunen, O, -H., Petäjä, J., Hast, J., Maaninen, A., Kopola, Fujita, H., & Toshiyoshi, H. (2007). *Novel Printing Processes for MEMS Fabry-perot Display Pixel,* Proceedings of The 14th Int. Display Workshop, pp. 1337-1340, ISBN 9781605603919, Sapporo, Japan, December 5-7, 2007

Marques, A., Moreno, I., Campos, J., & Yzuel, M, J. (2006). *Analysis of Fabry-Perot Interference Effects on The Modulation Properties of Liquid Crystal Displays,* Optics Communications, Vol. 265, Iss. 1, pp. 84-94, ISSN 0030-4018

Matsumoto, S. (1990). *Electronic Display Devices,* John Wiley & Sons, Inc., ISBN 0-471-92218-8, New York, USA

Obata, K., Sugioka, K., Shimazawa, N., & Midorikawa, K. (2006). *Fabrication of Microchip Based on UV Transparent Polymer for DNA Electrophoresis by F_2 Laser Ablation,* Applied physics A, Vol. 84, pp. 251-255, ISSN 0947-8396

Oh, H, -Y., Lee, C., & Lee, S. (2009). *Efficient Blue Organic Light-Emitting Diodes Using Newly-Developed Pyrene-Based Electron Transport Materials,* Organic Electronics, Vol. 10, Iss. 1, pp. 163-169, ISSN 1566-1199

Pollack, M, G., Fair, R, B., & Shenderov, A, D. (2000). *Electrowetting-based Actuation of Liquid Droplets for Microfluidic Applications,* Applied Physics Letters, Vol. 77, No. 11, (September 2000), pp. 1725-1726, ISSN 0003-6951

Puetz, J., & Aegerter, M. A. (2008). *Direct Gravure Printing of Indium Tin Oxide Nanoparticle Patterns on Polymer Foils,* Thin Solid Films, Vol. 561, pp. 4495-4501, ISSN 0040-6090

Schanda, J. (2007). *Colorimetry,* John Wiley & Sons, Inc., ISBN 978-0-470-04904-4, New Jersey, USA

Senda, K., Bae, B, S., & Esashi, M. (2008). *MEMS Membrane Switches Backplane for Matrix Driven Large Sign Display,* Proceedings of The 15th International Display Workshops, Vol. 2, pp. 1349-1352, Niigata, Japan, December 3-5, 2008

Senturia, S, D. (2001). *Microsystem Design,* Kluwer Academic Publishers, ISBN 0-306-47601-0, Massachusetts, USA

Smith, F, G. (2007). *Optics and Photonics,* John Wiley & Sons, Ltd., ISBN 978-0470017845, Chichester, England

Taii, Y. (2006). Master's degree thesis, The university of Tokyo

Taii, Y., Higo, A., Fujita, H., & Toshiyoshi, H. (2006). *A Transparent Sheet Display by Plastic MEMS,* Journal of Society for Information Display, Vol. 14, Iss. 8, pp. 735-741, ISSN 1071-0922

Vidotti, M., & Córdoba de Torresi, S, I. *(2008). Nanochromics: Old Materials, New Structures and Architectures for High Performance Devices,* Journal of the Brazilian *Chemical Society,* Vol. 19, No. 7, pp. 1248-1257, ISSN 0103-5053

Wang, Q, D., Duan, Y, G., Ding, Y. C., Lu, B. H., Xiang, J, W., & Yang, L. F. (2009). *Investigation on LIGA-like Process Based on Multilevel Imprint Lithography*, Microelectronics Journal, Vol. 40, Iss. 1, pp. 149-155, ISSN 0026-2692

Wong, W, S., Chabinyc, M. L., Limb, S., Ready, S, E., Lujan, R., Daniel, J., & Street, R, A. (2007). *Digital Lithographic Processing for Large-area Electronics*, Journal of Society for Information Display, Vol. 15, Iss. 7, pp. 463-470, ISSN 1071-0922

Acreo, http://www.acreo.se

Corning, http://www.corning.com

E Ink, http://www.eink.com

Ioffe Physical Technical Institute, http://www.ioffe.rssi.ru

Liquavista, http://www.liquavista.com/

Pixtronix, http://www.pixtronix.com

Qualcomm, http://www.qualcomm.com

Sipix, http://www.sipix.com

Silicon Light Machine, http://www.siliconlight.com

Texas Instruments, http://www.ti.com

16

Thermal Microactuators

Leslie M. Phinney, Michael S. Baker and Justin R. Serrano
Sandia National Laboratories
USA

1. Introduction

This chapter discusses the design, fabrication, characterization, modeling, and reliability of thermal microactuators. Microelectromechanical systems (MEMS) devices contain both electrical and mechanical components and are in use and under development for applications in the consumer products, automotive, environmental sensing, defense, and health care industries. Thermal microactuators are standard components in microsystems and can be powered electrically through Joule heating or optically with a laser. Examples of MEMS designs containing thermal microactuators include optical switches (Cochran et al., 2004; Sassen et al., 2008) and nanopositioners (Bergna et al., 2005). Advantages of thermal microactuators include higher force generation, lower operating voltages, and less susceptibility to adhesion failures compared to electrostatic actuators. Thermal microactuators do require more power and their switching speeds are limited by cooling times.

Extensive work has been performed designing, fabricating, testing, and modeling thermal microactuators. Howell et al. (2007) has reviewed the fundamentals of thermal microactuator design. Designs of electrically powered MEMS thermal actuators include actuators fabricated from a single material (Comtois et al., 1998; Park et al., 2001; Que et al., 2001) and bimorphs (Ataka et al., 1993). Thermal actuator designs using a single material are both symmetric, referred to as bent-beam or V-shaped, structures (Baker et al., 2004; Park et al., 2001; Phinney et al., 2009) and asymmetric (Comtois et al., 1998), which have a hot arm and a cold arm. Asymmetric actuators are also referred to as flexure actuators. Some studies investigated both bent-beam and flexure actuators (Hickey et al., 2003; Oliver et al., 2003). In addition to electrical heating, powering thermal microactuators optically using laser irradiation has been demonstrated (Oliver et al., 2003; Phinney & Serrano, 2007; Serrano & Phinney, 2008). Modeling efforts have focused on bent-beam microactuators (Baker et al., 2004; Enikov et al., 2005; Howell et al., 2007; Lott et al., 2002; Wong and Phinney, 2007) and flexure actuators (Mankame and Ananthasuresh, 2001).

This chapter focuses on bent-beam and flexure microactuators. In order for thermal actuators to operate, sufficient heating and thermal expansion of the components must occur. However, device temperatures that are too high result in permanent deformation, damage, and degradation in performance. In addition, packaging processes and conditions affect the performance and reliability of microsystems devices motivating studies on the effects of surrounding gas pressure and mechanical stress on thermal MEMS.

2. Thermal microactuator designs

Figure 1 shows schematics of three thermal microactuator designs: bimaterial, bent-beam, and flexure. Bimaterial actuators consist of materials with different coefficients of thermal expansion and function similarly to a bimetallic thermostat (Ataka et al., 1993). When the temperature changes due to an embedded heater, the microactuator moves due to the difference in the expansion associated with the temperature change (Fig. 1a). Bent-beam

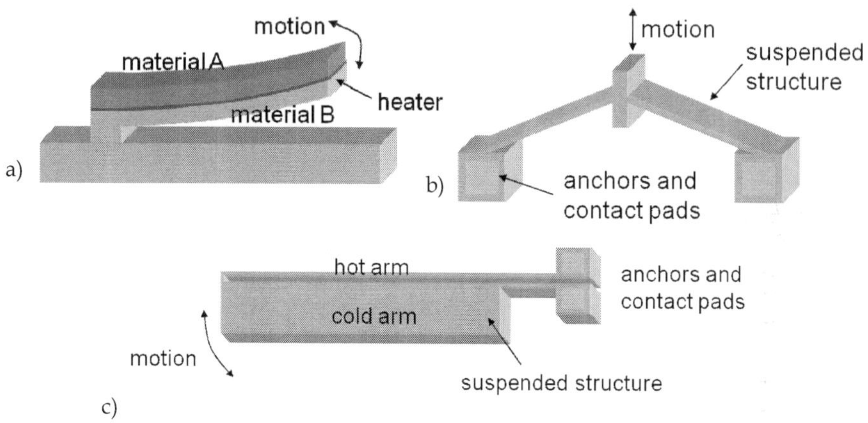

Fig. 1. Schematics of thermal microactuators: a) bimaterial, b) bent-beam, and c) flexure

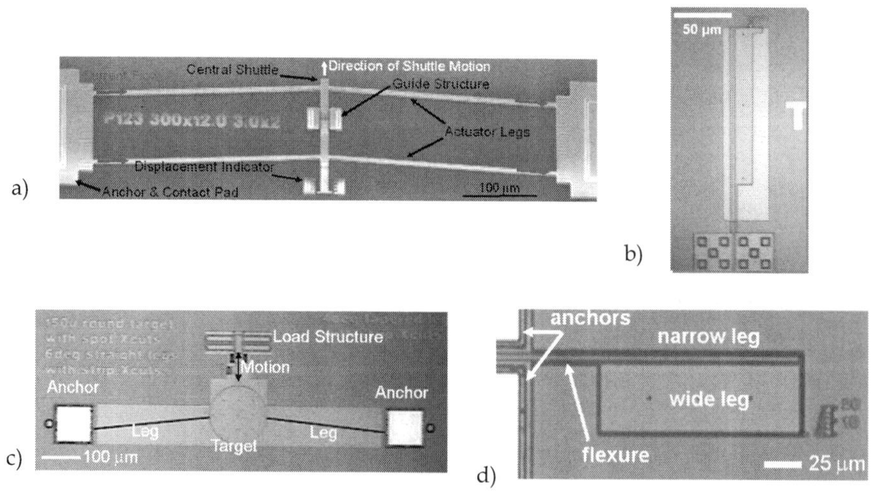

Fig. 2. Microscope pictures of thermal microactuators: a) electrically powered bent-beam, b) electrically powered flexure, c) laser powered bent-beam, and d) laser powered flexure

actuators have angled legs that expand when heated, providing force and displacement output as shown in Fig. 1b (Park et al., 2001; Que et al., 2001). The flexure actuator in Fig. 1c contains asymmetric legs, for example of unequal width, that flex to the side due to differential expansion when heated (Comtois et al., 1998). Figure 2 has pictures of electrically and optically powered bent-beam and flexure thermal microactuators.

3. Fabrication

Thermal microactuators are created using various microfabrication techniques including surface micromachining and silicon on insulator (SOI) processing which will be reviewed. Particular designs for surface micromachined thermal microactuators are presented in detail as characterization data for these designs are reported in Section 4.

3.1 Surface micromachining

Surface micromachining involves the sequential growth or deposition of thin films, patterning of features, and etching of the films to create multilayer structures and devices. Surface micromachining results in devices with in-plane dimensions from a few microns to millimeters and thicknesses of microns to 10 microns so they have low aspect ratios, i.e., thickness divided by length or width. Typical surface micromachining processes use polycrystalline silicon (polysilicon) for the structural layers and silicon dioxide for the sacrificial layers.

The surface micromachined thermal microactuators for which characterization data will be reported were fabricated using the SUMMiT V™ (Sandia Ultra-planar Multilevel MEMS Technology) process (Sniegowski and de Boer, 2000; SUMMiT V, 2008). The SUMMiT V process uses four structural polysilicon layers with a fifth layer as a ground plane. These layers are separated by sacrificial oxide layers that are etched away during the final release step. The two topmost structural layers, Poly3 and Poly4, are nominally 2.25 μm in thickness, while the bottom two, Poly1 and Poly2, are nominally 1.0 μm and 1.5 μm in thickness, respectively. The ground plane, Poly0, is 300 nm in thickness and lies above an 800 nm layer of silicon nitride and a 630 nm layer of silicon dioxide. The sacrificial oxide layers between the structural layers are each around 2.0 μm thick (Sniegowski and de Boer, 2000; SUMMiT V, 2008).

Figure 3 pictures schematics of an electrically heated bent-beam thermal microactuator with two legs and the cross-sectional area of an actuator leg with the width and thickness dimensions labeled. The SUMMiT V processing constraints on the sacrificial oxide cut between two polysilicon layers result in an I-beam shape for the thermal actuator legs (SUMMiT V, 2008). In this chapter, mechanical, electrical, and thermal characterization results are presented for bent-beam thermal microactuators with two actuator legs (Phinney et al., 2009). The thermal microactuator designs have the actuator legs fabricated from three laminated structural polysilicon layers: Poly1, Poly2, and Poly3 (Figure 4). This actuator design is referred to as the P123 actuator throughout this chapter. The second thermal actuator design is the same thermal actuator as the first design with a force gauge attached to the actuator shuttle (Figure 5) and is referred to as the P123F actuator. The force gauge consists of a linear bi-fold spring attached to the shuttle of the actuator using the Poly3 layer. Table 1 summarizes the geometries of the thermal microactuators with nominal dimensions specified according to the SUMMiT V Design Manual (SUMMiT V, 2008). The shuttle that connects beams at the center is 10 µm wide, 100 µm long and its thickness is the sum of t_1, t_2, and t_3.

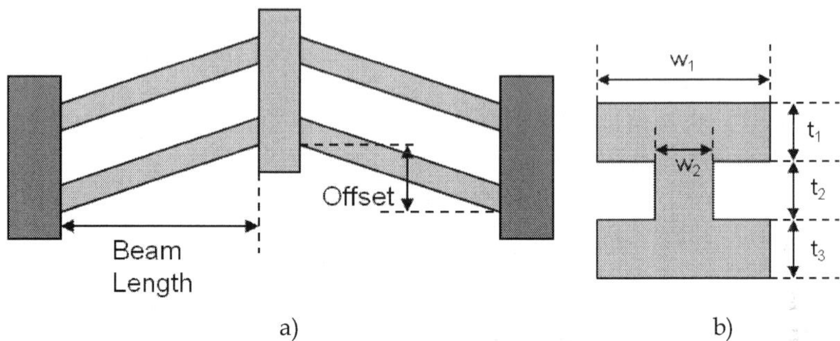

Fig. 3. Schematics of a) bent-beam thermal microactuator with two legs and b) actuator leg cross section showing dimensions that are specified in Table 1

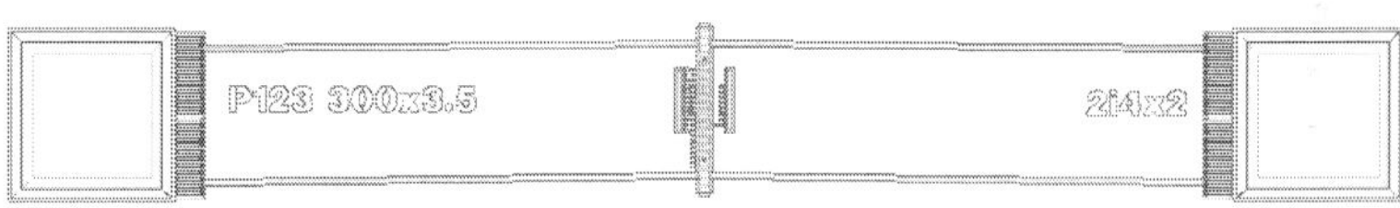

Fig. 4. P123 thermal microactuator schematic

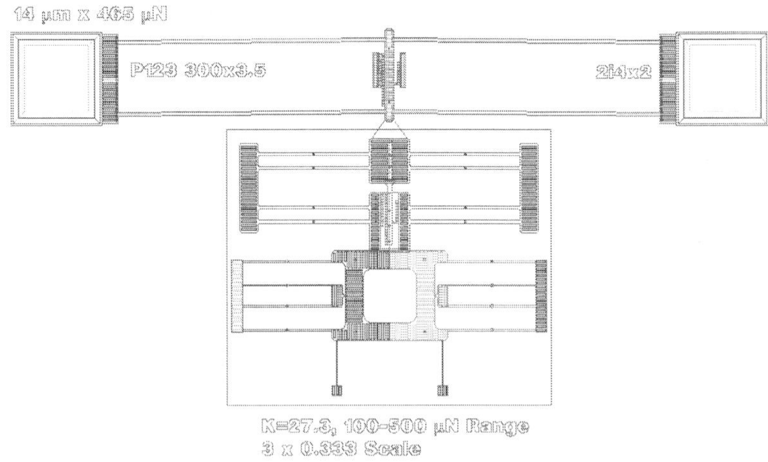

Fig. 5. P123F microactuator with an attached force gauge

Actuator	Gap to Substrate [μm]	Length [μm]	Offset [μm]	w_1 [μm]	w_2 [μm]	t_1 [μm]	t_2 [μm]	t_3 [μm]	Force Gauge
P123	2.0	300	3.5	4.0	2.0	2.5	2.0	2.25	No
P123F	2.0	300	3.5	4.0	2.0	2.5	2.0	2.25	Yes

Table 1. P123 and P123F thermal microactuator geometries

3.2 Silicon-on-insulator processing

A wide variety of microsystems devices such as microactuators, optical switches, accelerometers, and nanopositioners are fabricated with deep reactive ion etching (DRIE) using SOI (silicon on insulator) materials due to the high aspect ratios that can be achieved (Herrera et al., 2008). DRIE silicon etching is commonly referred to as Bosch etching and was patented by Lärmer and Schlip (1992). A thorough review of DRIE high aspect ratio silicon etching is presented by Wu et al. (2010). In SOI MEMS fabrication, the initial wafer has three layers: a single crystal silicon substrate wafer, a thin thermally grown silicon dioxide layer referred to as the buried oxide, and a mechanically thinned single crystal silicon layer called the device layer. A DRIE process enables high-aspect ratio, deep etching of features in silicon wafers using repeated cycles of conformal polymer deposition, ion sputtering, and chemical etching of the silicon. DRIE can be performed on both the device and substrate layers in order to pattern thermal microactuators from the device layer and remove the substrate underneath the microactuators (Milanović, 2004) to reduce heat loss and required power during operation (Skinner et al., 2008). Typically a metal layer is deposited on top of the device layer to improve electrical connections when the parts are packaged.

Example SOI thermal microactuator designs are pictured in Fig. 6 (Phinney et al., 2011). SOI thermal microactuators were fabricated from a wafer with: a 550 μm thick substrate, a 2 μm buried oxide layer, and a 125 μm thick device layer. Three bent-beam thermal microactuators were fabricated with four actuator legs having lengths from the anchor to the shuttle of 5500 μm or 7000 μm and leg widths of 50, 65, or 85 μm. During packaging, wires were bonded to the 0.7 μm aluminum layer that is deposited on top of the bond pad. Figure 6 shows a packaged die with the three thermal microactuators and bond wires visible.

Fig. 6. Picture of SOI thermal microactuators. Two wires bonded to each bond pad are visible in the image. The square bond pads are 900 μm x 900 μm. The connections to the package are outside of the image.

4. Characterization

During operation of an electrically powered thermal microactuator, a current is applied to heat the actuator and thereby create displacement or force output. Displacement and total actuator electrical resistance measurements as a function of input current are easily obtained and standard metrics of thermal microactuator performance that are used for design comparison and model validation. Output force as a function of position and spatially resolved temperature measurements are additional performance and reliability metrics that are more challenging to obtain.

The displacement, electrical resistance, and force measurements in sections 4.1, 4.2, and 4.3 were performed according to methods described Baker et al. (2004). Displacement and total electrical resistance results were measured on a probe station using a National Instruments Vision software package that performs sub-pixel image tracking. A displacement measurement error of ±0.25 μm was achieved by using 200X magnification. Force measurements were made using the P123F actuator design, in which a linear bi-fold spring is attached to the movable shuttle of the actuator. Force is applied manually to the actuator with a probe tip through the pull-ring attached to the spring. The displacement for a given force is determined from the vernier scale with ±1/6 μm resolution. The applied force is determined from the measured displacements and calculated spring stiffness. This method of force measurement was used due to the lack of other methods viable for force measurements at this scale. The displacement, electrical resistance, and force measurements are compared to the results from a model which will be described in Section 5.

Temperature measurements were obtained using Raman thermometry (Kearney et al., 2006a; Kearney et al., 2006b; Phinney et al., 2009; Phinney et al., 2010a; Serrano et al., 2006). In the Raman process, photons from the incident probe light source interact with the optical phonon modes of the irradiated material and are scattered to higher (anti-Stokes) or lower (Stokes) frequencies from the probe line frequency. In the case of silicon and polysilicon, the scattered Raman light arises from the triply degenerate optical phonon at the Brillouin zone center. The resulting spectrum for the Stokes (lower frequency) Raman response has a single narrow peak at approximately 520 cm^{-1} from the laser line frequency at room temperature. Increases in temperature affect the frequency, lifetime, and population of the phonon modes coupled to the Raman process, leading to changes in the Raman spectra, including shifting the peak positions, broadening of the Stokes Raman peak, and increasing the ratio of the anti-Stokes to Stokes signal. These changes in the Raman spectra are metrics for temperature mapping of MEMS. Peak width is sensitive only to surface temperature, and peak position is sensitive to both stress and temperature (Kearney et al., 2006a; Beechem et al., 2007). The ratio of the anti-Stokes to Stokes signal tends to require the longest data collection time for quality signals. Since the thermal microactuators are free to expand and relieve stress that would affect the Raman signal prior to the measurement, Raman peak position is used for the Raman thermometry measurements in this section.

4.1 Displacement

Figure 7 shows the displacement versus applied current for the P123 thermal microactuator. The positive displacement from the designed zero location at zero current is due to compressive residual stress resulting from fabrication processes. The model results shown on the figures are for the thermomechanical model presented by Baker et al. (2004) and summarized in Section 5. When a bias is specified after "Model" in the legend, the bias represents an edge bias which is subtracted from each side of thermal actuator leg nominal width. If a bias is not specified, the nominal width, 4.0 μm, is used in the model calculations. As the current is increased, the displacement versus current data exhibits an inflection point and roll-off in the curve. This is attributed to the maximum temperature in the thermal actuator legs becoming hot enough, above 550°C, that the polysilicon is softened or even melts (Baker et al., 2004). The thermal actuator legs have been observed to glow red under these conditions.

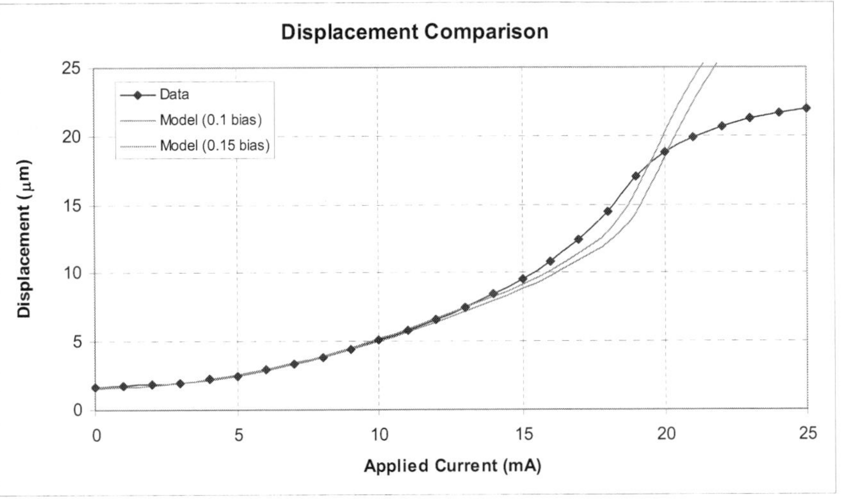

Fig. 7. Displacement versus current for the P123 thermal actuator

4.2 Resistance

Figure 8 shows the total electrical resistance for the actuators versus applied current for the P123 thermal microactuator. The resistance curve exhibits an inflection point, followed by a maximum, and then a decrease in resistance as the current is increased.

4.3 Force

Figure 9 shows the force for the P123F thermal actuator versus displacement when actuated at a constant current and voltage, 15 mA and 6.1V (Baker et al., 2004). For this test, the thermal actuator was held at a constant applied current and allowed to displace to its maximum unloaded position, which corresponds to the point on the graph where the curve intersects with the X-axis. Then using a probe tip, the force gauge was pulled away from the actuator, stretching the folded-beam spring in series between the probe tip and the actuator and applying a force to the thermal actuator center shuttle. The spring elongation was used to calculate the applied force and was recorded along with the actuator displacement. As the actuator is pulled back, the force increased to a maximum of 205 μN at ~6.75 μm. When pulled beyond this, the force begins to decrease due to buckling of the actuator legs. It is important to understand that this force curve represents the available output force of this single actuator design at this single applied power level. To fully characterize the force output of an actuator design, a force curve would need to be measured at several different power levels. This family of curves would then map out the full force versus displacement behavior. The error bars shown for each force level were determined based on an uncertainty analysis performed on the spring design, taking into account the uncertainty in beam width, length, thickness, and Young's Modulus, as well as the measurement uncertainty in the spring elongation.

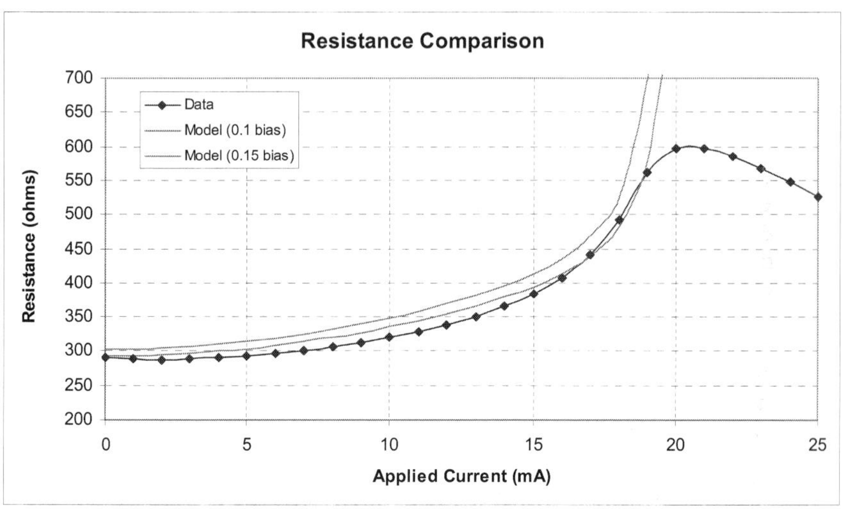

Fig. 8. Resistance versus current for the P123 thermal actuator

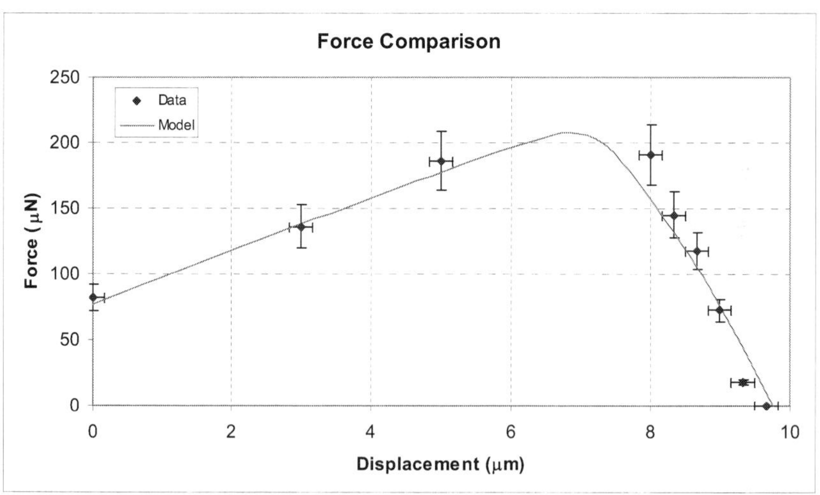

Fig. 9. Force versus position of the actuator at 15 mA for the P123F thermal actuator

4.4 Temperature measurements

Due to the challenges associated with obtaining spatially resolved temperature measurements on MEMS, thermal microactuator models have often been validated primarily from displacement and electrical measurements. Thermal microactuator performance depends on the temperatures of the microactuator legs. Experimentally measured temperatures are invaluable for understanding and improving thermal microactuator performance, model validation, and design optimization. Raman thermometry techniques were used to measure temperatures on electrically powered bent-beam thermal microactuators and laser powered flexure thermal microactuators.

4.4.1 Raman thermometry methods

Raman thermometry has been used to measure temperature profiles along the actuator legs of bent-beam and flexure thermal microactuators (Kearney et al., 2006a and 2006b; Serrano et al., 2006). Raman thermometry was performed using a Renishaw inVia Raman microscope. The microscope uses a 180° backscattering geometry and a 488 nm Ar+ laser as the probe that produces a diffraction-limited spot of 560 nm in diameter when focused by a 50×, 0.50-numerical-aperture objective. The actual measurement diameter within the sample is larger, 1.20 μm, because of spreading of the probe laser within the sample. The Raman signal from the sample surface is collected through the objective, dispersed by a grating spectrograph, and detected with a back-side illuminated, thermoelectrically cooled CCD camera (Princeton Instruments Pixis). Dispersion of the Raman signal at the CCD is 0.57 cm^{-1}/pixel.

Laser power at the sample is attenuated to minimize localized heating of the sample that would otherwise introduce a bias into the temperature measurement. Minimal heating of the sample is confirmed by obtaining Raman spectra at decreasing laser powers from a room-temperature sample until no change in the Raman peak position was observed. The 1.2 μm in-plane resolution of the Raman probe is capable of resolving widths of 2 μm to 4 μm for thermal microactuator legs. A detailed uncertainty analysis reveals that the reported Raman-measured temperatures are reliable to within ±10 to 11 K (Kearney et al., 2006a). These experimental results show that high-quality, reliable temperature measurements can be obtained. Most Raman thermometry measurements are performed when the devices are operating in a steady state in order to allow sufficient time for data collection with sufficient signal strength. Typical data collection times are on the order of tens of seconds to a few minutes. Transient measurements using periodic excitation are mentioned in Section 4.4.4.

4.4.2 Steady state electrically powered bent-beam thermal microactuator results

The temperature profiles reported in this section were taken using the Raman thermometry techniques reported by Kearney et al. (2006a, 2006b) and summarized in Section 4.4.1 on the surface micromachined actuators described in Section 3.1. Temperature measurements are made along one leg of the thermal microactuators starting from an anchor and ending at the center shuttle since the design and performance are symmetric. The chips with the P123 and P123F thermal microactuators were die attached and wire bonded in 24-pin Dual-in-Line Packages (DIP) that were inserted into a zero insertion force (ZIF) socket for the testing in laboratory air. The devices were powered with a Keithley 2400 Source Meter with a single lead on each anchor of the thermal microactuator.

Raman thermometry was used to measure temperatures along the lower left leg of P123 (four cases) and P123F (one case) thermal microactuators (Figure 10, Table 2). P123 microactuators on two packages, P5 and P6, were tested at two currents, 12 mA and 15 mA. The agreement between the temperature profiles for the P5 and P6 microactuators is within the experimental uncertainty of ±10 to 11°C. Thus, the observed device-to-device variation is within the measurement uncertainty. As the current is increased from 12 mA to 15 mA, the maximum temperature increases significantly from 210°C to 377°C. The maximum temperatures along the microactuator legs occur at about two-thirds of the distance from the anchor to the shuttle. Since these tests were conducted at laboratory air pressures, heat transport from the shuttle to the cooler underlying substrate results in the shuttle acting as a heat sink. The temperature profile along a P123F microactuator leg (P5 F) at 12 mA decreases even more at the shuttle than for P123 microactuators tested at 12 mA due to the connection to the force gauge providing another pathway for energy transport away from the shuttle.

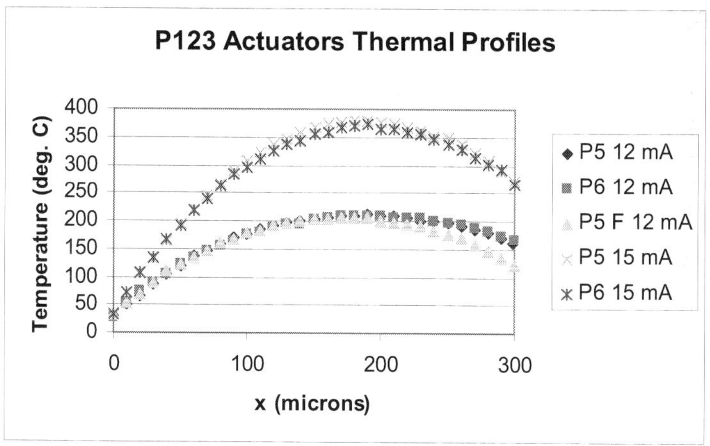

Fig. 10. Temperature profiles for the lower left leg of the P123 and P123F thermal actuators

Actuator	Current [mA]	Voltage [V]	Displacement [μm]	Maximum Temperature [°C]
P123	12	3.89	6.67	210
P123	15	5.58	9.61	377

Table 2. Average actuator displacements, electrical measurements, and maximum temperatures at laboratory air pressure

4.4.3 Steady state laser powered flexure thermal microactuator results

Temperature measurements have also been made on laser heated polysilicon MEMS surfaces using Raman thermometry (Serrrano and Phinney, 2008; Serrano et al., 2009; Serrano and Phinney, 2009). The Renishaw inVia microscope system was adapted to accommodate a stage holding a 1:1 relay lens through which an 808 nm continuous wave fiber-coupled laser with a 100 μm core fiber could be focused on a MEMS part. This enabled temperature measurements on a thermal microactuator surface during laser heating. The heating laser was at an angle of incidence of 60°, yielding a 200 μm × 100 μm elliptical spot on the surface. To avoid damage during measurement collection, a laser power of 314 mW was chosen. This power provides sufficient power to operate the device in a reliable fashion and avoid damaging the surface (Serrano and Phinney, 2008).

The flexure thermal microactuator was fabricated from using the SUMMiT V process and is pictured in Fig. 11a. The thermal microactuator was 200 μm long with a 2.5 μm wide × 200 μm long narrow leg. The opposite side consisted of the 100 μm wide by 150 μm long Poly4 target and a 2.5 μm wide × 50 μm long flexure element. The distance between the two legs was 5.0 μm. The full temperature profile of a 100 μm-wide Poly4 actuator was taken at 10 μm steps starting at the base of the narrow leg and down the near edge (the edge closest to the narrow leg) of the target surface and up to the base of the flexure element. The profile, shown in Fig. 11b, reveals that the temperature along both narrow elements (the thin leg and the flexure) increases linearly from the substrate temperature at the bond pads to the target surface temperature. On the target surface, the temperature in the near edge remains somewhat uniform from the point nearest the narrow leg up to the mid-length of the surface. Beyond this point, the temperature decreases as the flexure element is approached.

The temperature profile in Fig. 11b differs from that for an electrically heated flexure thermal microactuator which has the highest temperatures in the narrow leg (Serrano et al., 2006). Since the target of the laser powered thermal actuator achieves the highest temperatures, the wide leg expands more than the narrow leg and the actuator curls in the direction of the narrow leg when powered. For an electrically heated flexure thermal actuator, the narrow leg will expand more than the wide leg and the actuator will move in the direction of the wide leg.

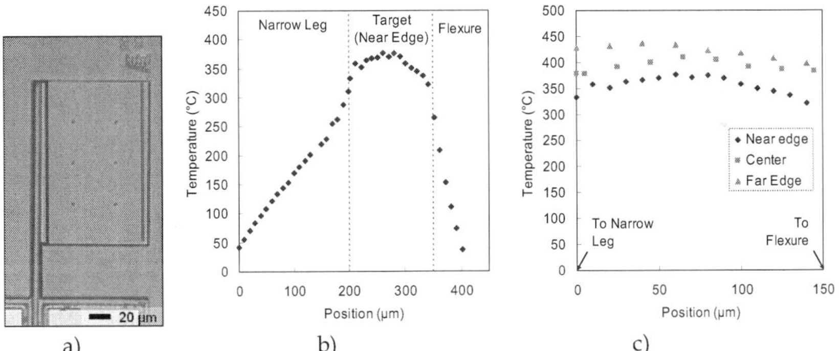

Fig. 11. Temperature profiles along a laser heated flexure actuator a) picture of flexure thermal microactuator and measurement paths, b) temperature profile along the near edge of the flexure actuator (blue line), and c) temperature profiles along the near edge, center, and far edge of the thermal microactuator target.

Comparing the profiles taken along the near, center and far edges of the target surface (Fig. 11c) reveals that the impact of the flexure element and the thin leg is most pronounced on the near edge. The presence of these elements results, on average, in average temperatures

~35°C and ~65°C higher along the center and the far edge, respectively, than at the near edge. Moreover, temperature measurements reveal that the top-right corner of the target surface attains the highest temperature, ~425°C. Subsequent measurements at varying laser powers revealed the effects of optical interference and temperature dependent optical properties on the peak temperatures (Serrano and Phinney, 2008; Serrano et al., 2009; Serrano and Phinney, 2009).

4.4.4 Transient bent-beam measurements

Raman thermometry usually requires data collection times on the order of tens of seconds to minutes to acquire sufficient signals from silicon and polysilicon thermal microactuators precluding transient measurements. Serrano and Kearney (2008) collected time resolved Raman thermometry measurements on polysilicon thermal microactuators using a phase-locked technique. They were able to achieve 100 μs temporal resolution for a polysilicon two leg thermal microactuator design similar to a P123 microactuator but with w_1 = 3.0 μm and an offset of 12 μm (Figure 3). Their measurements revealed that when the thermal microactuator is powered at 3.9 V it achieved a maximum temperature of ~150°C. The heating process took about 2 ms, and the thermal microactuator cooled in about 1.5 ms after turning off the power.

5. Modeling

Numerous research groups have developed numerical models of thermal microactuator performance (Baker et al., 2004; Bergna et al., 2005; Enikov et al., 2005; Howell et al., 2007; Lott et al., 2002; Mankame and Ananthasuresh, 2001; Serrano et al., 2006; and Wong and Phinney, 2007). These models include electrical, thermal, and mechanical effects and are implemented through finite difference as well as finite element approaches.

An example of a model for thermal microactuators is the coupled electro-thermo-mechanical model that was developed to predict actuator performance (displacement, temperature and output force) as a function of the geometry and applied current (Baker et al. (2004)). The model utilizes a finite-difference thermal model to predict the total thermal strain at a given input current, accounting for temperature dependent material properties including thermal conductivity, electrical resistivity, and coefficient of thermal expansion. Heat conduction through the air gap into the substrate is included through the use of a conduction shape factor that is determined from a two-dimensional thermal analysis using the commercial finite-element analysis software ANSYS. Shape factors were determined to be 1.9856 for a 2.0 μm gap between a P123 actuator and the substrate and 2.2336 for a gap of 2.8 μm. ANSYS is then used to model the structural response for the given thermal strain. This model is described in more detail by Baker et al. (2004).

The material properties used in the model include an electrical resistivity given by the following curve fit

$$\text{If } T<300 \quad \rho = (2.9713\times10^{-2})T + 20.858$$

$$\text{If } T>300 \text{ and } T<700 \quad \rho = (6.1600\times10^{-5})T^2 - (7.2473\times10^{-3})T + 26.402 \tag{1}$$

$$\text{If } T>700 \quad \rho = (8.624\times10^{-2})T - 8.8551$$

where the temperature is in degrees Celsius and the resistivity is in units of ohm-microns. The thermal conductivity is defined using the equation

$$k_p = \frac{1}{(-2.2\times10^{-11})T^3 + (9.0\times10^{-8})T^2 - (1.0\times10^{-5})T + 0.014} \tag{2}$$

where the temperature is in degrees Celsius and the thermal conductivity is in W/m/°C. At room temperature the thermal conductivity of polysilicon is 72 W/m/°C, and it decreases with increasing temperature. A value of 164 GPa was used for the Young's Modulus of polysilicon. The model curves in Figures 7-9 were calculated using this model. As seen in these figures, the predicted and measured displacements, electrical resistance, and force are in good agreement.

6. Reliability

The reliability of thermal microactuators depends on the packaging and environment as well as the initial design. Thermal microactuators have operated successfully for tens of millions of cycles; however, performance degradation mechanisms have been observed including plastic deformation of actuator legs, wear debris generation, void formation when operated in vacuum, changes in the grain structure, out of plane displacement, oxide growth, fracture of actuator legs, and die stress effects (Baker et al., 2004; Chu et al., 2006; Phinney et al., 2010b; Plass et al., 2004). Temperature measurements in reduced pressure environment and the effects of die stress are described in detail to illustrate the impact of environment and packaging on thermal microactuator performance.

6.1 Reduced pressure environments

Microsystems devices are often packaged at pressures lower than atmospheric, which dramatically affects the thermal performance of the parts since energy transfer to the environment is substantially reduced as the pressure is reduced (Phinney et al., 2010a). Thus, temperature measurements of thermal microactuators in varying pressures are crucial to optimizing device and package design as well as model validation.

Raman measurements were performed at nitrogen pressures varying from 0.05 Torr to 630 Torr using the Raman thermometry methods described in Section 4.3.1 and Torczynski et al. (2008). For reduced pressure measurements inside a Linkam thermal stage for which the pressure was controllable, a SUMMiT die with a P123 microactuator was packaged on a printed circuit board (PCB) to which wire leads were soldered. Each bond pad on the beam structure is wire-bonded to two separate connections on the PCB to allow for four-point sensing of the voltage. Quick-disconnect connectors were used inside the Linkam thermal stage to allow for easy exchange of parts. The PCB was placed in the center of a quartz crucible inside the stage and held in place with vacuum compatible carbon tape. The heating ability of the stage was used to heat the sample to a temperature of 300-310 K to ensure a consistent substrate temperature for the measurements. The devices were powered with a Keithley 2400 Source Meter in a four-point sensing configuration, where the current is sourced through the outside connections and the voltage is measured across the inner ones.

Raman thermometry was used to measure the temperature profiles for a P123 microactuator leg at pressures ranging from 0.05 Torr to 630 Torr (Figure 12). In order to maintain similar maximum temperatures in the P123 microactuator as the pressure reduced, the power applied to the microactuator was reduced as the pressure was decreased. The P123 microactuator power versus pressure is plotted in Figure 13. At pressures below about 5 Torr, the rate of decrease in the power to maintain the maximum temperature is less than at higher pressures as seen in Figure 13.

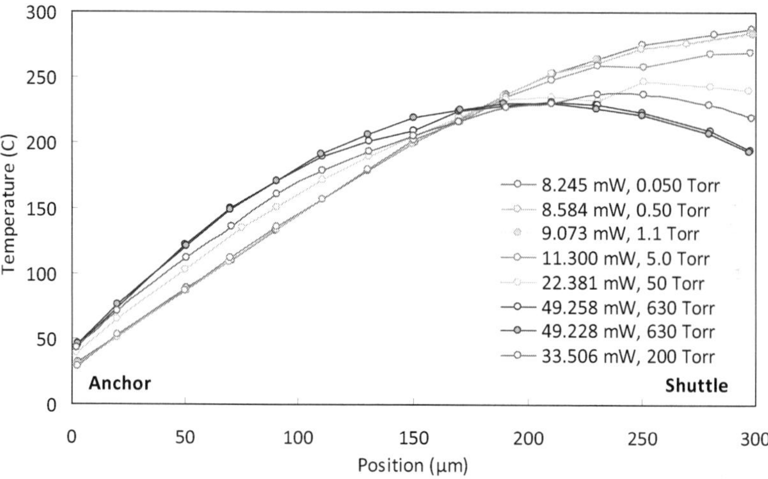

Fig. 12. Temperature profiles for a P123 microactuator leg at 0.05 to 630 Torr

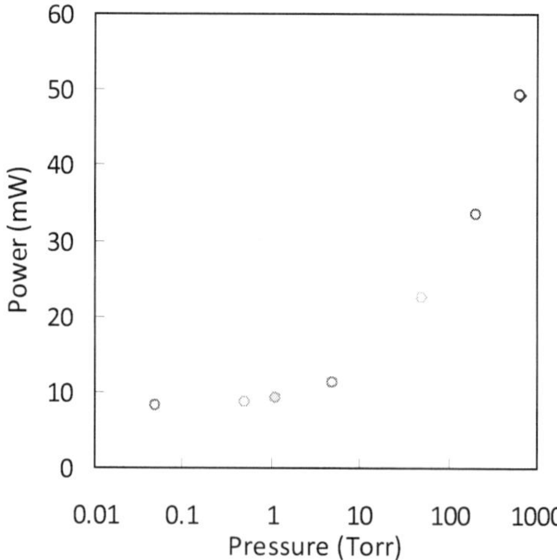

Fig. 13. Operating power for P123 thermal microactuator as a function of pressure for a constant actuator resistance

Pressure [Torr]	Current [mA]	Voltage [V]	Resistance [Ω]	Power [mW]	Displacement [μm]	Maximum Temp. [°C]	Location of Maximum [μm]	Average Temp. [°C]
0.050	4.910	1.67925	342.006	8.245	5.01	287	298.3	186
0.50	5.010	1.7133	341.976	8.584	5.01	284	297.9	185
1.1	5.150	1.76167	342.072	9.073	5.01	283	298.6	184
5.0	5.750	1.9653	341.791	11.300	5.01	269	297.6	182
50.0	8.090	2.7665	341.965	22.381	5.01	248	250.4	182
200	9.900	3.3844	341.859	33.506	5.01	238	230.4	182
630	12.000	4.1048	342.067	49.258	5.01	231	210.2	183
630	12.000	4.1023	341.858	49.228	5.01	230	210.2	184

Table 3. Actuator displacements, electrical measurements, location and magnitude of the temperature maxima, and average temperature under various nitrogen pressures

As seen in Figure 12, the location of the maximum temperature along the microactuator leg moves from around two-thirds of the distance from the anchor to the shuttle to the shuttle as the pressure is reduced. Heat transfer through the underlying gas from the microactuator to the cooler underlying substrate is significantly reduced as the pressure is decreased. At lower pressures, sufficient energy is not transferred from the shuttle to the substrate to allow the shuttle to maintain a lower temperature and act as a heat sink. The location of maximum temperature therefore moves from being on the microactuator legs at high pressures to being at the shuttle at the lower pressures. Table 3 summarizes the operating conditions used (pressure and current) as well as the voltage, resistance, power, displacement, magnitude and location of the temperature maximum, and the average temperature at the various pressures. It is important to note that, although the location and magnitude of the temperature maximum varies with pressure the length-averaged temperature, $\overline{T} = \frac{1}{L}\int T(x)dx$, remains fairly constant for all pressures, thus yielding the similar resistance and displacement values observed.

6.2 Effects of die stress

The effects of die stress were investigated by measuring thermal microactuator displacement as a function of applied current on a four-point bending stage for stresses ranging from -250 MPa compressive to 200 MPa tenisile (Phinney et al., 2010b). Displacement as a function of both input current and applied external stress is shown in Figure 14 for three stress conditions corresponding to -18.7 MPa (the residual stress due to fabrication), 208 MPa tensile, and -261 MPa compressive. Increasing tensile stress decreases the initial displacement of the thermal microactuator, and the amount of displacement that occurs due to an applied current decreases when the device under test is subject to stress from the four-point bending stage.
The numerical model predictions using the model described in Section 5 agree qualitatively with the average of the experimentally measured displacements. The initial displacement decreases and the displacement curves flatten with increasing applied stress. Additionally, the calculated displacements agree with the experimental data for currents up to 25 mA. In Fig. 14, the predicted displacements were calculated for distances between the bottom of the thermal microactuator and the substrate of 2.0 μm and 2.8 μm. At 35 mA, the predicted

displacements using the nominal distance between the bottom of the thermal microactuator and the substrate of 2.0 μm are significantly lower than the measured displacements for all three stress conditions. For the -18.7 MPa case with a gap of 2.0 μm, the predicted displacement is 14 μm and the average of the measured values is 19 μm. Interferometric measurements at stresses of -261 MPa, -122 MPa, -18.7 MPa, 115 MPa, and 212 MPa showed that at currents starting at 20 mA the thermal microactuator experiences upward out-of-plane displacement as well as forward displacement. The maximum measured out-of-plane displacement was 0.8 μm at 35 mA and occurred for the actuator at -18.7 MPa. The upward deflection increases the underlying gap size under the center of the thermal microactuator. When the gap is increased to 2.8 μm in the model, the predicted and measured thermal microactuator displacements are in good agreement. The remaining discrepancies are likely due to slight variations in the actuator geometry or material properties from nominal values.

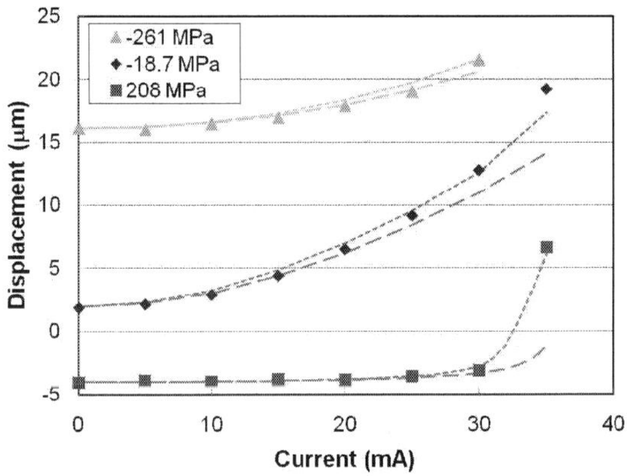

Fig. 14. Predicted and measured thermal microactuator displacement as a function of applied current. The symbols are the average of the experimentally measured displacements. The dotted lines represent the predictions for a gap between the bottom of the thermal microactuator and the substrate of 2.0 μm (long dashes) and 2.8 μm (short dashes).

7. Summary and conclusions

Thermal microactuators function due to thermal expansion of heated members and are versatile components for MEMS designs. Thermal microactuators are part of MEMS devices such as optical microswitches, nanopositioners, and microsensors. Two common thermal microactuator designs are the bent-beam actuator with angled legs that expand when heated and the flexure actuator with asymmetric legs that flex to the side due to differential expansion when heated. Thermal microactuators can be powered, heated, electrically or optically. Both surface micromachining and silicon on insulator (SOI) processing are used to fabricate MEMS thermal microactuators. In order to optimize the design, performance, and reliability of thermal microactuators, both experimental characterization and predictive modeling are necessary. Experimental measurements include displacement, electrical resistance, force output, and temperature. Raman thermometry is a useful technique for acquiring spatially resolved temperature profiles along microactuator legs. The environment and packaging impact thermal microactuator performance and lifetime.

Thermal, electrical, and mechanical measurements for bent-beam polycrystalline silicon thermal microactuators are reported, including displacement, overall actuator electrical resistance, force, and temperature profiles along microactuator legs in standard laboratory air pressures and reduced pressures down to 50 mTorr. Typical displacements are up to around 15 μm with output forces of about 200 μN. For such devices, electrical resistances are in the 100s of Ohms and temperatures will increase by 100s of degrees Celsius. When operated in laboratory air, heating and cooling times are around 1-2 ms. Decreasing the pressure in which the thermal microactuators are operated moves the location of maximum temperature from about two-thirds of the distance between the anchor and shuttle to being on the shuttle. At low pressures, the shuttle does not function as a heat sink since it is not able to transfer the energy to substrate. Laser heated flexure actuators flex in the opposite direction as electrically heated flexure actuators and have the highest temperature at the corner of the wide leg target that is furthest from the flexure and narrow leg. Increasing tensile stress on bent-beam thermal microactuator samples decreases the initial displacement, and applied stress reduces the displacement output for applied currents.

8. Acknowledgment

Sandia National Laboratories is a multi-program laboratory managed and operated by Sandia Corporation, a wholly owned subsidiary of Lockheed Martin Corporation, for the U.S. Department of Energy's National Nuclear Security Administration under contract DE-AC04-94AL85000.

9. References

Ataka, M.; Omodaka, A.; Takeshima, N. & Fujita, H. (1993) Fabrication and Operation of Polyimide Bimorph Actuators for a Ciliary Motion System. *Journal of Microelectromechanical Systems*, Vol. 2, pp. 146-150

Baker, M. S. ; Plass, R. A. ; Headley, T. J. & Walraven, J. A. (2004) Final Report: Compliant Thermomechanical MEMS Actuators LDRD #52553, Sandia Report SAND2004-6635, Sandia National Laboratories, Albuquerque, NM

Beechem, T.; Graham, S.; Kearney, S. P.; Phinney, L. M. & Serrano, J. R. (2007) Simultaneous Mapping of Temperature and Stress in Microdevices Using Micro-Raman Spectroscopy. *Review of Scientific Instruments*, Vol. 78, No. 6, Paper No. 061301, 9 pp.

Bergna, S.; Gorman, J. J. & Dagalakis, N. G. (2005) Design and Modeling of Thermally Actuated MEMS Nanopositioners, *Proceedings of the 2005 ASME International Mechanical Engineering Congress and Exposition*, Paper No. IMECE2005-82158, 8 pp., Orlando, Florida, USA, November 5-11, 2005

Chu, L. L.; Que, L.; Oliver, A. D. & Gianchandani, Y. B. (2006) Lifetime Studies of Electrothermal Bent-Beam Actuators in Single-Crystal Silicon and Polysilicon. *Journal of Microelectromechanical Systems*, Vol. 15, No. 3, pp. 498-506

Cochran, K. R.; Fan, L. & DeVoe, D. L. (2004) Moving Reflector Type Micro Optical Switch for High-Power Transfer in a MEMS-Based Safety and Arming System. *Journal of Micromechanics and Microengineering*, Vol. 14, No. 1, pp. 138-146

Comtois, J. H.; Michalicek, M. A. & Barron, C. C. (1998) Electrothermal Actuators Fabricated in Four-Level Planarized Surface Micromachined Polycrystalline Silicon. *Sensors and Actuators A*, Vol. 70, pp. 23-31

Enikov, E. T.; Kedar, S. S. & Lazarov, K. V. (2005) Analytical Model for Analysis and Design of V-Shaped Thermal Microactuators. *Journal of Microelectromechanical Systems*, Vol. 14, pp. 788-798

Herrera, G. V.; Bauer, T.; Blain, M. G.; Dodd, P. E.; Dondero, R.; Garcia, E. J.; Galambos, P. C.; Hetherington, D. L.; Hudgens, J. J.; McCormick, F. B.; Nielson, G. N.; Nordquist, C. D.; Okandan, M.; Olsson, R. H.; Ortiz, K.; Platzbecker, M. R.; Resnick, P. J.; Shul, R. J.; Shaw, M. J.; Sullivan, C. T. & Watts, M. R. (2008) SOI-Enabled MEMS Processes Lead to Novel Mechanical, Optical, and Atomic Physics Devices, *Proceedings of the 2008 IEEE SOI Conference*, pp. 5-8., New Paltz, New York, USA, October 6-9, 2008

Hickey, R.; Sameoto, D.; Hubbard, T. & Kujath, M. (2003) Time and Frequency Response of Two-Arm Micromachined Thermal Actuators. *Journal of Micromechanics and Microengineering*, Vol. 13, pp. 40-46

Howell, L. L.; McLain, T. W.; Baker, M. S. & Lott, C. D. (2007). Techniques in the Design of Thermomechanical Microactuators, In: *MEMS/NEMS Handbook: Techniques and Applications,* C.T. Leondes, (Ed.), 187-200, Springer US, ISBN 978-0-387-24520-1, New York, New York

Kearney, S. P.; Serrano, J. R.; Phinney, L. M.; S. Graham, S.; Beecham, T. & Abel, M. R.; (2006a) Noncontact Surface Thermometry for Microsystems: LDRD Final Report, Sandia Report SAND2006-6369, Sandia National Laboratories, Albuquerque, NM

Kearney, S. P.; Phinney, L. M. & Baker, M. S. (2006b) Spatially Resolved Temperature Mapping of Electrothermal Actuators by Surface Raman Scattering. *Journal of Microelectromechanical Systems*, Vol. 15, pp. 314-321

Lärmer, F. & Schlip, A. (1992) Method of Anisotropically Etching Silicon, German Patent DE4241045 (1992) US Patent 5501893 (1996)

Lott, C.D.; McLain, T. W.; Harb, J. N. & Howell, L. L. (2002) Modeling the Thermal Behavior of a Surface-Micromachined Linear-Displacement Thermomechanical Microactuator. *Sensors and Actuators A*, Vol. 101, pp. 239-250

Mankame, N. D. & Ananthasuresh, G. K. (2001) Comprehensive Thermal Modelling and Characterization of an Electro-Thermal-Compliant Microactuator. *Journal of Micromechanics and Microengineering*, Vol. 11, pp. 452-462

Milanović, V. (2004) Multilevel Beam SOI-MEMS Fabrication and Applications. *Journal of Microelectromechanical Systems*, Vol. 13, No. 1, pp. 19-30

Oliver, A. D.; Vigil, S. R. & Gianchandani, Y. B. (2003) Photothermal Surface-Micromachined Actuators. *IEEE Transactions on Electron Devices*, Vol. 50, pp. 1156-1157

Park, J.-S.; Chu, L. L.; Oliver, A. D. & Gianchandani, Y. B. (2001) Bent-Beam Electrothermal Actuators – Part II: Linear and Rotary Microengines. *Journal of Microelectromechanical Systems*, Vol. 10, pp. 255-262

Phinney, L. M.; Epp, D. S.; Baker, M. S.; Serrano, J. R. & Gorby, A. D. (2009) Thermomechanical Measurements on Thermal Microactuators, Sandia Report SAND2009-0521, Sandia National Laboratories, Albuquerque, NM

Phinney, L. M.; Lu, W.-Y. & Serrano, J. R. (2011) Raman and Infrared Thermometry for Microsystems, *Proceedings of ASME/JSME 2011 8th Thermal Engineering Joint*

Conference Paper No. AJTEC2011-41462, 7 pp., Honolulu, Hawaii, USA, March 13-17, 2011

Phinney, L. M. & Serrano, J. R. (2007) Influence of Target Design on the Damage Threshold for Optically Powered MEMS Thermal Actuators. *Sensors and Actuators A*, Vol. 134, pp. 538-543

Phinney, L. M.; Serrano, J. R.; Piekos, E. S.; Torczynski, J. R.; Gallis, M. A. & Gorby, A. D. (2010a) Raman Thermometry Measurements and Thermal Simulations for MEMS Bridges at Pressures from 0.05 Torr to 625 Torr. *Journal of Heat Transfer*, Vol. 132, Article ID 0072402, 9 pp.

Phinney, L. M.; Spletzer, M. A.; Baker, M. S. & Serrano, J. R. (2010b) Effects of Mechanical Stress on Thermal Microactuator Performance. *Journal of Micromechanics and Microengineering*, Vol. 20, Article ID 095011, 7 pp.

Plass, R.; Baker, M. S. & Walraven, J. A. (2004) Electrothermal Actuator Reliability Studies, *Proceedings of SPIE: Reliability, Testing, and Characterization of MEMS/MOEMS III*, Eds. D. M. Tanner & R. Ramesham, Vol. 5343, pp. 15-21

Que, L.; Park, J.-S. & Gianchandani, Y. B. (2001) Bent-Beam Electrothermal Actuators - Part I: Single Beam and Cascaded Devices. *Journal of Microelectromechanical Systems*, Vol. 10, pp. 247-254

Sassen, W. P.; Henneken, V. A.; Tichem, M. & Sarro, P. M. (2008) An Improved In-Plane Thermal Folded V-Beam Actuator for Optical Fibre Alignment. *Journal of Micromechanics and Microengineering*, Vol. 18, No. 1, Article No. 075033, 9 pp.

Serrano, J. R. & Kearney, S. P. (2008) Time-Resolved Micro-Raman Thermometry for Microsystems in Motion. *Journal of Heat Transfer*, Vol. 130, Article No. 122401, 5 pp.

Serrano, J. R. & Phinney, L. M. (2008) Displacement and Thermal Performance of Laser-Heated Asymmetric MEMS Actuators. *Journal of Microelectromechanical Systems*, Vol. 17, No. 1, pp. 166-174

Serrano, J. R. & Phinney, L. M. (2009) Effects of Layers and Vias on Continuous-Wave Laser Heating and Damage of Surface-Micromachined Structures. *Journal of Micro/Nanolithography, MEMS, and MOEMS*, Vol. 8, No. 4, Article ID 043030, 7 pp.

Serrano, J. R.; Phinney, L. M. & Kearney, S. P. (2006) Micro-Raman Thermometry of Thermal Flexure Actuators. *Journal of Micromechanics and Microengineering*, Vol. 16, pp. 1128-1134

Serrano, J. R.; Phinney, L. M. & Rogers, J. W. (2009) Temperature Amplification during Laser Heating of Polycrystalline Silicon Microcantilevers due to Temperature-Dependent Optical Properties. *International Journal of Heat and Mass Transfer*, Vol. 52, pp. 2255-2264

Skinner, J. L.; Dentinger, P. M.; Strong, F. W. & Gianoulakis, S. E. (2008) Low-Power Electrothermal Actuation for Microelectromechanical Systems. *Journal of Micro/Nanolithography, MEMS, and MOEMS*, Vol. 7, No. 4, Article Number 043025, 7 pp.

Sniegowski, J. J. & de Boer, M. P. (2000) IC-Compatible Polysilicon Surface Micromachining. *Annual Review of Materials Science*, Vol. 30, pp. 299-333

SUMMiT V™ Five Level Surface Micromachining Technology Design Manual, Version 3.1a (2008) Sandia Report SAND2008-0659P, Sandia National Laboratories, Albuquerque, NM

Torczynski, J. R.; Gallis, M. A.; Piekos, E. S.; Serrano, J. R.; Phinney, L. M. & Gorby, A. D. (2008) Validation of Thermal Models for a Prototypical MEMS Thermal Actuator, Sandia Report SAND2008-5749, Sandia National Laboratories, Albuquerque, NM

Wong, C. C. & Phinney, L. M. (2007) Computational Analysis of Responses of Micro Electro-Thermal Actuators, *Proceedings of ASME 2007 International Mechanical Engineering Congress & Exposition* Paper No. IMECE2007-41462, 9 pp., Seattle, Washington, USA, November 11-15, 2007

Wu, B.; Kumar, A. & Pamarthy, S. (2010) High Aspect Ratio Silicon Etch: A Review. *Journal of Applied Physics*, Vol. 108, No. 5, Article Number 051101, 20 pp.

17

Standalone Tensile Testing of Thin Film Materials for MEMS/NEMS Applications

Arash Tajik and Hamid Jahed
University of Waterloo
Canada

1. Introduction

The microelectronics industry has been consistently driven by the scaling roadmap, colloquially referred to as the Moore's law. Consequently, during the past decades, integrated circuits have scaled down further. This shrinkage could have never been possible without the efficient integration and exploitation of thin film materials.

Thin film materials, on the other hand, are the essential building blocks of the micro- and nano-electromechanical systems (MEMS and NEMS). Utilization of thin film materials provides a unique capability of further miniaturizing electromechanical devices in micro- and nano-scale. These devices are the main components of many sensors and actuators that perform electrical, mechanical, chemical, and biological functions. In addition to the wide application of thin film materials in micro- and nano-systems, this class of materials has been historically utilized in optical components, wear resistant coatings, protective and decorative coatings, as well as thermal barrier coatings on gas turbine blades.

In some applications, thin film materials are used mainly as the load-bearing component of the device. Microelectromechanical systems (MEMS) are the example of these applications. Thin film materials carry mechanical loads in thermal actuators, switches and capacitors in RF MEMS, optical switches, micro-mirror hinges, micro-motors, and many other miniaturized devices. In these applications, one of the main criteria to choose a specific material is its ability to perform the mechanical requirements. Therefore, a clear understanding of the mechanical behavior of thin film materials is of great importance in these applications. This understanding helps better analyze the creep in thermal actuators (Tuck et al., 2005; Paryab et al., 2006), to investigate the fatigue of polysilicon (Mulhstein et al., 2001; Shrotriya et al., 2004) and metallic micro-structures (Eberl et al., 2006; Larsen et al., 2003), to scrutinize the relaxation and creep behavior of switches made of aluminum (Park et al., 2006; Modlinski et al., 2004) and gold films (Gall et al., 2004), to study the hinge memory effect (creep) in micro-mirrors (Sontheimer, 2002), and to address the wear issues in micro-motors. (van Spengen, 2003)

In some other applications, the thin film material is not necessarily performing a mechanical function. However, during the fabrication process or over the normal life, the device experiences mechanical loads and hence may suffer from any of the mechanical failure issues. Examples of these cases are the thermal fatigue in IC interconnects (Gudmundson & Wikstrom, 2002), strain ratcheting in passivated films (Huang et al., 2002; He et al., 2000), the fracture and delamination of thin films on flexible substrates (Li & Suo, 2006), the fracture of porous low-k dielectrics (Tsui et al., 2005), electromigration (He et al., 2004), the chip-package-interaction (CPI) (Wang & Ho, 2005), and thin film buckling and delamination (Sridhar et al., 2001).

In order to address the above-mentioned failure issues and to design a device that has mechanical integrity and material reliability, an in-depth knowledge of the mechanical behavior of thin film materials is required. This information will help engineers integrate materials and design devices that are mechanically reliable and can perform their specific functions during their life-time without any mechanical failure.

In addition to the tremendous industrial and technological driving force that was mentioned earlier, there is a strong scientific motivation to study the mechanical behavior of thin film materials. The mechanical behavior of thin film structures have been known to drastically differ from their bulk counterparts. (Xiang, 2005) This discrepancy that has been referred to as the length-scale effect has been one of the main motivations in the scientific society to study the mechanical behavior of thin film materials. In order to provide fundamental mechanistic understanding of this class of materials, old problems and many of the known physical laws in materials science and mechanical engineering have to be revisited from a different and multidisciplinary prospective. These investigations will not be possible unless a concrete understanding of the mechanical behavior of thin film materials is achieved through rigorous experimental and theoretical research in this area.

2. Mechanical testing methods for thin film materials

In order to probe the mechanical behavior of thin film materials different approaches have been used by researchers. Tensile testing, nano-indentation (Olivier & Pharr, 2004), bulge test (Vlassak & Nix, 1992; Xiang et al., 2005), curvature method, micro-beam bending (Freund & Suresh, 2003), and a few other techniques were used to measure the mechanical properties of thin films on substrate and free-standing thin films. Among these methods, the first four techniques were more popular among the researchers and different measurements have been carried out using these methods.

In tensile testing, the film is patterned into a dog-bone shaped specimen and is then loaded in tension. By monitoring the load and strain across the gage length, the mechanical properties of the film in different load conditions namely monotonic, fatigue, creep, and relaxation can be found. Although tensile testing has been the primary method of experimental research in macro-scale applications, it has not been traditionally as popular among the thin film researchers, due in part to the difficulties in specimen handling, gripping, and strain measurements. However, the tensile testing has a unique advantage that all of the properties of the material can directly be extracted from the measurement data and no calibration model is required. On the other hand and in contrary to other methods, all loading conditions can be applied to the specimen and both free-standing films and films on substrate can be tested using this technique.

In bulge test method, a thin film specimen is loaded by a hydrostatic pressure and the deflection of the specimen is monitored. The pressure-deflection data is then correlated to the actual plain stress-strain behavior of the material through a correlation model. Free-standing single layer and multi-layer materials can be tested by this technique. The only application of this method that is reported in the literature is the monotonic static testing of thin film materials.

Nano-indentation is the advanced version of the classical hardness test method. In this technique, the specimen is loaded by a sharp indenter and the load-displacement (*P-h*) of the indenter is monitored during loading and unloading. The reduced elastic modulus and hardness are the two material parameters that can be extracted from the *P-h* data. This method can only be used for thin films on substrate.

Curvature method is one of the early methods that was used to probe the mechanical behavior of the thin film materials on substrate. In this method, the initial curvature of a substrate is measured and then the film material is deposited on the substrate. The variations in the curvature of the substrate before and after the deposition of the film are a good measure of the residual stresses in the film. This method can also be used to investigate the mechanical behavior of thin film materials on substrate under temperature cycling.

Among the aforementioned experimental methods, tensile testing technique is the only technique that can be used to extract the mechanical behavior of thin film materials under different loading conditions. In this method, all material parameters can be directly measured from the experimental data and it provides a straight-forward approach to the measurement. However, this method faces its own challenges in sample preparation, handling, and gripping and involves uncertainties in the measured strains. In the following sections, these challenges are discussed and different approaches to tackle these problems are presented.

3. Tensile testing techniques for thin films

The early efforts in the tensile testing of thin films were the concurrent research work of Ruud et al., 1993, Koskinen et al., 1993, and Read & Dally, 1993 in the early 1990's. Ruud et al., 1993 introduced a tensile testing technique to test free standing thin film specimens with gage section area of 10 mm long by 3.3 mm wide. They sandwiched the specimen ends between polished aluminum grippers using 5μm thick copper films and used a motor-driven micrometer for loading. Strain was measured by monitoring the displacement of laser spots diffracted from a series of lithography patterned photoresist islands. With this technique, they managed to determine the Young's modulus, Poisson's ratio, and yield strength of free-standing Cu, Ag, and Ni films (Ruud et al., 1993) and Ag/Cu multilayers (Huang & Spaepen, 2000), and to study the yield strength (Yu & Spaepen, 2004) and anelastic behavior (Yu, 2003) of thin Cu films on Kapton substrate.

Koskinen et al., 1993 used a relatively simple technique to test LPCVD polysilicon films. They introduced a gripping setup that could hold an array of 20 samples and was capable of loading individual specimens. Specimen ends were glued to the gripper and loaded by a motor driven stage. Gripper displacement was measured and used to calculate the strain.

While both of these techniques suffered from a reliable gripping and load train alignment, Read & Dally, 1993 and Read, 1998a, 1998b developed a sample fabrication procedure that could meet the demanding gripping and alignment issues, simultaneously. In their method, films were deposited on silicon substrate and after patterning the film to a dog-bone shape, the substrate was etched from the backside to open a window frame under the film, leaving it free-standing. After mounting the specimen in grippers, the frame edges were cut so that only the film is carrying the load. In this way, since the thick substrate is mounted in gripper jaws, there would be less slip and alignment will be an easier task. The concept of free standing film on supporting frame was used by other researchers to overcome the alignment issues (Cornella, 1999; Sharp et al., 1997; Emry & Pvirk, 2004a,2004b). This setup was later on improved by adding laser speckle interferometry (Read, 1998) and Digital Image Correlation (DIC) (Cheng et al., 2005) to measure in-plane strains. E-beam evaporated Ti-Al-Ti multilayer (Read & Dally, 1993), polySi, aluminum and its alloys, and electrodeposited copper (Cheng et al., 2005) were tested using this technique in the temperature range of 25-200°C.

William Sharpe (Sharpe et al., 1997; Yuan & Sharpe, 1997; Sharpe et al., 2004; Edwards et al., 2004; Oh & Sharpe, 2004) used interferometric strain displacement gage (ISDG) technique to measure strains in free-standing films under tensile loading. The ISDG was originally developed in the late 1980's for strain measurement in a non-contact mode (Sharpe, 1968, 1982, 1989) and macro-scale high-temperature applications (Li & Sharpe, 1996). This technique is based on Young's two-slit interference (Born & Wolf, 1983) generated from the diffraction of a laser beam from two sufficiently separate markers. Tensile behavior of polysilicon (Sharpe et al., 1997) and silicon nitride (Edwards et al., 2004) were studied using this technique. Oh & Sharpe, 2004 used this technique to investigate thermal expansion and creep behavior of polysilicon films, while Zupan (Zupan & Hemker, 2002; Zupan et al., 2001) studied the high temperature properties of γ-TiAl micro samples.

The tensile behavior of free standing gold films was studied by Emery and Povirk, 2003a, 2003b. They used the procedure of Read & Dally, 1993 for sample preparation and measured the cross-head displacement for strain calculations. Bravman group (Cornella, 1999; Lee et al., 2000) used the same concept to study the mechanical behavior of thin films with an emphasis on time dependent behavior of Al films. (Zhang et al., 2001; Lee et al., 2000, 2003, 2004, 2005)

Allameh et al., 2003, 2004 investigated fatigue behavior of LIGA Ni thin films under tensile loading. They used Focused Ion Beam (FIB) to mill 1μm deep markers on the specimen surface and monitored the motion of these markers under an optical microscope to calculate strain. Since LIGA films are relatively thick, i.e. a few tens of micrometers, they used common mounting methods for specimen gripping.

The advent and wide-spread availability of high resolution microscopy techniques led some researchers to use *in situ* tensile testing methods to characterize the mechanical behavior of thin films. Atomic Force Microscopy (AFM), Transmission Electron Microscopy (TEM), and Scanning Electron Microscopy (SEM) were among the instruments that were used for *in situ* studies. These techniques were utilized either to measure strain or to study the micro-structural deformations during specimen loading. Chasiotis and Knauss (Chasiotis & Knauss, 2002; Chasiotis, 2004; Knauss, et al., 2003) used AFM to measure the changes in surface topography during the loading and correlated this measurement to strain field using Digital Image Correlation (DIC). They also revised the electrostatic gripping technique, originally proposed by Tsuchiya et al., 1997, 1998, to prevent specimen slipping during the long-time AFM scans for each measurement point. They used this technique to study the influence of surface conditions (Chasiotis & Knauss, 2003a) and the size effect of elliptical and circular perforations (Chasiotis & Knauss, 2003b) on the mechanical strength of polysilicon. Chasiotis et al., 2007 used the same setup to study the strain rate effect on the mechanical behavior of Au films. However, due in part to the slow scan rate of AFM, they used cross-head displacement for strain measurements. Zhu et al., 2003 integrated the specimen with the loading system in a MEMS based device and used AFM to measure strains in polysilicon films under uniaxial tensile loading.

Haque & Saif, 2003, 2004 proposed a quantitative technique to study the deformation mechanisms in Al and Au nano-scale thin films. They used a MEMS device to load samples in TEM and SEM. Using the same technique, Samuel & Haque, 2006 studied the relaxation of Au films and Rajagopalan et al., 2007 reported the plastic deformation recovery in Al and Au thin films.
The above-mentioned methods were among the main research activities that used tensile testing to study the mechanical characteristics of thin film materials. In the following sections, the different parts of the current tensile testing devices, including gripper, loading system, and strain measurement subsystem along with the sample preparation process is discussed in more details. Future researchers can use the information provided in the following sections to choose appropriate solutions to their specific requirements in tensile testing of thin film materials.

4. Sample preparation and microfabrication techniques

Sample preparation is one of the main challenges in tensile testing of thin film specimens. Thin film materials are usually fabricated using one of the deposition techniques. In order to utilize any of these techniques to fabricate free-standing thin film "dog-bone" specimens, a designated microfabrication process has to be developed. This process depends on the specific requirements defined by the choice of gripping and sample handling method, the film material and deposition technique, and the availability of the specific procedures in any fabrication laboratory.
Ruud et al., 1993 used a relatively simple technique to fabricate free-standing films. They evaporated Cu and Ag films on glass substrate and after patterning the film to a dog-bone shaped specimen, they took the films off the substrate by sliding a razor blade underneath them while submerged in water. For Ni films, glass substrate was first coated with a layer of photoresist and Ni was then sputter deposited on it. The film was then released by etching the resist in acetone. Although both processes developed by them are relatively simple, films are prone to be damaged and wrinkle while releasing.
The concept of using a window frame in the substrate which was originally introduced by Read & Dally, 1993 was among the most popular methods that was used and further developed by other researchers. In this process, double-sided polished (DSP) <100> silicon wafers, were first coated by a thin layer of silicon oxide. Oxide layer on front-side was patterned and etched at specimen locations. Thin film material of interest was then deposited by e-beam evaporation and patterned to dog-bone shape specimens. The oxide layer on both sides was then patterned and etched in HF to form a hard mask for silicon substrate etching. Silicon was then etched in hydrazine to open window frames. Sharpe et al., 2003 utilized this technique to test thin polysilicon films. Figure 1-a shows a silicon carbide specimen that was fabricated by Edwards et al., 2004 using this concept. Emery & Povirk, 2003a, 2003b used the same process to fabricate e-beam evaporated gold. The main issue with this technique is the long Si substrate etching times that may cause the specimen film be attacked during etching process and special care is required in this regard.
Cornella, 1999 improved this concept by using dry etching processes rather than wet etching processes to fabricate specimens with higher film quality and process yield. In their process, Si substrate was first coated on front side with 1μm thick LPCVD silicon nitride to be used as an etch stop. Aluminum was then sputter deposited on front side and patterned to the dog-bone shape. Backside of the substrate was coated with thick photoresist to act as the

etching mask during substrate etching. Silicon substrate was entirely dry-etched until it reached silicon nitride layer. This layer was then removed in RIE to release the aluminum specimen. The specimen fabricated through this process is shown in Figure 1-b.

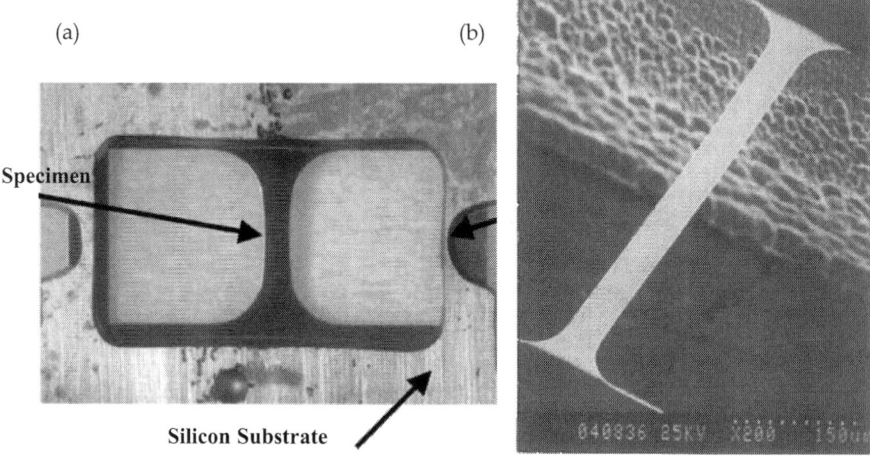

Fig. 1. (a) A free-standing silicon carbide specimen on substrate window frame (Sharpe et al., 2003) and (b) SEM image of an Al free-standing film on window frame. (Zhang et al., 2001)

A few researchers, who mainly worked on the mechanical behavior of polysilicon, used factory processes like polyMUMPS to fabricate their specimens. The specimen is then released by etching the oxide sacrificial layer. Although these processes are well developed and are readily available, they are limited to a few thin film materials, most of which are silicon based. When metallic materials are used as the structural layer, traditional silicon-based films are not good choices for sacrificial layer. A common practice in the fabrication of free-standing metallic devices in RF MEMS devices is to use polymers as the sacrificial layer. Chasiotis et al., 2007 used this technique to fabricate tensile specimens of Au films. As shown in Figure 2, they used PMMA and AZ 4110 photoresist as the sacrificial material for electroplated and evaporated Au films, respectively. For electroplated Au films, a molding process was utilized to pattern gold on PMMA and sacrificial layer was then etched to release the film. Evaporated Au films, however, were lithography patterned and the photoresist sacrificial layer was then stripped to release the structure. Although polymeric sacrificial layers are easier to remove and hence result in less attack to the metallic film, they are less applicable when high temperature processes are involved. In fact in high temperatures two problems arise; above the glass transition temperature, polymer layer starts a significant flow which causes deformation and wrinkling in the metallic film; on the other hand, the thermal mismatch between the polymer and the metallic film causes significant stresses on the film that, in high temperatures, may result in creep and permanent deformation. (Stance et al., 2007) Tajik, 2008 optimized this process in order to realize free-standing thin film specimens that are free of wrinkle and warpage.

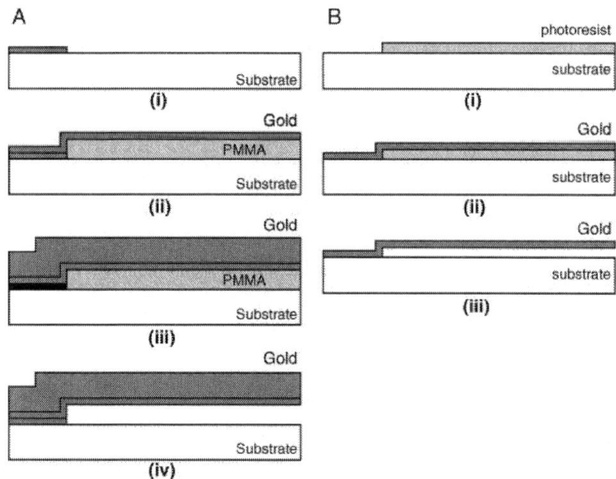

Fig. 2. Microfabrication process for electroplated (A), and evaporated (B) dog-bone specimens. (Chasiotis et al., 2007) In brief, for electroplated specimens (A), Ti/Au anchor is deposited via lift-off process (i), PMMA is spun coated and patterned through photolithography and RIE etching, and then, a thin Ti/Au seed layer is evaporated and patterned to the dog bone shape specimen (ii), a thick Au layer is deposited via electroplating to realize the final dog bone specimen (iii), PMMA sacrificial layer is etched to release the structure (iv). In the case of evaporated films (B), photoresists is used as the sacrificial layer and patterned via photolithography (i), Au film is evaporated and patterned (ii), and the sacrificial layer is then removed in stripper (iii).

5. Gripping

Gripping a film that usually has smaller thickness than even the surface roughness of the macro-machined grippers is a tough challenge. Under these circumstances, the film may slip or experience high stresses at the gripping location due to stress concentrations. On the other hand, aligning the two grippers is, in fact, a demanding task. Therefore, many researchers have designed and utilized a variety of gripping techniques to overcome this issue.

Ruud et al., 1993 sandwiched the free-standing thin film specimens between polished aluminum grippers using 5µm thick Cu foils. The concept of window frame in substrate (Read & Dally, 1993; Cornella, 1999) has made gripping much easier and common macro-machined grippers can be used to mount thick end grips of the specimen which is basically the thick silicon substrate rather than the thin film. As shown in Figure 3, Greek & Johnson, 1997 and Greek et al., 1997 used a connecting ring as a gripper. They inserted a probe connected to the load-train setup in the ring and loaded the specimen. Buchheit et al., 2003 used the same concept to pull micromachined silicon films. A cylindrical sapphire nano-indenter tip was inserted into the so-called "pull-tab" and utilizing the lateral loading capability of a nano-indenter, samples were loaded in tension. Emry & Povirk, 2003a, 2003b used the same technique for pulling tensile specimens on a substrate window frame. This methods, however, is only useful when tension-tension loading scenario is used. In cases where loading direction is changed or set to zero, backlash and rigid displacements cannot be avoided.

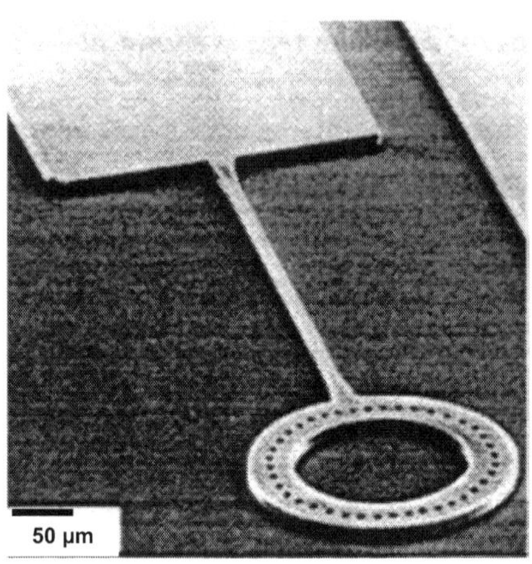

Fig. 3. Tensile testing specimen with a ring at one end for gripping and loading (Greek & Johnson, 1997)

Tsuchiya et al., 1997, 1998 introduced a novel technique to grip tensile testing specimens using electrostatic force. This technique is schematically shown in Figure 4-a. In this technique, a free-standing specimen with large end grip (puddle) is fabricated and is fixed to the gripper by electrostatic force. Specimen can be easily fixed to and released from the gripper by changing the polarity of the applied voltage. Sharpe & Bagdahn, 2004 argued that although the electrostatic gripping is very useful in static tensile tests, it fails during tension-tension fatigue testing. To overcome this issue, they glued a silicon carbide fiber to the puddle using viscous UV curable adhesive.

Chasiotis & Knauss, 2002 also reported that specimens mounted by electrostatic gripping slip during long-time static loadings and they experienced rigid-body motion of the specimen during their long-time AFM scans for deformation measurement. It was shown that the electrostatic gripping is only reliable when the applied tensile loads are below 0.1 N for their specimen geometry. They improved the technique by combining the electrostatic actuation with UV adhesive to meet their demanding requirements for a no-slip reliable gripper. (Figure 4-b)

In order to avoid slipping of the film at the gripper, Tajik, 2008 used a novel gripping method that could reliably grip the specimen for different modes of loading. In this method, the conventional serrated jaw macro-machined gripper was mounted on a double action arm. This mechanism provides a tight gripping of thin film micro-specimens, though the gripper itself is of macro-scale size. (Figure 5)

In conclusion, the application of substrate frame window concept makes griping much easier in the expense of having a more complicated specimen fabrication process. The electrostatic gripping, although seems straight forward, is only applicable for static low-load (<0.1 N) tests. The utilization of adhesive layer is necessary when conducting time-dependent tests or applying a dynamic load that requires a reliable no-slip gripper.

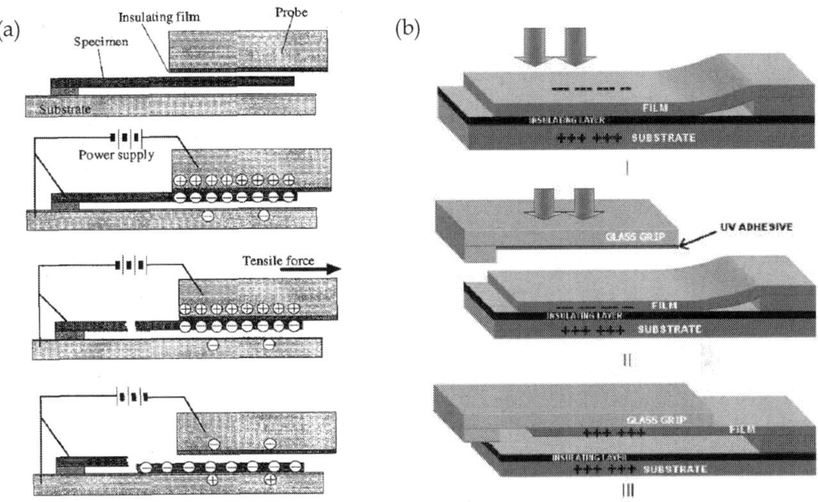

Fig. 4. (a) Schematic representation of electrostatic gripping technique (Tsuchiya et al., 1997) and (b) Combination of the electrostatic and UV adhesive gripping. (Chasiotis & Knauss, 2002)

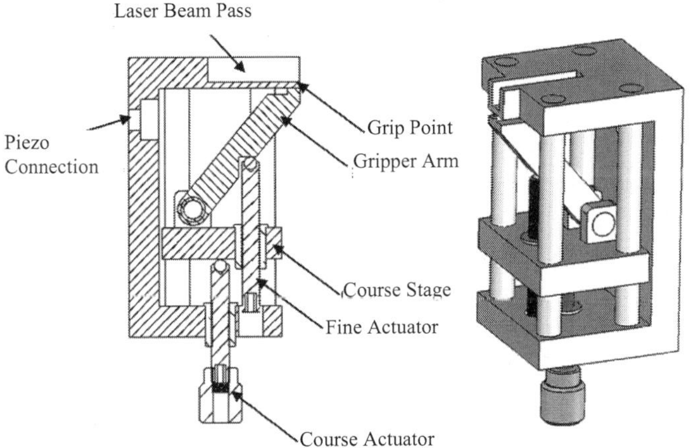

Fig. 5. The double action gripper (Tajik, 2008)

6. Load actuation and measurement

Having prepared the samples and mounted them in grippers, they had to be loaded to the required load level and the load value has to be measured. With the availability of different types of commercial actuators and load cells, this part of a tensile testing setup is not as challenging as the other parts. Piezo driven actuators were among the popular tools for loading. These actuators provide the capability of loading the specimen with different waveforms and frequencies. If they are fitted with any type of displacement sensor, e.g. strain gage, LVDT, or capacitive sensors, they can be controlled in a close-loop system in order to compensate for hysteresis and drift. On the other hand, this displacement feedback can be recorded as the cross-head displacement and be used for the measurement of strain. Many of the research works discussed earlier in this chapter are equipped with this type of actuators (Read & Dally, 1993; Cornella, 1999; Sharpe et al., 1997; Zupan & Hemker, 2001; Allameh et al., 2004). Inchworm (Chasiotis & Knauss, 2002) and motor-driven micrometer (Ruud et al., 1993) actuators are also among the type of actuators that were used for specimen loading. Sharpe & Bagdahn, 2004 used a loud-speaker operating at 20 kHz in their early fatigue tests on polysilicon to dynamically load their specimens at high frequencies. Almost all research groups used strain gage-based load cells to measure the applied load and hence the stress. Tajik, 2008 also used piezoelectric actuators along with precision strain gage load cells that can provide static and dynamic loading capabilities to test thin film specimens as thin as 300 nm.

A specific group of tensile test setups are those that have integrated the load-actuation and measurement with the specimen itself on a MEMS-based tensile testing device. These devices provide much higher resolution for load actuation and load measurement, making them a versatile tool to study the mechanical properties of nano-scale structures like carbon nanotubes and nano-wires (Zhu et al., 2007), and films that are substantially thin or have very small gage section areas. (Haque & Saif, 2004) On the other hand, because of their small size, they can be used for *in situ* study of deformation in Scanning Electron Microscope (SEM) or Transmission Electron Microscope (TEM). (Zhu et al., 2003; Haque & Saif, 2003, 2004; Samuel & Haque, 2006; Rajagopalan et al., 2007) However, these devices are not so applicable at the length scale where most of the thin films are usually fabricated and used in MEMS and microelectronics applications.

7. Strain measurement

The measurement of strain is the most challenging part of the tensile testing of thin film specimens. Due to the small size of thin films, none of the current macro-scale methods of strain measurement are applicable to tensile testing of thin films. Thin film specimens are at the same size scale of resistive strain gages and are too small for LVDT-based extensometers. Technically, any method of strain measurement that is used in contact with the specimen is not useful. Therefore, many researchers have developed or adapted non-contact strain measurement techniques to measure the strain during tensile testing. These techniques can be categorized into four different groups, including cross-head displacement; optical imaging; interferometry-based methods; and advanced microscopy techniques like AFM, SEM, and TEM for *in situ* strain measurement. In what follows, these methods are discussed in detail.

Read & Dally, 1993 monitored the cross-head displacement and used it as a measure for strain. There are many sources of error involved in this technique. Specimen may slip at the gripper. On the other hand, the gripper itself may have clearances that cause backlash during the changes in load direction. Compliance of the test setup is the other source that deteriorates the accuracy of the method. The cross-head displacement is a combination of all of the deformations in the load train, i.e. the deformations in load-cell, load actuator, the test rig, grippers and albeit in the specimen itself. Therefore, this measurement will not provide an accurate measure of stain in the gage section of the sample.

Cornella, 1999 measured the compliance of the test setup by compressing the load actuator to the load cell in the absence of specimen and subtracted this compliance from the actual measurements to find the deformation in the specimen. They reported that 76% of the measured displacement accounts for the actual deformation in the specimen (Zhang et al., 2006). In order to validate the strain relaxation measurements and to show that the drop in the stress level over time is the actual behavior of the specimen itself and not the test setup, they used iridium specimens. Iridium, due to its high melting point, has a very low relaxation at room temperature. Since these tests revealed no relaxation, they argued that their test setup is stiff enough and that the relaxation behavior that they monitored during the tensile testing of Al films is the actual material behavior.

Chasiotis et al., 2007 used the cross-head displacement to study the relaxation in gold thin films. They measured the deformation of the load-cell and the apparatus compliance and subtract it from the results. Due to the high compliance of their specimens compared to the setup, 99% of the cross-head displacement was due to the deformation in the specimen. They verified the accuracy of the crosshead displacement method by testing brittle materials with known elastic modulus.

Greek & Johnson, 1997 cancelled out the effect of the compliance of the test setup by testing specimens with identical gages section areas and different gage lengths. Assuming that the compliance of the test setup is constant for any test, they calculated the deformations caused by test setup and subtract it from the test results. This method is only applicable to the cases that the compliance of the specimens are sufficiently different at different gage lengths.

Emry & Povirk, 2003a also measured the displacement in grippers by monitoring the displacement of two markers using a video camera. They argued that their method has the limitation that the measured strain is not the actual strain in the gage section. They also reported a non-linearity in the stress-strain curve in low loads. Figure 6 shows this non-linearity which is technically an experimental error. They extrapolated the linear portion of the results to find the zero point of stress-strain curve. Due to this non-linearity in the curve, calculating the yield stress using the 0.2% offset rule was erroneous. Therefore they found the yield point by defining it as the point where the slope of the stress-strain curve drops to one tenth of the elastic modulus.

Due to the uncertainties involved in the application of cross-head displacement for strain measurements, a number of techniques have been introduced to measure the strain directly on the gage section. An inexpensive way of measuring strain is to put markers on the specimen's gage section and monitor their displacement using a camera. Allameh et al., 2004 used a video camera and monitored the deformation of two markers milled by Focused Ion Beam (FIB) on LIGA Ni specimens. Markers were 300μm apart and were located by block matching in a series of images captured by a camera during tensile testing. They have not discussed the accuracy of their setup; however, they reported that the strains that were measured as such were only used in the plastic deformation regime. In this method, one way to achieve higher displacement measurement resolution is to use higher optical magnification. However, at high magnifications, the field of view (FOV) of the objective is so small that the two markers cannot be fit in a single image, simultaneously. To overcome this issue, two different approaches have been used. Cheng et al., 2005 used a low magnification (350X) to fit 180μm-apart markers in a single image. Utilizing the digital image correlation (DIC) method, they post-processed the data to the resolution of 0.02 pixel. For their optical setup, this resolution is the equivalent of 0.01μm displacement on the specimen which translates to 55μstrain resolution for a 180μm gage length. Since compared to other techniques, usually less or no preparation is required on the surface of the specimen and inexpensive optical imaging equipment can be used for this purpose, this method is becoming more popular among researchers. The only disadvantage of DIC is that this method is computationally expensive and therefore cannot be used in real-time and strain-controlled measurements.

In another attempt to tackle the small FOV issue in high magnifications, Ogawa et al., 1997 proposed a double field of view approach. As shown in Figure 7, a low magnification objective is used to view the two markers which were 1-1.4mm apart. Image of each marker position was then magnified on a separate CCD and their displacement was monitored. In their technique, they could measure displacements to better than 1μm corresponding to 0.1% strain for a 1mm gage length.

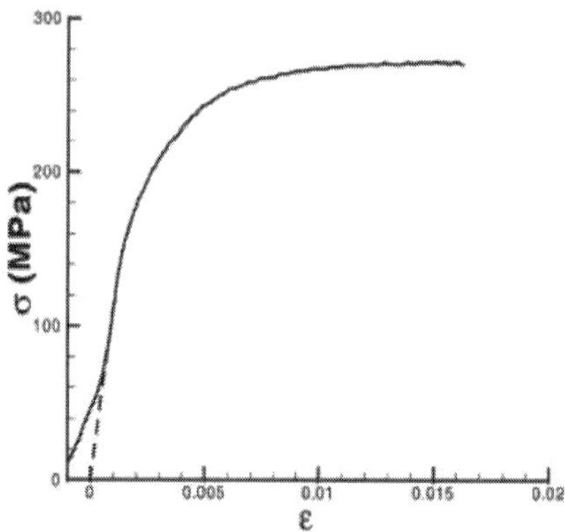

Fig. 6. Stress-strain curve of gold thin films, showing non-linearity at low loads. The dashed line is fitted to the curve to find the zero point. (Emry & Povirk, 2003a)

Ruud et al., 1993 patterned a two-dimensional area of photoresist islands on the specimen and monitored the displacement of the diffracted spots from these islands to directly measure the strain on the gage section. As shown in Figure 8, presence of the photoresist islands results in diffraction patterns when a laser beam is illuminated on the surface of the sample. Diffracted spots are then detected using a two-dimensional position sensor. By monitoring the displacement of these spots, the relative displacement of the islands and hence the specimen deformation can be measured. In this setup, system works in the third diffraction order and the distance between the sample and position detector serves as an optical lever for magnification. They reported that the resolution of the system is limited by the signal-to-noise ratio of the detector and provides strain resolution of 50μstrains.

The main advantage of this technique is that since axial and lateral strains can be measured simultaneously, not only Young's modulus but also the Poisson's ratio can be calculated. Since the modulus of the photoresist islands is sufficiently lower than that of the film's, their presence has very negligible effect on the mechanical behavior of the specimen material.

The aforementioned techniques were all based on optical imaging. The main issue with imaging techniques is that their resolution is limited by the optical setup, and more specifically by the magnification and the CCD resolution. An advanced method of improving the resolution of optical devices is to use light interference. This approach which is the basis of interferometry techniques has been used by a few researchers for strain measurement in thin film materials.

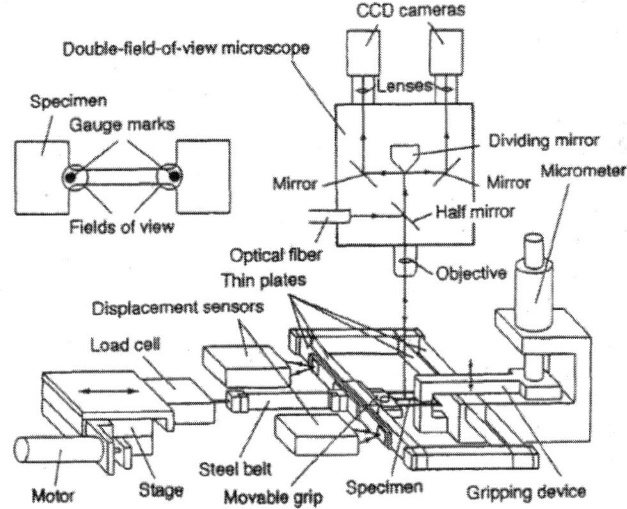

Fig. 7. Double field of view microscope setup for strain measurement (Ogawa et al., 1997)

Sharpe et al., 1997 used Interferometric Strain/Displacement Gage (ISDG) method to measure axial and lateral strains in thin film specimens. ISDG was originally proposed by Sharpe (Sharpe, 1968, 1982, 1986, 1989) in the late 1980s to measure strains in macro-specimens in non-contact mode. The principle of this technique is based on the diffraction and interference of light from two slits, i.e. Young's two slit interference. In this method, two markers are put on the specimen. They can be fabricated either by nano-indentation or by FIB-assisted deposition. When the laser beam is illuminated on them the beam is diffracted. The diffracted beams interfere to form interference fringes. The frequency of the fringes is directly proportional to the distance between the two markers. When the specimen is elongated under the applied load, the distance between the two markers varies, resulting in a change in the fringe frequency. By monitoring this frequency the strains can be directly calculated on the specimen. Since the fringe frequency is also affected by the rigid-body motion of the specimen, two separate detectors have to be used to cancel out the effect of this motion. A schematic of this setup is shown in Figure 9. In the original setup, in order to measure the strains from the fringe data, the location of the fringe minimum was isolated at the beginning of the test and was followed through a complex algorithm. In this algorithm only a small part of the optical signal was used and most of it which contained a lot of information was omitted in calculations. With this algorithm, the strain resolution was 5μstains with uncertainty of ±30μstrains. Zupan & Hemker, 2002 used Fourier Transforms on the whole optical signal and improved the uncertainty of the technique to ±15μstrains. An advantage of ISDG is that if markers are placed along the width of the specimen, the lateral strains and hence the Poisson's ratio can also be measured with this technique. However, since the markers have to be at least 300μm apart, (Hemker & Sharpe, 2007) wide specimens have to be utilized.

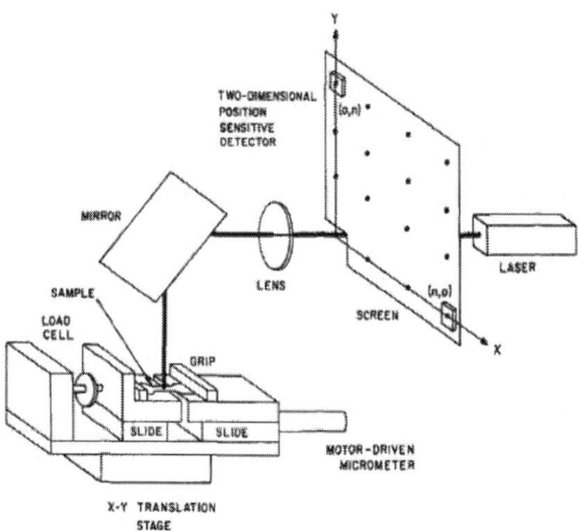

Fig. 8. Schematic diagram of the setup to use diffraction spots for strain measurement (Ruud et al., 1993)

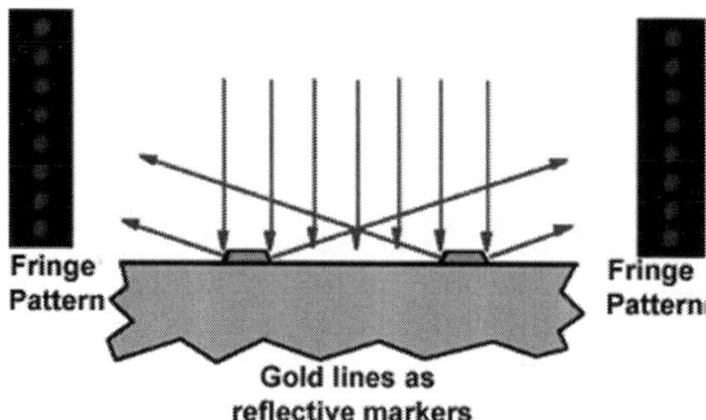

Fig. 9. Schematic representation of ISDG technique. (Sharpe et al., 1997)

In another effort to use interferometric measurement techniques, Read, 1998 used speckle interferometry to measure in-plane strains in thin film specimens. The main advantage of speckle interferometry is that since it uses the speckles caused by the surface topography, no surface preparation or marker fabrication is required. However, it is usually computationally expensive and has low signal-to-noise ratio. In this technique, instead of the strain, the strain rate was measured in the elastic deformation regime and was used to calculate the modulus of elasticity. The uncertainty of the calculated modulus was reported to be 5%.

Tajik, 2008 used another interferometric technique to measure the strain fields in a thin film specimen. This method is schematically represented in Figure 10-a. Diffraction gratings are milled on the freestanding thin film specimen using Focused Ion Beam (FIB). (Figure 10-b and c) Gratings are then illuminated by two laser beams and the interferogram thus formed is captured via a CCD camera. Deformation of the specimen and hence the gratings will result in changes in the interferogram. Intensity of the interferogram can be correlated to the displacement field, such that

$$I(x,y) = I_0(x,y) + \gamma(x,y)\cos(2\pi(2f_s)U(x,y)) \tag{1}$$

where $I(x,y)$ is the interferogram intensity, and $I_0(x,y)$ and $\gamma(x,y)$ are the background intensity and fringe visibility, respectively. In this equation, f_s is the spatial frequency of the gratings and $U(x,y)$ represents in the axial displacement field.

As shown in Equation (1), displacement field is essentially the phase of the interferogram. Therefore, in order to measure displacement field, one needs to calculate the phase value from the interferogram data. For this purpose, continuous wavelet transformation (CWT) has been used in this study. In brief, it has been shown that the spatial derivative of the displacement field, which in fact corresponds to strains, correlates with the value of the scaling parameter at the ridge of the transformation,

$$\varepsilon_{xx}(x) = \frac{1}{2f_s}\frac{f_c}{a_r(x)} \tag{2}$$

where $a_r(x)$ is the scale value on the ridge of the wavelet for any coordinate x and f_c is the center frequency of the Morlet wavelet. This method is then used to extract uniform and non-uniform strain fields. It is shown that wavelet transformations have exceptional capabilities in denoising the experimental interferogram, given appropriate wavelet parameters are chosen in the analysis. Figure 11 demonstrates the capabilities of this method to reconstruct the non-uniform strain field around a hole in a plate under axial tension.

The last group of strain measurement methods are those that used advanced microscopy techniques to measure the strain in thin films. The challenges involved in using optical microscopy led researchers to use other microscopy techniques like AFM, SEM, and TEM to measure the deformation of thin films specimens. These techniques, on the other hand, can provide an insight into the microstructural deformation of thin film materials during loading.

Chasiotis & Knauss, 2002 monitored the surface topography changes during loading using Atomic Force Microscopy (AFM). Correlating the AFM images of the deformed and undeformed surface using Digital Image Correlation (DIC), they calculated the strain field.

The process of imaging is very time intensive and usually takes about 10 minutes for each scan and is confined to an area of a few microns long. In addition to the time required for each AFM scan, there is also the post-processing time added for DIC. They reported 400 μstrains resolution in their strain measurements for a 512×512 pixel image and 1/8 pixel DIC resolution. However, the main advantage of this technique is that it provides a whole-field strain data which helped the analysis of the strain field around geometry inclusions and notches (Chasiotis & Knauss, 2003), as well as cracks. (Chasiotis et al., 2006; Cho et al., 2007). A typical displacement field at the vicinity of a crack in polysilicon that has been measured using AFM/DIC is shown in Figure 12.

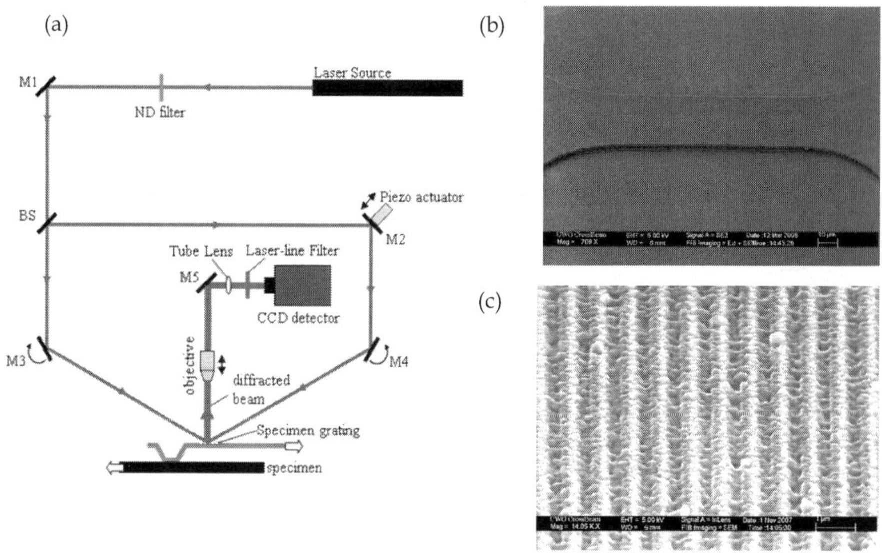

Fig. 10. (a) Schematic representation of the optical setup for moiré interferometry, (b) diffraction gratings fabricated using FIB milling on the test specimen, and (c) high magnification image of the gratings on the specimen. (Tajik, 2008)

A few researchers used MEMS based devices to study the mechanical properties of thin films in electron microscopes. If the thin film samples are electron transparent, i.e. have nanometer thickness, *in situ* studies in TEM are also possible which provides more information on the microstructural deformations during loading. Haque & Saif, 2003, 2004 used the technique to study the mechanical behavior of Al and Au thin film specimens under SEM and TEM. The resolution of the measured strain and stress depends on the magnification of the microscope and the size of the specimen. At 100nm microscope resolution, the strain resolution was 0.05% for 200nm thick and 185μm long Al films and was 0.03% for 100nm thick and 275μm long specimens. Rajagopalan et al., 2007 reported lower strain resolutions of 0.005% and 0.01% for Al and Au films, respectively. Although this method of strain measurement has a high resolution and provides extra information on the microstructural deformations, its force and displacement scales are within the limits of nanostructures rather than common thin film materials. A stress-strain curve along with respective microstructural observations produced by this method is shown in Figure 13.

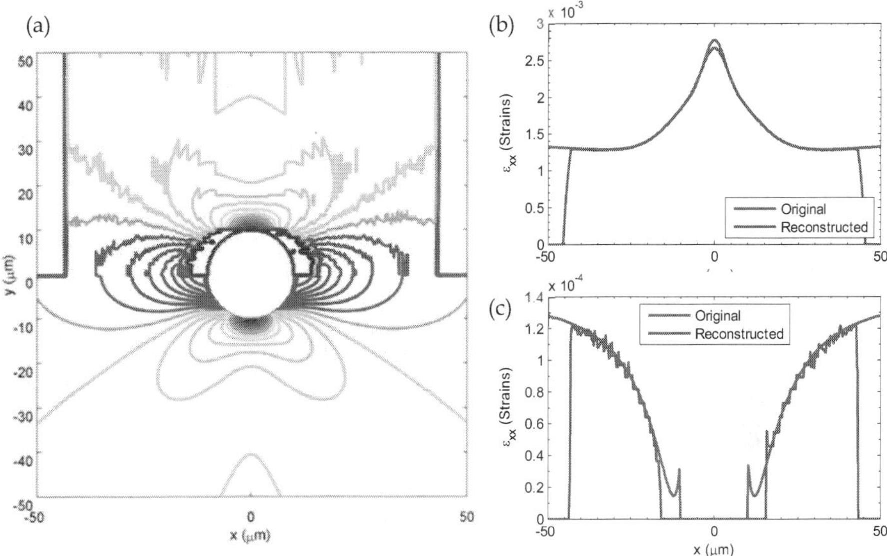

Fig. 11. (a) Reconstructed (top) and original (bottom) strain field using continuous wavelet transformations, (b) reconstructed strains at top of the hole (y=12μm), (c) through the hole (y=0). (Tajik, 2008)

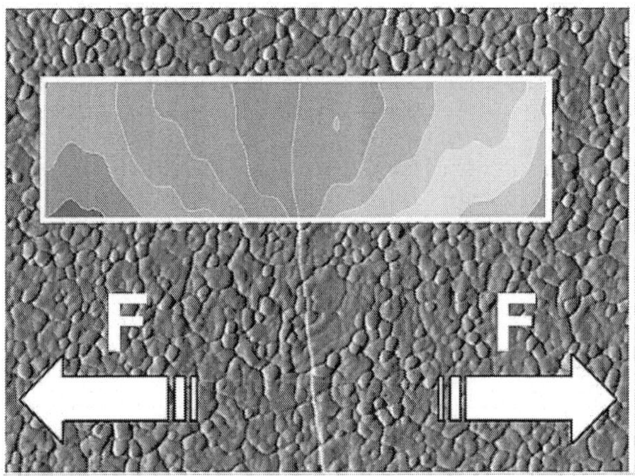

Fig. 12. Displacement field in the vicinity of a crack in polysilicon film measured by AFM/DIC. (Chasiotis, 2004)

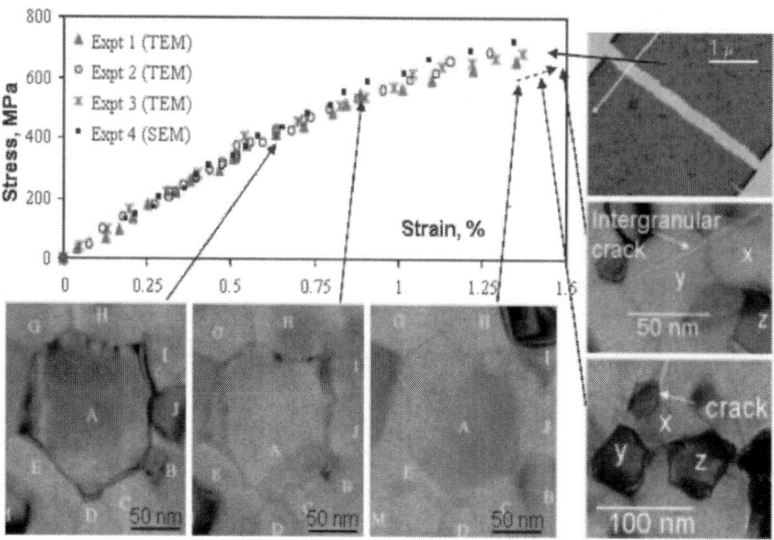

Fig. 13. *In situ* TEM test results for Al thin film specimens. Microstructural deformations corresponding to the tensile test is also presented. (Haque & Saif, 2004)

8. Conclusion

In this chapter, the major research activities that used a tensile testing method to study the mechanical behavior of thin film materials were reviewed. For this purpose, most of the research groups designed and used their custom made test setups. A historical overview of the development of these test devices is presented. Early research in this filed started in early 90's and has continuously been pursued since then. Tensile testing research is categorized into four major steps, namely sample preparation, gripping, load actuation and measurement, and strain measurement. Each research group has mainly tackled one of these challenges and tried to implement innovative designs to address these requirements. Each one of these steps is discussed in detail in their designated sections 5-7. Overall, among all the sample preparation techniques reviewed here, the window frame approach and polymeric sacrificial layer seems to be the most reliable fabrication processes. In terms of gripping, depending on the sample geometry, fabrication technique, and loading requirements, any of the methods presented by Tajik, 2008, Cornella, 1999, or Chasiotis & Knauss, 2002 can be utilized. A combination of window frame specimen and rough macro-gripping seems to be basis of future tensile testing techniques. Piezo actuators can be utilized as the loading actuator. They can provide enough resolution and their maximum load is within the range that is required for tensile loading of thin film specimens. Also, they can provide static and dynamic loads which can be used to test specimens under tensile, fatigue, creep, and relaxation experiments. Strain measurement still seems to be the most challenging part of the tensile testing. Different methods have been traditionally used in these experiments; however, depending on the type of application, each method has its own advantages and disadvantage. Methods based on

interferometry have much higher resolution and can be used to extract strain values as well as strain fields. However, their main disadvantage is the hardware complexity and cost. On the other hand, optical imaging methods in combination with Digital Image Correlation (DIC) can provide about the same resolution; however, they are computationally expensive and cannot be used in real-time and strain-controlled experiments. Therefore, depending on the specific type of application, one needs to choose either method to measure strains or strain fields across the gage length.
The information obtain through this review provide a detailed understanding of the challenges involved in tensile testing of thin film materials and different approaches that were used to tackled these issues. The results will help design and implement a device that can meet these challenges toward a reliable and precise study of the mechanical behavior of thin film materials.

9. Acknowledgment

The financial support from the Natural Sciences and Engineering Council of Canada (NSERC) is appreciated. Center for Integrated RF Engineering (CIRFE) and WatLabs of the University of Waterloo, Nanofabrication Laboratory of the University of Western Ontario and the Canadian Centre for Electron Microscopy in McMaster University are acknowledged for providing research facilities. A great portion of this research is conducted in Laboratory for the Mechanical Properties of Thin Film Materials at the University of Waterloo.

10. References

Allameh, S. M. (2003). An introduction to mechanical properties related issues in MEMS structures, *Journal of Materials Science*, Vol. 38, pp. 4115-4123.

Allameh, S. M., Lou, J., Kavishe, F., Buchheir, T., & Soboyejo, W. O. (2004). An investigation of fatigue in LIGA Ni MEMS thin films, *Materials Science and Engineering A*, Vol. 371, pp. 256-266.

Born, M., & Wolf, E. (1983). Principles of Optics: Electromagnetic Theory of Propagation, Interference, and Diffraction of Light, Cambridge University Press, ISBN 0521642221, Cambridge, UK.

Buchheit, T. E., Glass, S. J., Sullivan, J. R., Mani, S. S., Lavan, D. A., Friedmann, T. a., & Janek, R., (2003). Micromechanical testing of MEMS materials, *Journal of Materials Science*, Vol. 38, pp. 4081-4086.

Chassiotis, I., & Knauss, W. G. (2002). A new microtensile tester for the study of MEMS materials with the aid of Atomic Force Microscopy, *Experimental Mechanics*, Vol. 42, No. 1, pp. 51-57.

Chasiotis, I., & Knauss, W. G. (2003a). The mechanical strength of polysilicon films: Part 1. The influence of fabrication governed surface conditions, *Journal of the Mechanics and Physics of Solids*, Vol. 51, pp. 1533-1550.

Chasiotis, I., & Knauss, W. G. (2003b). The mechanical strength of polysilicon films: Part 2. Size effects associated with elliptical and circular perforations, *Journal of the Mechanics and Physics of Solids*, Vol. 51, pp. 1551-1572.

Chasiotis, I. (2004). Mechanics of thin films and microdevices, *IEEE Transactions on Device and Materials Reliability*, Vol. 4, No. 2, pp. 176-188.

Chasiotis, I., Cho, S. W., & Jonnalagadda, K. (2006). Fracture toughness and subcritical crack growth in polycrystalline silicon, *Journal of Applied Mechanics*, Vol. 73, No. 5, pp. 714-722.

Chasiotis, I., Bateson, C., Timpano, K., McCarty, A. S., Barker, N. S., & Stanec, J. R. (2007). Strain rate effects on the mechanical behavior of nanocrystalline Au films, *Thin Solid Films*, Vol. 515, No. 6, pp. 3183-3189.

Cheng, Y. W., Read, T. D., McCloskey, J. D., & Wright, J. E. (2005). A tensile-testing technique for micrometer-sized free-standing thin films, *Thin Solid Films*, Vol. 484, pp. 426-432.

Cho, S. W., Jonnalagadda, K., & Chasiotis, I. (2007). Mode I and mixed mode fracture of polysilicon for MEMS, *Fatigue and Fracture of Engineering Materials and Structures*, Vol. 30, No. 1, pp. 21-31.

Cornella, G. (1999). Monotonic and cyclic testing of thin film materials for MEMS applications, *PhD Dissertation*, Stanford University.

Eberl, C., Spolenak, R., Arzt, E., Kubat, F., Leidl, A., Ruile, W., & Kraft, O. (2006). Ultra high-cycle fatigue in pure Al films and line structures, *Materials Science and Engineering A*, Vol. 421, pp. 68-76.

Edwards, R. L., Coles, G., & Sharpe, W. N. (2004). Comparison of tensile and buldge tests for thin film silicon nitride, *Experimental Mechanics*, Vol. 44, No. 1, pp. 49-54.

Emry, R. D., & Povirk, G. L. (2003a). Tensile behavior of free-standing gold films. Part I. Coarse grained films, *Acta Materialia*, Vol. 51, pp.2067-2078.

Emry, R. D., & Povirk, G. L. (2003b). Tensile behavior of free-standing gold films. Part II. Fine grained films, *Acta Materialia*, Vol. 51, pp.2079-2087.

Freund, L. B., & Suresh, S. (2003). *Thin Film Materials, Stress, Defect Formation and Surface Evolution*, Cambridge University Press, ISBN 0521822815 Cambridge, UK.

Gall, K., West, N., Spark, K., Dunn, M. L., & Finch, D. S. (2004). Creep of thin film Au on biomaterial Au/Si microcantilevers, *Acta Materialia*, Vol. 52, pp. 2133-2146.

Greek, S., & Johnson, S. (1997). Tensile Testing of thin film microstructures, *Proceedings of the SPIE*, Vol. 3224, pp. 344-351.

Greek, S., Ericson, F., Johnsson, S., & Schweitz, J. A. (1997). In situ tensile strength measurement and Weibull analysis of thin film and thin film micromachined polysilicon structures, *Thin Solid Films*, Vol. 292, pp. 247-254.

Gudmundson, P., & Wikstrom, A. (2002). Stresses in thin films and interconnect lines, *Microelectronics Engineering*, Vol. 60, p. 17-29.

Haque, M. A., & Saif, M. T. A. (2003). A review of MEMS based microscale and nanoscale tensile and bending testing, *Experimental Mechanics*, Vol. 43, No.3, pp. 248-255.

Haque, M. A., & Saif, M. T. A. (2004). Deformation mechanisms in free-standing nano-scale thin films: A quantitative *in situ* transmission electron microscope study, *Proceedings of the National Academy of Science*, Vol. 101, No. 17, pp. 6335-6340.

He, J., Suo, Z., Marieb, T. N., & Maiz, J. A. (2004). Electromigration lifetime and critical void volume, *Applied Physics Letters*, Vol. 85, pp. 4639-4641.

He, M. Y., Evans, A. G., & Hutchinson, J. W. (2000). The ratcheting of compressed thermally grown thin films on ductile substrate, *Acta Materialia*, Vol. 48, pp. 2593-2601.

Hemker, K. J. & Sharpe, W. N. (2007). Microscale characterization of mechanical properties, *Annual Review of Materials Research*, Vol. 37, pp. 93-126.

Huang, H., & Spaepen F. (2000). Tensile testing of free-standing Cu, Ag, and Al thin films and Ag/Cu multilayers, *Acta Materialia*, Vol. 48, pp. 3261-3269.

Huang, M., Suo, Z., & Ma, Q. (2002). Plastic ratcheting induced cracks in thin film structures, *Journal of the Mechanics and Physics of Solids*, Vol. 50, pp. 1079-1098.

Knuass, W. G., Chasiotis, I., & Huang, Y. (2003). Mechanical measurements at the micron and nanometer scales, *Mechanics of Materials*, Vol. 35, pp. 217-231.

Koskinen, J., Steinwall, J. E., Soave, R., & Johnson, H. H. (1993). Microtensile testing of free-standing polysilicon fibers of various grain sizes, *Journal of Micromechanics and Microengineering*, Vol. 3, pp. 13-17.

Larsen, k. P., Rasmussen, A. A., Ravnkilde, J. T., Ginnerup, M., & Hansen, O. (2003). MEMS device for bending test: measurements of fatigue and creep of electroplated nickel, *Sensors and Actuators A*, Vol. 103, pp. 156-164.

Lee, H. J., Cornella, G., & Bravman J. C. (2000). Stress relaxation of free-standing aluminum beams for microelectromechanical systems applications, *Applied Physics Letters*, Vol. 76, No. 23, pp. 3415-3417.

Lee, H. J., Zhang, P., & Bravman, J. C. (2003). Tensile failure by grain thinning in micromachined aluminum thin films, *Journal of Applied Physics*, Vol. 93, No. 3, pp. 1443-1451.

Lee, H. J., Zhang, P., & Bravman, J. C. (2004). Study on the strength and elongation of free-standing Al beams for microelectromechanical systems applications, *Applied Physics Letters*, Vol. 84, No. 6, pp. 915-917.

Lee, H. J., Zhang, P., & Bravman J. C. (2005). Stress relaxation in free-standing aluminum beams, *Thin Solid Films*, Vol. 476, pp. 118-124.

Li, K., & Sharpe, W. N. (1996). Viscoplastic behavior of a notch root at 650°C: ISDG measurement and finite element modeling, *Journal of Engineering Materials and Technology*, Vol. 118, pp. 88-93.

Li, T., & Suo, Z. (2006). Deformability of thin metal films on elastomer substrates, *International Journal of Solids and Structures*, Vol. 43, pp. 2351-2363.

Modlinski, R., Witvrouw, A., Ratchev, P., Jourdain, A., Simons, V., Tilmans, H. A. C., Toonder, J. M. J., Puers, R., & De Wolf, I. (2004). Creep as a reliability problem in MEMS, *Microelectronics Reliability*, Vol. 44, pp. 1733-1738.

Mulhstein, C. L., Brown, S. B., & Ritchie, R. O. (2001). High cycle fatigue of single crystal silicon thin films, *Journal of Microelectromechanical Systems*, Vol. 10, No. 4, pp. 593-600.

Ogawa, H., Suzuki, K., Kaneko, S., Nakano, Y., Ishikawa, Y., & Kitahara, T. (1997). Measurements of mechanical properties of microfabricated thin films, *Proceedings of IEEE Microelectromechanical Systems Workshop*, pp. 430-435.

Oh, C. S., & Sharpe, W. N. (2004). Techniques for measuring thermal expansion and creep of polysilicon, *Sensors and Actuators A*, Vol. 112, pp. 66-73.

Oliver, W. C., & Pharr, G. M. (2004). Measurement of hardness and elastic modulus by instrumented indentation: Advances in understanding and refinements to methodology, *Journal of Materials Research*, Vol. 19, pp. 3-20.

Park, J. H., Kim, Y. J., & Choa, S. H. (2006). Mechanical properties of Al-3%Ti thin films for reliability analysis of RF MEMS switch, *Key Engineering Materials*, Vol. 306-308, pp. 1319-1324.

Paryab, N., Jahed, H., & Khajepour, A. (2006). Failure mechanisms of MEMS thermal actuators, *Proceedings of the ASME IMECE*, Paper no. IMECE2006-15128, pp. 397-406

Rajagopalan, J., Han, J. H., Saif, M. T. H. (2007). Plastic deformation recovery in freestanding nanocrystalline aluminum and gold thin films, *Science*, Vol. 315, pp. 1831-1834.

Read, D. T., & Dally, J. W. (1993). A new method for measuring the strength and ductility of thin films, *Journal of Materials Research*, Vol. 8, No. 7, pp. 1542-1549.

Read, D. T. (1998a). Piezo-actuated microtensile test apparatus, *Journal of Testing and Evaluation*, Vol. 26, No. 3, pp. 255-259.

Read, D. T. (1998b). Young's modulus of thin film by speckle interferometry, *Measurement Science and Technology*, Vol. 9, pp. 676-685.

Ruud, J. A., Josell, D., Spaepen, F., & Greer, A. L. (1993). A new method for tensile testing of thin films, *Journal of Materials Research*, Vol. 8, No.1, pp. 112-117.

Samuel, B. A., & Haque, M. A. (2006). Room temperature relaxation of freestanding nanocrystalline gold films, *Journal of Micromechanics and Microengineering*, Vol. 16, pp. 929-934.

Sharpe, W. N. (1968). The interferometric strain gage, *Experimental Mechanics*, Vol. 8,No. 4, pp. 164-170.

Sharpe, W. N. (1982). Application of the interferometric strain/displacement gage, Optical Engineering, Vol. 21, No. 3, pp. 483-488.

Sharpe, W. N. (1989). An interferometric strain/displacement measurement system, *NASA Technical Memorandum*, 101638.

Sharpe, W. N., Yuan, B., & Edwards, R. L. (1997). A new technique for measuring the mechanical properties of thin films, *Journal of Microelectromechanical Systems*, Vol. 6, No. 3, pp. 193-199.

Sharpe, W. N., Bagdahn, J., Jackson K., & Coles , G. (2003). Tensile testing of MEMS materials - recent progress, *Journal of Materials Science*, Vol. 38, pp. 4075-4079.

Sharpe, W. N., & Bagdahn, J. (2004). Fatigue testing of polysilicon - A review, *Mechanics of Materials*, Vol. 36, No. 1-2, pp. 3-11.

Shrotriya, P., Allameh, S. M., & Soboyejo, W. O. (2004). On the evolution of surface morphology of polysilicon MEMS structures during fatigue, *Mechanics of Materials*, Vol. 36, pp. 35-44.

Sontheimer, A. B. (2002). Digital Micromirror Device (DMD) hinge memory lifetime reliability modeling, *Proceedings of IEEE International Reliability Physics Symposium*, pp. 118-121.

Sridhar, N., Srolovitz, D. J., & Suo, Z. (2001). Kinetics of buckling of a compressed film on a viscous substrate, *Applied Physics Letters*, Vol. 78, pp. 2482-2484.

Stance, J. R., Smith, C. H., Chasiotis, I., & Barker, N. S. (2007). Realization of low-stress Au cantilever beams, *Journal of Micromechanics and Microengineering*, Vol. 17, N7-N10.

Tajik, A. (2008). An Experimental Technique for the study of the Mechanical Behavior of Thin Film Materials at Micro- and Nano-Scale, *MASc Thesis*, University of Waterloo.

Tsuchiya, T., Tabata, O., Sakata, J., & Taga, Y. (1997). Specimen size effect on tensile strength of surface micromachined polycrystalline silicon thin films, *Proceedings of IEEE Microelectromechanical Systems*, pp. 529-534.

Tsuchiya, T., Tabata, O., Sakata, J., & Taga, Y. (1998). Specimen size effect on tensile strength of surface micromachined polycrystalline silicon thin films, *Journal of Microelectromechanical Systems*, Vol. 7, No. 1, pp. 106-113.

Tsui, T. Y., McKerrow, A. J., & Vlassak, J. J. (2005). Constraint effect on thin film channel cracking behavior, *Journal of Materials Research*, Vol. 20, pp. 2266-2273.

Tuck, K., Jungen A., Geisberger, Ellis, M., & Skidmore, G. (2005). A study of creep in polysilicon MEMS devices, *Journal of Engineering Materials and Technology*, Vol. 127, pp. 90-96.

van Spengen, W. M. (2003). MEMS reliability from a failure mechanisms perspective, *Microelectronics Reliability*, Vol. 43, pp. 1049-1060.

Vlassak, J. J., & Nix, W. D. (1992). A new buldge test technique for the determination of Young's modulus and Poisson's ratio of thin films, *Journal of Materials Research*, Vol. 7, pp. 3242-3249.

Wang, G., Ho, P. S., & Groothuis, S. (2005). Chip-packaging interaction: a critical concern for Cu/Low-k packaging, *Microelectronics Reliability*, Vol. 45, pp. 1079-1093.

Xiang, Y. (2005). Plasticity in Cu thin films: an experimental investigation of the effect of microstructure, *PhD Dissertation*, Harvard University.

Xiang, Y., Chen, X., & Vlassak, J. J. (2005). Plane-strain buldge test for thin films, *Journal of Materials Research*, Vol. 20, pp. 2360-2370.

Yu, D. Y. W. (2003). Microtensile testing of free-standing and supported metallic thin films, *PhD Dissertation*, Harvard University.

Yu, D. Y. W., & Spaepen, F. (2004). The yield strength of thin copper films on Kapton, *Journal of Applied Physics*, Vol. 95, No. 6, pp. 2991-2997.

Yuan, B., & Sharpe, W. N. (1997). Mechanical testing of polysilicon thin films with the ISDG, *Experimental Techniques*, Vol. 21, No. 2, pp. 32-35.

Zhang, P., Lee, H. J., & Bravman, J. C. (2001). Mechanical testing of free-standing aluminum microbeams for MEMS application, in *Mechanical Properties of Structural Films*, Ed. Mulstein, C. and Brown, S. B., ASTM STP 1413.

Zhu Y., Barthelat, F., Labossiere, P. E., Moldovan, N., & Espinosa, H. D. (2003). Nanoscale displacement and strain measurement, *Proceedings of the SEM Annual Conference on Experimental and Applied Mechanics*, paper 155.

Zhu, Y., Ke, C., & Espinosa, H. D. (2007). Experimental techniques for the mechanical characterization of one-dimensional nanostructures, *Experimental Mechanics*, Vol. 47, No. 1, pp. 7-24.

Zupan, M. & Hemker, K. J. (2001). High temperature microsample tensile testing of γ-TiAl, *Materials Science and Engineering A*, Vol. 319-321, pp. 810-814.

Zupan, M., Hayden, C., Boehlert, C. J., & Hemker, K. J. (2001). Development of high temperature microsample testing, *Experimental Mechanics*, Vol. 41, No. 3, pp. 1-6.

Zupan, M., & Hemker, K. J. (2002). Application of Fourier analysis to the laser based interferomteric strain/displacement gage, *Experimental Mechanics*, Vol. 42, No. 2, pp. 214-220.

18

Diamond, Diamond-Like Carbon (DLC) and Diamond-Like Nanocomposite (DLN) Thin Films for MEMS Applications

T. S. Santra[1], T. K. Bhattacharyya[2], P. Patel[3], F. G. Tseng[1] and T. K. Barik[4]

[1]*Institute of Nanoengineering and Microsystems (NEMS), National Tsing Hua University, Hsinchu, Taiwan*

[2]*Department of Electronics and Electrical Communication Engineering, Indian Institute of Technology, Kharagpur, West Bengal,*

[3]*Department of Electrical and Computer Engineering, University of Illinois at Urbana Champaign,*

[4]*School of Applied Sciences and Humanities, Haldia Institute of Technology, Haldia, Purba Medinipur, West Bengal,*

[1]*Republic of China*

[2,4]*India*

[3]*USA*

1. Introduction

Amorphous carbon films have been utilized in many types of engineering systems and adapted to fulfill a wide variety of applications. The uses of surface coatings are mainly to protect structural materials from high temperature environments or to confine the electric charge largely on interfaces between materials with differing electronic properties mainly for enormous commercial significance. Diamond, Diamond-like Carbon (DLC) and Diamond-like Nanocomposite (DLN) thin films are based on amorphous carbon films. Diamond, DLC and DLN thin films has generated a great interest in the academia due to its fundamental and technological importance. Presently, researchers have given much attention to fabricate the Micro- or Nano- electromechanical systems (MEMS/NEMS) with different types of materials. The characteristics lengths of these technologies are micrometer to nanometer range. MEMS are defined as miniature devices which combining with mechanical, electrical, optical, and biological fields to fabricate integrated circuits (IC) or other similar manufacturing devises. The applications of these MEMS technologies are in different vast areas, like biomedical, environmental, transportation, manufacturing, robotics, space sciences, computing systems etc [1-5]. Researchers have much expectation of these new frontier technologies after silicon-based microelectronic technologies. For excellent MEMS devices, the coating materials should have the properties like high hardness, high modulus of elasticity, high thermal conductivity and tensile strength, high fracture toughness, low surface roughness, very low coefficient of friction, low thermal expansion, high band gap energy, high transmission capability etc. All of these unique and attractive properties present in Diamond, Diamond-like Carbon (DLC) and Diamond-like Nanocimposite (DLN) based thin films [6-14]. The amorphous carbon films based MEMS are fully dominated the silicon-based MEMS technologies. The silicon-based MEMS with mechanical loading have lack of high fracture toughness facing with high reliability. Under some extreme conditions like very high temperature or very high particle radiation, silicon may fail to sustain these properties. However, silicon have very large coefficient of friction, high surface energy, high wear rate and small band gap energy, which cannot fulfill the all material properties of MEMS [15-18]. To overcome these drawbacks of silicon materials, researchers are continuously trying to look for new materials for MEMS applications. Ceramics (wide band gap), semiconductors (such as SiC), Polymers (PDMS, PMMA), can play important role for MEMS fabrications. Except these materials, diamond, diamond-like carbon (DLC) and diamond-like nanocomposite (DLN) etc are promising materials for MEMS applications. High elasticity and tensile strength of DLC and DLN films can suitable for high frequency MEMS devices. The temperature withstanding capability of both DLC and DLN films is up to 600 ^{0}C or slightly more. The biocompatibility of DLC and DLN films is strongly effective for biosensors in diagnostics and therapies, surface coatings for surgical instruments, prosthetic replacements etc. Chemically modified DLC and DLN surfaces can act as sensing trace of gases to detect biomolecules in biological research. We have presented a brief review about the latest properties of different amorphous carbon based diamond, Diamond-like Carbon (DLC) and Diamond-like Nanocomposite (DLN) thin films and their application in MEMS/NEMS devices.

2. Preperation of diamond films

Diamond, Diamond-like carbon (DLC) and Diamond-like Nanocomposite (DLN) exist in different form of amorphous carbon based thin films have generated a great interest in the academia due to its fundamental and technological importance. The carbon materials which arises from the strong dependence of their physical properties of the ratio of

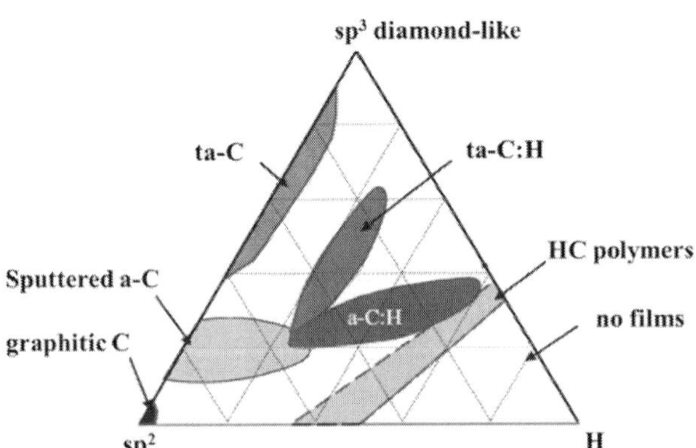

Fig. 1. Ternary phase diagram of amorphous carbons. The three corners correspond to diamond, graphite and hydrocarbons respectively, ([21], permission to reprint obtained from Royal Society London).

sp^2 (graphite-like) to sp^3 (diamond-like) bonds. The amorphous carbon is a mixture of sp^2, sp^3 and sp^1 sites with the presence of nitrogen and hydrogen. The nitrogen free carbon films are shown in Fig. 1 on ternary phase diagram. In this figure, the phase diagram defines the regions of pure carbon (designated a-C), tetrahedral amorphous carbon (ta-C), and hydrogenerated amorphous carbon (a-C:H) with the corresponding extent of hydrogenation [19-21]. To increase the degree of sp^3 carbon bonding, better amorphous carbon (a-C) films can be produced by any kind of deposition systems. If sp^3 carbon bonding is very high, then this a-C can be denoted as a tetrahedral amorphous carbon (ta-C) [22]. Fig. 1 shows amorphous hydrocarbon (a-C:H) or diamond like films, but it is not higher order due to large hydrogen content. To achieve less hydrogen content with much more sp^3 bond, plasma enhanced chemical vapour deposition (PECVD) technique is ideal to generate tetrahedral amorphous carbon films [20]. The sp^3 content influence the mechanical properties of the films. The mechanical and wear resistance properties are more prominent with increase of hydrogen content into the films. On the other hand, surface energy and coefficient of friction decreases with greater hydrogen passivation into the films. Again, the sp^2 content influences the electronic properties of the films.

Diamond, Diamond-Like Carbon (DLC) and Diamond-Like Nanocomposite (DLN) thin films can be deposited by different chemical vapor deposition technique like plasma enhanced CVD, plasma assisted CVD, microwave plasma CVD or a hot filament [23-27], ion beam deposition, pulsed laser ablations, filtered cathodic arc deposition, magnetron sputtering etc. The DC plasma jet chemical vapor deposition can be used for Diamond like carbon films deposition also [28]. Table 1 shows the different properties of diamond films [29-31, 6].

	a-C:H (DLC)	ta-C (DLC)	UNCD	Diamond	DLN
H (atomic %)	30	0	0	<0.1	---
Sp^3 fraction	<0.5	>0.8	>0.9	~1.0	0.5-0.8
Density (Kg/m^{-3})	2350	3260	3500	3515	---
E (GPa)	300	757	300	1050	90-160
Hardness (GPa)	<15	>20	>45	45	9-17
Residual stress (GPa)	1-2	8-10	0	0	---

Table 1. Different parameters of Diamond, Diamond-Like Carbon (DLC) and Diamond-Like Nanocomposite (DLN) thin films.

In this section, we describe diamond-like carbon deposition by plasma which consists of argon (99.998%), hydrogen (99.9%) and CH_4 (99.5%) is used as a carbon source and is mixed into the plasma jet. The plasma jet is sprayed onto a substrate fixed on a water-cooled substrate holder. The hydrocarbon species in the gas phase for the CH_4-Ar-H_2 gas system, temperatures (500 ^{0}C to 6000 ^{0}C) and a total pressure of 0.25 atm (25 KPa) has been computed using the thermodynamic computer program. The deposition was performed on Si(111) surface with the growth rate 80 μm/hr The CH_4/H_2 gas ratio and substrate temperature influence the properties of diamond. Fig. 2 shows the Scanning Electron Microscopy (SEM) morphology of DLC thin films [28]. Diamond-like Carbon (DLC) and Diamond-like Nanocomposite (DLN) are is basically amorphous carbon based films. In amorphous carbon structure, there is a possibility to form both threefold coordinate (sp^2 -site) as in graphite and fourfold coordinate (sp^3 -site) as in diamond [32]. Each of the four valance electron lies in the sp^3 -site forms σ-bonds with neighbors [33]. In sp^2-site, only three electrons are used in σ-bonds and the forth electron forms a π-bond, which lies normal to the σ -bonding plane. In sp^2 -site, only the π-bond is weakly bonded, and hence, it usually lies closest to the Fermi level and controls the electronic properties of the films. On the other hand, in sp^3 -site, the σ-bond controls the mechanical properties of the films [34]. These electrical and mechanical properties are very important parameters for every DLC and DLN based materials.

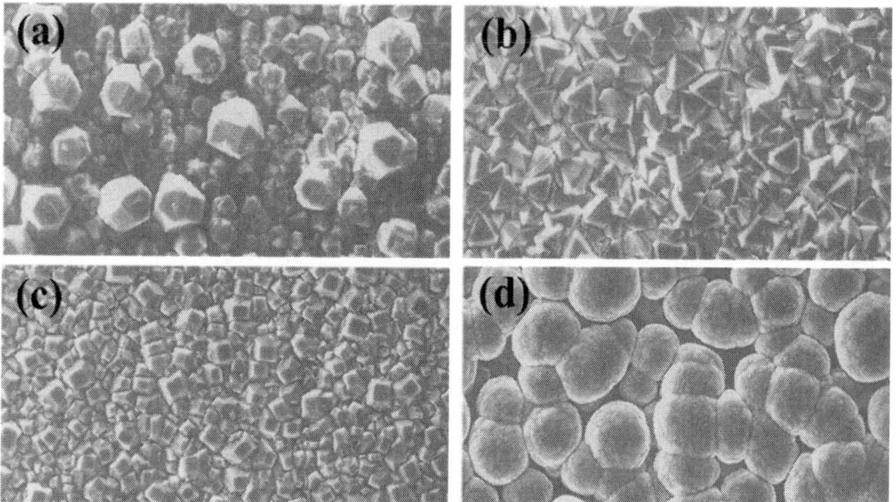

Fig. 2. SEM micrographs of diamond film surfaces deposited at different CH_4/H_2 ratios. (a) 1% (b) 2%, (c) 5% and (d) 8%. (— Scale bar 20 μm for all figures), ([28], permission to reprint obtained from Elsevier).

3. Properties of diamond films

3.1 Structural properties

In this section we analyze the details about High Resolution Transmission Electron Microcopy (HRTEM), Fourier Transform Infrared (FTIR) spectroscopy and Raman spectroscopy of DLN films. These DLN films are prepared by using liquid gas precursor as hexamethyldisilane (HMDS), hexamethyldisiloxane (HMDSO) and hexamethyldisilazane (HMDSN) with argon and nitrogen as a source gas by Plasma Enahanced Chemical Vapor Deposition (PECVD). The HRTEM image in Fig. 3 of DLN films for HMDSN precursor on silicon substrate confirm the nucleation and growth of Si_3N_4 nanoparticles in the amorphous matrix of sizes 6–30 nm. On the other hand, SiC and SiOx nanoparticles having same sizes were found in the DLN films using HMDS and HMDSO precursors, respectively.

The FTIR analysis of DLN films shows that the films predominantly consist of C-C, C-H, Si-C and Si-H bonding. FTIR and Raman spectroscopic results conform to a large extent with structural model [35-36]. DLN films are consisting of mostly two interpenetrating networks of a-Si:O and a-C:H. FTIR spectroscopy is a well known method for investigating the bonding structure of atoms by using the IR absorption spectrum which is related to vibration of atoms [37]. DLN films are deposited in same bias voltages. FTIR spectra of DLN films are given in Fig. 4 The main absorption band is the Si-C stretching in 750-800 cm^{-1} due to Si-$(CH_3)_3$ vibration. Strong Si-O (Si-O-H) stretching in the range of 850-1000 cm^{-1} is due to Si-$(CH_3)_2$ vibration. A very weak C=C stretching peak appears in the range of 1560 cm^{-1}, which indicates non graphite bonding of carbon [38]. The Si-H absorbance band appears in the range 2200 cm^{-1} region. C-H stretching band appears in 2850 cm^{-1} -3100 cm^{-1} region. This type of stretching is very important for DLN films. In DLN films CO_2 vibration appears due to atmospheric carbon present during experiment, and N-H vibration in 3450 cm^{-1} region is due to presence of nitrogen in the precursors. Here C-H stretching and Si-O stretching mainly comprise of the a-C:H and a-Si:O networks.

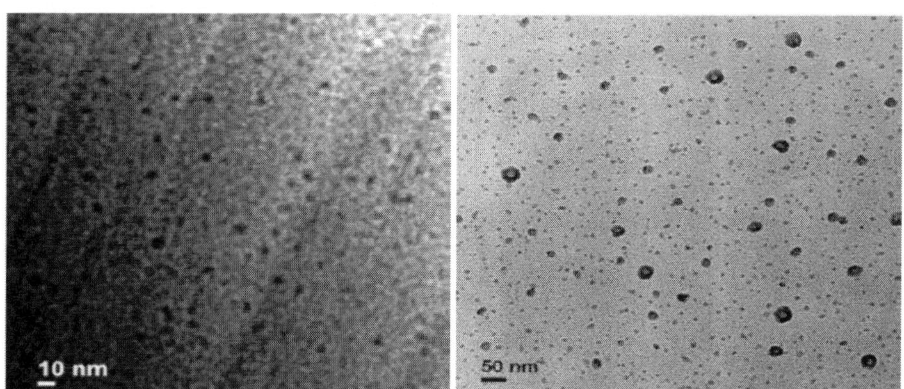

Fig. 3. HRTEM image of DLN films on silicon substrate (a) HMDSO precursor (Left fig.) (b) HMDSN precursor (right fig.), ([6], permission to reprint obtained from American Institute of Physics (AIP)).

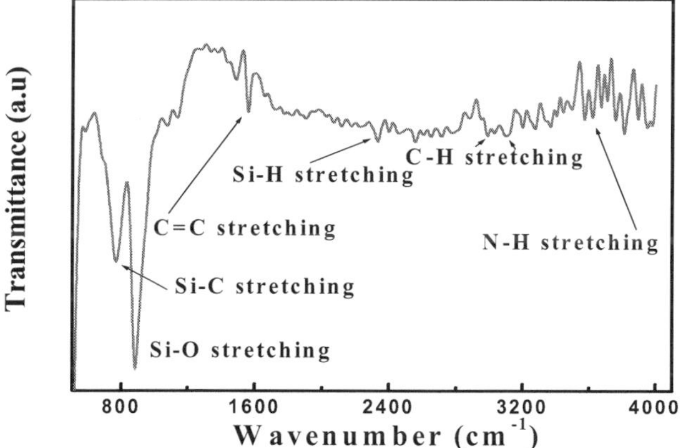

Fig. 4. FTIR spectra of DLN films.

The first order Raman spectra of DLN films as shown in Fig. 5, which is excited by visible light, is usually decomposed into two bands, the D and G bands. Broad asymmetric diamond-like peaks in the region 1000-1800 cm^{-1} are typical characteristics of amorphous carbon films. Raman spectra were deconvoluted into Gaussians D and G peaks by curve fitting. DLN films are example of amorphous carbon phase, much like DLC films and probably dominated by sp^3 bonding [39].

The shape of the spectra varies with substrate material composition. The position of D and G peak widths can be correlated to the film properties such as hardness, wear and electrical characteristics for conventional diamond like carbon films [40]. The position of D and G peak can be shifted due to film structure, light source of Raman spectroscopy, Gaussian fitting method and so on. Rosenblatt and Vairs have suggested the existence of new structural type of diamond-like form of carbon in which phonon frequency is around (1540±20) cm^{-1} depending on the distortion from the graphite structure. The D band (which is around 1330 cm^{-1}) corresponds to breathing mode of sp^2 atoms in hexagonal ring formed by graphite structure, which means disorder of bond angle resulting due to disappearance of the long range translation symmetry of polycrystalline graphite and amorphous carbon films, while G peak (located around 1535 cm^{-1}) is related to C-C bond stretching vibration of all pair of sp^2 atoms in both ring and chains of graphite layer for single crystalline graphite structure [41-44]. Here D means disorder G means graphite.

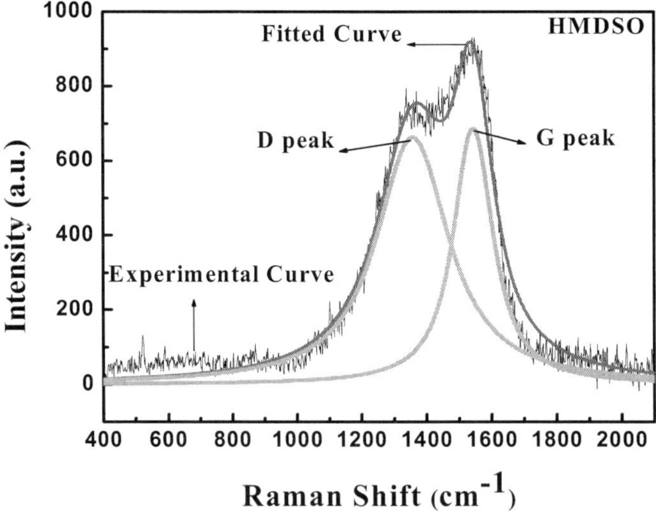

Fig. 5. Raman spectra of DLN films.

3.2 Mechanical properties

The method of measuring hardness and elastic modulus of thin films by nanoindentation test is explained by Oliver and Pharr [45-46]. This method is widely adopted to characterize the mechanical behavior of low dimensional materials, while the numerous refinements have been made to further improvement of its accuracy. The curve for DLN samples of loading and unloading forces versus displacement into the films, at maximum load up to 20 mN are shown in Fig. 6. This figure shows a good reproducibility of the nanoindentation test [47]. The average hardness of the DLN films is measured using three indents with 20 mN load which is around 9-17 GPa. The average reduced elastic modulus of the DLN films is measured under the 20 mN loading force with 300 nm displacement which is around 90-160 GPa.

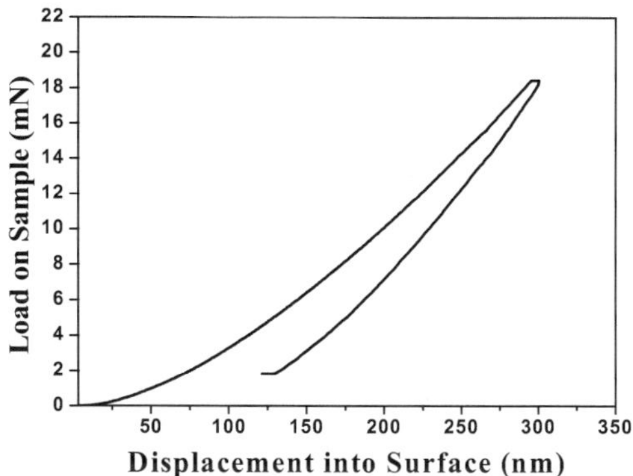

Fig. 6. Loading unloading curve of DLN sample film by Nanoindentation test for DLN films, ([6], permission to reprint obtained from Elsevier).

3.3 Tribological properties

To measure the friction coefficient of the DLN films by scratch test method, the loading rate was 16 N for both normal force and tractional force and the scratch length was about 3 mm. The friction coefficient of the DLN films is estimated by taking the ratio of normal force to the lateral force. The variation in normal load and tractional force with respect to stroke length, and the corresponding variation in coefficient of friction against normal load, for DLN films are shown in Fig. 7.

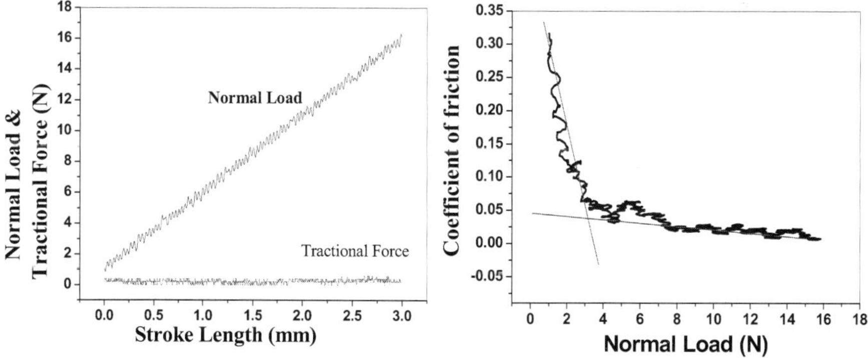

Fig. 7. Scratch Behavior for DLN coating sample (a) Variation of normal load and tractional force with stroke length (left fig), (b) Corresponding coefficient of friction variation with normal load (right fig.).

The average friction coefficient of DLN films using conical diamond tip is estimated which is nearly 0.03–0.05. The tribological properties of DLN films are most important for their use as protective coatings in MEMS/NEMS technology. Recently, the DLC films have been used as rigid disk for microelectromechanical or nanoelectromechanical devices. These protective coatings must have excellent wear and tear resistance, high adhesiveness and very low friction coefficient. For DLC films, the friction coefficient is around 1 but for DLN films, friction coefficient is around 0.03–0.05 as stated above. Hence, for modern microsystems or nanosystems *i.e.* MEMS or NEMS, we can use the films as protective coatings compared to DLC films. The surface morphologies of DLN films are analyzed by using AFM. Fig. 8 shows the AFM image of DLN films in two dimensional (2D) and three dimensional (3D) views.

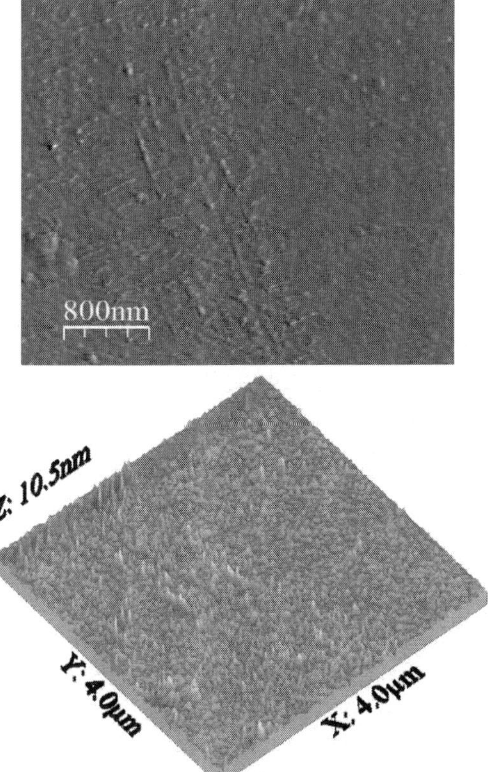

Fig. 8. The Surface morphology of DLN films deposited on silicon substrate:
2D view (top)…view (bottom).

From AFM analysis, we have estimated the mean surface roughness (R_a) and maximum peak-to-valley height (R_{max}) of the DLN films, which are 0.292–3.2 nm and 6.1–33 nm, respectively. From this analysis, it is also confirmed that all the DLN films have no surface defects such as macroparticles and pinholes. Again, very less surface roughness influences the mechanical and tribological performances of the films for microscale and nanoscale devices. Hence, DLN films could provide the better performance for the applications in microelectromechanical systems (MEMS) or nanoelectromechanical systems (NEMS) devices.

4. Diamond, DLC, DLN for MEMS technology

Recently, the microelectromechanical systems (MEMS) and nanoelectromechanical systems (NEMS) technology are fully dominated by Si based materials for their fabrication. These materials have good mechanical and electrical properties for fabrication of MEMS/NEMS based sensors and actuators. Also the silicon materials have large surface area to fabricate the device. However, these materials have some limitation like high temperature withstanding capability, aggressive media, high energy particle radiation etc. For these limitations, diamond films would be good choice for fabrication of MEMS/NEMS device. Some advantageous properties of different materials including diamond and silicon are given in Table 2.

	Diamond	**Silicon**	**3C-SiC**	**AlN**	**Ni**
Young Modulus (GPa)	1050	165	307	331	210
Coefficient of thermal expansion at RT (ppm/^{0}C)	1	4.2	3.8	4.6	13.4
Density (g/cm^3)	3.52	2.33	3.21	3.26	8.91

Table 2. Different material properties compared with diamond.

Recently researchers are concentrating for ceramic based materials as well as diamond to fabricate the MEMS devices instead of silicon materials. As structural properties, diamond has much more sp^3 phase content, which improves the very good mechanical properties. Also much more sp^2 content DLC can improve the electronic properties of materials compared to silicon material. Finally, very high hardness, high modulus of elasticity, high tensile strength, low surface roughness, low coefficient of friction, good wear rate of diamond, DLC and DLN can act as promising materials for MEMS/NEMS devices.

4.1 Microfabrication, pattern transfer and diamond film patterning

4.1.1 Microfabrication and pattern transfer

The standard process for microfabrication is to deposit of thin films into whole over the wafer and then need to remove the unwanted part by etch or polishing of thin films from the wafer. The microfabrication process can come in two ways, one is the directional process and another is the diffusion process. Fig. 9 shows the directional and diffuse process. The directional process which include electron or ion , photons, beam of atom which impinges into the whole wafer (such as lithography, e-beam evaporation, ion implantation etc.). The diffuse process which include the immersion process where the whole wafer surrounded by vapor, liquid or gases (By CVD or oxidation). To deposit the specific region in both process, need to use the mask in which the unwanted portion will be cover by mask and the open portion of the mask will be deposited metal or ions. The masking of the substrate can prevent the ions or atoms to react with the substrate material.

The another process is called the localized process by where the beam energy can falls into specific region of the substrate. The localized process can be divided in to focused beam

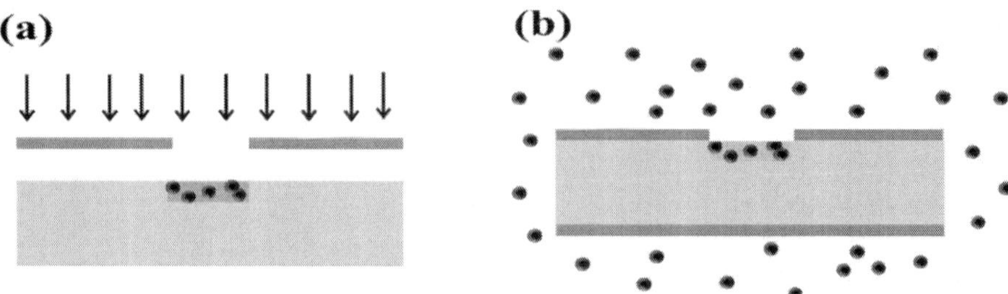

Fig. 9. (a) Directional process and (b) diffusion process

processing and microstructure assisted processing. By this process the reaction will occurred when beam of microstructure will provide energy. Fig. 10 shows the localized processing of focused beam supplies energy and microstructure provide energy.

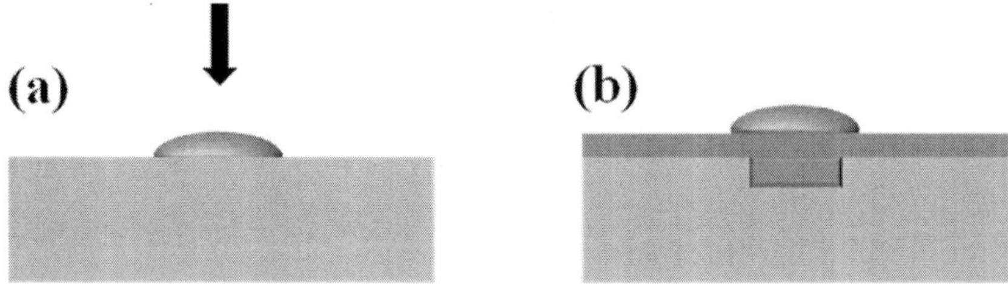

Fig. 10. Localized processing (a) focused ion beam supplies energy and (b) microstructure provide anergy.

The photolithography process is important for film patterning. By this process in the beginning the surface preparation of the films is more important. This process start from cleaning of the wafer for remove the moisture, baked the wafer, increase the adhesion promotion of the wafer by apply hexamethyldisiloxane (HMDS), $(CH_3)_3$-Si-NH-Si-$(CH_3)_3$. The HMDS treatment can reduce the pressure to form the monolayer onto the surface of the wafer, causes the more hydrophobic of the wafer which prevents the moisture condensation. After that spin coating, soft bake, UV exposure with mask alignment, development and hard bake technique is required for whole process. After this lithographic process of the photoresist on to the wafer, the wet etching or dry etching techniques is important to pattern the any type of substrate materials. For this photolithographic pattern, the photoresist uses basically polymeric resist. This resist can dissolve or it can insoluble in the developer solution according to the positive or negative photoresist to react with UV light. The pattern is form by this polymeric resist only use of mask in front of UV light. To pattern the polymeric resist on to the substrate , different process can be applicable. This process can be divided into additive or subtractive process. By which we can add some material into the substrate or we can remove the material from the substrate by using the photoresist with mask pattern. Fig. 11 shows the different additive and subtractive process. Fig 11(a) shows the lift off process. By this process we pattern the photoresist to fall the UV light on to the mask. Then deposit the metal on the top of the wafer and finally we remove the photoresist from the wafer by development process. Fig. 11(b) is the electroplating process by which, we can deposit the metal on top of the pattern photoresist on to the wafer and finally remove the photoresist. So required metal can deposit on to the wafer and reaming metal will be remove from the wafer. The electroplating technique can provide high aspect ratio structure compared to lift off technique.

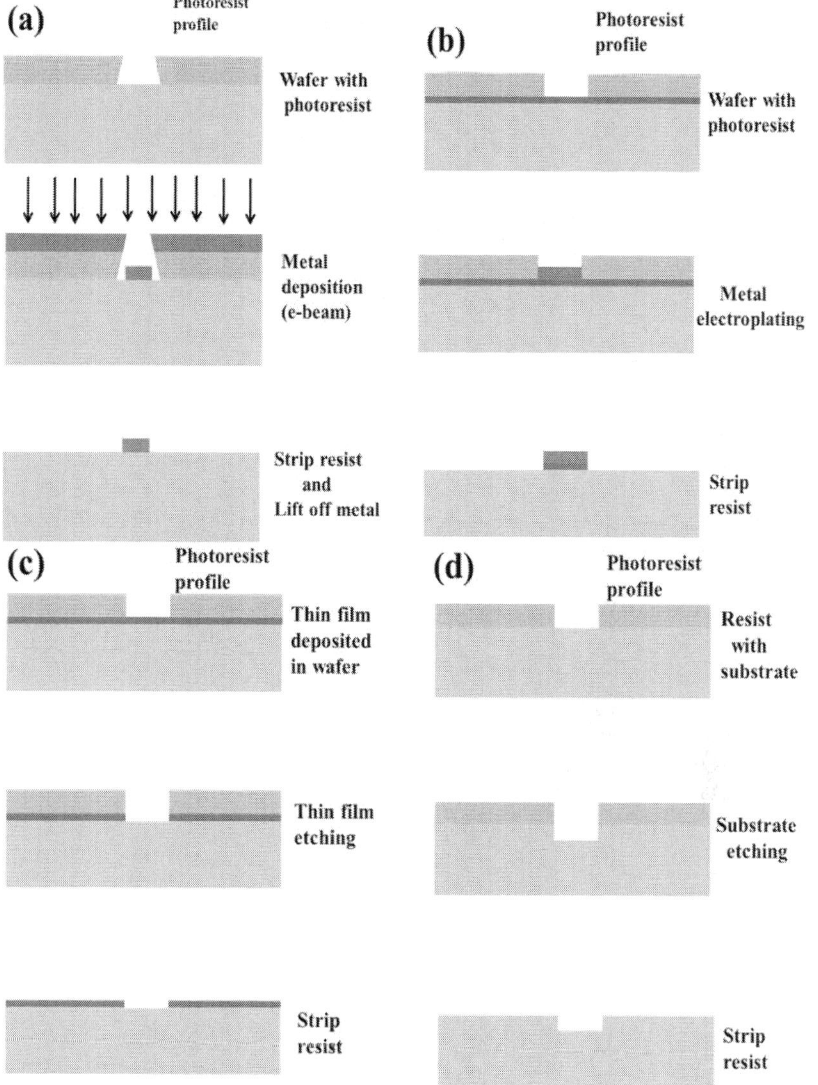

Fig. 11. Pattern transfer process, (a) lift-off technique, (b) electroplating , (c) etching of thin film, (d) etching of substrate material

In lift off technique the deposited metal film thickness should be one third of the photoresist thickness in order to make the discontinuity of the films . But electroplating technique can deposited the large thickness of metal with photoresist pattern compare to lift-off technique. The lift-off technique and electroplating technique is the additive process. Where as in Fig.11 (c) demonstrate the thin film etching technique by photoresist pattern, which indicate the subtractive process. Fig. 11(d) demonstrate the etching technique of substrate material by photoresist pattern (subtractive process)

4.1.2 Diamond film patterning

To make the pattern of diamond or diamond films is not easy task for fabrication of microstructure. The authors of this chapter are continuously trying to etch the DLN films for MEMS structure. But till now we are unable to get the success for patterning [48]. Hence, some of the diamond films (DLN) also applicable as coating materials in MEMS devices. But the other diamond films are able to pattern for MEMS device. Diamond films have very high chemical inertness. So wet chemical etching is almost impossible for diamond etching. The dry etching is possible for diamond films. The plasma based etching like Reactive Ion etching (RIE) is possible for diamond pattern. RIE has higher anisotropy, better uniformity and control, better selectivity compared to wet etching [49]. Generally RIE consists two electrode which creates the electric field to accelerate the ions towards the surface of the samples. Diamond pattern also possible by Inductive Couple Plasma (ICP). The plasma density of ICP is two times higher in order of magnitude compare to RIE plasma. In ICP, the RF power are capacitivly coupled and magnetic coil surrounded into the chamber for active species generation. The AC field is induced by RF coil which located in front of RF transducer helps to generate high density plasma due to confinement of electrons. In ICP process the plasma densities is very high around 10^{11} ion/cm^2 [50]. By this ICP process the diamond patterning is extremely fast with good shape under low self-bias condition [51]. The oxygen based diamond etching as shown in Fig. 12, where diamond pattern with 1.7 µm thick films etch by ICP [52].

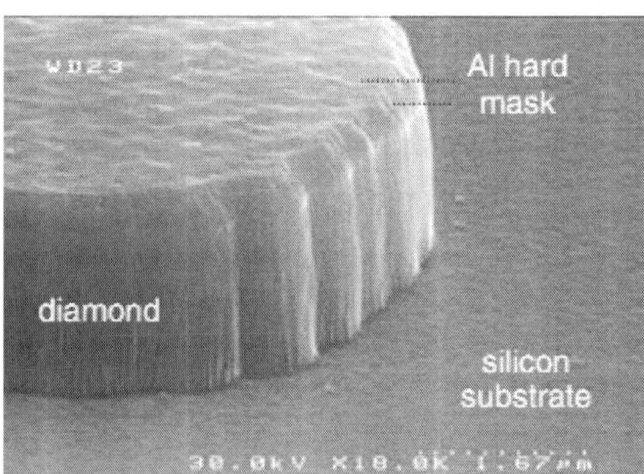

Fig. 12. Nano Crystalline Diamond (NCD) patterned by ICP etching using an Al hard mask. The etching time was 5 minutes, ([52], permission to reprint obtained from John Wiley & Sons, Ltd.).

Another diamond pattern technique is known as Focused Ion Beam (FIB). The FIB process commercially very expensive but we can etch the material very preciously even in nm level.

By this process focused beam of ions concentrate on a surface in order to create very small structures. This technique are widely used in microelectronics industry also. This Ion beam technique can cut the materials in very precise way. By this process, the materials can be milling to accelerate the concentrated gallium ions source on a specific site. The gallium ion source react with the surface and metallic precursor gases to produce the precise cut of conductive lines. In the FIB systems, where liquid metal ion sources are capable to form very small probes with high current densities [53]. Fig. 13 shows the FIB based diamond film pattern.

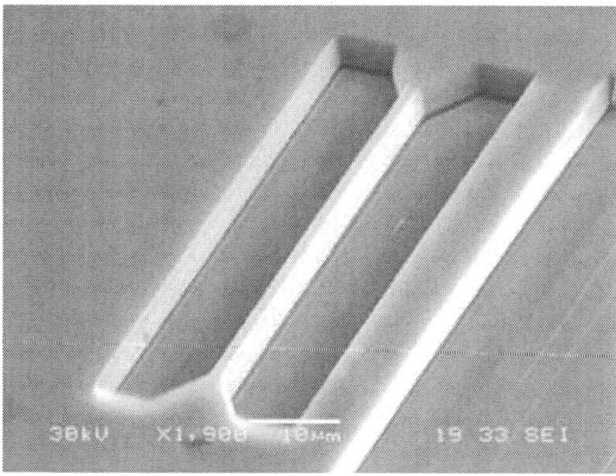

Fig. 13. Micrometer-scale bridge structure machined in single-crystal diamond released by chemical etching of graphitic layer. The graphitic layer produced by ion implantation into the diamond substrate using FIB, ([52], permission to reprint obtained from John Wiley & Sons, Ltd.).

4.2 Diamond, DLC, DLN films as coating material for MEMS

Diamond, DLC and DLN films have high hardness, high modulus of elasticity, low thermal expansion coefficient, low surface roughness and low coefficient of friction and good wear properties. This all properties are very important to apply this materials as a anti-stiction coating or tribological coating on MEMS/NEMS devices. The moving parts of any microactuators, like microgears, microengine or bearing have surface contact of their won functions during operations. Due to very low coefficient of friction and surface roughness of DLC and DLN films, the films can be act as a excellent solid lubricants of microgears or microengine based MEMS devices. Very small surface roughness can reduce the friction and wear rate in microstructure component during operations [54-56]. DLC coating which can improve the lifetime of linear motor actuated by electrostatic force which drive by sliding component laterally [55].

4.3 Diamond, DLC, DLN films for sensors and actuators applications

Diamond, Diamond-like carbon(DLC) and Diamond-like Nanocomposite(DLN) have wide band gap, good thermal conductivity and very good piezoelectric properties, high chemical stability, high fracture strength and very high stiffness. These key properties of diamond are potentially applicable for sensing and actuating application in harsh environments [57]. For piezoresistive diamond materials mainly the nanocryastalline diamond materials are used for minimize the grain boundary influence. The piezoresistive boron doped diamond materials, the gauge factor must be need as small as possible to achieve the reasonable activation and for low sheet resistance. The gauge factors of the diamond films has been found in the range of 70-4000 and even its remain unchanged in high temperatures [58-59]. As a sensor of diamond material, the aim to measurement of mechanical effect like acceleration, pressure from piezoelectric diamond or piezoresistive diamond transducers elements. These diamond materials have been utilize for different type of sensors like pressure or UV [60-61], temperature sensors [62-63] up to 700 ^{0}C. The pressure sensors of diamond materials are based on hermitically sealed membrane structure, that can detect the bowing pressure difference. The diamond pressure sensors can be made by doped diamond as a piezoresistor or undoped diamond as a flexible diaphragm. Fig. 14, shows the boron doped piezoresistive diamond pressure sensors [64]. The undoped and boron doped diamond deposited on silicon substrate by CVD process. To improve the sensitivity of this pressure sensors, the diaphragm thickness and boron doped was varied from 62 to 5 μm and 2 to 0.5 μm respectively. This sensors evaluated from room temperature to 250 ^{0}C and pressure between 0-0.07 MPa.

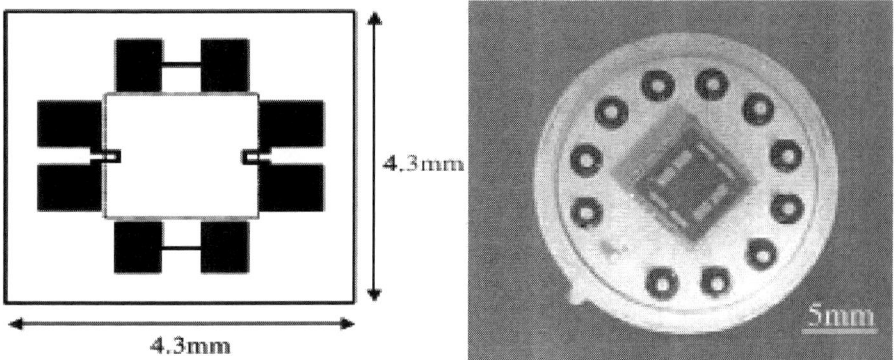

Fig. 14. Boron doped polycrystalline diamond based pressure sensors. (a) Chip layout (left fig.) (b) Pressure sensors (right fig), ([64], permission to reprint obtained from Elsevier).

Diamond films can act as a high acceleration sensors also. Diamond films have high fracture strength and high stiffness. So these films can withstand in high acceleration and shock also. Fig. 15, shows the diamond based acceleration sensors. In this figure the piezoresistive transducers elements located on the four suspended bridge which looks like whetstone bridge [65]. The sensors structure realized based on the movement of center seismic mask which suspended by four arms.

The diamond material can be act as a actuators. The microthermal actuators based on bilayer structure which delivers an out of plane motion [66] and the shape bimorph structure made by single material [67]. The bilayer which consists two layer of materials with different thermal expansion of coefficient. The deflection of bilayer actuators depends upon coefficient of thermal expansion of the materials. The diamond, DLC, and DLN have have very low coefficient of thermal expansions around 10^{-6} K^{-1}. Hence, diamond and DLC materials can be used as bi-layer thermal actuators. The high tensile strength and very high elastic modulus of diamond, DLC, and DLN materials is the potential candidate to fabricate the contact mode micro-actuators. Also these materials used to fabricate the micro-engine, gears, rotors etc. Fig. 16, shows the diamond based micro-gears and accelerometers [68].

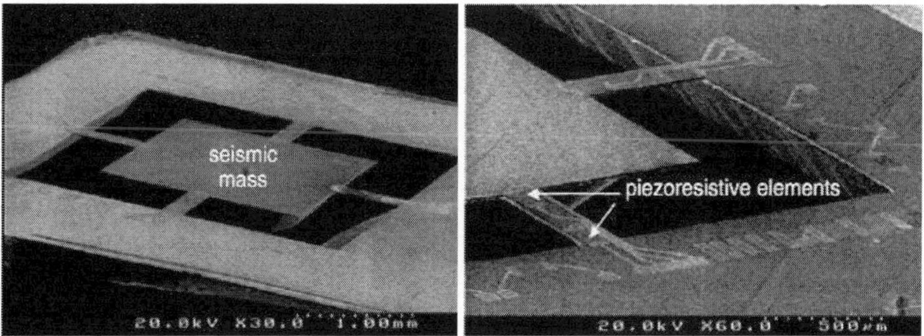

Fig. 15. Diamond material based acceleration sensors including a piezoresistive elements and seismic mask, ([52], permission to reprint obtained from John Wiley & Sons, Ltd.).

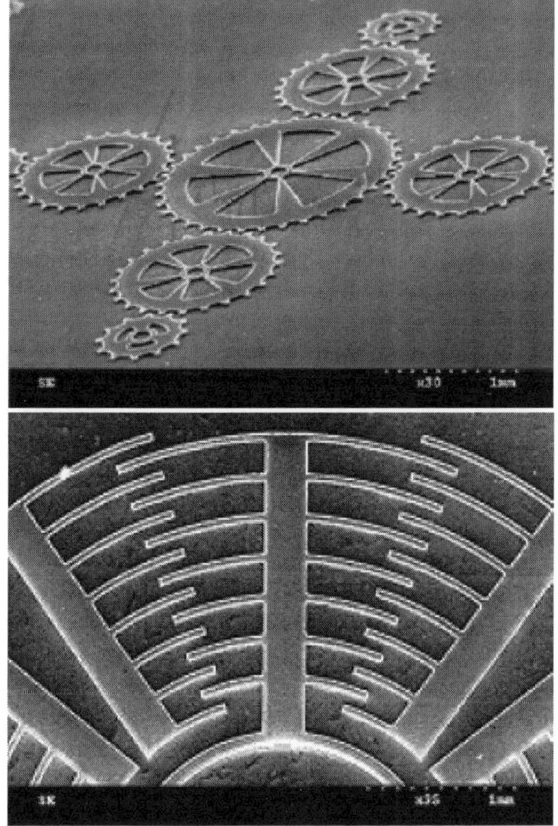

Fig. 16. Diamond based microgears (top fig.) and accelerometer (bottom fig.), ([68], permission to reprint obtained from Elsevier).

4.4 Diamond films for RF-MEMS application

Diamond, DLC and DLN are potential materials for RF-MEMS applications. The RF-MEMS switches have potential for RF communications because of their high isolation and low power consumption. The RF switches, which usually based on electrostatic actuators. This RF switch can be used as diamond switch [69-73]. The boron doped diamond can be considered as semiconductor with wideband gap for fabricate the electronic device. Diamond materials are ideal candidate for RF power electronics where the important factors are like high speed, high power density, effective thermal management and passive matching components with low loss at microwave frequencies. So surface area is needed for passive component, active device, waveguide circuits and heat dissipation. The substrate needs for monolithic integrated RF power electronics in gigahertz frequency range. Leaky diamond films used in RF capacitive MEMS switches with low RF loss up to 65 GHz have excellent electronic properties of the diamond layers [74]. Leaky diamond mode can trapped the charges and eliminating the charge injection and increases the switches reliability. Fig. 17 shows the single anchored cantilever in coplanar waveguide for ON and OFF switching of microwave signals. Where the centre line is bridge by the switch need to content the contact in series. In the first figure (left side) the switch was staying perpendicular way on to the waveguide. This switching is quite low because of air damping [75-76]. The very large surface area can provide the high switching speed.

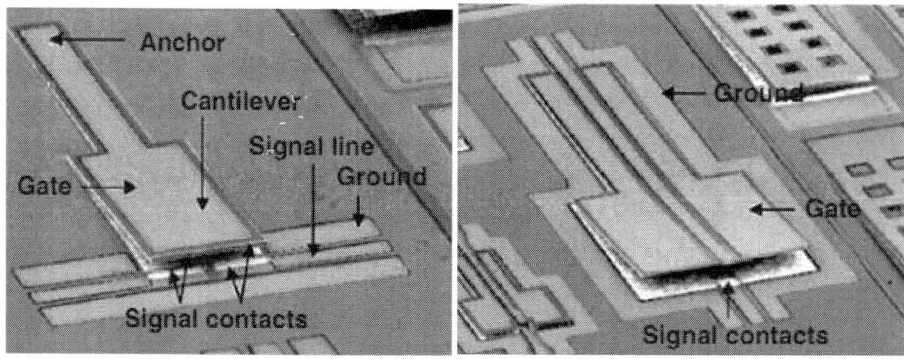

Fig. 17. Electrostatic actuators with coplanar RF waveguide. (a)The cantilever actuator closes a gap in a signal line from the side (left fig.) (b) The signal line is part of the beam itself (right fig.), ([52], permission to reprint obtained from John Wiley & Sons, Ltd.).

In right side of the figure where cantilever is the part of coplanar waveguide pattern. The RF line of ground plane act as a electrode of the parallel plate actuator [63].

5. Diamond, DLC, DLN for biomedical applications

Diamond, Diamond-like Carbon(DLC) and Diamond-like Nanocomposite (DLN) thin films have high hardness, high wear resistance, high corrosion resistance and chemical inertness, low coefficient of friction, very low surface roughness, very good infrared light transparency. All of these properties are ideal for any material as a biocompatible materials. As a biocompatible material it's can be applicable for orthopedic, cardiovascular, contact lenses, catheter, prosthetic replacement etc. The biomaterial can be determine by the reaction of materials in biological environment [77]. The biomaterial, to apply our human body, should be chemically and biologically inert to the surrounding cells and our body fluids. In these diamond structure, they have two phase. One is sp^2 hybridized carbon atoms in hexagonal ring formed by graphite structure. It is the disorder of bond angle, resulting due to disappearance of the long range translation symmetry of polycrystalline graphite and amorphous carbon films. On the other hand, the C-C stretching vibration of sp^3 hybridized carbon atoms in both the rings and chains. Which indicate the disorder diamond phase. In the films if percentage of sp^3 carbon is much more, then films should be good mechanical properties. The diamond and DLC films have good orthopedic applications. In our human body, the hip and knee joint are subjected the friction and wear, as a result which forms the polyethylene debris at a rate of 10^{10} particulates per year [78]. These phagocytosed particulates forms osteolysis, granulomatosis lesions and bone resorptions which causes the aseptic loosening of the prosthesis and pain. In last couple of years researchers are trying to apply this diamond, DLC, and DLN films coating for knee and hip replacement to decrease the wear rate. The DLC with Co-Cr alloy can reduce the significant wear rates of both sliding surfaces [79]. The amorphous diamond coatings which can improve the wear and corrosion resistance by a factor of millions compare to conventional materials [80].

The biocompatibility of DLC and DLN films have determined by the interaction of cells with DLC and DLN surface. The biocompatibility of the films can be performed by characterization of cytotsicity, protein adsorption or microphase adhesion property of the films. In the study of cell adhesion on different substrates, the increase of number of cell adhesion onto the substrate indicates that the cell have greater chance to adhere onto the substrate.

Fig. 18 shows fluoresce microscopic image of the different cells growth on different surface and Fig. 19 shows the corresponding scanning electron microscope (SEM) image of cell growth on different surface. From both of these figures, the PC 12 cells growing characteristic in neuronal process. This process interact with platinum and ultrananocrystalline diamond (UNCD) diamond surface. In silicon surface the cells formation as like closed packed islands formation.

The PC 12 cells growth on the platinum and UNCD surface exhibiting distinctive outgrowth of axons and dendrites on the surface. The growth of the PC 12 cells on the surfaces are quite less compare to HeLa cells. The figure shows that, PC 12 cells growth on UNCD surface most suitably compared to other surfaces. The area cover by PC 12 cells is almost highest in UNCD surface compared to another material surface. Which indicates that the UNCD surface have much more biocompatibility compared to other surfaces. Finally the maximum cells attachment, cell spreading and nuclear area coverage of the cell in UNCD surface is much more compered to other surfaces [81].

In medical application DLC can coat over the stent. Where a stent is a metal tube that inserted permanently into an artery. Stent helps to open an artery for blood circulation through it. Recently, the use of cardiovascular implantation of stent increasing in the world. The side effect of the artery stent lies in its release of metal ions and thrombogenecity. So it is desirable in such a material that can prevent the release of metal ions and that materials also will be hemocompatible, very high corrosion resistance and long lasting in human blood environment [82]. DLC coat stent are suitable for this pour pose. Some of the companies are made the DLC coted stent for medical purpose use. Fig. 20 shows the multilayer nanocomposite coated with DLC as an intermediate layer under the name biodiamond stent [83].

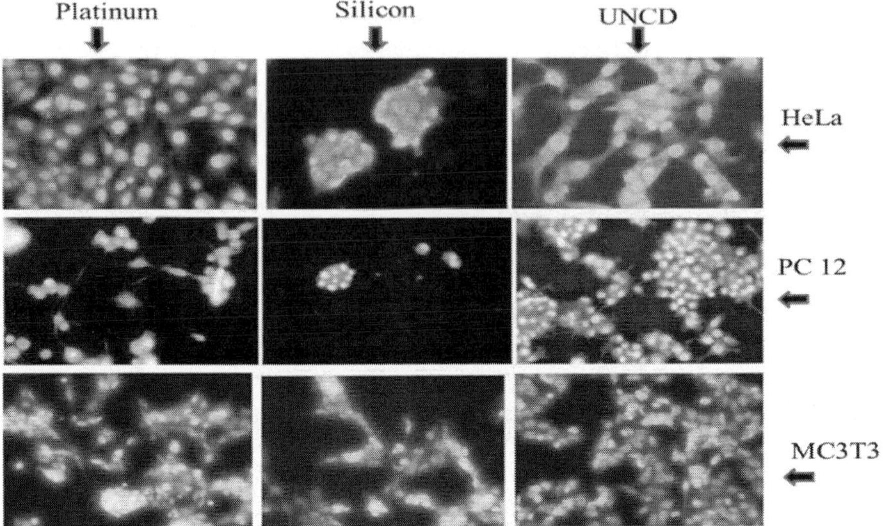

Fig. 18. Florescence microscopic image of cell attachment in platinum (first column), silicon (second column) and UNCD (third column) surfaces, ([81], permission to reprint obtained from Springer).

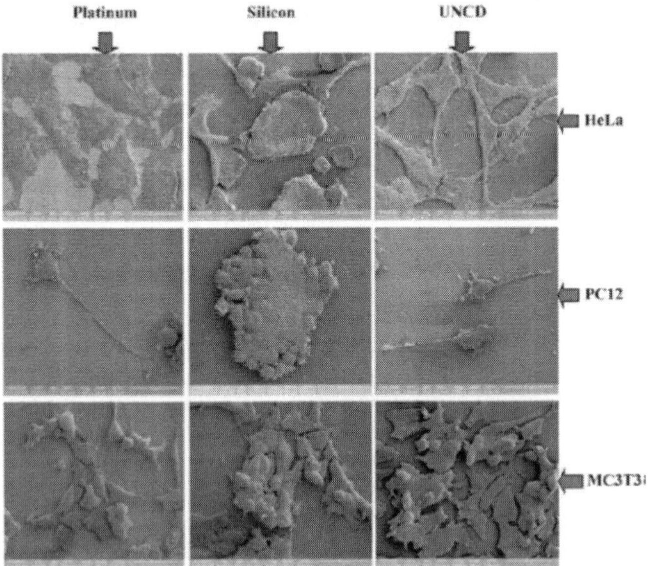

Fig. 19. Scanning Electron Microscope image of cell attachment on platinum (first column), Silicon (second column) and UNCD (third column) surfaces, ([81], permission to reprint obtained from Springer).

The variety of studied of biocompatibility of Diamond, DLC, DLN films shows that, the materials is potentially applicable in biomedical purpose. The characterization of these materials and determination of its surface properties are necessary to correlate the different in *vitro* and in *vivo* results. The diamond coating, due to use of its long term commercially in medical purpose needs more *vivo* and *vitro* experiments.

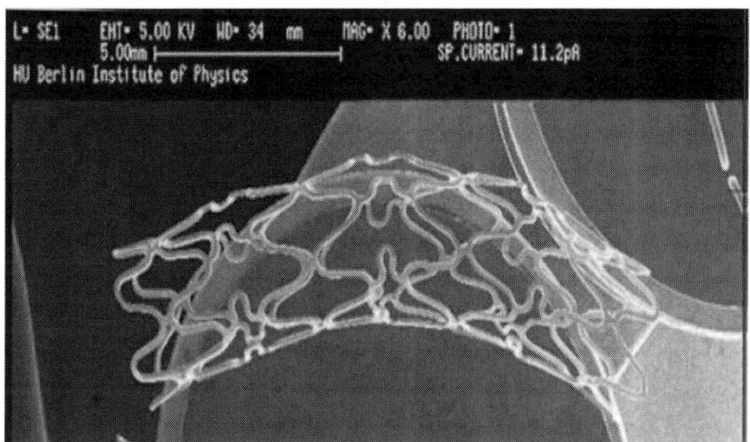

Fig. 20. A bio-diamond stent ([83], permission to reprint obtained from Elsevier).

6. Conclusions

In this chapter, we have described mainly the characterization technique of diamond, Diamond-Like Carbon (DLC) and Diamond-Like Nanocomposite (DLN) thin films and their application in MEMS devices. From HRTEM, FTIR, Raman Spectroscopy analysis, we conclude that the hydrocarbon groups are bonded with two interpenetrating networks (a-C:H and a-Si:O) of DLN films. And also from HRTEM analysis, DLN films contain Si_3N_4, SiC and SiO_x nanoparticles within amorphous matrix, which help to reduce the compressive stress of the films. Raman Spectroscopy shows that the DLN films should have higher concentrations of sp^3 carbon than the conventional DLC films. High sp^3 contents influence the mechanical properties of the films as a result, hardness and elastic modulus will increase due to higher sp^3 content. From our all characterize technique we can conclude that diamond, Diamond-Like Carbon (DLC) and Diamond-Like Nanocomposite (DLN) thin films have very high hardness, high modulus of elasticity, high tensile strength, high thermal conductivity, very less surface roughness, low coefficient of friction low thermal expansion and good wear properties. All of these properties are unique material properties for application in MEMS/NEMS device. The excellent tribological properties of the diamond films are very useful to improve the stiction, friction and wear resistance of MEMS/NEMS based microcomponent. The chemical inertness and high temperature withstanding capability of this films can be useful for biosensor and microfluidic devices. In our microfabrocation part we discussed different thin film deposition technique with different pattern like lift-off, electroplating, thin film etching and substrate etching technique. Finally we have discussed the diamond film patterning and their application in

MEMS/NEMS technologies like pressure sensor, accelerometer and different actuators. As a biocompatible material, diamond films have lot of applications in biomedical purpose such as orthopedic, cardiovascular, contact lenses, catheter, prosthetic replacement etc. Also the UNCD films have much more cells attachment, cell spreading and nuclear coverage area compared to another materials. Which confirms the biocompatibility issue of UNCD materilas. The chemical inertness, biocompatibility and excellent thermal conductivity of diamond, DLC and DLN films can be exploited in the development of biosensors, microfluidic device for lab-on-a-chip and implantable medical devices applications.

7. Acknowledgments

The authors T.S. Santra, P. Patel and T.K. Barik are gratefully acknowledge Dr. Arnab Bhattacharyya, National Metallurgical Laboratory (NML), Jamshedpur, India, for kind help to write this chapter.

8. References

[1] Verpoorte et al.(2003). Proc. IEEE 91 930
[2] Vellekoop.M.J & Kostner.S. (September 2006). Proc. 20th Eurosensors, Vol 1 pp 28 (Goteborh, Sweden)
[3] Jacob-Cook et al. (1996). J. Micromech. Microeng, Vol .6, pp.148
[4] Tsai.N.C & Sue.C.Y .(2007). Sensors Actuators, Vol.136, pp. 178
[5] Brown et al. (1998). IEEE Trans. Microw. Theory Technol. Vol.46, pp. 1868
[6] Santra et al.(2010). J. Appl. Phys,Vol.107, No. 124320-1 – 124320-9 and Santra et al. (2011). Surf. Coat. Technol., Vol. 206, pp. 228-233
[7] Morshed et al. (2003). J. Surface and Coating Technology, Vol.541, pp.163-164
[8] Silva et al .(1988). Amorphous carbon: State of the Art (World Scientific, Singapore)
[9] Lifshitz et al.(1999). Diamond Relat. Mater, Vol.8, pp.1659
[10] Bushan et al. (1999). Diamond Relat. Mater, Vol.8, pp.1985
[11] Tillmann et al. (2007). Thin Solid Films, Vol.516, pp.262
[12] Shirakura et al. (2006). Thin Solid Film, Vol.494, pp.84
[13] Liu et al .(2003). Thin Solid Films, Vol.36, pp.244
[14] Das et al. (2007). Materials in Medicine, Vol.18, No 3, pp.493
[15] Spengen et al.(2003). J.Adhes. Sci. Technol, Vol.17, pp.563
[16] Tanner et al.(1999). Microelectron. Reliab, Vol 39, pp.401
[17] Tanner et al. (April 2000). Proc. IEEE Int. Reliability Phys, Symp.(San Jose) pp. 139
[18] http://www.sandia.gov/media.NewsRel/NR2000/diamond.htm.
[19] Jacob, W & Moller, w. (1993). Appl. Phy. Lett., Vol. 63, pp. 1771-1773.
[20] Weiler et al. (1994). Appl. Phy. Lett., Vol. 64(21), pp. 2797-2799.
[21] Ferrari .A.C & Robertson, J. (2004). Phil. Trans. R. Soc. Lond., Vol. A362, pp. 2477-2512.
[22] Mekenzie, D.R. (1996). Rep. Prog. Phys. Vol. 59, p. 1611.
[23] Blormen et al. (1999). Sensors Actuators, Vol.73, p.24.
[24] Blormen et al. (1999). Sensors Actuators, Vol. 78, p. 41.
[25] Fu et al. (2000). Mater. Sci. Eng, A., Vol. 282, pp. 38-48.
[26] Fu et al. (1999). J. Mater. Sci., Vol. 34, pp. 2269-83.
[27] Fu et al. (2003). Thin Solid Films, Vol. 424, pp. 107-114.
[28] Boundina et al. (1992). Diamond and Related Materials I, pp. 380-387.

[29] Grill. A, (1997). Surf. Coat.Technol, Vol. 507, pp. 94-95.
[30] Robertson. J. (2002). Mater. Sci. Eng. Rep, Vol. R37, p. 129.
[31] Luo et al. (2007). J. Micromech. Microeng., Vol. 17, pp. S147-S163.
[32] Tamor, M.A. & Vassell, W.C. (1994). J. Appl. Phys, Vol. 76, p. 3823.
[33] Robertson, J. (1983). Adv. Phys, Vol. 32, p.361.
[34] Robertson, J. (1994). Pure Appl. Chem., Vol. 66, p.1789.
[35] Dorfman, V.F. & Pypkin, B.M. (1991). Surf. Coat. Technol., Vol. 48, p.193.
[36] Dorfman. V.F. & Pypkin, B.N. (1994). U.S. Patent, Patent Number: 5, 352, 493.
[37] Skoog et al. (1998). Principles of Instrumental Analysis, 4th ed. (Harcourt Brace & Company, Philadelphia), Chaps. 16.
[38] Nadler et al. (1984). Thin Solid Films, Vol. 116, p. 241.
[39] Tallant et al. (1995). Daimond Relat. Mater., Vol. 4, p. 191.
[40] Marchon et al. (19991). J. Appl. Phys., Vol. 69, p. 5748.
[41] Tuinstra, F. & Koenig, J.L. (1970). J. Chem. Phys., Vol. 53, p. 1126.
[42] Dillion, R.O. & Woollam, J.A. (1984). Phys. Rev., Vol. B29, p. 3482.
[43] Yang et al. (2003). Materials Letters, Vol. 57, p. 3305.
[44] Wu et al. (2008). Thin Solid Films, Vol. 517, pp. 1141–1145.
[45] Oliver, W.C. & Pharr. G.M. (1992). J. Mater. Res., Vol. 7, p. 1564.
[46] Oliver, W.C. & Pharr, G.M. (2004). J. Mater. Res., Vol. 19, p. 3.
[47] Savvides, N. & Bell, T.J. (1992). J. Appl. Phys., Vol. 72, p. 2791.
[48] Kundu et al. (2009). Advanced Material Research, Vol. 74, pp. 269-272.
[49] Rahman et al. (2001). J. Appl. Phys., Vol. 89, p. 5.
[50] http://www.axinst.com
[51] Enlund et al. (2005). Carbon, Vol. 43, pp.1839–1842.
[52] Sussmann, R.S. (2009). CVD diamond for electronic device and sensors, Wiley & Sons, ISBN-978-0-470-06532-7.
[53] Olivero et al. (2005). Adv. Mat., Vol. 17(20), pp. 2427-2430.
[54] Beerschwinger et al. (1994). J. Micromech. Microeng., Vol. 4, p. 95.
[55] Smallwood et al. (2006). Wear, Vol. 260, p. 1179.
[56] Liu, H.W. & Bhushan, B. (2003). J. Vac. Sci. Technol., Vol. A 21, p. 1528.
[57] Kohn et al. (2007). (Eds), Elsevier pp. 131–181.
[58] Tang et al. (2006). Diam. Rel. Mater., Vol. 15, p. 199.
[59] Sahli, S. & Aslam, D.M. (1998). Sensors Actuators, Vol. A 71, p. 193.
[60] Davidson et al. (1999). Diam. Rel. Mater., Vol. 8, p. 1741.
[61] Ciancaglioni et al. (2005). Diam. Rel. Mater., Vol. 14, p. 526.
[62] Ran et al. (1993). Diam. Rel. Mater., Vol. 2, p. 793.
[63] Yang et al. (1997). Diam. Rel. Mater., Vol. 6, p. 394.
[64] Yamamoto et al. (2005). Diam. Rel. Mater., Vol. 14, pp. 657–660.
[65] Kohn et al. (1999). Diam. Rel. Mater., Vol. 8, pp. 934–940.
[66] Baltes et al. (1998). Proc. IEEE, Vol. 86, p. 1660.
[67] Hickey et al. (2002). Aci. Technol., Vol. A 23, p. 71.
[68] Fu et al. (2003). J. Mat. Proc. Technol., Vol. 132, p. 73.
[69] Adamschik et al. (2002). Diam. Rel. Mater., Vol. 11, pp. 672–676.
[70] Kohn et al. (12-15 May, 2002). 23rd International Conference on Microelectronics (MIEL), Nis, Yugoslavia, Proceedings, pp. 59–66.

[71] Schmid et al. (12-15 May, 2002). 23rd International Conference on Microelectronics (MIEL), Nis, Yugoslavia.
[72] Kohn et al. (8-13 June, 2003). Evaluation of CVD diamond for heavy duty microwave switches, IEEE MTT-S International Microwave Symposium, Philadelphia, PA, USA, Digest, pp. 1625–1628.
[73] Schmid et al. (2003). Semicond. Sci. Technol., Vol. 18, pp. S72–S76.
[74] Chee et al. (2005). Proc. 13th GAAS Symp. (Paris).
[75] Kusterer et al. (2006). New Diamond and Frontier Carbon Technology, Vol. 16 (6), pp. 295–321.
[76] Schmid. P. (2006). Ph. D thesis, University of Ulm, Germany.
[77] Von Recum, A.F. (1999). Editor. Handbook of Biomaterials Evaluation.Philadelphia: Taylor and Francis.
[78] Dearnaley et al. (2005). Surf. Coat. Technol., Vol. 200, pp. 2518-2526.
[79] Sheeja et al. (2005). Surf. Coat. Technol, Vol. 190, p. 231.
[80] Lappalainen, R .(May, 2005). Intl. Conf. on Metallurgical Coatings and Thin Films, San Diego, CA.
[81] Bajaj et al. (2007), Biomed. Microdevices, Vol. 9, pp. 787-794.
[82] Roy, R.K. & Lee. K.R. (2006). J. Biomedical Material Research, Part B, Applied Biomaterials, DOI 10.1002/jbmb
[83] Choi et al. (2006). Diamond Related Material, Vol. 15, pp. 38-43.